INDUSTRIAL ENGINEERING

FE Review Manual

Executive Director of Engineering Education: Brian S. Reitzel, PE

INDUSTRIAL ENGINEERING FE REVIEW MANUAL

Published by Kaplan Engineering Education

1-877-884-0828

www.kaplanengineering.com

Printed in the United States of America.

13 14 15 10 9 8 7 6 5 4 3 2 1

ISBN: 978-1-4277-4533-0 / 1-4277-4533-1
PPN: 3200-3318

CONTENTS

PERMISSIONS

"Rules of Professional Conduct," Chapter 4, reprinted by permission of NCEES.
Source: Model Rules, National Council of Examiners for Engineering and Surveying, 2007. www.ncees.org

Chapter 3 Exhibit 31 and Figures 3.16a-b, 3.17a-b, 3.18a, 3.21, 3.22, 3.30, 3.32, 3.33, and 3.34 reprinted with permission of John Wiley & Sons, Inc.
Source: Callister, William D., Jr. Materials Science and Engineering: An Introduction, 6/e. J. Wiley & Sons. 2003.

Figures 3.16c, 3.17c, and 3.18b reprinted by permission of the estate of William G. Moffatt.
Source: Moffatt, William G. The Structure and Property of Materials, Volume 1. J. Wiley & Sons. 1964.

Tables 3.4 and 3.5 reprinted by permission of McGraw-Hill Companies.
Source: Fontana, M., Corrosion Engineering. McGraw-Hill Companies.

Figure 3.25 used by permission of ASM International.
Source: Mason, Clyde W., Introductory Physical Metallurgy: p. 33. 1947.

Figure 3.37 used by permission of ASM International.
Source: Rinebolt, J.A., and W. J. Harris, Jr., "Effect of Alloying Elements on Notch Toughness of Pearlitic Steels." Transactions of ASM, Volume 43: pp. 1175–1201. 1951

CHAPTER 1

Introduction

OUTLINE

HOW TO USE THIS BOOK

Fundamentals of Engineering FE Exam Preparation is designed to help you prepare for the Fundamentals of Engineering exam. The book covers the full breadth and depth of topics covered by the new Fundamentals of Engineering exams. Each chapter of this book covers a major topic on the exam, reviewing important terms, equations, concepts, analysis methods, and typical problems. Solved examples are provided throughout each chapter to help you apply the concepts and to model problems you may see on the exam. After reviewing the topic, you can work the end-of-chapter problems to test your understanding. The problems are typical of what you will see on the exam, and complete solutions are provided so that you can check your work and further refine your solution methodology.

The following sections provide you with additional details on the process of becoming a licensed professional engineer and on what to expect at the exam.

BECOMING A PROFESSIONAL ENGINEER

To achieve registration as a Professional Engineer, there are four distinct steps: (1) education, (2) the Fundamentals of Engineering (FE) exam, (3) professional experience, and (4) the Professional Engineer (PE) exam. These steps are described in the following sections.

Education

Generally, no college degree is required to be eligible to take the FE exam. The exact rules vary, but all states allow engineering students to take the FE exam before they graduate, usually in their senior year. Some states, in fact, have no education requirement at all. One merely needs to apply and pay the application fee. Perhaps the best time to take the exam is immediately following completion of related coursework. For most engineering students, this will be the end of the senior year.

Fundamentals of Engineering Examination

This six-hour, multiple-choice examination is known by a variety of names— Fundamentals of Engineering, Engineer-in-Training (EIT), and Intern Engineer— but no matter what it is called, the exam is the same in all states. It is prepared and graded by the National Council of Examiners for Engineering and Surveying (NCEES).

Experience

States that allow engineering seniors to take the FE exam have no experience requirement. These same states, however, generally will allow other applicants to substitute acceptable experience for coursework. Still other states may allow a candidate to take the FE exam without any education or experience requirements.

Typically, several years of acceptable experience is required before you can take the Professional Engineer exam—the duration varies by state, and you should check with your state licensing board for details.

Professional Engineer Examination

The second national exam is called Principles and Practice of Engineering by NCEES, but many refer to it as the Professional Engineer exam or PE exam. All states, plus Guam, the District of Columbia, and Puerto Rico, use the same NCEES exam. Review materials for this exam are found in other engineering license review books.

FUNDAMENTALS OF ENGINEERING EXAMINATION

Laws have been passed that regulate the practice of engineering in order to protect the public from incompetent practitioners. Beginning in 1907, the individual states began passing title acts regulating who could call themselves engineers and offer services to the public. As the laws were strengthened, the practice of engineering was limited to those who were registered engineers, or to those working under the supervision of a registered engineer. Originally the laws were limited to civil engineering, but over time they have evolved so that the titles, and sometimes the practice, of most branches of engineering, are included.

There is no national licensure law; licensure is based on individual state laws and is administered by boards of registration in each state. You can find a list of contact information for and links to the various state boards of registration at the Kaplan Engineering Web site: *www.kaplanengineering.com*. This list also shows the exam registration deadline for each state.

Examination Development

Initially, the states wrote their own examinations, but beginning in 1966 NCEES took over the task for some of the states. Now the NCEES exams are used by all states. Thus it is easy for engineers who move from one state to another to achieve licensure in the new state. About 50,000 engineers take the FE exam annually. This represents about 65% of the engineers graduated in the United States each year.

The development of the FE exam is the responsibility of the NCEES Committee on Examination for Professional Engineers. The committee is composed of people from industry, consulting, and education, all of whom are subject-matter experts. The test is intended to evaluate an individual's understanding of mathematics, basic sciences, and engineering sciences obtained in an accredited bachelor degree of engineering. Every five years or so, NCEES conducts an engineering task analysis survey. People in education are surveyed periodically to ensure the FE exam specifications reflect what is being taught. This was last done in 2012-2013, and the survey results drove the format change for the 2014 exam. Previously a general engineering portion of the exam was given in the morning and a discipline specific in the afternoon. Now the entire exam is discipline specific.

The exam questions are prepared by the NCEES committee members, subject matter experts, and other volunteers. All people participating must hold professional licensure. When the questions have been written, they are circulated for review in workshop meetings and by mail. You will see mostly metric units (SI) on the exam. Some problems are posed in U.S. customary units (USCS) because the topics typically are taught that way. All problems are four-way multiple choice.

Examination Structure

The FE exam will be six hours in length, which includes a tutorial, a break, the exam, and a brief survey at the conclusion of the exam. There are 110 questions total on the exam.

The exam will be divided into two sections with a 25-minute break in the middle. Examinees will be given 5 hours and 20 minutes to complete approximately 55 questions prior to the scheduled break and the remaining questions afterward.

Seven different exams are in the test booklet, including one for each of the following six branches: civil, mechanical, electrical, chemical, industrial, environmental. An Other Disciplines exam is included for those examinees not covered by the six engineering branches. If you are taking the FE as a graduation requirement, your school may compel you to take the exam that matches the engineering discipline in which you are obtaining your degree. Otherwise, you can choose the exam you wish to take.

Examination Dates

Beginning in January 2014, the FE will be administered during four testing windows throughout the year: January–February, April–May, July–August, and October–November. Registration will be open year-round. Those wishing to take the exam must apply to their state board several months before the exam date.

Examination Procedure

Registration for the computer-based FE exam is scheduled to open on November 4, 2013, and will be open year-round. You will register and schedule your appointment through your My NCEES account on the NCEES Web site. You will first select your exam location, and then you will be presented with a list of available exam dates for your appointment. If you are not happy with the choices, you can browse through the available dates at another NCEES-approved testing center.

The examination is closed book. You may not bring any reference materials with you to the exam. To replace your own materials, NCEES has prepared a *Fundamentals of Engineering (FE) Supplied-Reference Handbook*. The handbook contains engineering, scientific, and mathematical formulas and tables for use in the examination. Examinees will receive the handbook from their state registration board prior to the examination. The *Fundamentals of Engineering Supplied-Reference Handbook* is also included in the exam materials distributed at the beginning of each exam period.

Examination-Taking Suggestions

Those familiar with the psychology of examinations have several suggestions for examinees:

1. There are really two skills that examinees can develop and sharpen. One is the skill of illustrating one's knowledge. The other is the skill of familiarization with examination structure and procedure. The first can be enhanced by a systematic review of the subject matter. The second, exam-taking skills, can be improved by practice with sample problems—that is, problems that are presented in the exam format with similar content and level of difficulty.
2. Examinees should answer every problem, even if it is necessary to guess. There is no penalty for guessing. The best approach to guessing is to try to eliminate one or two of the four alternatives. If this can be done, the chance of selecting a correct answer obviously improves from 1 in 4 to 1 in 2 or 3.
3. Plan ahead with a strategy and a time allocation. There are problems in 13 subject areas. Compute how much time you will allow for each of the subject areas. You might allocate a little less time per problem for the areas in which you are most proficient, leaving a little more time in subjects that are more difficult for you. Your time plan should include a reserve block for especially difficult problems, for checking your scoring sheet, and finally for making last-minute guesses on problems you did not work. Your strategy might also include time allotments for two passes through the exam—the first to work all problems for which answers are obvious to you, and the second to return to the more complex, time-consuming problems and the ones at which you might need to guess.
4. Read all four multiple-choice answer options before making a selection. All distractors (wrong answers) are designed to be plausible. Only one option will be the best answer.
5. Do not change an answer unless you are absolutely certain you have made a mistake. Your first reaction is likely to be correct.
6. If time permits, check your work.
7. Do not sit next to a friend, a window, or other potential distraction.

License Review Books

To prepare for the FE exam, you need one or two review books.

1. You need this book, to provide a review of the discipline-specific examination.

2. You will need *Fundamentals of Engineering (FE) Supplied-Reference Handbook.* At some point this NCEES-prepared book will be provided to applicants by their state registration board. You may want to obtain a copy sooner so you will have ample time to study it before the exam. Pay close attention to the *Fundamentals of Engineering Supplied-Reference Handbook* and the notation used in it because it is the only book you will have at the exam.

Textbooks

If you still have your university textbooks, they can be useful in preparing for the exam, unless they are out of date. To a great extent the books will be like old friends with familiar notation. You probably need both textbooks and license review books for efficient study and review.

Examination Day Preparations

The exam day will be a stressful and tiring one. You should take steps to eliminate the possibility of unpleasant surprises. If at all possible, visit the examination site ahead of time to determine the following:

1. How much time should you allow for travel to the exam on that day? Plan to arrive about 15 minutes early. That way you will have ample time, but not too much time. Arriving too early, and mingling with others who are also anxious, can increase your anxiety and nervousness.

2. Where will you park?

3. How does the exam site look? Will you have ample workspace? Will it be overly bright (sunglasses), or cold (sweater), or noisy (earplugs)? Would a cushion make the chair more comfortable?

4. Where are the drinking fountain and lavatory facilities?

5. What about food? Most states do not allow food in the test room (exceptions for ADA). Should you take something along for energy in the exam? A light bag lunch during the break makes sense.

Items to Take to the Examination

Although you may not bring books to the exam, you should bring the following:

- *Calculator*—NCEES has implemented a more stringent policy regarding permitted calculators. For a list of permitted models, see the NCEES Web site *(www.ncees.org).* You also need to determine whether your state permits pre-programmed calculators. Bring extra batteries for your calculator just in case, and many people feel that bringing a second calculator is also a very good idea.

- *Clock*—You must have a time plan and a clock or wristwatch. You will not be allowed to use your phone as a clock.

- *Exam Assignment Paperwork*—Take along the letter assigning you to the exam at the specified location to prove that you are the registered person. Also bring something with your name and picture (driver's license or identification card).
- *Items Suggested by Your Advance Visit*—If you visit the exam site, it will probably suggest an item or two that you need to add to your list.
- *Clothes*—Plan to wear comfortable clothes. You probably will do better if you are slightly cool, so it is wise to wear layered clothing.

Special Medical Condition

If you have a medical situation that may require special accommodation, notify the licensing board well in advance of exam day.

Examination Scoring and Results

Examinees will be notified via e-mail when their results are available for viewing in My NCEES. The process is still being finalized, but most examinees should receive their results within 7 to 10 business days.

Errata

The authors and publisher of this book have been careful to avoid errors, employing technical reviewers, copyeditors, and proofreaders to ensure the material is as flawless as possible. Any known errata and corrections are posted on the product page at our Web site, *www.kaplanengineering.com*. If you believe you have discovered an inaccuracy, please notify Customer Service at *Kaplanaeinfo@kaplan.com*.

CHAPTER 2

Mathematics: Analytic Geometry, Calculus, Matrix Operations, Vector Analysis, and Linear Algebra

OUTLINE

ALGEBRA

Factorials

Definition. The factorial of a non-negative integer, n, is defined as $n!$ $n! = n(n-1)(n-2)(n-3)\ldots$ and so forth.

For example, $6! = 6(5)(4)(3)(2)(1) = 720$.

Also, $1! = 1$ and $0! = 1$.

Factorials can be written as multiples of other factorials:

$$6! = 6(5!) = 6(5)(4!) = 6(5)(4)(3!) = 6(5)(4)(3)(2!) = 6(5)(4)(3)(2)(1!) = 720$$

Factorials can be multiplied and divided.

$$\frac{n!}{(n-1)!} = \frac{n(n-1)!}{(n-1)!} = n$$

Exponents

Definition. Any number defined as base, b, can be multiplied by itself x number of times, which is denoted as b^x.

For example, $b^4 = b(b)(b)(b)$.

Properties of Exponents

$$b^0 = 1; \quad b^1 = b; \quad b^{-1} = \frac{1}{b}; \quad b^{-x} = \frac{1}{b^x}$$

$$b^{x+y} = \left(b^x\right)\left(b^y\right); \quad b^{x-y} = \frac{b^x}{b^y}; \quad b^{x \times y} = \left(b^x\right)^y$$

$$b^{\frac{1}{2}} = \sqrt{b}; \quad b^{\frac{1}{x}} \sqrt[x]{b}; \quad b^{\frac{x}{y}} = \left(\sqrt[y]{b}\right)^x$$

Logarithms

Definition. If b is a finite positive number, other than 1, and $b^x = N$, then x is the logarithm of N to the base b, or $\log_b N = x$. If $\log_b N = x$, then $b^x = N$.

Properties of Logarithms

$$\log_b b = 1; \ \log_b 1 = 0; \ \log_b 0 = \begin{cases} +\infty, \text{ when } b \text{ lies between 0 and 1} \\ +\infty, \text{ when } b \text{ lies between 1 and } \infty \end{cases}$$

$$\log_b \left(M \circ N\right) = \log_b M + \log_b N \quad \log_b \ M/N = \log_b \ M - \log_b N$$

$$\log_b N^P = p \log_b N \quad \log_b \sqrt[r]{N^P} = \frac{p}{r} \log_b N$$

$$\log_b N = \log_a N / \log_a b; \quad \log_b b^N = N; \quad b^{\log_b N} = N$$

Systems of Logarithms

Common (Briggsian)—base 10.

Natural (Napierian or hyperbolic)—base 2.7183 (designated by e or ε).

The abbreviation of *common logarithm* is log, and the abbreviation of *natural logarithm* is ln.

Example 2.1

(i) Solve for a if $\log_a 10 = 0.25$ and (ii) find $\log\left(\frac{1}{x}\right)$, if $\log x = 0.3332$.

Solution

(i) If $\log_a N = x$, then $N = a^x$ or $a = N^{(1/x)}$

Here, $N = 10$ and $x = 0.25$; then $a = 10^{\frac{1}{0.25}} = 10{,}000$

(ii) Since $\log \frac{M}{N} = \log M - \log N$, $\log\left(\frac{1}{x}\right) = \log 1 - \log x = -0.3332$

Example 2.2

Solve the equation $\log x + \log (x - 3) - \log 4 = 0$.

Solution

Since $(\log M + \log N - \log P) = \log \left(\frac{MN}{P}\right)$, $\log x + \log (x - 3) - \log 4$

$= \log \left[\frac{x(x-3)}{4}\right] = 0$

$\left[\frac{x(x-3)}{4}\right] = 10^0 = 1$; simplifying, $x^2 - 3x - 4 = 0$. Finding the roots,

$x = 4$ and -1; then $x = 4$ (-1 is not an answer because the log of a negative number is undefined).

The Solution of Algebraic Equations

Definition. A root, x, is any value such that $f(x) = 0$.

The Quadratic Equation

If $ax^2 + bx + c = 0$, then

$$x = \frac{-b \pm \sqrt{b^2 - 4ac}}{2a}$$

If $b^2 - 4ac > 0$, the two roots are real and unequal; if $b^2 - 4ac = 0$, the two roots are real and equal; if $b^2 - 4ac < 0$, the two roots are imaginary.

Example 2.3

Find the root(s) of the following equations (i) $x + 4 = 0$, (ii) $x^2 + 4x + 3 = 0$, (iii) $x^2 - 4x + 4 = 0$, (iv) $x^2 + 4 = 0$, and (v) $3x^3 + 3x^2 - 18x = 0$.

Solution

For the quadratic equations in (ii), (iii), and (iv), use either the equation or simply a scientific calculator to find the roots. The results yield:

(i) -4, (ii) -1 and -3 (real and distinct roots), (iii) 2 and 2 (real and equal roots),

(iv) $+i2$ and $-i2$ (complex roots; always occur in pairs called *conjugates*)

For (v), factoring, $3x(x^2 + x - 6) = 0$. Roots are 0, -3, and 2.

Example 2.4

Find the equation whose roots are 3 and -2.

Solution

If the roots are x_1, x_2, x_3, etc., the equation is $(x - x_1)(x - x_2)(x - x_3)\ldots = 0$; here, $(x - 3)(x - (-2)) = 0$; $(x - 3)(x + 2) = 0$; $x^2 - x - 6 = 0$.

Progressions

Arithmetic Progression

An arithmetic progression is a, $a + d$, $a + 2d$, $a + 3d$,. . ., where d = common difference.

The nth term is $t_n = a + (n - 1)d$

The sum of n terms is $S_n = \frac{n}{2}[2a + (n-1)d] = \frac{n}{2}(a + t_n)$

Geometric Progression

A geometric progression is a, ar, ar^2, ar^3,. . ., where r = common ratio.

The nth term is $t_n = ar^{n-1}$

The sum of n terms is $S_n = a\left(\frac{1-r^n}{1-r}\right)$

If $r^2 < 1$, S_n approaches a definite limit as n increases indefinitely, and

$$S_\infty = \frac{a}{1-r}$$

Example 2.5

Consider the arithmetic progression 1, 3, 5, 7, 9, 11, 13, . . . (i) Find the sum of the first seven terms and (ii) the 18th term of the progression.

Solution

(i) First term $a = 1$, number of terms $n = 7$, and the common difference $d = 2$

sum $S_n = \frac{n}{2}[2a + (n-1)d] = \frac{7}{2}[2 + 6(2)] = 49$

(ii) Number of terms $n = 18$, first term $a = 1$, and the common difference $d = 2$

The last term or the 18th term is $= a + (n - 1)\,d = 1 + (18 - 1)2 = 35$.

Example 2.6

Find the sum of the series 1, 0.5, 0.25, 0.125, 0.0625, . . .

Solution

This geometric series is convergent. First term $a = 1$ and the common ratio $r = 0.5$.

As the number of terms tend to infinity, sum $S = \frac{a}{1-r} = \frac{1}{1-0.5} = 2$

Example 2.7

Consider the geometric progression 2, 4, 8, 16, 32, 64, 128, . . . (i) Find the sum of the first seven terms and (ii) the 20th term of the series.

Solution

(i) First term $a = 2$, common ratio $r = 2$, and the number of terms $n = 7$

sum $S = a\frac{(1-r^n)}{(1-r)} = \frac{2(1-2^7)}{(1-2)} = 256$

(ii) The 20th term of the series is $= ar^{(n-1)} = 2(2)^{(20-1)} = 1{,}048{,}576$.

COMPLEX QUANTITIES

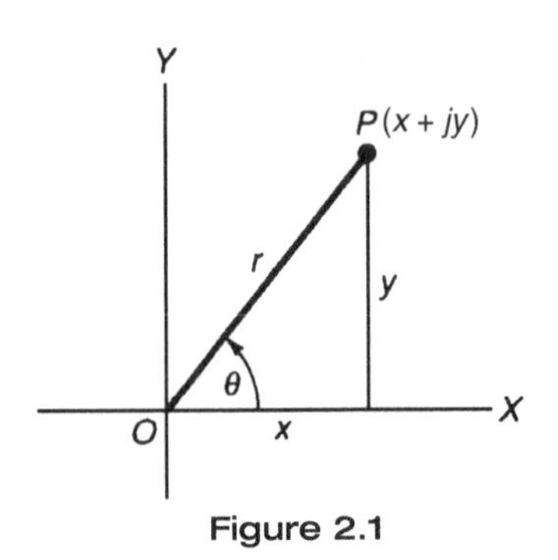

Figure 2.1

Definition and Representation of a Complex Quantity

If $z = x + jy$, where $j = \sqrt{-1}$ and x and y are real, z is called a complex quantity and is completely determined by x and y.

If $P(x, y)$ is a point in the plane (Figure 2.1), then the segment OP in magnitude and direction is said to represent the complex quantity $z = x + jy$.

If θ is the angle from OX to OP and r is the length of OP, then $z = x + jy = r(\cos\theta + j\sin\theta) = re^{j\theta}$, where $\theta = \tan^{-1} y/x$, $r = +\sqrt{x^2 + y^2}$ and e is the base of natural logarithms. The pair $x + jy$ and $x - jy$ are called complex conjugate quantities.

Properties of Complex Quantities

Let z, z_1, and z_2 represent complex quantities; then

Sum or difference: $z_1 \pm z_2 = (x_1 \pm x_2) + j(y_1 \pm y_2)$

Equation: If $z_1 = z_2$, then $x_1 = x_2$ and $y_1 = y_2$

Periodicity: $z = r(\cos\theta + j\sin\theta) = r[\cos(\theta + 2k\pi) + j\sin(\theta + 2k\pi)]$, or $z = re^{j\theta} = re^{j(\theta + 2k\pi)}$ and $e^{j2k\pi} = 1$, where k is any integer.

Exponential-trigonometric relations: $e^{jz} = \cos z + j\sin z$, $e^{-jz} = \cos z - j\sin z$,

$$\cos z = \frac{1}{2}\left(e^{jz} + e^{-jz}\right), \quad \sin z = \frac{1}{2j}\left(e^{jz} - e^{-jz}\right)$$

TRIGONOMETRY

Definition of an Angle

An angle is the amount of rotation (in a fixed plane) by which a straight line may be changed from one direction to any other direction. If the rotation is counterclockwise, the angle is said to be positive; if clockwise, negative.

Measure of an Angle

A degree is $\frac{1}{360}$ of the plane angle about a point, and a radian is the angle subtended at the center of a circle by an arc equal in length to the radius. One complete circle contains 180 degrees or 2π radians; 1 radian = $\pi/180$ degrees.

Figure 2.2

Trigonometric Functions of an Angle

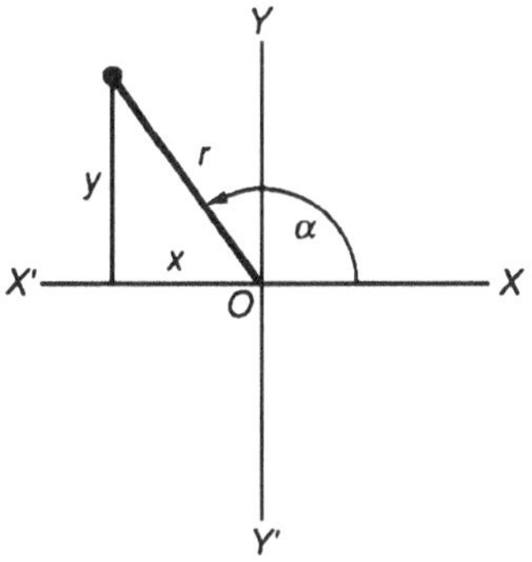

sine (sin) $\alpha = y/r$ cosecant (csc) $\alpha = r/y$

cosine (cos) $\alpha = x/r$ secant (sec) $\alpha = r/x$

tangent (tan) $\alpha = y/x$ cotangent (cot) $\alpha = x/y$

The variable x is positive when measured along OX and negative along OX'. Similarly, y is positive when measured parallel to OY, and negative parallel to OY'.

$$\sin 0° = 0;\ \sin 90° = 1;\ \sin 180° = 0;\ \sin 270° = -1$$

$$\cos 0° = 1;\ \cos 90° = 0;\ \cos 180° = -1;\ \cos 270° = 0$$

Fundamental Relations among the Functions

$$\sin\alpha = \frac{1}{\csc\alpha};\quad \cos\alpha = \frac{1}{\sec\alpha};\quad \tan\alpha = \frac{1}{\cot\alpha} = \frac{\sin\alpha}{\cos\alpha}$$

$$\csc\alpha = \frac{1}{\sin\alpha};\quad \sec\alpha = \frac{1}{\cos\alpha};\quad \cot\alpha = \frac{1}{\tan\alpha} = \frac{\cos\alpha}{\sin\alpha}$$

$$\sin^2\alpha + \cos^2\alpha = 1;\quad \sec^2\alpha - \tan^2\alpha = 1;\quad \csc^2\alpha - \cot^2\alpha = 1$$

Functions of Multiple Angles

$$\sin 2\alpha = 2\sin\alpha\cos\alpha$$

$$\cos 2\alpha = 2\cos^2\alpha - 1 = 1 - 2\sin^2\alpha = \cos^2\alpha - \sin^2\alpha$$

$$\tan 2\alpha = (2\tan\alpha)/(1 - \tan^2\alpha)$$

$$\cot 2\alpha = (\cot^2\alpha - 1)/(2\cot\alpha)$$

Functions of Half Angles

$$\sin\frac{1}{2}\alpha = \sqrt{\frac{1-\cos\alpha}{2}};\quad \cos\frac{1}{2}\alpha = \sqrt{\frac{1+\cos\alpha}{2}}$$

$$\tan\frac{1}{2}\alpha = \frac{1-\cos\alpha}{\sin\alpha} = \frac{\sin\alpha}{1+\cos\alpha} = \sqrt{\frac{1-\cos\alpha}{1+\cos\alpha}}$$

Functions of Sum or Difference of Two Angles

$$\sin(\alpha \pm \beta) = \sin\alpha\cos\beta \pm \cos\alpha\sin\beta$$

$$\cos(\alpha \pm \beta) = \cos\alpha\cos\beta \mp \sin\alpha\sin\beta$$

$$\tan(\alpha \pm \beta) = \frac{\tan\alpha \pm \tan\beta}{1 \mp \tan\alpha\tan\beta}$$

Sums, Differences, and Products of Two Functions

$$\sin\alpha + \sin\beta = 2\sin\frac{1}{2}(\alpha+\beta)\cos\frac{1}{2}(\alpha-\beta)$$

$$\sin\alpha - \sin\beta = 2\cos\frac{1}{2}(\alpha+\beta)\sin\frac{1}{2}(\alpha-\beta)$$

$$\cos\alpha + \cos\beta = 2\cos\frac{1}{2}(\alpha+\beta)\cos\frac{1}{2}(\alpha-\beta)$$

$$\cos\alpha - \cos\beta = 2\sin\frac{1}{2}(\alpha+\beta)\sin\frac{1}{2}(\alpha-\beta)$$

$$\tan\alpha \pm \tan\beta = \frac{\sin(\alpha+\beta)}{\cos\alpha\cos\beta}$$

$$\sin^2\alpha - \sin^2\beta = \sin(\alpha+\beta)\sin(\alpha-\beta)$$

$$\cos^2\alpha - \cos^2\beta = \sin(\alpha+\beta)\sin(\alpha-\beta)$$

$$\cos^2\alpha - \sin^2\beta = \cos(\alpha+\beta)\cos(\alpha-\beta)$$

$$\sin\alpha\sin\beta = \frac{1}{2}\cos(\alpha-\beta) - \frac{1}{2}\cos(\alpha+\beta)$$

$$\cos\alpha\cos\beta = \frac{1}{2}\cos(\alpha-\beta) + \frac{1}{2}\cos(\alpha+\beta)$$

$$\sin\alpha\cos\beta = \frac{1}{2}\sin(\alpha+\beta) + \frac{1}{2}\sin(\alpha-\beta)$$

Example 2.8

Simplify the following expressions to trigonometric functions of angles less than 90 degrees. Note: There is more than one correct answer for each.

i) cos 370°
ii) sin 120°

Solution

Typical strategies for simplifying the functions are shown below:

i) cos 370° = cos (360° + 10°) = (cos 360°)(cos 10°) – (sin 360°)(sin 10°) = (1)(cos 10°) – (0)(sin 10°) = cos 10°

ii) sin 120° = sin (90° + 30°) = (sin 90°)(cos 30°) + (cos 90°)(sin 30°) = (1)(cos 30°) + (0)(sin 30°) = cos 30°

or

sin 120° = sin (180° – 60°) = (sin 180°)(cos 60°) – (cos 180°)(sin 60°) = (0)(cos 60°) – (–1)(sin 60°) = sin 60°

Properties of Plane Triangles

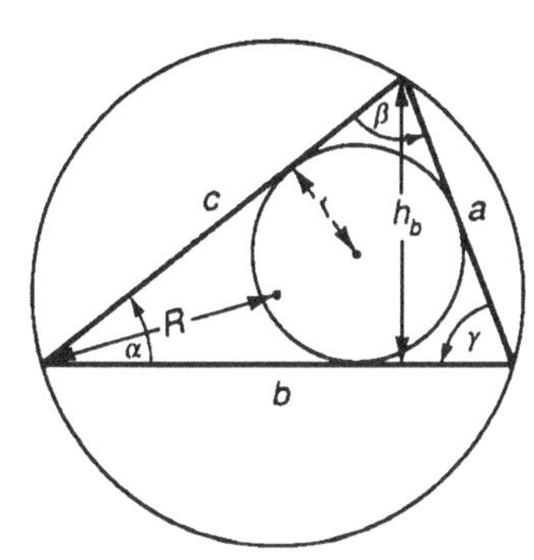

Figure 2.3

Notation. α, β, γ = angles; a, b, c = sides; A = area; h_b = altitude on b; $s = \frac{1}{2}(a+b+c)$; r = radius of inscribed circle; R = radius of circumscribed circle

$$\alpha + \beta + \gamma = 180° = \pi \text{ radians}$$

$$\frac{a}{\sin\alpha} = \frac{b}{\sin\beta} = \frac{c}{\sin\gamma}$$

$$\frac{a+b}{a-b} = \frac{\tan\frac{1}{2}(\alpha+\beta)}{\tan\frac{1}{2}(\alpha-\beta)}$$

$$a^2 = b^2 + c^2 - 2bc\cos\alpha \qquad a = b\cos\gamma + c\cos\beta$$

$$\cos\alpha = \frac{b^2+c^2-a^2}{2bc} \qquad \sin\alpha = \frac{2}{bc}\sqrt{s(s-a)(s-b)(s-c)}$$

$$\sin\frac{\alpha}{2}\sqrt{\frac{(s-b)(s-c)}{bc}} \qquad \cos\frac{\alpha}{2} = \sqrt{\frac{s(s-a)}{bc}}$$

$$\tan\frac{\alpha}{2} = \sqrt{\frac{(s-b)(s-c)}{s(s-a)}} = \frac{r}{s-a}$$

$$h_b = c\sin\alpha = a\sin\gamma = \frac{2}{b}\sqrt{s(s-a)(s-b)(s-c)}$$

$$r = \sqrt{\frac{(s-a)(s-b)(s-c)}{s}} = (s-a)\tan\frac{\alpha}{2}$$

$$R = \frac{a}{2\sin\alpha} = \frac{abc}{4A}$$

$$A = \frac{1}{2}bh_b = \frac{1}{2}ab\sin\gamma = \frac{a^2\sin\beta\sin\gamma}{2\sin\alpha} = \sqrt{s(s-a)(s-b)(s-c)} = rs$$

Example 2.9

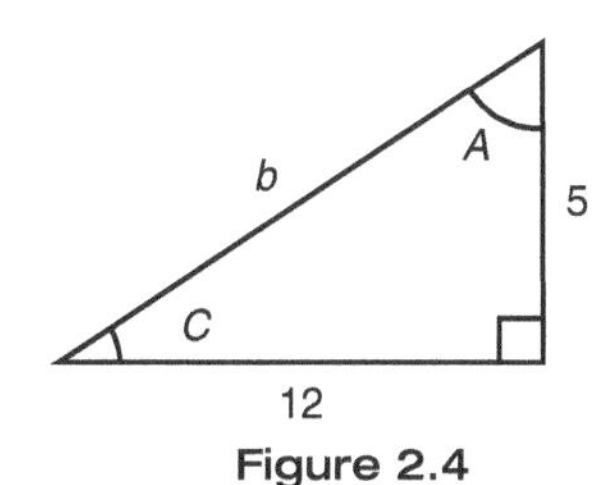

Figure 2.4

Find the side b and the angles A and C for the triangle in Figure 2.4.

Solution

$\tan(A) = 12/5$; then, Angle $A = \tan^{-1}(12/5) = 67.38°$

Since the sum $(A + C + 90°) = 180°$; $C = 90° - A = 22.62°$

Now, $\cos(A) = \frac{5}{b}$; then, $b = \frac{5}{\cos(A)} = 13$

Example 2.10

Find the side c and the angles A and B for the triangle in Figure 2.5.

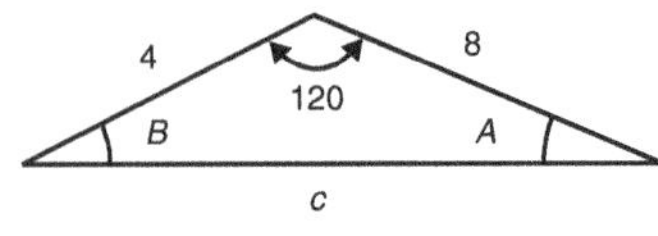

Figure 2.5

Solution

Since two sides and an included angle are given, use the law of cosines.

$c^2 = 4^2 + 8^2 - 2(4)(8)\cos 120$; solving $c = 10.583$

Now use law of sines to find the remaining angles.

$\frac{10.583}{\sin 120} = \frac{4}{\sin A} = \frac{8}{\sin B}$; solving, $A = 19.1°$ and $B = 40.89°$

(Check: sum of the angles = 180°)

Example 2.11

Simplify: (i) $(\sec^2\theta)(\sin^2\theta)$ (ii) $\sin(A + B) + \sin(A - B)$
(iii) $2\sin^2\theta + 1 + \cos^2\theta$

Solution

(i) $(1/\cos^2\theta)\sin^2\theta = \tan^2\theta$ (ii) $2\sin A\cos B$

(iii) $2\sin^2\theta + 1 + (2\cos^2\theta - 1) = 2(\sin^2\theta + \cos^2\theta) = 2$

GEOMETRY AND GEOMETRIC PROPERTIES (MENSURATION)

Notation. a, b, c, d, and s denote lengths, A denotes area, V denotes volume

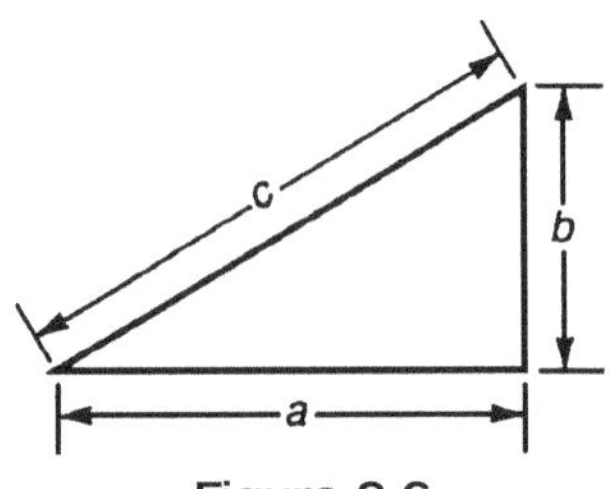

Figure 2.6

Right Triangle

$$A = \frac{1}{2}ab$$

$$c = \sqrt{a^2 + b^2},\quad a = \sqrt{c^2 - b^2},\quad b = \sqrt{c^2 - a^2}$$

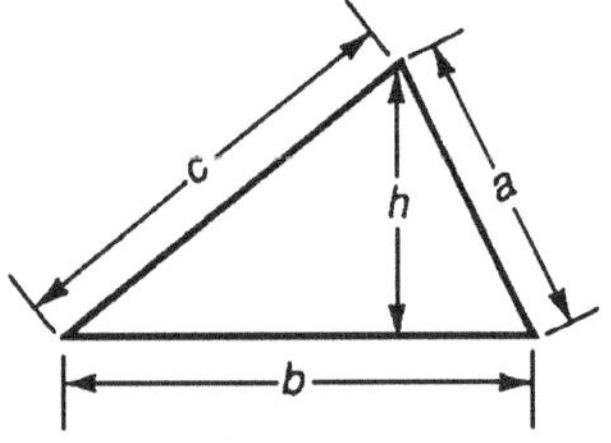

Figure 2.7

Oblique Triangle

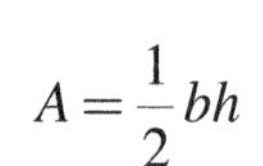

$$A = \frac{1}{2}bh$$

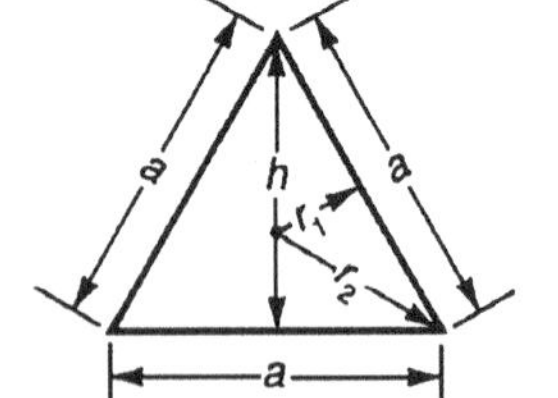

Figure 2.8

Equilateral Triangle

All sides are equal and all angles are 60°.

$$A = \frac{1}{2}ah = \frac{1}{4}a^2\sqrt{3},\qquad h = \frac{1}{2}a\sqrt{3},\qquad r_1 = \frac{a}{2\sqrt{3}},\qquad r_2 = \frac{a}{\sqrt{3}}$$

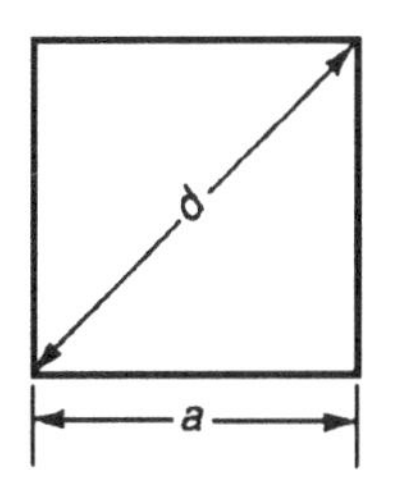

Figure 2.9

Square

All sides are equal, and all angles are 90°.

$$A = a^2, \qquad d = a\sqrt{2}$$

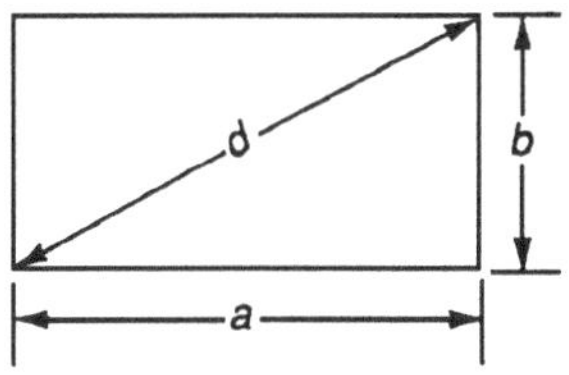

Figure 2.10

Rectangle

Opposite sides are equal and parallel, and all angles are 90°.

$$A = ab, \qquad d = \sqrt{a^2 + b^2}$$

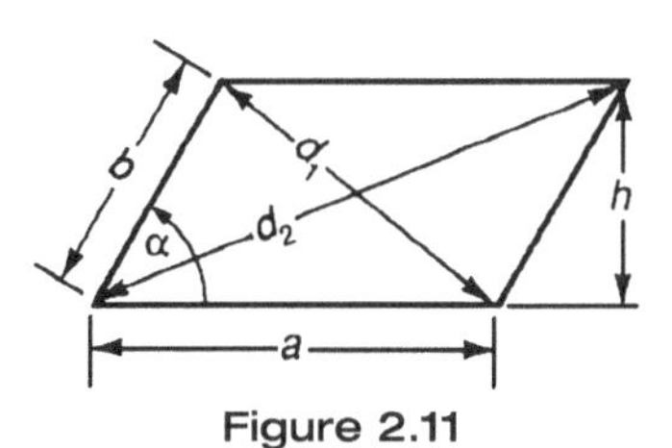

Figure 2.11

Parallelogram

Opposite sides are equal and parallel, and opposite angles are equal.

$$A = ah = ab \sin\alpha, \; d_1 = \sqrt{a^2 + b^2 - 2ab\cos\alpha}, \; d_2 = \sqrt{a^2 + b^2 + 2ab\cos\alpha}$$

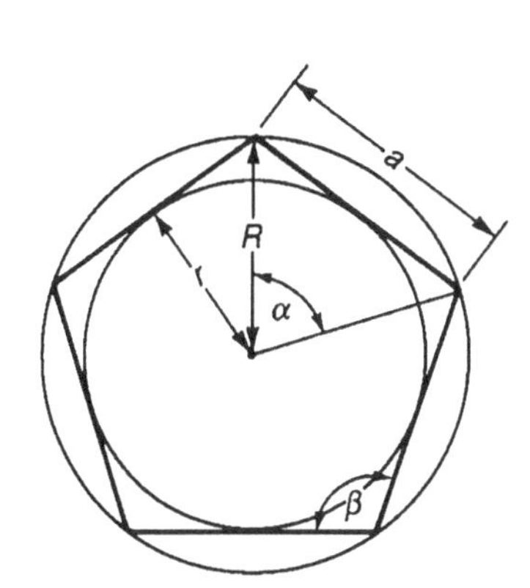

Figure 2.12

Regular Polygon of *n* Sides

All sides and all angles are equal.

$$\beta = \frac{n-2}{n} 180° = \frac{n-2}{n} \pi \text{ radians}, \qquad \alpha \frac{360°}{n} = \frac{2\pi}{n} \text{ radians}, \qquad A = \frac{nar}{2}$$

Circle

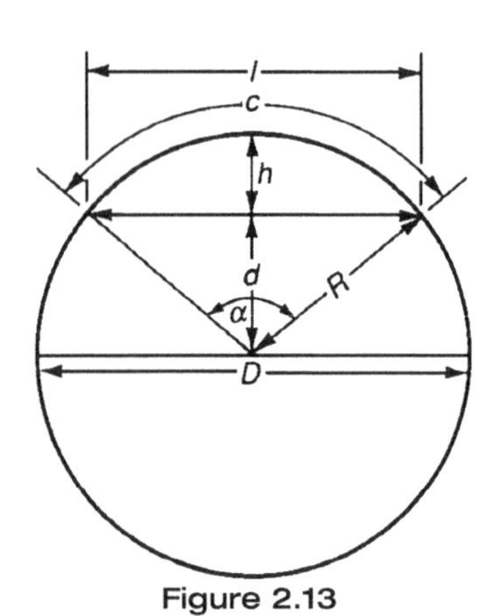

Figure 2.13

Notation. C = circumference, α = central angle in radians

$$C = \pi D = 2\pi R$$

$$c = R\alpha = \frac{1}{2}D\alpha = D\cos^{-1}\frac{d}{R} = D\tan^{-1}\frac{1}{2d}$$

$$l = 2\sqrt{R^2 - d^2} = 2R\sin\frac{\alpha}{2} = 2d\tan\frac{\alpha}{2} = 2d\tan\frac{c}{D}$$

$$d = \frac{1}{2}\sqrt{4R^2 - l^2} = \frac{1}{2}\sqrt{D^2 - l^2} = R\cos\frac{\alpha}{2}$$

$$h = R - d$$

$$\alpha = \frac{c}{R} = \frac{2c}{D} = 2\cos^{-1}\frac{d}{R}$$

$$A_{(circle)} = \pi R^2 = \frac{1}{4}\pi D^2 = \frac{1}{2}RC = \frac{1}{4}DC$$

$$A_{(sector)} = \frac{1}{2}Rc = \frac{1}{2}R^2\alpha = \frac{1}{8}D^2\alpha$$

Ellipse

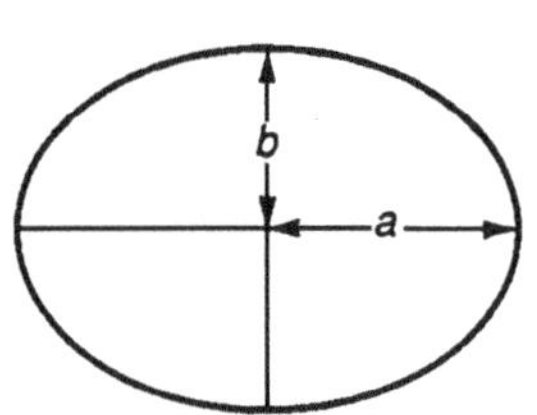

Figure 2.14

$$A = \pi ab$$

$$\text{Perimeter } (s) = \pi(a+b)\left[1 + \frac{1}{4}\left(\frac{a-b}{a+b}\right)^2 + \frac{1}{64}\left(\frac{a-b}{a+b}\right)^4 + \frac{1}{256}\left(\frac{a-b}{a+b}\right)^6 + \cdots\right]$$

$$\text{Perimeter } (s) \approx \pi\frac{a+b}{4}\left[3(1+\lambda) + \frac{1}{1-\lambda}\right], \quad \text{where } \lambda = \left[\frac{a-b}{2(a+b)}\right]^2$$

Parabola

Figure 2.15

$$A = \frac{2}{3}ld$$

Cube

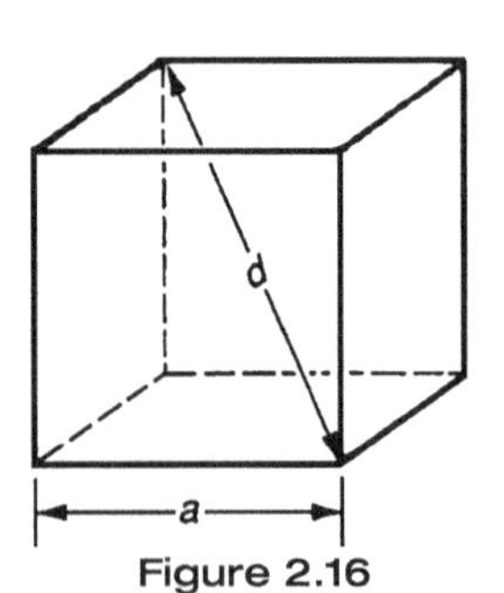

Figure 2.16

$$V = a^3 \qquad d = a\sqrt{3}$$

Total surface area = $6a^2$

Prism or Cylinder

$V =$ (area of base) (altitude, h)

Lateral area = (perimeter of right section)(lateral edge, e)

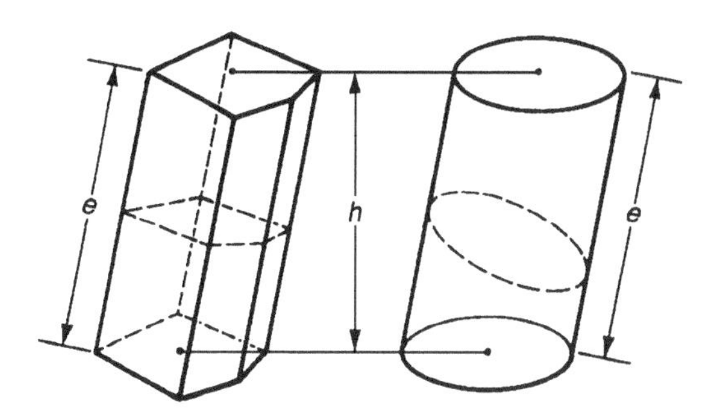

Figure 2.17

Pyramid or Cone

$V = \frac{1}{3}$ (area of base) (altitude, h)

Lateral area of regular figure $= \frac{1}{2}$ (perimeter of base)(slant height, s)

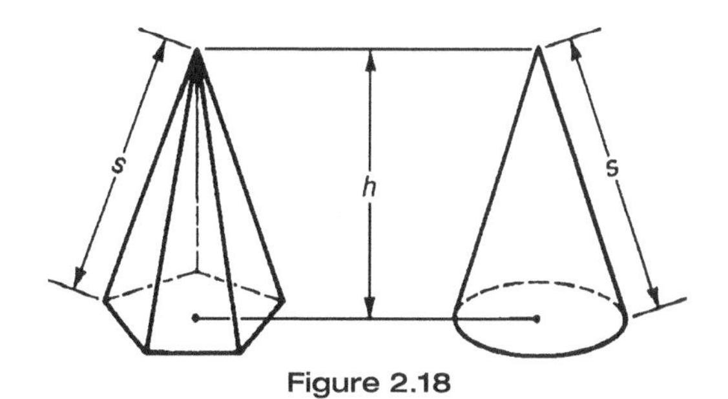

Figure 2.18

Sphere

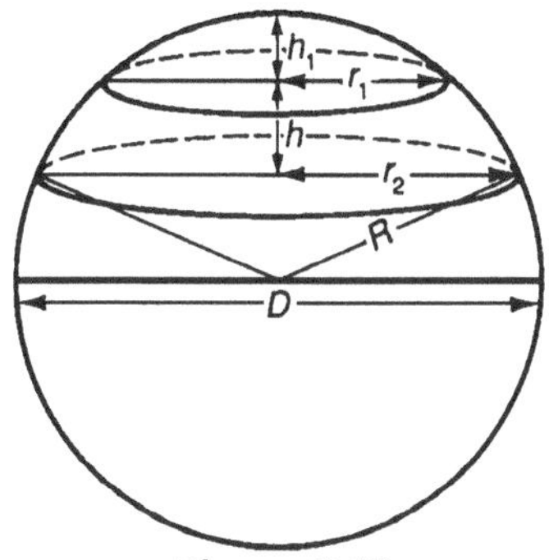

Figure 2.19

$$A_{(sphere)} = 4\pi R^2 = \pi D^2$$

$$A_{(zone)} = 2\pi Rh = \pi Dh$$

$$V_{(sphere)} = \frac{4}{3}\pi R^3 = \frac{1}{6}\pi D^3$$

$$V_{(spherical\ sector)} = \frac{2}{3}\pi R^2 h = \frac{1}{6}\pi D^2 h$$

PLANE ANALYTIC GEOMETRY

Rectangular Coordinates

Let two perpendicular lines, $X'X$ (x-axis) and $Y'Y$ (y-axis) meet at a point O (origin). The position of any point $P(x, y)$ is fixed by the distances x (abscissa) and y (ordinate) from $Y'Y$ and $X'X$, respectively, to P. Values of x are positive to the right and negative to the left of $Y'Y$; values of y are positive above and negative below $X'X$.

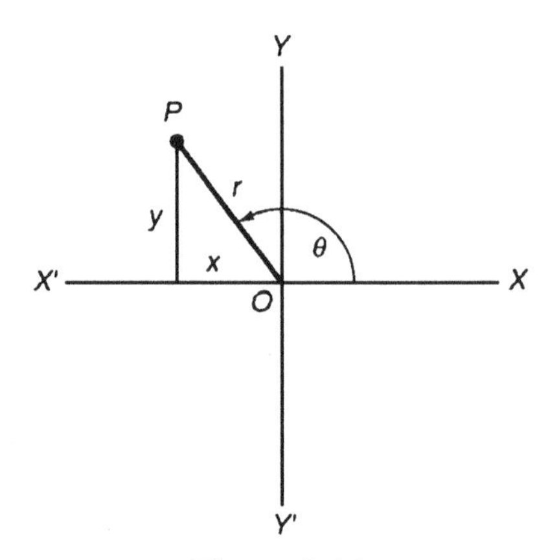

Figure 2.20

Polar Coordinates

Let O (origin or pole) be a point in the plane and OX (initial line) be any line through O. The position of any point $P(r, \theta)$ is fixed by the distance r (radius vector) from O to the point and the angle θ (vectorial angle) measured from OX to OP (Figure 2.20).

A value for r is positive and is measured along the terminal side of θ; a value for θ is positive when measured counterclockwise and negative when measured clockwise.

Relations Connecting Rectangular and Polar Coordinates

$$x = r\cos\theta, \quad y = r\sin\theta$$

$$r = \sqrt{x^2 + y^2}, \quad \theta = \tan^{-1}\frac{y}{x}, \quad \sin\theta = \frac{y}{\sqrt{x^2 + y^2}},$$

$$\cos\theta = \frac{x}{\sqrt{x^2 + y^2}}, \quad \tan\theta = \frac{y}{x}$$

Points and Slopes

Let $P_1(x_1, y_1)$ and $P_2(x_2, y_2)$ be any two points, and let α_1 be the angle from the x axis to P_1P_2, measured counterclockwise.

The length P_1P_2 is $d = \sqrt{(x_2 - x_1)^2 + (y_2 - y_1)^2}$.

The midpoint of P_1P_2 is $\left(\frac{x_1 + x_2}{2}, \frac{y_1 + y_2}{2}\right)$.

The point that divides P_1P_2 in the ratio $n_1{:}n_2$ is $\left(\frac{n_1x_2 + n_2x_1}{n_1 + n_2}, \frac{n_1y_2 + n_2y_1}{n_1 + n_2}\right)$.

The slope of P_1P_2 is $\tan\ \alpha = m = \frac{y_2 - y_1}{x_2 - x_1}$.

The angle between two lines of slopes m_1 and m_2 is $\beta = \tan^{-1}\frac{m_2 - m_1}{1 + m_1m_2}$.

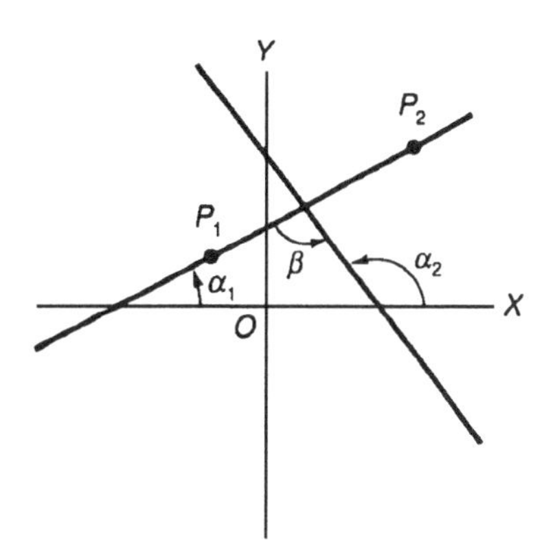

Figure 2.21

Two lines of slopes m_1 and m_2 are perpendicular if $m_2 = -\frac{1}{m_1}$.

Example 2.12

Find the distance between the points (1, –2) and (–4, 2).

Solution

The distance is $d = \sqrt{(x_2 - x_1)^2 + (y_2 - y_1)^2} = \sqrt{[1-(-4)]^2 + [-2-2]^2}$

$= \sqrt{5^2 + 4^2} = 6.4$

Locus and Equation

The collection of all points that satisfy a given condition is called the locus of that condition; the condition expressed by means of the variable coordinates of any point on the locus is called the equation of the locus.

The locus may be represented by equations of three kinds: (1) a rectangular equation involves the rectangular coordinates (x, y); (2) a polar equation involves the polar coordinates (r, θ); and (3) parametric equations express x and y or r and θ in terms of a third independent variable called a parameter.

The following equations are generally given in the system in which they are most simply expressed; sometimes several forms of the equation in one or more systems are given.

Straight Line

$Ax + By + C = 0$ $[-A/B = \text{slope}]$

$y = mx + b$ [m = slope, b = intercept on OY]

$y - y_1 = m(x - x_1)$ [m = slope, $P_1(x_1, y_1)$ is a known point on the line]

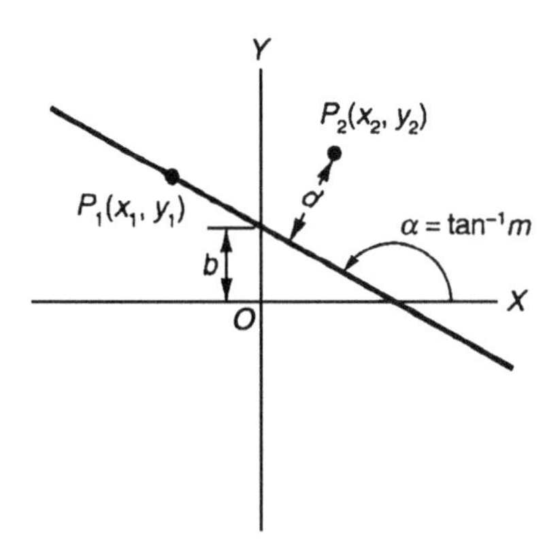

Figure 2.22

Example 2.13

Find the equation of the line passing through the points (2, 1) and (3, –3).

Solution

Slope can be found as $m = \dfrac{y_2 - y_1}{x_2 - x_1} = \dfrac{1-(-3)}{2-3} = -4.$

The point-slope form is $(y - 1) = -4(x - 2)$ or $(y + 3) = -4(x - 3)$.

Simplifying, either equation yields $y = -4x + 9$ or equivalently $4x + y - 9 = 0$.

Example 2.14

Find the equation of the straight line passing through the point (3, 1) and perpendicular to the line passing through the points (3, –2) and (–3, 7).

Solution

Slope of the line passing through (3, –2) and (–3, 7) is

$$m_1 = \frac{y_2 - y_1}{x_2 - x_1} = \frac{7-(-2)}{-3-3} = -\frac{3}{2}$$

Slope of the line passing through (3, 1) is $m_2 = -\dfrac{1}{m_1} = \dfrac{2}{3}$, since the two lines are perpendicular to each other.

Equation is $(y-1) = \dfrac{2}{3}(x - 3)$ or $2x - 3y - 3 = 0$.

Circle

The locus of a point at a constant distance (radius) from a fixed point C (center) is a circle.

$(x-h)^2+(y-k)^2=a^2$	$C(h, k)$, radius $=a$	
$r^2+b^2\pm 2\,br\cos(\theta-\beta)=a^2$	$C(b,\beta)$, radius $=a$	[Figure 2.23(a)]
$x^2+y^2=2ax$	$C(a, 0)$, radius $=a$	
$r=2a\cos\theta$	$C(a, 0)$, radius $=a$	[Figure 2.23(b)]
$x^2+y^2=2ay$	$C(0, a)$, *radius* $=a$	
$r=2a\sin\theta$	$C(0, a)$, *radius* $=a$	[Figure 2.23(c)]
$x^2+y^2=a^2$	$C(0, 0)$, radius $=a$	
$r=a$	$C(0, 0)$, radius $=a$	[Figure 2.23(d)]
$x=a\cos\phi$, $y=a\sin\phi$	ϕ = angle from OX to radius	

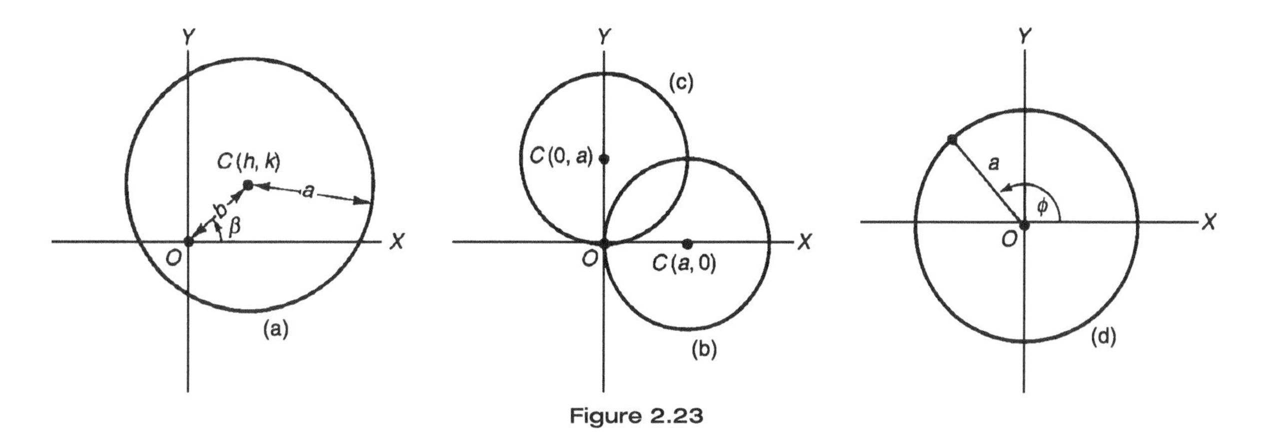

Figure 2.23

Example 2.15

Find the equation of the circle (i) with center at (0, 0) and radius 3, and (ii) with center at (1, 2) and radius 4.

Solution

Equation of a circle with center at (h, k) and radius r is $(x-h)^2+(y-k)^2=r^2$.

i) Here, $h=0$, $k=0$, and $r=3$; equation of the circle is $x^2+y^2=3^2$ or $x^2+y^2-9=0$
ii) Here, $h=1$, $k=2$, and $r=4$; equation of the circle is $(x-1)^2+(y-2)^2=4^2$
Simplifying, $x^2-2x+y^2-4y-11=0$

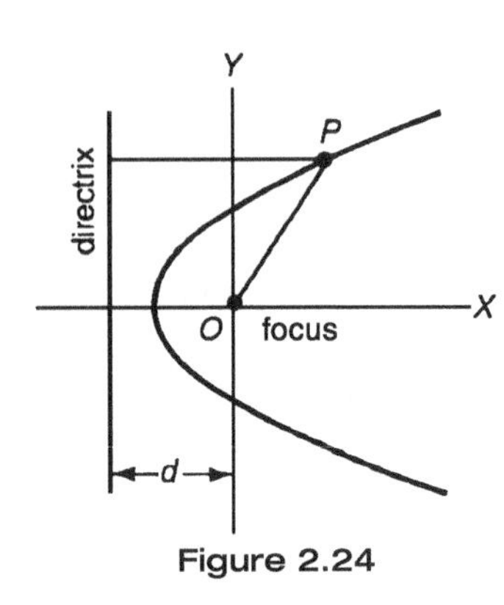

Figure 2.24

Conic

A **conic** is the locus of a point whose distance from a fixed point (focus) is in a constant ratio e, called the eccentricity, to its distance from a fixed straight line (directrix).

$$x^2 + y^2 = e^2(d + x)^2 \qquad d = \text{distance from focus to directrix}$$

$$r = \frac{de}{1 - e\cos\theta}$$

The conic is called a parabola when $e = 1$, an ellipse when $e < 1$, and a hyperbola when $e > 1$.

Example 2.16

Which conic section is represented by each of the following equations: (i) $x^2 + 4xy + 4y^2 + 2x = 10$ and (ii) $x^2 + y^2 - 2x - 4y - 11 = 0$.

Solution

The general equation of a conic section is $Ax^2 + 2Bxy + Cy^2 + 2Dx + 2Ey + F = 0$, where both A and C are not zeros. If $B^2 - AC > 0$, a *hyperbola* is defined; if $B^2 - AC = 0$, a *parabola* is defined; if $B^2 - AC < 0$, an *ellipse* is defined. (Note: If B is zero and A = C, a *circle* is defined.)

If A = B = C = 0, a *straight line* is defined.

If B = 0, A = C, a *circle* is defined with equation $x^2 + y^2 + 2ax + 2by + c = 0$.

Center is at (–a, –b) and radius = $\sqrt{a^2 + b^2 - c}$ provided $a^2 + b^2 - c > 0$.

(i) Here, A = 1, B = 2, and C = 4. Then, $B^2 - AC = (2)^2 - (1)(4) = 0$. The equation represents a parabola.

(ii) Here, A = 1, B = 0, C = 1. Because A = C and B = 0, the equation represents a circle.

[Note that the center is at (1, 2), and the radius is $\sqrt{(-1)^2 + (-2)^2 - (-11)} = 4$.]

Parabola

A parabola is a special case of a conic where $e = 1$.

$(y-k)^2 = a(x-h)$ Vertex (h, k), axis $\parallel OX$

$y^2 = ax$ Vertex $(0, 0)$, axis along OX [Figure 2.25(a)]

$(x-h)^2 = a(y-k)$ Vertex (h, k), axis $\parallel OY$

$x^2 = ay$ Vertex $(0, 0)$, axis along OY [Figure 2.25(b)]

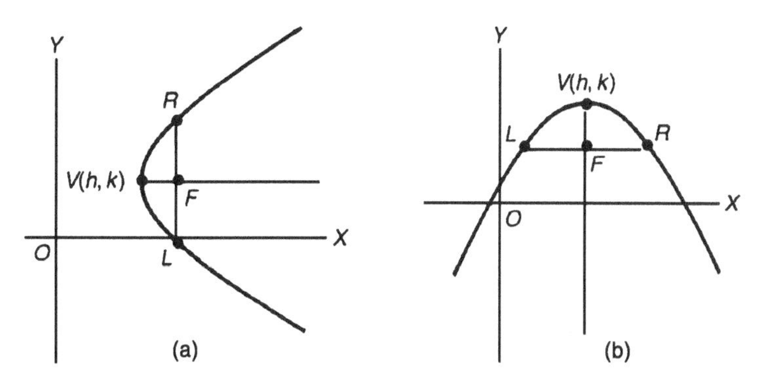

Figure 2.25

Distance from vertex to focus $= VF = \frac{1}{4}a$. Latus rectum $= LR = a$.

Example 2.17

Find the equation of a parabola (i) with center at (0, 0) and focus at (4, 0) and (ii) with center at (4, 2) and focus at (8, 2).

Solution

The equation of a parabola with center at (h, k) and focus at $(h + p/2, k)$ is given as

$(y-k)^2 = 2p(x-h)$

(i) $h = 0$; $k = 0$; $h + p/2 = 4$; then, $p = 8$ and the equation is $y^2 = 16x$.

(ii) $h = 4$; $k = 2$; $h + p/2 = 8$; then, $p = 8$ and the equation is $(y-2)^2 = 16(x-4)$.

Ellipse

This is a special case of a conic where $e < 1$.

$$\frac{(x-h)^2}{a^2} + \frac{(y-k)^2}{b^2} = 1 \qquad \text{Center } (h, k), \text{ axes } \parallel OX, OY$$

$$\frac{x^2}{a^2} + \frac{y^2}{b^2} = 1 \qquad \text{Center } (0, 0), \text{ axes along } OX, OY$$

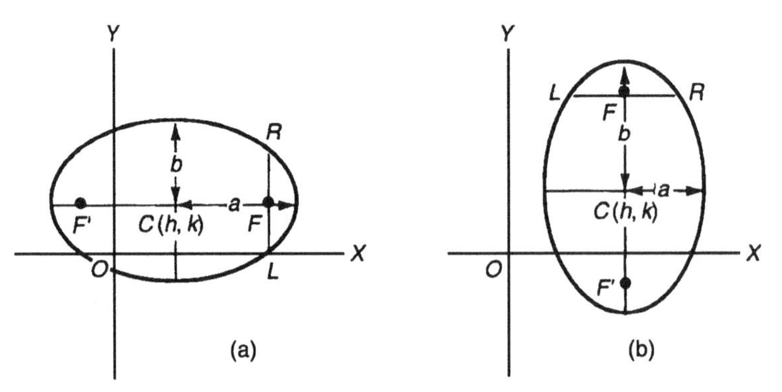

Figure 2.26

	$a > b$, Figure 2.26(a)	$b > a$, Figure 2.26(b)
Major axis	$2a$	$2b$
Minor axis	$2b$	$2a$
Distance from center to either focus	$\sqrt{a^2 - b^2}$	$\sqrt{b^2 - a^2}$
Latus rectum	$\frac{2b^2}{a}$	$\frac{2a^2}{b}$
Eccentricity, e	$\sqrt{\frac{a^2 - b^2}{a}}$	$\sqrt{\frac{b^2 - a^2}{b}}$
Sum of distances of any point P from the foci, $PF' + PF$	$2a$	$2b$

Example 2.18

Find the equation of an ellipse with center at origin, x-axis intercept (4, 0), and y-axis intercept (0, 2).

Solution

Equation of an ellipse with center at origin and x-axis intercept of $(a, 0)$ and y-axis intercept of $(0, b)$ is $\frac{x^2}{a^2} + \frac{y^2}{b^2} = 1$. Here, $a = 4$ and $b = 2$; then, the equation is $\frac{x^2}{4^2} + \frac{y^2}{2^2} = 1$.

Hyperbola

This is a special case of a conic where $e > 1$.

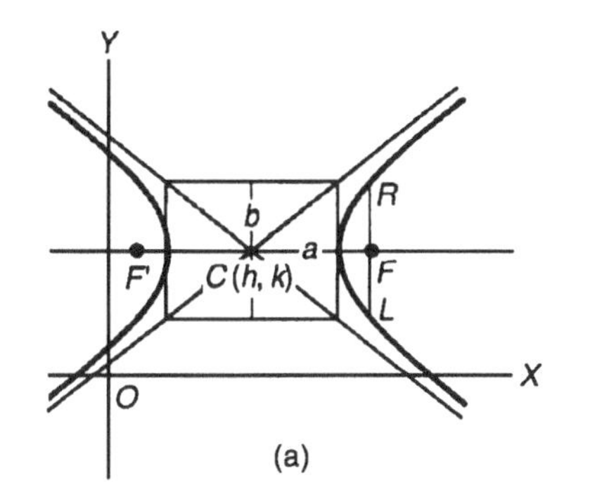

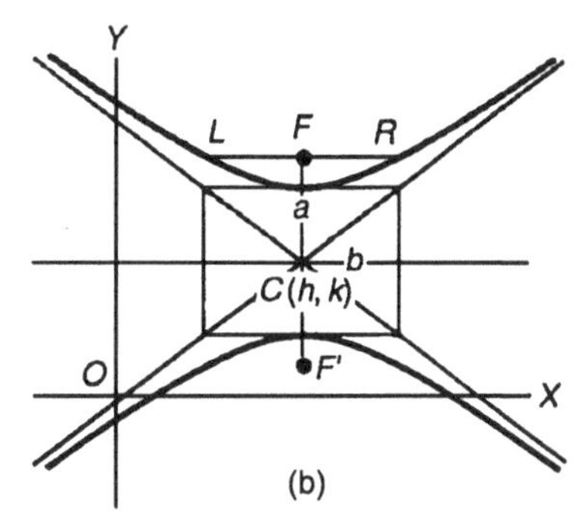

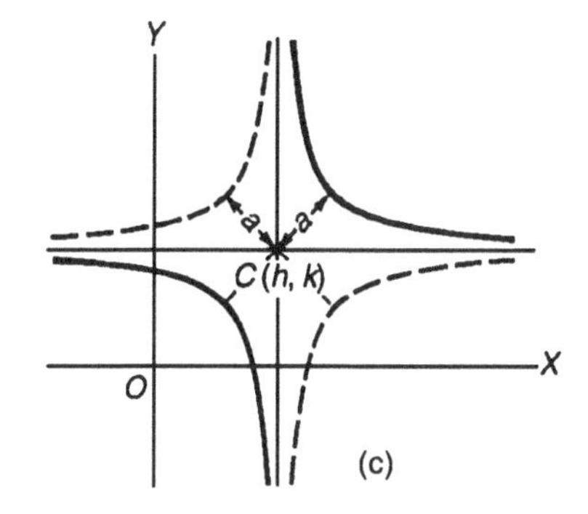

Figure 2.27

$$\frac{(x-h)^2}{a^2} - \frac{(y-k)^2}{b^2} = 1 \qquad C(h, k), \text{ transverse axis } \| \ OX$$

$$\frac{x^2}{a^2} - \frac{y^2}{b^2} = 1 \qquad C(0, 0), \text{ transverse axis along } OX$$

$$\frac{(y-k)^2}{a^2} - \frac{(x-h)^2}{b^2} = 1 \qquad C(h, k), \text{ transverse axis } \| \ OY$$

$$\frac{y^2}{a^2} - \frac{x^2}{b^2} = 1 \qquad C(0, 0), \text{ transverse along } OY$$

Transverse axis = $2a$; conjugate axis = $2b$

Distance from center to either focus = $\sqrt{a^2 + b^2}$

Latus rectum = $\dfrac{2b^2}{a}$

Eccentricity, $e = \dfrac{\sqrt{a^2 + b^2}}{a}$

Difference of distances of any point from the foci = $2a$.

The asymptotes are two lines through the center to which the branches of the hyperbola approach arbitrarily closely; their slopes are $\pm b/a$ [Figure 2.27(a)] or $\pm a/b$ [Figure 2.27(b)].

The rectangular (equilateral) hyperbola has $b = a$. The asymptotes are perpendicular to each other.

$$(x-h)(y-k) = \ \pm e = \sqrt{2} \qquad \text{Center } (h, k), \text{ asymptotes } \| \ OX, OY$$

$$xy = \ \pm e = \sqrt{2} \qquad \text{Center } (0, 0), \text{ asymptotes along } OX, \ OY$$

The + sign gives the solid curves in Figure 2.27(c); the – sign gives the dotted curves in Figure 2.27(c).

Example 2.19

What is the equation of the hyperbola with center at origin, passing through (±2, 0), and an eccentricity of $\sqrt{10}$?

Solution

Equation of a hyperbola with center at origin, x-axis intercepts of (± a, 0), and eccentricity e is $\frac{x^2}{a^2} - \frac{y^2}{b^2} = 1$, where $b = a\sqrt{e^2 - 1}$. Here, $a = 2$; $e = \sqrt{10}$; then, $b = a\sqrt{e^2 - 1} = 2\sqrt{10 - 1} = 6$; equation is $\frac{x^2}{2^2} - \frac{y^2}{6^2} = 1$.

VECTORS

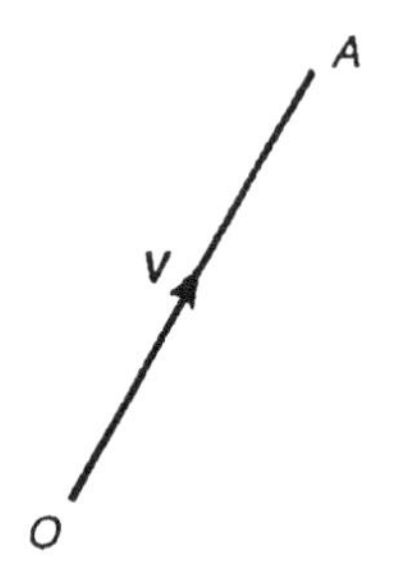

Figure 2.28

Definition and Graphical Representation of a Vector

A vector (**V**) is a quantity that is completely specified by magnitude *and* a direction. A scalar (s) is a quantity that is completely specified by a magnitude *only*.

The vector (**V**) may be represented geometrically by the segment $\overrightarrow{OA}$, the length of OA signifying the magnitude of **V** and the arrow carried by OA signifying the direction of **V**. The segment $\overrightarrow{AO}$ represents the vector –**V**.

Graphical Summation of Vectors

If $\mathbf{V}_1$ and $\mathbf{V}_2$ are two vectors, their graphical sum $\mathbf{V} = \mathbf{V}_1 + \mathbf{V}_2$ is formed by drawing the vector $\mathbf{V}_1 = \overrightarrow{OA}$, from any point O, and the vector $\mathbf{V}_2 = \overrightarrow{AB}$ from the end of $\mathbf{V}_1$ and joining O and B; then $\mathbf{V} = \overrightarrow{OB}$. Also, $\mathbf{V}_1 + \mathbf{V}_2 = \mathbf{V}_2 + \mathbf{V}_1$ and $\mathbf{V}_1 + \mathbf{V}_2 - \mathbf{V} = 0$ (Figure 2.29(a)).

Similarly, if $\mathbf{V}_1, \mathbf{V}_2, \mathbf{V}_3, \ldots, \mathbf{V}_n$ are any number of vectors drawn so that the initial point of one is the end point of the preceding one, then their graphical sum $\mathbf{V} = \mathbf{V}_1 + \mathbf{V}_2 + \ldots + \mathbf{V}_n$ is the vector joining the initial point of $\mathbf{V}_1$ with the end point of $\mathbf{V}_n$ (Figure 2.29(b)).

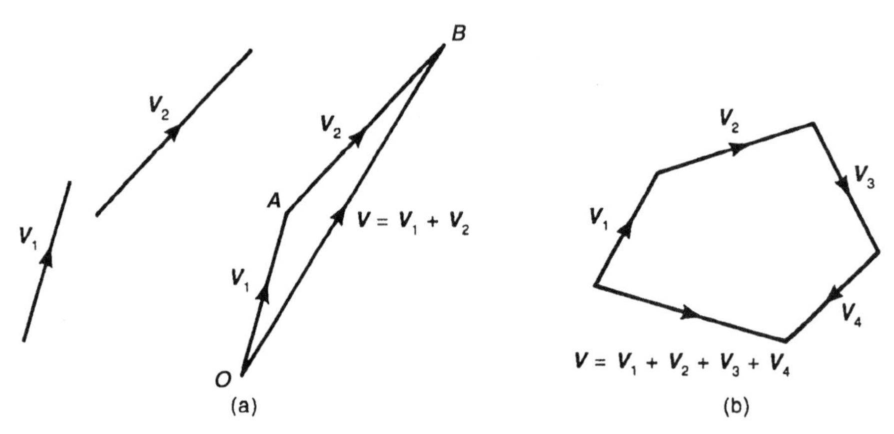

Figure 2.29

Analytic Representation of Vector Components

A vector **V** that is considered as lying in the *x*-*y* coordinate plane (Figure 2.30(a)) is completely determined by its horizontal and vertical components *x* and *y*. If **i** and **j** represent vectors of unit magnitude along *OX* and *OY*, respectively, and *a* and *b* are the magnitude of *x* and *y*, then **V** may be represented by $\mathbf{V} = a\mathbf{i} + b\mathbf{j}$, its magnitude by $|\mathbf{V}| = +\sqrt{a^2 + b^2}$, and its direction by $\alpha = \tan^{-1} b/a$.

A vector **V** in three-dimensional in space is completely determined by its components *x*, *y*, and *z* along three mutually perpendicular lines *OX*, *OY*, and *OZ*, directed as shown in Figure 2.30(b). If **i**, **j**, and **k** represent vectors of unit magnitude along *OX*, *OY*, *OZ*, respectively, and *a*, *b*, and *c* are the magnitudes of the components *x*, *y*, and *z*, respectively, then **V** may be represented by $\mathbf{V} = a\mathbf{i} + b\mathbf{j} + c\mathbf{k}$, its magnitude by, $|\mathbf{V}| = +\sqrt{a^2 + b^2 + c^2}$, and its direction by $\cos\alpha : \cos\beta : \cos\gamma = a : b : c$.

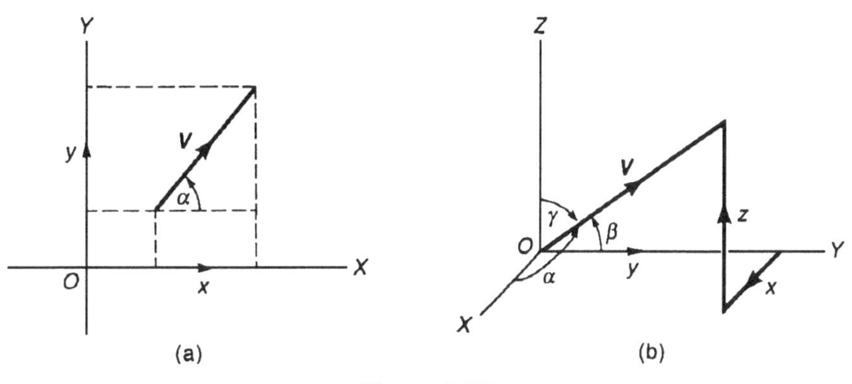

Figure 2.30

Properties of Vectors

$$\mathbf{V} = a\mathbf{i} + b\mathbf{j} \quad \text{or} \quad \mathbf{V} = a\mathbf{i} + b\mathbf{j} + c\mathbf{k}$$

Vector Sum V of any Number of Vectors, V_1, V_2, V_3,...

$$\mathbf{V} = \mathbf{V}_1 + \mathbf{V}_2 + \mathbf{V}_3 + \ldots = (a_1 + a_2 + a_3 + \ldots)\,\mathbf{i} + (b_1 + b_2 + b_3 + \ldots)\,\mathbf{j} + (c_1 + c_2 + c_3 + \ldots)\,\mathbf{k}$$

Product of a Vector V and a Scalar *s*

The product $s\mathbf{V}$ has the same direction as **V**, and its magnitude is *s* times the magnitude of **V**.

$$s\mathbf{V} = (sa)\,\mathbf{i} + (sb)\,\mathbf{j} + (sc)\,\mathbf{k}$$

$$(s_1 + s_2)\,\mathbf{V} = s_1\mathbf{V} + s_2\mathbf{V} \qquad (\mathbf{V}_1 + \mathbf{V}_2)s = \mathbf{V}_1 s + \mathbf{V}_2 s$$

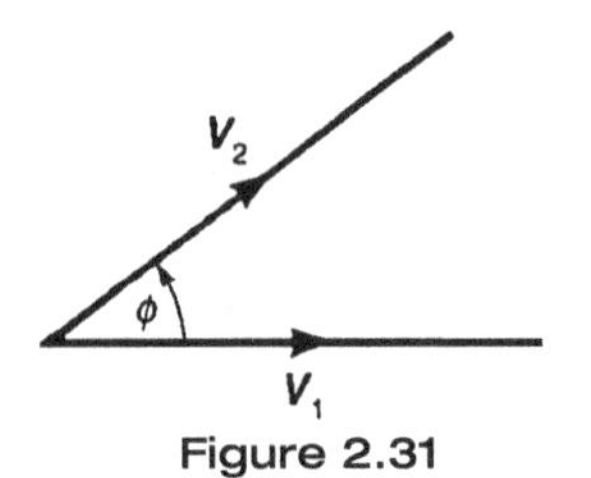

Figure 2.31

Scalar Product or Dot Product of Two Vectors: $\mathbf{V}_1 \bullet \mathbf{V}_2$

$$\mathbf{V}_1 \bullet \mathbf{V}_2 = |\mathbf{V}_1||\mathbf{V}_2|\cos\phi, \text{ where } \phi \text{ is the angle between } \mathbf{V}_1 \text{ and } \mathbf{V}_2$$

$$\mathbf{V}_1 \bullet \mathbf{V}_2 = \mathbf{V}_2 \bullet \mathbf{V}_1; \quad \mathbf{V}_1 \bullet \mathbf{V}_1 = |\mathbf{V}_1|^2; \quad (\mathbf{V}_1 + \mathbf{V}_2) \bullet \mathbf{V}_3 = \mathbf{V}_1 \bullet \mathbf{V}_3 + \mathbf{V}_2 \bullet \mathbf{V}_3$$

$$(\mathbf{V}_1 + \mathbf{V}_2) \bullet (\mathbf{V}_3 + \mathbf{V}_4) = \mathbf{V}_1 \bullet \mathbf{V}_3 + \mathbf{V}_1 \bullet \mathbf{V}_4 + \mathbf{V}_2 \bullet \mathbf{V}_3 + \mathbf{V}_2 \bullet \mathbf{V}_4$$

$$\mathbf{i} \bullet \mathbf{i} = \mathbf{j} \bullet \mathbf{j} = \mathbf{k} \bullet \mathbf{k} = 1; \quad \mathbf{i} \bullet \mathbf{j} = \mathbf{j} \bullet \mathbf{k} = \mathbf{k} \bullet \mathbf{i} = 0$$

In a plane, $\mathbf{V}_1 \bullet \mathbf{V}_2 = a_1 a_2 + b_1 b_2$; in space, $\mathbf{V}_1 \bullet \mathbf{V}_2 = a_1 a_2 + b_1 b_2 + c_1 c_2$.

The scalar product of two vectors $\mathbf{V}_1 \bullet \mathbf{V}_2$ is a scalar quantity and may physically represent the work done by a constant force of magnitude $|\mathbf{V}_1|$ on a particle moving through a distance $|\mathbf{V}_2|$, where ϕ is the angle between the direction of the force and the direction of motion.

Vector Product or Cross Product of Two Vectors: $\mathbf{V}_1 \times \mathbf{V}_2$

The vector product is $\mathbf{V}_1 \times \mathbf{V}_2 = \mathbf{l}\ |\mathbf{V}_1||\mathbf{V}_2| \sin\phi$, where ϕ is the angle from $\mathbf{V}_1$ to $\mathbf{V}_2$ and $\mathbf{l}$ is a unit vector perpendicular to the plane of the vectors $\mathbf{V}_1$ to $\mathbf{V}_2$ and so directed that a right-handed screw driven in the direction of $\mathbf{l}$ would carry $\mathbf{V}_1$ into $\mathbf{V}_2$.

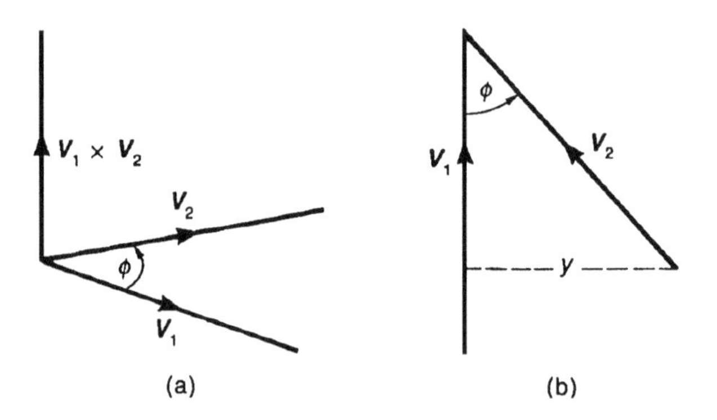

Figure 2.32

$$\mathbf{V}_1 \times \mathbf{V}_2 = -\mathbf{V}_2 \times \mathbf{V}_1; \quad \mathbf{V}_1 \times \mathbf{V}_1 = 0$$

$$(\mathbf{V}_1 + \mathbf{V}_2) \times \mathbf{V}_3 = \mathbf{V}_1 \times \mathbf{V}_3 + \mathbf{V}_2 \times \mathbf{V}_3$$

$$\mathbf{V}_1 \bullet (\mathbf{V}_2 \times \mathbf{V}_3) = \mathbf{V}_2 \bullet (\mathbf{V}_3 \times \mathbf{V}_1) = \mathbf{V}_3 \bullet (\mathbf{V}_1 \times \mathbf{V}_2)$$

$$\mathbf{i} \times \mathbf{i} - \mathbf{j} \times \mathbf{j} = \mathbf{k} \times \mathbf{k} = 0; \quad \mathbf{i} \times \mathbf{j} = \mathbf{k}; \quad \mathbf{j} \times \mathbf{k} = \mathbf{i}; \quad \mathbf{k} \times \mathbf{i} = \mathbf{j}$$

In the x-y plane, $\mathbf{V}_1 \times \mathbf{V}_2 = (a_1 b_2 - a_2 b_1)\mathbf{k}$.

In space, $\mathbf{V}_1 \times \mathbf{V}_2 = (b_2 c_3 - b_3 c_2)\mathbf{i} + (c_3 a_1 - c_1 a_3)\mathbf{j} + (a_1 b_2 - a_2 b_1)\mathbf{k}$.

The vector product of two vectors is a vector quantity and may physically represent the moment of a force $\mathbf{V}_1$ about a point O placed so that the moment arm is $y = |\mathbf{V}_2| \sin\phi$ (see Figure 2.32(b)).

Example 2.20

Vectors **A** and **B** are defined as: $\mathbf{A} = i - 2j + 3k$, $\mathbf{B} = 2i + j - 2k$

Find (i) **A** + **B**, (ii) **A** – **B**, (iii) 2**A**, (iv) |**A**|, (v) dot product **A**• **B**, and (vi) the cross product **A** × **B**.

Solution

(i) $\mathbf{A} + \mathbf{B} = 3i - j + k$ (ii) $\mathbf{A} - \mathbf{B} = -i - 3j + 5k$ (iii) $2\mathbf{A} = 2i - 4j + 6k$

(iv) $|\mathbf{A}| = \sqrt{1^2 + (-2)^2 + 3^2} = \sqrt{14}$

(v) Dot product $\mathbf{A} \bullet \mathbf{B} = (2) + (-2) + (-6) = -6$

(vi) Cross product $\mathbf{A} \times \mathbf{B} = \begin{vmatrix} i & j & k \\ 1 & -2 & 3 \\ 2 & 1 & -2 \end{vmatrix} = i + 8j + 5k$

LINEAR ALGEBRA

Matrix Operations

Matrices are rectangular arrays of real or complex numbers. Their great importance arises from the variety of operations that may be performed on them. Using the standard convention, the across-the-page lines are called **rows** and the up-and-down-the-page lines are **columns**. Entries in a matrix are **addressed** with double subscripts (always row first, then column). Thus the matrix

$$\mathbf{A} = \begin{bmatrix} 1 & 2 & 3 \\ 0 & 9 & -3 \end{bmatrix}$$

is 2 × 3, and the "9" is a_{22}. The "2" is a_{12} and the "0" is a_{21}. One also can refer to entries with square brackets, the "9" being $[A]_{22}$ and the "3" $[A]_{13}$.

If two matrices are the same size, they may be added: $[A + B]_{ij} = [A]_{ij} + [B]_{ij}$. Thus,

$$\begin{bmatrix} 1 & 2 & 3 \\ 0 & 9 & -3 \end{bmatrix} + \begin{bmatrix} 1 & 3 \\ 2 & 4 \end{bmatrix}$$

is not defined, but

$$\begin{bmatrix} 1 & 2 & 3 \\ 0 & 9 & -3 \end{bmatrix} + \begin{bmatrix} 1 & 3 & 5 \\ 2 & 4 & 6 \end{bmatrix} = \begin{bmatrix} 2 & 5 & 8 \\ 2 & 13 & 3 \end{bmatrix}$$

is proper.

Any matrix may be multiplied by a **scalar** (a number): $[c\mathbf{A}]_{ij} = c[\mathbf{A}]_{ij}$, so that

$$5\begin{bmatrix} 1 & 5 \\ 0 & 6 \end{bmatrix} = \begin{bmatrix} 5 & 25 \\ 0 & 30 \end{bmatrix}$$

The most peculiar matrix operation (and the most useful) is matrix multiplication. If **A** is $m \times n$ and **B** is $n \times p$, then $\mathbf{A} \bullet \mathbf{B}$ (or **AB**) is of size $m \times p$, and

$$[AB]_{ij} = \sum_{k=1}^{n} a_{jk} \bullet b_{kj}$$

The **dot product** (scalar product) of the *i*th row of **A** with the *j*th column of **B**, as in

$$\begin{bmatrix} 1 & 2 & 3 \\ 0 & 9 & -3 \end{bmatrix} \bullet \begin{bmatrix} 1 & 5 \\ 0 & 6 \end{bmatrix}$$

is not defined (owing to the mismatch of row and column lengths), but

$$\begin{bmatrix} 1 & 2 & 3 \\ 0 & 9 & -3 \end{bmatrix} \bullet \begin{bmatrix} 1 & 5 \\ 0 & 6 \\ 7 & 8 \end{bmatrix} = \begin{bmatrix} 1\,(1)+2\,(0)+3\,(7) & 1\,(5)+2\,(6)+3\,(8) \\ 0\,(1)+9\,(0)-3\,(7) & 0\,(5)+9\,(6)-3\,(8) \end{bmatrix} = \begin{bmatrix} 22 & 41 \\ -21 & 30 \end{bmatrix}$$

is correct.

A matrix with only one row or one column is called a vector, so a matrix times a vector is a vector (if defined). Thus $\mathbf{A}\,(m \times n) \bullet \mathbf{X}\,(n \times 1) = \mathbf{Y}(m \times 1)$, so a matrix can be thought of as an **operator** that takes vectors to vectors.

Another useful way of working with matrices is transposition: If **A** is $m \times n$, $\mathbf{A}^t$ is $n \times m$ and is the result of interchanging rows and columns. Hence

$$\begin{bmatrix} 1 & 2 & 3 \\ 0 & 9 & -3 \end{bmatrix}^t = \begin{bmatrix} 1 & 0 \\ 2 & 9 \\ 3 & -3 \end{bmatrix}$$

These various operations interact in the usual pleasant ways (and one decidedly unpleasant way); the standard convention is that all of the following combinations are defined:

$$\mathbf{A} + \mathbf{B} = \mathbf{B} + \mathbf{A}$$
$$\mathbf{A} + (\mathbf{B} + \mathbf{C}) = (\mathbf{A} + \mathbf{B}) + \mathbf{C}$$
$$c(\mathbf{A} + \mathbf{B}) = c\mathbf{A} + c\mathbf{B}$$
$$(c + d)\,\mathbf{A} = c\mathbf{A} + d\mathbf{A}$$
$$(-1)\mathbf{A} + \mathbf{A} = (0\mathbf{A})$$
$$\mathbf{A} \bullet \mathbf{B} \neq \mathbf{B} \bullet \mathbf{A} \text{ (in general)}$$
$$\mathbf{A} \bullet (\mathbf{B} \bullet \mathbf{C}) = (\mathbf{A} \bullet \mathbf{B}) \bullet \mathbf{C}$$
$$\mathbf{A} \bullet (\mathbf{B} + \mathbf{C}) = \mathbf{A} \bullet \mathbf{B} + \mathbf{A} \bullet \mathbf{C}$$
$$(\mathbf{A} + \mathbf{B}) \bullet \mathbf{C} = \mathbf{A} \bullet \mathbf{C} + \mathbf{B} \bullet \mathbf{C}$$
$$(\mathbf{A} + \mathbf{B})^t = \mathbf{A}^t + \mathbf{B}^t$$
$$(\mathbf{A} \bullet \mathbf{B})^t = \mathbf{B}^t \bullet \mathbf{A}^t$$

In addition, matrices **I**, which are $n \times n$ and whose entries are 1 on the diagonal $i = j$ and 0 elsewhere, are multiplicative identities: $\mathbf{A} \bullet \mathbf{I} = \mathbf{A}$ and $\mathbf{I} \bullet \mathbf{A} = \mathbf{A}$. Here the two **I** matrices may be different sizes; for example,

$$\begin{bmatrix} 1 & 2 & 3 \\ 0 & 9 & -3 \end{bmatrix} \bullet \begin{bmatrix} 1 & 0 & 0 \\ 0 & 1 & 0 \\ 0 & 0 & 1 \end{bmatrix} = \begin{bmatrix} 1 & 2 & 3 \\ 0 & 9 & -3 \end{bmatrix}$$

but

$$\begin{bmatrix} 1 & 0 \\ 0 & 1 \end{bmatrix} \bullet \begin{bmatrix} 1 & 2 & 3 \\ 0 & 9 & -3 \end{bmatrix} = \begin{bmatrix} 1 & 2 & 3 \\ 0 & 9 & -3 \end{bmatrix}$$

I is called the identity matrix, and the size is understood from context.

Example **2.21**

Verify that the transpose of $\mathbf{A} + \mathbf{BC}$ is $\mathbf{C}'\mathbf{B}' + \mathbf{A}'$ if

$$\mathbf{A} = \begin{bmatrix} 1 & 1 \\ 2 & 3 \end{bmatrix} \quad \mathbf{B} = \begin{bmatrix} 1 & 2 & 3 \\ 0 & 9 & -3 \end{bmatrix} \quad \mathbf{C} = \begin{bmatrix} 1 & 5 \\ 0 & 6 \\ 7 & 8 \end{bmatrix}$$

Solution

$$\mathbf{BC} \begin{bmatrix} 22 & 41 \\ -21 & 30 \end{bmatrix}, \text{ so } \mathbf{A} + \mathbf{BC} = \begin{bmatrix} 23 & 42 \\ -19 & 33 \end{bmatrix} \text{ and } [\mathbf{A} + \mathbf{BC}]' = \begin{bmatrix} 23 & -19 \\ 42 & 33 \end{bmatrix}$$

On the other hand, $\mathbf{C}'\mathbf{B}' = \begin{bmatrix} 1 & 0 & 7 \\ 5 & 6 & 8 \end{bmatrix} \bullet \begin{bmatrix} 1 & 0 \\ 2 & 9 \\ 3 & -3 \end{bmatrix} = \begin{bmatrix} 22 & -21 \\ 41 & 30 \end{bmatrix}$ and

$$\mathbf{A}' = \begin{bmatrix} 1 & 2 \\ 1 & 3 \end{bmatrix}, \text{ so } \mathbf{A}' + \mathbf{C}'\mathbf{B}' = \mathbf{C}'\mathbf{B}' + \mathbf{A}' = \begin{bmatrix} 23 & -19 \\ 42 & 33 \end{bmatrix}$$

Types of Matrices

Matrices are classified according to their appearance or the way they act. If **A** is square and $\mathbf{A}' = \mathbf{A}$, then **A** is called symmetric. If $\mathbf{A}' = -\mathbf{A}$, then it is skew-symmetric.

If **A** has complex entries, $\mathbf{A}^*$ then is called the Hermitian adjoint of **A**. If $\mathbf{A}^* = \mathbf{A}'$ (complex conjugate), then

$$\begin{bmatrix} 1+i & i \\ 3 & 4-i \end{bmatrix}^* = \begin{bmatrix} 1-i & -i \\ 3 & 4+i \end{bmatrix}' = \begin{bmatrix} 1-i & 3 \\ -i & 4+i \end{bmatrix}$$

If $\mathbf{A} = \mathbf{A}^*$, then A is called Hermitian. If $\mathbf{A}^* = -\mathbf{A}$, the name is skew-Hermitian.

If **A** is square and $a_{ij} = 0$ unless $i = j$, **A** is called diagonal. If **A** is square and zero below the diagonal ($[\mathbf{A}]_{ij} = 0$ if $i > j$), **A** is called upper triangular. The transpose of such a matrix is called lower triangular.

If A is square and there is a matrix $\mathbf{A}^{-1}$ such that $\mathbf{A}^{-1} \bullet \mathbf{A} = \mathbf{A} \bullet \mathbf{A}^{-1} = \mathbf{I}$, **A** is nonsingular. Otherwise, it is singular. If **A** and **B** are both nonsingular $n \times n$ matrices, then **AB** is nonsingular and $(\mathbf{AB})^{-1} = \mathbf{B}^{-1}\mathbf{A}^{-1}$, because $(\mathbf{AB})(\mathbf{B}^{-1}\mathbf{A}^{-1}) = \mathbf{A}(\mathbf{B}\,\mathbf{B}^{-1})\mathbf{A}^{-1} = \mathbf{AIA}^{-1} = \mathbf{AA}^{-1} = \mathbf{I}$, as does $(\mathbf{B}^{-1}\mathbf{A}^{-1}) \bullet (\mathbf{AB})$.

If $\mathbf{A}'\mathbf{A} = \mathbf{AA}' = \mathbf{I}$ and **A** is real, it is called orthogonal (the reason will appear below). If $\mathbf{A}^*\mathbf{A} = \mathbf{AA}^* = \mathbf{I}$ (**A** complex), **A** is called unitary. If **A** commutes with $\mathbf{A}^*$, so that $\mathbf{AA}^* = \mathbf{A}^*\mathbf{A}$, then **A** is called normal.

Elementary Row and Column Operations

The most important tools used in dealing with matrices are the elementary operations: R for row, C for column. If **A** is given matrix, performing $R(i \leftrightarrow j)$ on **A** means interchanging Row i and Row j. $R_i(c)$ means multiplying Row i by the number c (except $c = 0$). $R_j + cR_i$ means multiply Row i by c and add this result into Row j $(i \neq j)$. Thus, if

$$\mathbf{A} = \begin{bmatrix} 1 & 2 & 3 \\ 4 & 5 & 6 \\ 7 & 8 & 0 \end{bmatrix}$$

then

$$R(2 \leftrightarrow 3)\,(\mathbf{A}) = \begin{bmatrix} 1 & 2 & 3 \\ 7 & 8 & 0 \\ 4 & 5 & 6 \end{bmatrix} \qquad C_1(2)(\mathbf{A}) = \begin{bmatrix} 2 & 2 & 3 \\ 8 & 5 & 6 \\ 14 & 8 & 0 \end{bmatrix}$$

$$R_1 - R_2(\mathbf{A}) = \begin{bmatrix} -3 & -3 & -3 \\ 4 & 5 & 6 \\ 7 & 8 & 0 \end{bmatrix}$$

These operations are used in reducing matrix problems to simpler ones.

Example 2.22

Solve $\mathbf{AX} = \mathbf{B}$ where

$$\mathbf{A} = \begin{bmatrix} 1 & 2 & 3 \\ 4 & 5 & 6 \\ 7 & 8 & 9 \end{bmatrix}, \quad \mathbf{X} = \begin{bmatrix} x \\ y \\ z \end{bmatrix}, \quad \mathbf{B} = \begin{bmatrix} 1 \\ 1 \\ 1 \end{bmatrix}$$

Solution

Form the "augmented" matrix

$$[\mathbf{A}|\mathbf{B}] = \begin{bmatrix} 1 & 2 & 3 & 1 \\ 4 & 5 & 6 & 1 \\ 7 & 8 & 9 & 1 \end{bmatrix}$$

and perform elementary row operations on this matrix until the solution is apparent:

$$\begin{bmatrix} 1 & 2 & 3 & 1 \\ 4 & 5 & 6 & 1 \\ 7 & 8 & 9 & 1 \end{bmatrix} \begin{matrix} R_2 - 4R_1 \\ R_3 - 7R_1 \end{matrix} \begin{bmatrix} 1 & 2 & 3 & 1 \\ 0 & -3 & -6 & -3 \\ 7 & -6 & -12 & -6 \end{bmatrix} \begin{matrix} R_2\left(-\frac{1}{3}\right) \\ \\ R_3\left(-\frac{1}{6}\right) \end{matrix} \begin{bmatrix} 1 & 2 & 3 & 1 \\ 0 & 1 & 2 & 1 \\ 0 & 1 & 2 & 1 \end{bmatrix}$$

$$R_3 - R_2 \begin{bmatrix} 1 & 2 & 3 & 1 \\ 0 & 1 & 2 & 1 \\ 0 & 0 & 0 & 0 \end{bmatrix}$$

The answer is now apparent: $y + 2z = 1$ and $x + 2y + 3z = 1$, or, z arbitrary, $y = 1 - 2z$, $x = 1 - 2(1 - 2z) - 3z = -1 + z$. This system of equations has an infinite number of solutions.

Example 2.23

Solve the system of equations

$$\begin{aligned} x + y - z &= a \\ 2x - y + 3z &= 2 \\ 3x + 2y + z &= 1 \end{aligned}$$

for x, y, and z in terms of a.

Solution

Strip off the variables x, y, and z:

$$\begin{bmatrix} 1 & 1 & -1 & a \\ 2 & -1 & 3 & 2 \\ 3 & 2 & 1 & 1 \end{bmatrix} \begin{matrix} R_2 - 2R_1 \\ R_3 - 3R_1 \end{matrix} \begin{bmatrix} 1 & 1 & -1 & a \\ 0 & -3 & 5 & 2-2a \\ 0 & -1 & 4 & 1-3a \end{bmatrix}$$

$$\begin{matrix} R_2(2 \leftrightarrow 3) \\ R_2(-1) \end{matrix} \begin{bmatrix} 1 & 1 & -1 & a \\ 0 & 1 & -4 & 3a-1 \\ 0 & -3 & 5 & 2-2a \end{bmatrix} \begin{matrix} R_1 - R_2 \\ R_3 + 3R_2 \end{matrix} \begin{bmatrix} 1 & 0 & 3 & 1-2a \\ 0 & 1 & -4 & 3a-1 \\ 0 & 0 & -7 & 7a-1 \end{bmatrix}$$

The solution is now clear:

$$\begin{aligned} z &= \frac{7a-1}{-7} = -a + \frac{1}{7} \\ y &= 3a - 1 + 4z = 3a - 1 - 4a + \frac{4}{7} = -a - \frac{3}{7} \\ x &= 1 - 2a - 3z = 1 - 2a + 3a - \frac{3}{7} = a + \frac{4}{7} \end{aligned}$$

Example 2.24

Find $\mathbf{A}^{-1}$ if

$$\mathbf{A} = \begin{bmatrix} 1 & 1 & -1 \\ 1 & 2 & 3 \\ 3 & 2 & 1 \end{bmatrix}$$

Solution

Since this amounts to solving $\mathbf{AX} = \mathbf{B}$ three times, with

$$\mathbf{B} = \begin{bmatrix} 1 \\ 0 \\ 0 \end{bmatrix} \quad \mathbf{B} = \begin{bmatrix} 0 \\ 1 \\ 0 \end{bmatrix} \quad \mathbf{B} = \begin{bmatrix} 0 \\ 0 \\ 1 \end{bmatrix}$$

form

$$[\mathbf{A}|\mathbf{I}] = \begin{bmatrix} 1 & 1 & -1 & 1 & 0 & 0 \\ 1 & 2 & 3 & 0 & 1 & 0 \\ 3 & 2 & 1 & 0 & 0 & 1 \end{bmatrix}$$

and perform row operations until a solution emerges.

$$[\mathbf{A}|\mathbf{I}] = \frac{R_2 - R_1}{R_3 - 3R_1} \begin{bmatrix} 1 & 1 & -1 & 1 & 0 & 0 \\ 0 & 1 & 4 & -1 & 1 & 0 \\ 0 & -1 & 4 & -3 & 0 & 1 \end{bmatrix}$$

$$\frac{R_1 - R_2}{R_3 + R_2} z \begin{bmatrix} 1 & 0 & -5 & 2 & -1 & 0 \\ 0 & 1 & 4 & -1 & 1 & 0 \\ 0 & 0 & 8 & -4 & 1 & 1 \end{bmatrix}$$

$$\begin{matrix} R_3\left(\frac{1}{8}\right) \\ \frac{R_2 - 4R_3}{R_1 + 5R_3} \end{matrix} \begin{bmatrix} 1 & 0 & 0 & -\frac{1}{2} & -\frac{3}{8} & \frac{5}{8} \\ 0 & 1 & 0 & 1 & \frac{1}{2} & -\frac{1}{2} \\ 0 & 0 & 1 & -\frac{1}{2} & \frac{1}{8} & \frac{1}{8} \end{bmatrix}$$

Thus,

$$\mathrm{A}^{-1} = \begin{bmatrix} -\frac{1}{2} & -\frac{3}{8} & \frac{5}{8} \\ 1 & \frac{1}{2} & -\frac{1}{2} \\ -\frac{1}{2} & \frac{1}{8} & \frac{1}{8} \end{bmatrix}$$

Example **2.25**

Verify that $\mathbf{A}^{-1}\mathbf{A} = \mathbf{I}$ in Example 2.24.

Solution

$$8\mathbf{A}^{-1}\mathbf{A} = \begin{bmatrix} -4 & -3 & 5 \\ 8 & 4 & -4 \\ -4 & 1 & 1 \end{bmatrix} \begin{bmatrix} 1 & 1 & -1 \\ 1 & 2 & 3 \\ 3 & 2 & 1 \end{bmatrix}$$

$$= \begin{bmatrix} -4-3+15 & -4-6+10 & 4-9+5 \\ 8+4-12 & 8+8-8 & -8+12-4 \\ -4+1+3 & -4+2+2 & 4+3+1 \end{bmatrix} = 8\begin{bmatrix} 1 & 0 & 0 \\ 0 & 1 & 0 \\ 0 & 0 & 1 \end{bmatrix} = 8\mathbf{I}$$

Example **2.26**

Describe the set of solutions of $\mathbf{AX} = \mathbf{B}$.

Solution

If $\mathbf{AX}_0 = \mathbf{B}$ is one solution, and $\mathbf{AY} = 0$, then $\mathbf{A}(\mathbf{X}_0 + \mathbf{Y})$ is a solution, so all solutions are of the form $\mathbf{X} = \mathbf{X}_0 + \mathbf{Y}$ where $\mathbf{AY} = 0$. Thus, if $\mathbf{N} = \{\mathbf{Y} : \mathbf{AY} = 0\}$ is the null space of $\mathbf{A}$, the set of solutions to $\mathbf{AX} = \mathbf{B}$ is $\mathbf{X}_0 + \mathbf{N} = \{\mathbf{X}_0 + \mathbf{Y} : \mathbf{Y} \in \mathbf{N}\}$.

Determinants

The determinant of a square matrix is a scalar representing the *volume* of the matrix in some sense. Matrices that are not square do not have determinants.

The determinant is frequently indicated by vertical lines, viz. $|A|$. It is a complicated formula, and one way to find it is by induction. The determinant of a 1×1 matrix is $|a| = a$. The determinant of a 2×2 matrix is

$$\begin{vmatrix} a & b \\ c & d \end{vmatrix} = ad - bc$$

The determinant of an $n \times n$ matrix is given in terms of n determinants, each of size $(n-1) \times (n-1)$. If $\mathbf{A}$ is $n \times n$ and $\mathbf{M}_{ij}$ is the matrix obtained by removing the ith row and the jth column from $\mathbf{A}$, then

$$|A| = \sum_{j=1}^{n} (-1)^{1+j} a_{1j} \, |M_{1j}|$$

Example 2.27

Find the determinant

$$\begin{vmatrix} 1 & 2 & 3 \\ 4 & 0 & 6 \\ 7 & 8 & 9 \end{vmatrix}$$

Solution

$$\begin{aligned} |A| &= (-1)^{1+1} a_{11} |M_{11}| + (-1)^{1+2} a_{12} |M_{12}| + (-1)^{1+3} a_{13} |M_{13}| \\ &= 1 \begin{vmatrix} 0 & 6 \\ 8 & 9 \end{vmatrix} - 2 \begin{vmatrix} 4 & 6 \\ 7 & 9 \end{vmatrix} + 3 \begin{vmatrix} 4 & 0 \\ 7 & 8 \end{vmatrix} \\ &= -48 - 2(36 - 42) + 3(32) = 60 \end{aligned}$$

Example 2.28

Find the determinant

$$\begin{vmatrix} 0 & 0 & 2 & 0 \\ 1 & 2 & 7 & 3 \\ 4 & 0 & 3 & 6 \\ 7 & 8 & -6 & 9 \end{vmatrix}$$

Solution

$$\begin{aligned} |A| &= a_{11} |M_{11}| - a_{12} |M_{12}| + a_{13} |M_{13}| - a_{14} |M_{14}| \\ &= 0 |M_{11}| - 0 |M_{12}| + 2 |M_{13}| - 0 |M_{14}| = 2\,(60) = 120 \end{aligned}$$

The last example provides a clue to the evaluation of large determinants, but the use of the first row of **A** in the definition of a determinant was arbitrary. For any row or column (fix i or j),

$$|A| = \sum_{j=1}^{n} (-1)^{1+j} a_{i1j} |M_{ij}|$$

The interaction of the determinant with elementary row or column operations is simple: Interchanging two rows changes the sign of the determinant; multiplying a row by a constant multiplies the determinant by that constant.

Example 2.29

Evaluate the determinant

$$\begin{vmatrix} 1 & 2 & 3 & 4 \\ 1 & 1 & 1 & 0 \\ 4 & 0 & 3 & 2 \\ 0 & 3 & 0 & 1 \end{vmatrix}$$

Solution

Choose a row or column with many zeroes and introduce still more:

$$|A| = C_2 - 3C_4 \, |A| = \begin{vmatrix} 1 & -10 & 3 & 4 \\ 1 & 1 & 1 & 0 \\ 4 & -6 & 3 & 2 \\ 0 & 0 & 0 & 1 \end{vmatrix} = (-1)^{4+4} a_{44} \begin{vmatrix} 1 & -10 & 3 \\ 1 & 1 & 1 \\ 4 & -6 & 3 \end{vmatrix}$$

$$\underset{R_3 - 4R_1}{\overset{R_2 - R_1}{=}} \begin{vmatrix} 1 & -10 & 3 \\ 0 & 11 & -2 \\ 0 & 34 & -9 \end{vmatrix} = (-1)^{1+1} a_{11} \begin{vmatrix} 11 & -2 \\ 34 & -9 \end{vmatrix} = -99 + 68 = -31$$

Example 2.30

Find which values, if any, of the number c make $\mathbf{A}$ singular if

$$\mathbf{A} = \begin{vmatrix} 1 & 2 & c \\ 4 & 5 & 6 \\ 1 & 1 & 1 \end{vmatrix}$$

Solution

$|\mathbf{A}| = (-1)^2(5-6) + (-1)^3(2)(4-6) + (-1)^4 c(4-5) = -1 + 4 - c = 0.$ Hence $\mathbf{A}$ is singular for only one value of c, $c = 3$.

Cramer's Rule is a consequence of adj(A): If $\mathbf{A}$ is nonsingular, the ith component of the solution of $\mathbf{A}\,\mathbf{X} = \mathbf{B}$ is $x_i = \dfrac{|A_i|}{|A|}$, where A_i is the result of replacing the ith column of $\mathbf{A}$ by $\mathbf{B}$.

Example 2.31

Find x_2 in $\mathbf{AX} = \mathbf{B}$ by Cramer's Rule if

$$\mathbf{A} = \begin{bmatrix} 1 & 2 & 1 & 1 \\ 3 & 4 & 5 & -2 \\ 6 & 7 & 1 & 5 \\ -1 & 0 & 2 & 0 \end{bmatrix} \quad \text{and} \quad \mathbf{B} = \begin{bmatrix} 1 \\ 2 \\ 3 \\ 4 \end{bmatrix}$$

Solution

First,

$$|A| = \begin{vmatrix} 1 & 2 & 3 & 1 \\ 3 & 4 & 11 & -2 \\ 6 & 7 & 13 & 5 \\ -1 & 0 & 0 & 0 \end{vmatrix} = (-1)^{4+1}(-1) \begin{vmatrix} 2 & 3 & 1 \\ 4 & 11 & -2 \\ 7 & 13 & 5 \end{vmatrix}$$

$$= \begin{vmatrix} 0 & 0 & 1 \\ 8 & 17 & -2 \\ -3 & -2 & 5 \end{vmatrix} = (-1)^{1+3}(1) \begin{vmatrix} 8 & 17 \\ -3 & -2 \end{vmatrix} = -16 + 51 = 35$$

Next, the numerator of x_2 is

$$\begin{vmatrix} 1 & 1 & 1 & 1 \\ 3 & 2 & 5 & -2 \\ 6 & 3 & 1 & 5 \\ -1 & 4 & 2 & 0 \end{vmatrix} = \begin{vmatrix} 1 & 0 & 0 & 0 \\ 3 & -1 & 2 & -5 \\ 6 & -3 & -5 & -1 \\ -1 & 5 & 3 & 1 \end{vmatrix} = \begin{vmatrix} -1 & 2 & -5 \\ -3 & -5 & -1 \\ 5 & 3 & 1 \end{vmatrix} = \begin{vmatrix} -1 & 2 & -5 \\ 0 & -11 & 14 \\ 0 & 13 & -24 \end{vmatrix}$$

$$= -\begin{vmatrix} -11 & 14 \\ 13 & -24 \end{vmatrix} = -\begin{vmatrix} -11 & 14 \\ 2 & -10 \end{vmatrix} = -(110-28) = -82, x_2 = -\frac{82}{35}$$

NUMERICAL METHODS

This portion of numerical methods includes techniques of finding roots of polynomials by the Routh-Hurwitz criterion and Newton methods, Euler's techniques of numerical integration and the trapezoidal methods, and techniques of numerical solutions of differential equations.

Root Extraction

Routh-Hurwitz Method (without Actual Numerical Results)

Root extraction, even for simple roots (i.e., without imaginary parts), can become quite tedious. Before attempting to find roots, one should first ascertain whether they are really needed or whether just knowing the area of location of these roots will suffice. If all that is needed is knowing whether the roots are all in the left half-plane of the variable (such as is in the s-plane when using Laplace transforms—as is frequently the case in determining system stability in control systems), then one may use the Routh-Hurwitz criterion. This method is fast and easy even for higher-ordered equations. As an example, consider the following polynomial:

$$p_n(x) = \prod_{m=1}^{n} (x - x_m) = x^n + a_1 x^{n-1} + a_2 x^{n-2} + \cdots + a_{n-1} \quad \textbf{(2.1)}$$

Here, finding the roots, x_m, for $n > 3$ can become quite tedious without a computer; however, if one only needs to know if any of the roots have positive real parts, one can use the Routh-Hurwitz method. Here, an array is formed listing the coefficients of every other term starting with the highest power, n, on a line, followed by a line listing the coefficients of the terms left out of the first row. Following rows are constructed using Routh-Hurwitz techniques, and after completion of the array, one merely checks to see if all the signs are the same (unless there is a zero coefficient—then something else needs to be done) in the first column; if none, no roots will exist in the right half-plane. In case of zero coefficient, a simple technique is used; for details, see almost any text dealing with stability of control systems. A short example follows.

$F(s) = s^3 + 3s^2 + 10$ Array:				Where the s^1 term is formed as
	s^3	1	2	
	s^2	3	10	
$= (s + ?)(s + ?)(s + ?)$	s^1	$-\frac{4}{3}$	0	$(3 \times 2 - 10 \times 1)/3 = -\frac{4}{3}$. For details, refer to any text on control systems or numerical methods.
	s^0	10	0	

Here, there are two sign changes: one from 3 to $-\frac{4}{3}$, and one from $-\frac{4}{3}$ to 10. This means there will be two roots in the right half-plane of the s-plane, which yield an unstable system. This technique represents a great savings in time without having to factor the polynomial.

Newton's Method

The use of Newton's method of solving a polynomial and the use of iterative methods can greatly simplify a problem. This method utilizes synthetic division and is based upon the remainder theorem. This synthetic division requires estimating a root at the start, and, of course, the best estimate is the actual root. The root is the correct one when the remainder is zero. (There are several ways of estimating this root, including a slight modification of the Routh-Hurwitz criterion.)

If a $P_n(x)$ polynomial (see Equation (2.1)) is divided by an estimated factor $(x - x_1)$, the result is a reduced polynomial of degree $n-1$, $Q_{n-1}(x)$, plus a constant remainder of b_{n-1}. Thus, another way of describing Equation (2.1) is

$$P_n(x)/(x - x_1) = Q_{n-1}(x) + b_{n-1}/(x - x_1) \quad \text{or} \quad P_n(x) = (x - x_1)Q_{n-1}(x) + b_{n-1} \qquad \textbf{(2.2)}$$

If one lets $x = x_1$, Equation (2.2) becomes

$$P_n(x = x_1) = (0)Q_{n-1}(x) + b_{n-1} = b_{n-1} \qquad \textbf{(2.3)}$$

Equation 2.3 leads directly to the remainder theorem: "The remainder on division by $(x - x_1)$ is the value of the polynomial at $x = x_1$, $P_n(x_1)$."[1]

Newton's method (actually, the Newton-Raphson method) for finding the roots for an nth-order polynomial is an iterative process involving obtaining an estimated value of a root (leading to a simple computer program). The key to the process is getting the first estimate of a possible root. Without getting too involved, recall that the coefficient of x^{n-1} represents the sum of all of the roots and the last term represents the product of all n roots; then the first estimate can be "guessed" within a reasonable magnitude. After a first root is chosen, find the rate of change of the polynomial at the chosen value of the root to get the next, closer value of the root x_{n+1}. Thus the new root estimate is based on the last value chosen:

$$x_{n+1} = x_n - P_n(x_n)/P_n'(x_n), \\ \text{where } P_n'(x_n) = dP_n(x)/dx \text{ evaluated at } x = x_n \qquad \textbf{(2.4)}$$

[1] Gerald & Wheatley, *Applied Numerical Analysis*, 3rd ed., Addison-Wesley, 1985.

NUMERICAL INTEGRATION

Numerical integration routines are extremely useful in almost all simulation-type programs, design of digital filters, theory of z-transforms, and almost any problem solution involving differential equations. And because digital computers have essentially replaced analog computers (which were almost true integration devices), the techniques of approximating integration are well developed. Several of the techniques are briefly reviewed below.

Euler's Method

For a simple first-order differential equation, say $dx/dt + ax = af$, one could write the solution as a continuous integral or as an interval type one:

$$x(t) = \int^{t} [-ax(\tau) + af(\tau)]d\tau \tag{2.5a}$$

$$x(kT) = \int^{kT-T} [-ax + af]d\tau + \int_{kT-T}^{kT} [-ax + af]d\tau = x(kT - T) + A_{rect} \tag{2.5b}$$

Here, A_{rect} is the area of $(-ax + af)$ over the interval $(kT - T) < \tau < kT$. One now has a choice looking back over the rectangular area or looking forward. The rectangular width is, of course, T. For the forward-looking case, a first approximation for x_1 is[2]

$$\begin{aligned} x_1(kT) &= x_1(kT - T) + T[ax_1(kT - T) + af(kT - T)] \\ &= (1 - aT)x_1(kT - T) + aTf(kT - T) \end{aligned} \tag{2.5c}$$

Or, in general, for Euler's forward rectangle method, the integral may be approximated in its simplest form (using the notation $t_{k+1} - t_k$ for the width, instead of T, which is kT–T) as

$$\int_{t_k}^{t_{k+1}} x(\tau)d\tau \approx (t_{k+1} - t_k)x(t_k) \tag{2.6}$$

Trapezoidal Rule

This trapezoidal rule is based upon a straight-line approximation between the values of a function, $f(t)$, at t_0 and t_1. To find the area under the function, say a curve, is to evaluate the integral of the function between points a and b. The interval between these points is subdivided into subintervals; the area of each subinterval is approximated by a trapezoid between the end points. It will be necessary only to sum these individual trapezoids to get the whole area; by making the intervals all the same size, the solution will be simpler. For each interval of delta t (i.e., $t_{k+1} - t_k$), the area is then given by

$$\int_{t_k}^{t_{k+1}} x(\tau)\, d\tau \approx (1/2)(t_{k+1} - t_k)[x(t_{k+1}) + x(t_k)] \tag{2.7}$$

[2] This method is as presented in Franklin & Powell, *Digital Control of Dynamic Systems*, Addison-Wesley, 1980, page 55.

This equation gives good results if the delta t's are small, but it is for only one interval and is called the "local error." This error may be shown to be – (1/12) (delta $t)^3 f''(t = \xi_1)$, where ξ_1 is between t_0 and t_1. For a larger "global error" it may be shown that

$$\text{Global error} = -(1/12)(\text{delta } t)^3\,[f''(\xi_1) + f''(\xi_2) + \cdots + f''(\xi_n)] \qquad \textbf{(2.8)}$$

Following through on Equation 2.8 allows one to predict the error for the trapezoidal integration. This technique is beyond the scope of this review or probably the examination; however, for those interested, please refer to pages 249–250 of the previously mentioned reference to Gerald & Wheatley.

NUMERICAL SOLUTIONS OF DIFFERENTIAL EQUATIONS

This solution will be based upon first-order ordinary differential equations. However, the method may be extended to higher-ordered equations by converting them to a matrix of first-ordered ones.

Integration routines produce values of system variables at specific points in time and update this information at each interval of delta time as T (delta $t = T = t_{k+1} - t_k$). Instead of a continuous function of time, $x(t)$, the variable x will be represented with discrete values such that $x(t)$ is represented by $x_0, x_1, x_2, \ldots, x_n$. Consider a simple differential equation as before as based upon Euler's method,

$$dx/dt + ax = f(t)$$

Now assume the delta time periods, T, are fixed (not all routines use fixed step sizes); then one writes the continuous equations as a difference equation where $dx/dt \approx (x_{k+1} - x_k)/T = -ax_k + f_k$ or, solving for the updated value, x_{k+1},

$$x_{k+1} = x_k - Tax_k + Tf_k \qquad \textbf{(2.9a)}$$

For fixed increments by knowing the first value of $x_{k=0}$ (or the initial condition), one may calculate the solution for as many "next values" of x_{k+1} as desired for some value of T. The difference equation may be programmed in almost any high-level language on a digital computer; however, T must be small as compared to the shortest time constant of the equation (here, $1/a$).

The following equation—with the "f" term meaning "a function of" rather than as a "forcing function" term as used in Equation (2.5a)—is a more general form of Equation (2.9a). This equation is obtained by letting the notation x_{k+1} become $y\,[k+1\;\Delta t]$ and is written (perhaps somewhat more confusingly) as

$$y[(k+1)\Delta t] = y(k\Delta t) + \Delta t f[y(k\Delta t), (k\Delta t)] \qquad \textbf{(2.9b)}$$

Reduction of Differential Equation Order

To reduce the order of a linear time-dependent differential equation, the following technique is used. For example, assume a second-order equation: $x'' + ax' + bx = f(t)$. If we define $x = x_1$ and $x' = x_1' = x_2$, then

$$x_2' + ax_2 + bx_1 = f(t)$$
$$x_1' = x_2 \text{ (by definition)}$$
$$x_2' = -b_{x1} - ax_2 + f(t)$$

This technique can be extended to higher-order systems and, of course, be put into a matrix form (called the state variable form). And it can easily be set up as a matrix of first-order difference equations for solving digitally.

PROBLEMS

2.1 The simplest value of $\frac{[(n+1)!]^2}{n!(n-1)!}$ is:
a. n^2
b. $n(n + 1)$
c. $n + 1$
d. $n(n + 1)^2$

2.2 If $x^{3/4} = 8$, x equals:
a. 6
b. 9
c. –9
d. 16

2.3 If $\log_a 10 = 0.250$, $\log_{10} a$ equals:
a. 4
b. 0.50
c. 2
d. 0.25

2.4 If $\log_5 x = -1.8$, $\log_x 5$ is:
a. 0.35
b. 0.79
c. –0.56
d. undefined

2.5 If $\log x + \log (x - 10) - \log 2 = 1$, x is:
a. –1.708
b. 5.213
c. 7.824
d. 11.708

2.6 A right circular cone, cut parallel with the axis of symmetry, reveals a(n):
a. circle
b. hyperbola
c. eclipse
d. parabola

2.7 The expression $\frac{6!}{3!0!}$ is equal to:
a. ∞
b. 120
c. 2!
d. 0

2.8 To find the angles of a triangle, given only the lengths of the sides, one would use:
a. the law of cosines
b. the law of tangents
c. the law of sines
d. the inverse-square law

2.9 If $\sin\alpha = \dfrac{a}{\sqrt{a^2+b^2}}$, which of the following equations is true?

a. $\tan^{-1}\dfrac{b}{a} = \dfrac{\pi}{2} - \alpha$

b. $\tan^{-1}\dfrac{b}{a} = -\alpha$

c. $\cos^{-1}\dfrac{b}{\sqrt{a^2+b^2}} = \dfrac{\pi}{2} - \alpha$

d. $\cos^{-1}\dfrac{a}{\sqrt{a^2+b^2}} = \alpha$

2.10 The sine of 840° equals:
a. –cos 30°
b. –cos 60°
c. sin 30°
d. sin 60°

2.11 One root of $x^3 - 8x - 3 = 0$ is:
a. 2
b. 3
c. 4
d. 5

2.12 Roots of the equation $3x^3 - 3x^2 - 18x = 0$ are:
a. –2, 3
b. 0, – 2, 3
c. $2 + 1i, 2 - 1i$
d. 0, 2, –3

2.13 The equation whose roots are $-1 + i1$ and $-1 - i1$ is given as:
a. $x^2 + 2x + 2 = 0$
b. $x^2 - 2x - 2 = 0$
c. $x^2 + 2 = 0$
d. $x^2 - 2 = 0$

2.14 Natural logarithms have a base of:
a. 3.1416
b. 2.171828
c. 10
d. 2.71828

2.15 $(5.743)^{1/30}$ equals:
a. 1.03
b. 1.04
c. 1.05
d. 1.06

2.16 The value of $\tan(A + B)$, where $\tan A = 1/3$ and $\tan B = 1/4$ (A and B are acute angles) is:
a. 7/12
b. 1/11
c. 7/11
d. 7/13

2.17 To cut a right circular cone in such a way as to reveal a parabola, it must be cut:
a. perpendicular to the axis of symmetry
b. at any acute angle to the axis of symmetry
c. at any obtuse angle to the axis of symmetry
d. none of these

2.18 The equation of the line perpendicular to $3y + 2x = 5$ and passing through $(-2, 5)$ is:
a. $2x = 3y$
b. $2y = 3x$
c. $2y = 3x + 16$
d. $3x = 2y + 8$

2.19 Equation of a line that has a slope of –2 and passes through (2, 0) is:
a. $y = -2x$
b. $y = 2x + 4$
c. $y = -2x + 4$
d. $y = 2x - 4$

2.20 Equation of a line that intercepts the x-axis at $x = 4$ and the y-axis at $y = -6$ is:
a. $3x - 2y = 12$
b. $2x - 3y = 12$
c. $x + y = 6$
d. $x - y = 4$

2.21 The x-axis intercept and the y-axis intercept of the line $x + 3y + 9 = 0$ are:
a. 0 and 3
b. –9 and –3
c. –3 and –9
d. 9 and 0

2.22 The distance between the points (1, 0, –2) and (0, 2, 3) is:
a. 3.45
b. 5.39
a. 6.71
b. 7.48

2.23 The equation of a parabola with center at (0, 0) and directrix at $x = -2$ is:

a. $y^2 = 4x$
b. $y = 2x^2$
c. $y^2 = 8x$
d. $y^2 = 2x$

2.24 The equation of the directrix of the parabola $y^2 = -4x$ is:

a. $y = 2$
b. $x = 1$
c. $x + y = 0$
d. $x = -2$

2.25 The equation of an ellipse with foci at $(\pm 2, 0)$ and directrix at $x = 6$ is:

a. $\frac{x^2}{12} + \frac{y^2}{8} = 1$

b. $\frac{x^2}{8} + \frac{y^2}{8} = 1$

c. $\frac{x^2}{12} + \frac{y^2}{12} = 1$

d. $\frac{x^2}{4} + \frac{y^2}{4} = 1$

2.26 The foci of the ellipse $\left(\frac{x}{3}\right)^2 + \left(\frac{y}{2}\right)^2 = 1$ are at:

a. $(\pm\sqrt{5}, 0)$
b. $(\pm\sqrt{2}, 0)$
c. $(0, \pm\sqrt{2})$
d. $(0, \pm\sqrt{5})$

2.27 The equation of a hyperbola with center at (0, 0), foci at $(\pm 4, 0)$, and eccentricity of 3 is:

a. $\frac{x^2}{16} - \frac{y^2}{16} = 1$

b. $\frac{x^2}{16} - \frac{y^2}{128} = 1$

c. $\frac{x^2}{64} - \frac{y^2}{48} = 1$

d. $\frac{x^2}{128} - \frac{y^2}{128} = 1$

2.28 The equation of a circle with center at (1, 2) and passing through the point (4, 6) is:

a. $x^2 + y^2 = 25$
b. $(x + 1)^2 + (y - 2)^2 = 25$
c. $x^2 + (y - 2)^2 = 25$
d. $(x - 1)^2 + (y - 2)^2 = 25$

2.29 The length of the tangent from (4, 8) to the circle $x^2 + (y - 1)^2 = 3^2$ is:
a. 3.81
b. 4.14
c. 5.66
d. 7.48

2.30 The conic section described by the equation $x^2 - 10xy + y^2 + x + y + 1 = 0$ is:
a. circle
b. parabola
c. hyperbola
d. ellipse

2.31 A triangle has sides of length 2, 3, and 4. The angle subtended by the sides of length 2 and 4 is:
a. 21.2°
b. 35.0°
c. 46.6°
d. 61.2°

2.32 Length *a* of one side of the triangle below is:

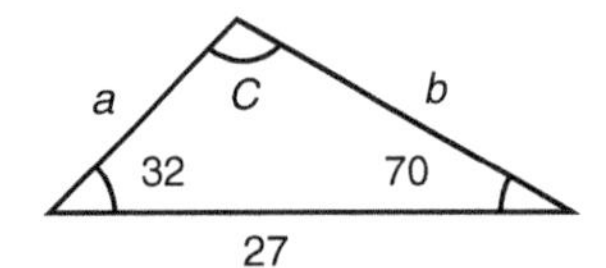

a. 25.9
b. 19.1
c. 12.7
d. 4.8

2.33 The relation $\sec\theta - (\sec\theta)(\sin^2\theta)$ can be simplified as:
a. $\sin\theta$
b. $\tan\theta$
c. $\cot\theta$
d. $\cos\theta$

2.34 If $\sin\theta = m$, $\cot\theta$ is:

a. $\dfrac{\sqrt{1-m^2}}{m}$

b. $\dfrac{m}{\sqrt{1-m^2}}$

c. $\sqrt{1-m^2}$

d. m

2.35 If vectors **A** = $3i - 6j + 2k$ and **B** = $10i + 4j - 6k$, their cross product **A** × **B** is:

a. $-12\mathbf{i} + 38\mathbf{j}$
b. $12\mathbf{i} + 24\mathbf{j} + 36\mathbf{k}$
c. $-12\mathbf{i} + 18\mathbf{j} + 24\mathbf{k}$
d. $28\mathbf{i} + 38\mathbf{j} + 72\mathbf{k}$

2.36 The sum of all integers from 10 to 50 (both inclusive) is:

a. 990
b. 1110
c. 1230
d. 1420

2.37 The 50th term of the series 10, 16, 22, 28, 34, 40. . . is:

a. 272
b. 304
c. 428
d. 584

2.38 Sum of the infinite series 4, 2, 1, 0.5, 0.25. . . is:

a. 8
b. 128
c. 10,400
d. ∞

2.39 If $\mathbf{A} = \begin{bmatrix} 1 & 2 & 3 \\ 1 & 2 & 9 \end{bmatrix}$ and $\mathbf{B} = \begin{bmatrix} 5 & 1 \\ 6 & 0 \\ 4 & 7 \end{bmatrix}$, the (2,1) entry of AB is:

a. 29
b. 53
c. 33
d. 64

2.40 The inverse of the matrix $\begin{bmatrix} 1 & 1 \\ 3 & 2 \end{bmatrix}$ is:

a. $\begin{bmatrix} 2 & -1 \\ -3 & 1 \end{bmatrix}$

b. $\begin{bmatrix} 2 & 3 \\ 1 & 1 \end{bmatrix}$

c. $\begin{bmatrix} 1 & 3 \\ 1 & 2 \end{bmatrix}$

d. $\begin{bmatrix} -2 & 1 \\ 3 & -1 \end{bmatrix}$

2.41 The determinant of the matrix $\begin{bmatrix} 1 & 2 & -1 \\ 3 & 0 & 2 \\ 2 & -2 & -1 \end{bmatrix}$ is:

a. 4
b. 16
c. 24
d. –16

2.42 In the system of equations

$$\begin{aligned} 3x_1 + 2x_2 - x_3 &= 5 \\ x_2 - x_3 &= 2 \\ x_1 + 2x_2 - 3x_3 &= -1 \end{aligned}$$

the value of $x_2 =$ is:
a. 2
b. –1
c. 4
d. 6

2.43 What is the determinant of M?

$$\boldsymbol{M} = \begin{bmatrix} 0 & 1 & 1 & 1 \\ 1 & 1 & 1 & 1 \\ 1 & 1 & 3 & 1 \\ 2 & 1 & 3 & 4 \end{bmatrix}$$

a. –6
b. 6
c. 0
d. 7

SOLUTIONS

2.1 d. The value (n + 1)! may be written as (n +1)(n) [(n –1)!]. It may be written also as n!(n + 1). Hence the given expression may be written as follows:

$$\frac{\{(n+1)(n)\,[(n-1)!]\}\,\{n\,!(n+1)\}}{n\,!(n-1)!} = (n+1)^2 n$$

2.2 d. Raise both sides of the equation to the 4/3 power:

$$[x^{3/4}]^{4/3} = 8^{4/3}$$
$$x = \sqrt[3]{8^4} = \sqrt[3]{(2^3)^4} = 2^{\frac{3\bullet 4}{3}} = 2^4 = 16$$

2.3 a. $\log_a 10 = 0.250$ can be written as $10 = a^{0.250}$. Taking $\log_{10}$, $\log_{10} 10 = \log_{10} a^{0.250}$, $1 = 0.250 \log_{10} a$ and

$$\log_{10} a = \frac{1}{0.250} = 4$$

2.4 c. Since $(\log_5 x)\,(\log_x 5)=1$, $\log_x 5 = \dfrac{1}{-1.8} = -0.556$

2.5 d. Since log (a) + log (b) – log(c) = log $\dfrac{ab}{c}$, the given equation

simplifies to $\log \dfrac{x(x-10)}{2} = 1$. Equivalently, $\dfrac{x(x-10)}{2} = 10$ or

$x^2 - 10x - 20 = 0$. The roots are 11.708 and –1.708. Since log is not defined for negative values, $x = 11.708$.

2.6 b.

2.7 b. $\dfrac{6!}{3!0!} = \dfrac{6(5)(4)(3)!}{3!(1)} = 120$

2.8 a. The law of cosines is $a^2 = b^2 + c^2 - 2bc \cos A$ for any plane triangle with angles A, B, C and sides a, b, c, respectively. This law can be applied to solve for the angles, given three sides in a plane triangle (Exhibit 2.8).

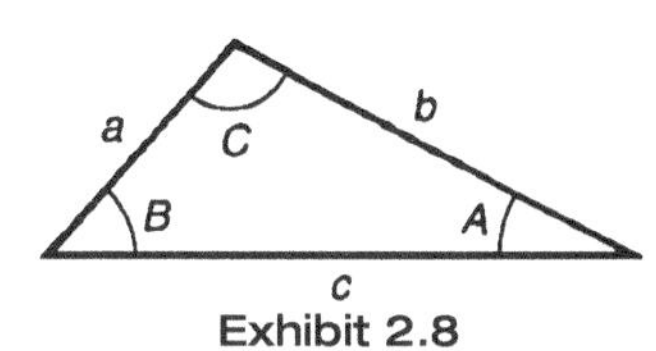

Exhibit 2.8

2.9 a. The triangle appears in Exhibit 2.9.

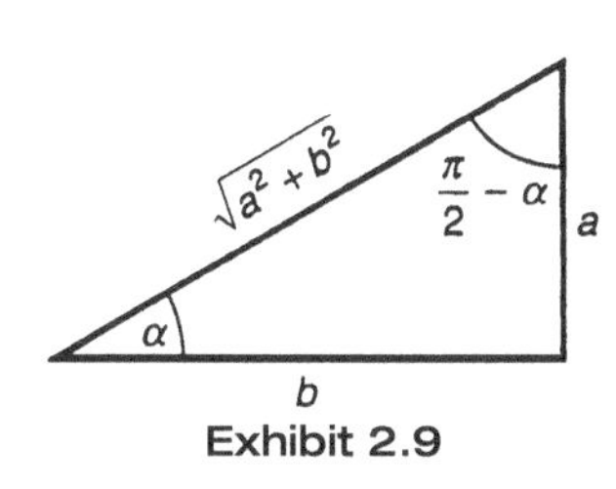

Exhibit 2.9

$$\tan\left(\frac{\pi}{2} - \alpha\right) = \frac{b}{a}$$
$$\tan^{-1}\frac{b}{a} = \frac{\pi}{2} - \alpha$$

2.10 d.

840° = 2(360) + 120 = 2(2π) rad + 120°
sin [2(2π) rad + 120°] = sin 120° = sin 60°

2.11 b. The solution is obtained by seeing which of the five answers satisfies the equation.

x	$x^3 - 8x - 3$
2	–11
3	0
4	29
5	82
6	165

2.12 b. Any scientific calculator can be used to find the roots as 0, 3, and –2.

2.13 a. If x_1 and x_2 are the roots, the equation is $(x - x_1)(x - x_2) = 0$. Since the roots are $-1 + i$ and $-1 - i$, the equation is $(x - (-1 + i))(x - (-1 - i)) = 0$ or $(x + 1 - i)(x + 1 + i) = 0$ or $x^2 + 2x + 2 = 0$.

2.14 d. Common logarithms have base 10. Natural, or napierian, logarithms have base $e = 2.71828$.

2.15 d. $\log\ (5.743)^{1/30} = \frac{1}{30}\log 5.743 = \frac{1}{30}(0.7592) = 0.0253$

The antilogarithm of 0.0253 is 1.06.

2.16 c.

$$\sin\ (A + B) = (\sin A \cos B) + (\cos A \sin B)$$

$$\cos\ (A + B) = (\cos A \cos B) - (\sin A \sin B)$$

$$\tan\ (A + B) = \frac{\sin(A + B)}{\cos(A + B)} = \frac{(\sin A \cos B) + (\cos A \sin B)}{(\cos A \cos B) - (\sin A \sin B)}$$

Dividing by $\cos A \cos B$,

$$\tan(A + B) = \frac{\frac{(\sin A \cos B)}{(\cos A \cos B)} + \frac{(\cos A \sin B)}{(\cos A \cos B)}}{\frac{(\cos A \cos B)}{(\cos A \cos B)} - \frac{(\sin A \sin B)}{(\cos A \cos B)}} = \frac{\tan A + \tan B}{1 - \tan A \tan B}$$

$$= \frac{\frac{1}{3} + \frac{1}{4}}{1 - \frac{1}{3} \times \frac{1}{4}} = \frac{\frac{4}{12} + \frac{3}{12}}{1 - \frac{1}{12}} = \frac{\frac{7}{12}}{\frac{11}{12}} = \frac{7}{11}$$

The problem could also be solved by determining angle A (whose tangent is 1/3) and angle B (whose tangent is 1/4). Then we could find the tangent of $(A + B)$.

$$\tan^{-1}\frac{1}{3} = 18.435° \quad \tan^{-1}\frac{1}{4} = 14.036°$$

$$\tan(18.435 + 14.036)° = \tan(32.471°) = 0.6364 = \frac{7}{11}$$

2.17 d. To reveal a parabola, a right circular cone must be cut parallel to an element of the cone and intersecting the axis of symmetry.

2.18 c. Rewriting the given line, $y = \frac{5}{3} - \frac{2}{3}x$. This line has slope $-\frac{2}{3}$, so a perpendicular line must have slope $\frac{3}{2}$. Using the point-slope form, $\frac{y-5}{x+3} = \frac{3}{2}$. Simplifying, $y = \frac{3}{2}x + 8$.

2.19 d. The equation of a straight line is $y = mx + b$ where the slope is m, (x, y) is any point on the line, and b is the y-axis intercept. Here, $m = -2$ and (2, 0) is a point. Substituting these values, $0 = -2(2) + b$ or $b = 4$. Then, the equation is $y = -2x + 4$ or $y + 2x = 4$.

2.20 a. (4,0) and (0, –6) are two points on the straight line. Then, the slope $m = \frac{y_2 - y_1}{x_2 - x_1} = \frac{0-(-6)}{4-0} = \frac{3}{2}$. Substituting one of the points, say, (0, –6) in the general equation $y = mx + b$; $-6 = 0 + b$. Then, $b = -6$ and the equation is $y = \left(\frac{3}{2}\right)x - 6$ or $3x - 2y = 12$.

2.21 b. For y-axis intercept, $x = 0$. Substituting this in the equation, $0 + 3y + 9 = 0$ or $y = -3$ is the y-axis intercept. For x-axis intercept, $y = 0$. Substituting this in the equation, $x + 0 + 9 = 0$ or $x = -9$ is the x-axis intercept.

2.22 b. Distance d between any two points (x_1, y_1, z_1) and (x_2, y_2, z_2) can be determined as $d^2 = (x_1 - x_2)^2 + (y_1 - y_2)^2 + (z_1 - z_2)^2$. Here, $d^2 = (1-0)^2 + (0-2)^2 + (-2-3)^2 = 30$; $d = 5.385$.

2.23 c. Directrix, $x = -p/2 = -2$ or $p = 4$. Equation of a parabola with center at origin is $y^2 = 2px$; as $p = 4$, $y^2 = 8x$.

2.24 b. Equation of a parabola with center at origin is $y^2 = 2px$. Since $y^2 = -4x$, $2p = -4$ or $p = -2$. Equation of a directrix is $x = (-p/2)$; since $p = 2$, the equation is $x = 1$.

2.25 a. Focus $ae = 2$ and directrix $\frac{a}{e} = 6$. Solving, $a^2 = 12$ and $e = 1/\sqrt{3}$. But $e = \sqrt{1 - \frac{b^2}{a^2}}$; solving, $b^2 = 8$. The equation of an ellipse is

$$\frac{x^2}{a^2} + \frac{y^2}{b^2} = 1 \text{ or } \frac{x^2}{12} + \frac{y^2}{8} = 1.$$

2.26 a. The equation of an ellipse is $\frac{x^2}{a^2} + \frac{y^2}{b^2} = 1$. Here, $a = 3$ and $b = 2$.

The eccentricity $e = \sqrt{1 - \frac{b^2}{a^2}} = \sqrt{\frac{5}{9}}$, and the foci $= (\pm ae, 0) = (\pm \sqrt{5}, 0)$.

2.27 b. Focus $ae = 4$ and eccentricity $e = 3$. Solving, $a = 4/3$.

Also, $b = a\sqrt{e^2 - 1} = \frac{4\sqrt{8}}{3}$. Equation for a hyperbola is

$$\frac{x^2}{a^2} - \frac{y^2}{b^2} = 1 \Rightarrow \frac{x^2}{16} - \frac{y^2}{128} = 1.$$

2.28 d. Radius of a circle with center at (h, k) and passing through a point (x, y) is $r^2 = (x - h)^2 + (y - k)^2$. Here, $r^2 = (4 - 1)^2 + (6 - 2)^2 = 25$. Equation of a circle with center at (h, k) is $(x - h)^2 + (y - k)^2 = r^2$; here, $(x - 1)^2 + (y - 2)^2 = 25$.

2.29 d. Length of the tangent from any point (x', y') outside a circle with center at (h, k) and radius r is given as $t^2 = (x' - h)^2 + (y' - k)^2 - r^2$. Here, $(x', y') = (4, 8)$, center $(h, k) = (0, 1)$, and $r^2 = 3^2$. Then, $t^2 = (4 - 0)^2 + (8 - 1)^2 - 3^2 = 56$ or $t = 7.483$.

2.30 c. General equation of a conic section is $Ax^2 + 2Bxy + Cy^2 + 2Dx + 2Ey + F = 0$, where both A and C are not zeros. Here, A = 1; B = –5; C = 1; then, $(B^2 - AC) > 0 \Rightarrow$ hyperbola.

2.31 c. Since three sides of a triangle are given, use the law of cosines; $3^2 = 2^2 + 4^2 - 2(2)(4)\cos(\theta)$ where θ is the angle opposite to side 3. Solving, $\theta = 46.6°$.

2.32 a. Angle $C = 180 - (70 + 32) = 78°$. Using the law of sines,

$\frac{a}{\sin 70} = \frac{b}{\sin 32} = \frac{27}{\sin 78}$. Solving, $a = 25.94$.

2.33 d. $(\sec\theta)(1 - \sin^2\theta) = \sec\theta\cos^2\theta = \frac{1}{\cos\theta}\cos^2\theta = \cos\theta$

2.34 a. $\sin\theta = \frac{m}{1} = \frac{\text{opposite side}}{\text{hypotenuse}}$; then, adjacent side $= \sqrt{1 - m^2}$

and $\cos\theta = \frac{\text{adjacent side}}{\text{hypotenuse}} = \frac{\sqrt{1-m^2}}{1}$; then, $\cot\theta = \frac{\cos\theta}{\sin\theta} = \frac{\sqrt{1-m^2}}{m}$

2.35 d. Cross product, $\mathbf{A} \times \mathbf{B} = \begin{vmatrix} i & j & k \\ 3 & -6 & 2 \\ 10 & 4 & -6 \end{vmatrix}$

Expanding, $i[(-6)(-6) - (2)(4)] - j[(3)(-6) - (2)(10)] + k[(3)(4) - (-6)(10)] = 28\mathbf{i} + 38\mathbf{j} + 72\mathbf{k}$

2.36 c. This is an arithmetic series; first term $a = 10$, last term $l = 50$, and common difference $d = 1$. The number of terms, n, can be calculated as, $l = a + (n - 1)d$; $50 = 10 + (n - 1)1$; solving, $n = 41$.

sum $S = \frac{n(a+1)}{2} = \frac{41}{2}(10 + 50) = 1230$

2.37 b. This is an arithmetic series; first term $a = 10$ and the common difference $d = 6$. Taking the number of terms n as 50, the nth term (last term) can be calculated as, $l = a + (n - 1)d$. In this case, $l = 10 + (50 - 1)6 = 304$.

2.38 a. This is a geometric series; first term $a = 4$ and common ratio $r = 0.5$. Since $r < 1$, the series is convergent and the sum as the number of terms n tend to infinity is $S = \frac{a}{1-r} = \frac{4}{1-0.5} = 8$.

2.39 b. To compute the (2,1) entry, take [1 2 9] • [5 6 4] = 5 + 12 + 36 = 53.

2.40 d. To invert a 2×2 matrix,

$$\begin{bmatrix} a & b \\ c & d \end{bmatrix}^{-1} = \frac{1}{ad-bc}\begin{bmatrix} d & -b \\ -c & a \end{bmatrix} = \frac{1}{2-3}\begin{bmatrix} 2 & -1 \\ -3 & 1 \end{bmatrix}$$

2.41 c. This 3×3 determinant can be computed quickly be expanding it in minors, especially around the second column:

$$\begin{vmatrix} 1 & 2 & -1 \\ 3 & 0 & 2 \\ 2 & -2 & 1 \end{vmatrix} = (-1)^{1+2}(2)\begin{vmatrix} 3 & 2 \\ 2 & -1 \end{vmatrix} + (-1)^{2+2}(0)\begin{vmatrix} 1 & -1 \\ 2 & -1 \end{vmatrix} + (-1)^{3+2}(-2)\begin{vmatrix} 1 & -1 \\ 3 & 2 \end{vmatrix}$$

$$= -2(-3-4) + 0 + 2(2+3) = 24$$

2.42 d. By Cramer's Rule,

$$x_2 = \frac{\text{Det}\begin{bmatrix} 3 & 5 & -1 \\ 0 & 2 & -1 \\ 1 & -1 & -3 \end{bmatrix}}{\text{Det}\begin{bmatrix} 3 & 2 & -1 \\ 0 & 1 & -1 \\ 1 & 2 & -3 \end{bmatrix}} = \frac{3\begin{vmatrix} 2 & -1 \\ -1 & -3 \end{vmatrix} + 1\begin{vmatrix} 5 & -1 \\ 2 & -1 \end{vmatrix}}{3\begin{vmatrix} 1 & -1 \\ 2 & -3 \end{vmatrix} + 1\begin{vmatrix} 2 & -1 \\ 1 & -1 \end{vmatrix}} = \frac{3(-7) + (-3) = -24}{3(-1) + (-1) = -4} = 6$$

2.43 a. To evaluate a 4×4 matrix, one must do some row or column operations and expand by minors:

$$\begin{bmatrix} 0 & 1 & 1 & 1 \\ 1 & 1 & 1 & 1 \\ 1 & 1 & 3 & 1 \\ 2 & 1 & 3 & 4 \end{bmatrix} \sim \begin{bmatrix} 0 & 1 & 1 & 1 \\ 1 & 1 & 1 & 1 \\ 0 & 0 & 2 & 0 \\ 0 & -1 & 1 & 2 \end{bmatrix}$$

Taking minors of column 1,

$$\text{Det}(M) = (1)(-1)^{2+1}\,\text{Det}\begin{bmatrix} 1 & 1 & 1 \\ 0 & 2 & 0 \\ -1 & 1 & 2 \end{bmatrix} = -(4+2) = -6$$

CHAPTER 3

Engineering Science

OUTLINE

KINEMATICS OF A PARTICLE

Consider a point P that moves along a smooth path as indicated in Figure 3.1. The position of the point may be specified by the vector $\mathbf{r}(t)$, defined to extend from an arbitrarily selected, fixed point O to the moving point P. The **velocity v** of the point is defined to be the derivative with respect to t of $\mathbf{r}(t)$, written as

$$\mathbf{v} = \frac{d\mathbf{r}}{dt} \tag{3.1}$$

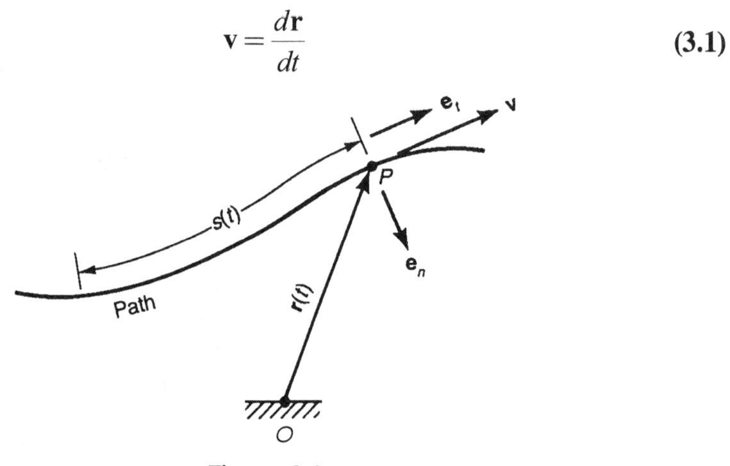

Figure 3.1

Although this definition is sometimes used for evaluation [that is, by differentiating a specific expression for $\mathbf{r}(t)$], it will often be more direct to use other relationships. It follows from the above definition that the velocity vector is tangent to the path of the particle; thus, upon introduction of a unit vector $\mathbf{e}_t$, defined to be tangent to the path, the velocity can also be expressed as

$$\mathbf{v} = v\mathbf{e}_t \tag{3.2}$$

The position of P can also be specified in terms of the distance $s(t)$ traveled along the path from an arbitrarily selected reference point. Then an incremental change in position may be approximated as $\Delta\mathbf{r} \approx \Delta s\mathbf{e}_t$, in which the accuracy increases as the increments Δt and $\Delta\mathbf{r}$ approach zero. This leads to still another way of expressing the velocity as

$$\mathbf{v} = \frac{ds}{dt}\mathbf{e}_t \tag{3.3}$$

The scalar

$$v = \frac{ds}{dt} = \dot{s} \tag{3.4}$$

can be either positive or negative, depending on whether the motion is in the same or the opposite direction as that selected in the definition of $\mathbf{e}_t$.

The **acceleration** of the point is defined as the derivative of the velocity with respect to time:

$$\mathbf{a} = \frac{d\mathbf{v}}{dt} \tag{3.5}$$

A useful relationship follows from application to Equation (3.2) of the rules for differentiating products and functions of functions:

$$\frac{d\mathbf{v}}{dt} = \dot{v}\mathbf{e}_t + \mathbf{v}\frac{ds}{dt}\frac{d\mathbf{e}_t}{ds}$$

As the direction of $\mathbf{e}_t$ varies, the square of its magnitude, $|\mathbf{e}_t|^2 = \mathbf{e}_t \bullet \mathbf{e}_t$, remains fixed and equal to 1, so that

$$\frac{d}{ds}|\mathbf{e}_t|^2 = \frac{d}{ds}(\mathbf{e}_t \bullet \mathbf{e}_t) = 2\mathbf{e}_t \bullet \frac{d\mathbf{e}_t}{ds} = 0$$

This shows that $d\mathbf{e}_t/ds$ is either zero or perpendicular to $\mathbf{e}_t$. With another unit vector $\mathbf{e}_n$ defined to be in the direction of $d\mathbf{e}_t/ds$, this vector may be expressed as

$$\frac{d\mathbf{e}_t}{ds} = \kappa\mathbf{e}_n$$

The scalar κ is called the local **curvature** of the path; its reciprocal, $\rho = 1/\kappa$, is called the local **radius of curvature** of the path. In the special case in which the path is straight, the curvature and hence $d\mathbf{e}_t/ds$ are zero. These lead to the following expression for the **acceleration** of the point:

$$\mathbf{a} = \dot{v}\mathbf{e}_t + \frac{v^2}{\rho}\mathbf{e}_n \tag{3.6}$$

The two terms express the **tangential** and **normal** (or **centripetal**) components of acceleration.

If a driver of a car with sufficient capability "steps on the gas," a positive value of $\dot{v}$ is induced, whereas if he "steps on the brake," a negative value is induced. If the path of the car is straight (zero curvature or "infinite" radius of curvature), the entire acceleration is $\dot{v}\mathbf{e}_t$. If the car is rounding a curve, there is an additional component of acceleration directed laterally, toward the center of curvature of the path. These components are indicated in Figure 3.2, a view of the plane of $\mathbf{e}_t$ and $d\mathbf{e}_t/ds$.

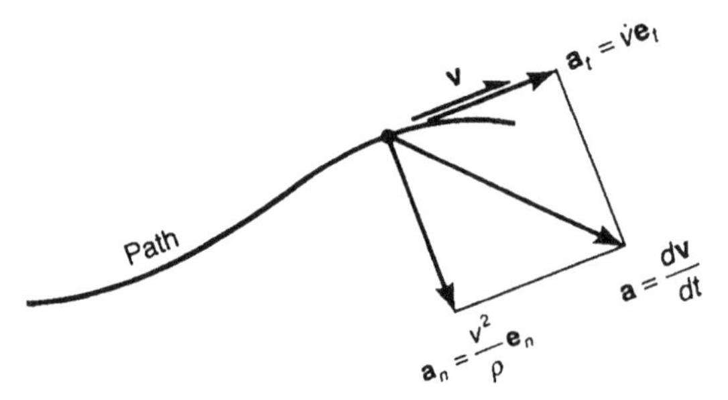

Figure 3.2

Example 3.1

At a certain instant, the velocity and acceleration of a point have the rectangular Cartesian components given by

$$\mathbf{v} = (3.5\mathbf{e}_x - 7.2\mathbf{e}_y + 9.6\mathbf{e}_z)\ \text{m/s}$$
$$\mathbf{a} = (-20\mathbf{e}_x + 20\mathbf{e}_y + 10\mathbf{e}_z)\ \text{m/s}^2$$

At this instant, what are the rate of change of speed dv/dt and the local radius of curvature of the path?

Solution

The rectangular Cartesian components of the unit tangent vector can be determined by dividing the velocity vector by its magnitude:

$$\mathbf{e}_t = \frac{\mathbf{v}}{|\mathbf{v}|} = \frac{3.5\mathbf{e}_x - 7.2\mathbf{e}_y + 9.6\mathbf{e}_z}{\sqrt{(3.5)^2 + (-7.2)^2 + (9.6)^2}} = 0.280\mathbf{e}_x - 0.576\mathbf{e}_y + 0.768\mathbf{e}_z$$

The rate of change of speed can then be determined as the projection of the acceleration vector onto the tangent to the path:

$$\dot{v} = \mathbf{e}_t \bullet \mathbf{a} = [(0.280)(-20) + (-0.576)(20) + (0.768)(10)]\ \text{m/s}^2$$
$$= -9.44\ \text{m/s}^2$$

The negative sign indicates the projection is opposite to $\mathbf{e}_t$ (which was defined by the above equation to be in the same direction as the velocity). This means that the speed is *decreasing* at 9.44 m/s. One sees from Figure 3.2 that the normal component of acceleration has magnitude

$$a_n = \sqrt{|\mathbf{a}|^2 - \dot{v}^2} = \sqrt{(-20)^2 + (20)^2 + (10)^2 - (-9.44)^2\,\text{m/s}^2} = 28.5\ \text{m/s}^2$$

which, from Equation (3.6), is related to the speed and radius of curvature by $a_n = v^2/\rho$. Rearrangement of this equation gives the radius of curvature as

$$\rho = \frac{v^2}{a_n} = \frac{[(3.5)^2 + (-7.2)^2 + (9.6)^2]\ \text{m}^2/\text{s}^2}{28.5\ \text{m/s}^2} = 5.48\ \text{m}$$

Relating Distance, Velocity, and the Tangential Component of Acceleration

The basic relationships among tangential acceleration a_t, velocity $v\mathbf{e}_t$, and distance s are

$$\frac{dv}{dt} = a_t \quad \text{or} \quad v = v_0 + \int a_t\, dt \tag{3.7}$$

$$\frac{ds}{dt} = v \quad \text{or} \quad s = s_0 + \int v\, dt \tag{3.8}$$

in which v_0 and s_0 are constants of integration. An alternative relationship comes from writing $dv/dt = (ds/dt)(dv/ds) = v\, dv/ds$:

$$v\frac{dv}{ds} = a_t \quad \text{or} \quad v^2 = v_0^2 + 2\int a_t\, ds \tag{3.9}$$

Equations (3.7) and (3.8) are useful in dealing with the *time* histories of acceleration, velocity, and distance, whereas Equation (3.9) is helpful in dealing with the manner in which velocity and acceleration vary with distance.

Example 3.2

The variation of tangential acceleration with time is given in Exhibit 1. If a point with an initial velocity of 24 m/s is subjected to this acceleration, what will be its velocity at $t = 6$ s, 10 s, and 15 s, and what will be the values of s at $t = 4$ s, 7.6 s, and 15 s?

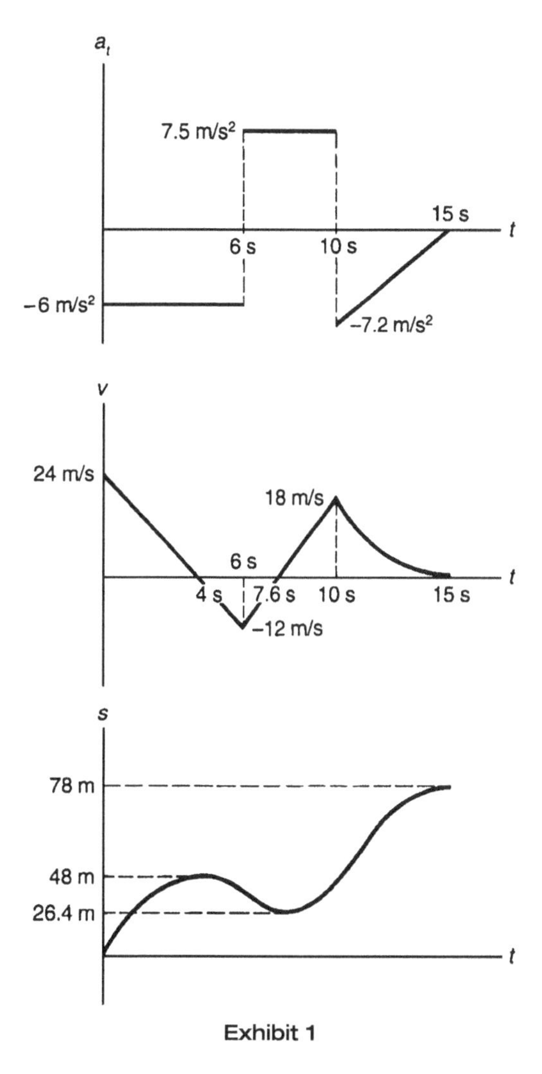

Exhibit 1

Solution

Equation (3.7) has the following graphical interpretations: At each point, the slope of the *v*-*t* curve is equal to the ordinate on the a_t-*t* curve. During any interval, the change in the value of *v* is equal to the area under the a_t-*t* curve for the same interval. With these rules and the given initial value of *v*, the variation of *v* with *t* can be plotted, and values of *v* can be calculated for each point.

The reader should use these rules to verify all details of the *v*-*t* curve shown. Equation (3.8) indicates that identical rules for slopes, ordinates, and areas relate the curve of distance *s* to that of velocity *v*, so the same procedure can be used to construct the *s*-*t* curve from the *v*-*t* curve. Again, the reader should verify all details of this curve.

Example **3.3**

The tangential acceleration of the pendulum bob shown in Exhibit 2 varies with position according to $a = -g\sin(s/l)$, in which *g* is the local acceleration of gravity. If a speed v_0 is imparted at the vertical position (where $s = 0$), what will be the maximum value of *s* reached?

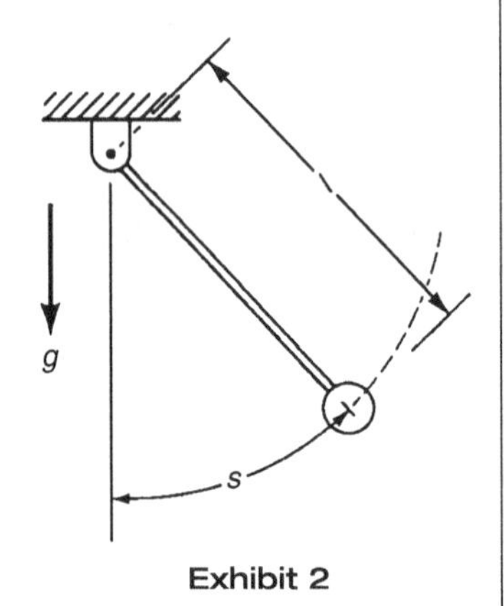

Exhibit 2

Solution

Because the relationship between tangential acceleration and *position* is given, Equation (3.9) will prove useful. The integrated form leads to

$$v^2 = v_0^2 + 2\int_0^s g\sin\left(\frac{s}{l}\right)ds$$

$$= v_0^2 + 2gl\left(1 - \cos\frac{s}{l}\right)$$

which gives the velocity *v* in the terms of any position *s*. Since $v = \dot{s}$, the maximum *s* will occur when $v = 0$, and the corresponding *s* is easily isolated from the above equation after setting $v = 0$:

$$s_{\max} = l\cos^{-1}\left(1 - \frac{v_0^2}{2gl}\right)$$

Observe that if $v_0^2 > 2gl$, no real value of $s_{\max}$ exists, because *v* never reaches zero in that case.

Constant Tangential Acceleration

When the tangential acceleration is constant, Equations (3.7) through (3.9) reduce to

$$v = v_0 + a_t t \tag{3.10}$$

$$s = s_0 + v_0 t + \frac{1}{2}a_t t^2 \tag{3.11}$$

$$v^2 = v_0^2 + 2a_t s \tag{3.12}$$

Rectilinear Motion

In the special case in which the path is a straight line, the unit tangent vector $\mathbf{e}_t$ is constant, and the curvature $1/\rho$ is zero throughout. The acceleration is then given by $(dv/dt)\mathbf{e}_t$, and the subscript on the symbol a_t may be dropped without ambiguity.

Example 3.4

A particle is launched vertically upward with an initial speed of 10 m/s and subsequently moves with constant downward acceleration of magnitude 9.8 m/s^2. What is the maximum height reached by the particle? How long does it take to return to the original launch position? And how fast is it traveling at its return to the launch position?

Solution

In this case the path will be straight and the acceleration is constant. With $\mathbf{e}_t$ defined as upward, the constant scalars appearing in Equations (3.10) through (3.12) have the values $v_0 = 10$ m/s and $a_t = a = -9.8$ m/s^2, so that these equations become

$$v = 10\text{ m/s} - (9.8\text{ m/s}^2)t \qquad \textbf{(i)}$$

$$s = (10\text{ m/s})t - \frac{1}{2}(9.8\text{ m/s}^2)t^2 \qquad \textbf{(ii)}$$

$$v^2 = (10\text{ m/s})^2 - 2(9.8\text{ m/s}^2)s \qquad \textbf{(iii)}$$

The maximum height reached can be obtained by setting $v = 0$ in (iii), which gives

$$s_{\max} = \frac{(10\text{ m/s})^2}{2(9.8\text{ m/s}^2)} = 5.1\text{ m}$$

The time required to reach this height can be obtained by setting $v = 0$ in (i), which gives

$$t_1 = \frac{10\text{ m/s}}{9.8\text{ m/s}^2} = 1.02\text{ s}$$

Finally, setting $s = 0$ in (iii) yields the two values of v that specify the velocity at the launch position:

$$v = \pm 10\text{ m/s}$$

The positive value gives the upward initial velocity, and the negative value gives the equal-magnitude, downward velocity of the particle when it returns to the launch position.

Rectangular Cartesian Coordinates

Multidimensional motion can be analyzed in terms of components associated with a set of fixed unit vectors $\mathbf{e}_x$, $\mathbf{e}_y$, and $\mathbf{e}_z$, which are defined to be mutually perpendicular. For some aspects of analysis, it is also important that they form a "right-handed" set, or $\mathbf{e}_z = \mathbf{e}_x \times \mathbf{e}_y$, $\mathbf{e}_x = \mathbf{e}_y \times \mathbf{e}_z$, and $\mathbf{e}_y = \mathbf{e}_z \times \mathbf{e}_x$. In terms of these unit vectors, the position, velocity, and acceleration can be expressed as

$$\mathbf{r} = x\mathbf{e}_x + y\mathbf{e}_y + z\mathbf{e}_z$$
$$\mathbf{V} = v_x\mathbf{e}_x + v_y\mathbf{e}_y + v_z\mathbf{e}_z$$
$$\mathbf{A} = a_x\mathbf{e}_x + a_y\mathbf{e}_y + a_z\mathbf{e}_z$$

with

$$v_x = \dot{x}$$
$$a_x = \dot{v}_x = \ddot{x}, \quad \text{etc.}$$

Example 3.5

A wheel rolls without slipping along a straight surface with the orientation of the wheel given in terms of the angle $\theta(t)$. See Exhibit 3. Express the velocity and acceleration of the point P on the rim of the wheel in terms of this angle, its derivatives, and the radius b of the wheel.

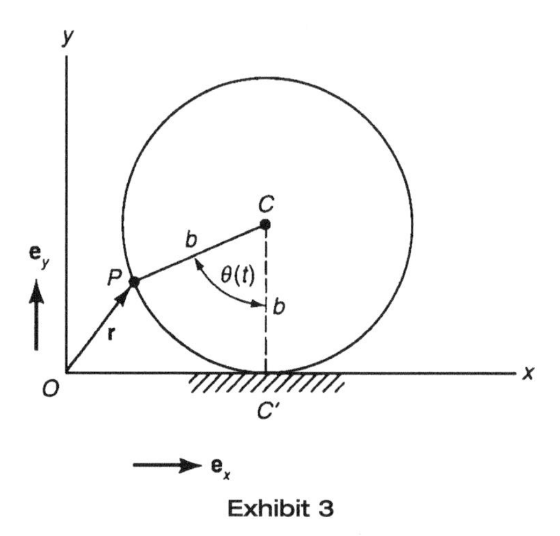

Exhibit 3

Solution

The origin for the x-y coordinates of P is the location of P when $\theta = 0$. Because the wheel rolls without slipping, the distance OC' is equal to the length of the circular arc PC'. The x-coordinate of P is then $OC' - b\sin\theta = b\theta - b\sin\theta$. The y-coordinate of P is that of C (i.e., b) minus $b\cos\theta$. In terms of these coordinates, the position vector from O to P may be expressed as

$$\mathbf{r} = b(\theta - \sin\theta)\mathbf{e}_x + b(1 - \cos\theta)\mathbf{e}_y$$

The velocity is then determined by differentiation of this expression:

$$\mathbf{v} = b\dot{\theta}[(1 - \cos\theta)\mathbf{e}_x + \sin\theta\mathbf{e}_y]$$

The acceleration is determined by another differentiation:

$$\mathbf{a} = b\ddot{\theta}[(1 - \cos\theta)\mathbf{e}_x + \sin\theta\mathbf{e}_y] + b\dot{\theta}^2(\sin\theta\mathbf{e}_x + \cos\theta\mathbf{e}_y)$$

These expressions may be simplified somewhat by rewriting them in terms of the unit vectors $\mathbf{e}_r$ and $\mathbf{e}_\theta$ as defined in Exhibit 4. These unit vectors are given in terms of the original horizontal and vertical unit vectors by

$$\mathbf{e}_r = -\sin\theta\mathbf{e}_x - \cos\theta\mathbf{e}_y$$
$$\mathbf{e}_\theta = -\cos\theta\mathbf{e}_x + \sin\theta\mathbf{e}_y$$

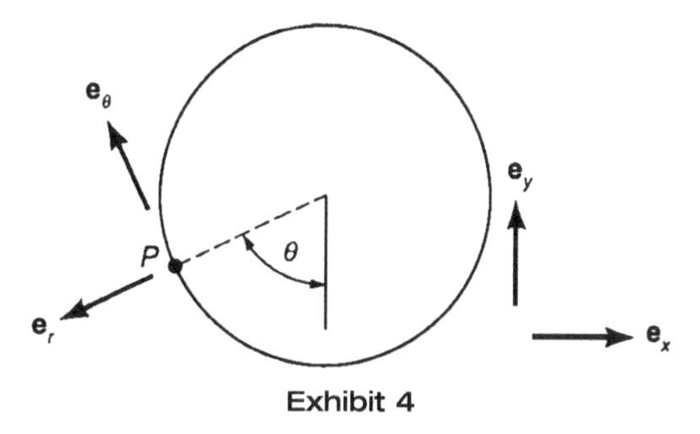

Exhibit 4

The previous expressions for velocity and acceleration can now be written as

$$\mathbf{v} = b\dot{\theta}(\mathbf{e}_x + \mathbf{e}_\theta)$$
$$\mathbf{a} = b\ddot{\theta}(\mathbf{e}_x + \mathbf{e}_\theta)b\dot{\theta}^2\mathbf{e}_r$$

Further simplification is possible upon examination of the sum $\mathbf{e}_x + \mathbf{e}_\theta$, shown in Exhibit 5. The magnitude of this sum is $2 \sin(\theta/2)$, and its direction is perpendicular

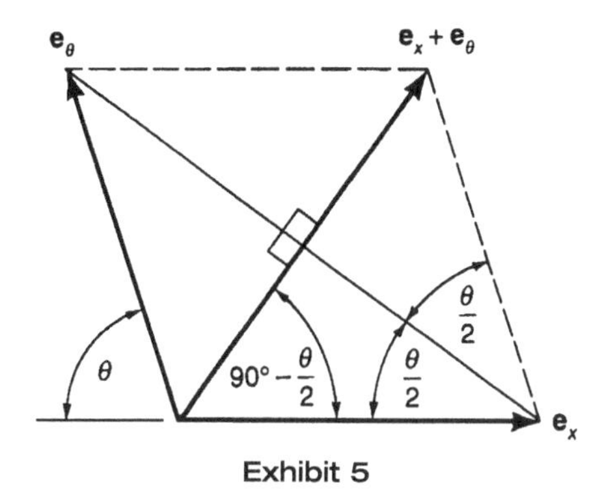

Exhibit 5

to the line connecting points P and C'. The velocity can thus be expressed as $\mathbf{v} = (2b \sin \theta/2\, \dot{\theta})\mathbf{e}_t$ in which the unit tangent vector is perpendicular to the line PC'. The acceleration can be simplified correspondingly to

$$\mathbf{a} = 2b \sin\frac{\theta}{2}\ddot{\theta}\mathbf{e}_t - b\dot{\theta}^2\mathbf{e}_r$$

Several steps were taken to reach the results in Example 3.5. The position vector was expressed in terms of the geometric constraints on the rolling of the wheel, differentiation led to expressions for the velocity and acceleration, and the introduction of auxiliary unit vectors and several trigonometric relationships simplified several expressions.

As mentioned earlier, direct use of the definitions expressed by Equations (3.1) and (3.5) may not be the easiest means of evaluating velocities and accelerations. Indeed, we will now review some kinematic relationships for rigid bodies that will make much shorter work of this example.

Circular Cylindrical Coordinates

Figure 3.3 shows a coordinate system that is useful for a number of problems in particle kinematics. The x and y coordinates of the rectangular Cartesian system are replaced with the distance r and the angle ϕ, while the definition of the z-coordinate remains unchanged. Two of the unit vectors associated with the rectangular

Cartesian system are also replaced with $\mathbf{e}_r = \cos\phi\,\mathbf{e}_x + \sin\phi\,\mathbf{e}_y$ and $\mathbf{e}_\phi = -\sin\phi\,\mathbf{e}_x + \cos\phi\,\mathbf{e}_y$.

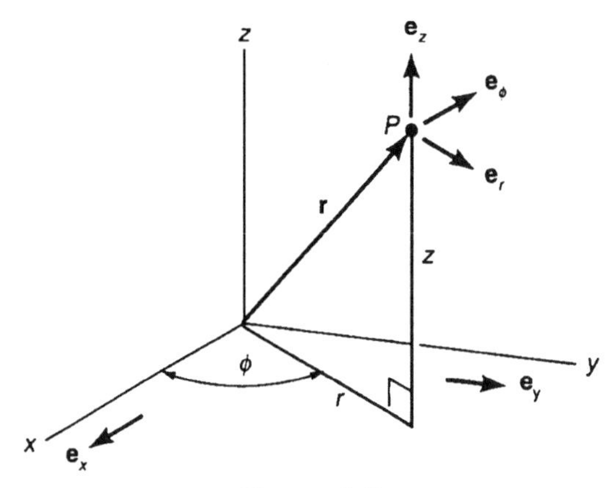

Figure 3.3

Since the angle ϕ varies, these two unit vectors also vary; their derivatives may be obtained by differentiating the above expressions:

$$\frac{d\mathbf{e}_r}{dt} = (-\mathbf{e}_x \sin\phi + \mathbf{e}_y \cos\phi)\frac{d\phi}{dt} = \dot{\phi}\mathbf{e}_\phi$$
$$\frac{d\mathbf{e}_\phi}{dt} = (-\mathbf{e}_x \cos\phi - \mathbf{e}_y \sin\phi)\frac{d\phi}{dt} = -\dot{\phi}\mathbf{e}_r$$

These are used along with the expression $\mathbf{r} = r\,\mathbf{e}_r + z\,\mathbf{e}_z$ for position to obtain expressions for velocity and acceleration:

$$\begin{aligned}\mathbf{v} = \dot{\mathbf{r}} &= \dot{r}\mathbf{e}_r + r\dot{\mathbf{e}}_r + \dot{z}\mathbf{e}_z \\ &= \dot{r}\mathbf{e}_r + r\dot{\phi}\mathbf{e}_\phi + \dot{z}\mathbf{e}_z\end{aligned} \tag{3.13}$$

$$\begin{aligned}\mathbf{a} = \dot{\mathbf{v}} &= \ddot{r}\mathbf{e}_r + \dot{r}\dot{\mathbf{e}}_r + (\dot{r}\dot{\phi} + r\ddot{\phi})\mathbf{e}_\phi + r\dot{\phi}\dot{\mathbf{e}}_\phi + \ddot{z}\mathbf{e}_z \\ &= (\ddot{r} - r\dot{\phi}^2)\mathbf{e}_r + (r\ddot{\phi} + 2\dot{r}\dot{\phi})\mathbf{e}_\phi + \ddot{z}\mathbf{e}_z\end{aligned} \tag{3.14}$$

Example 3.6

In Exhibit 6, the slider moves along the rod as it rotates about the fixed point *O*. At a particular instant, the slider is 200 mm from *O*, moving outward at 3 m/s relative to the rod; this relative speed is increasing at 130 m/s². At the same instant, the rod is rotating at a constant rate of 191 rpm. Evaluate the velocity and acceleration of the slider, and determine the rate of change of speed of the slider.

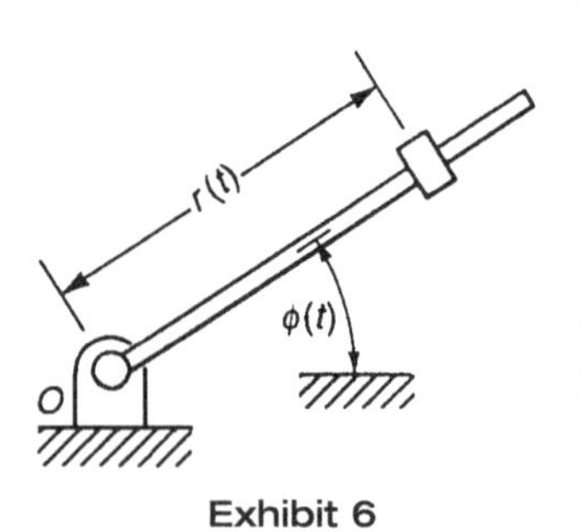

Exhibit 6

Solution

The angular speed of the rod is

$$\dot{\phi} = (191\text{ rpm})\frac{2\pi\text{ rad/rev}}{60\text{ s/min}} = 20.00\text{ rad/s}$$

and its angular acceleration $\ddot{\phi}$ is zero. Other values to be substituted into Equations (3.13) and (3.14) are r = 0.2 m, $\dot{r} = 3$ m/s, and $\ddot{r} = 130$ m/s².
Substitution into Equations (3.13) and (3.14) leads directly to the following radial and transverse components of velocity and acceleration:

$$\mathbf{v} = (3\,\mathbf{e}_r + 4\,\mathbf{e}_\phi)\text{ m/s}$$
$$\mathbf{a} = (50\,\mathbf{e}_r + 120\,\mathbf{e}_\phi)\text{ m/s}^2$$

Now the radial and transverse components of the unit vector tangent to the path can be obtained by dividing the velocity vector by its magnitude:

$$\mathbf{e}_t = \frac{3\mathbf{e}_r + 4\mathbf{e}_\phi}{\sqrt{(3)^2 + (4)^2}} = 0.6\mathbf{e}_r + 0.8\mathbf{e}_\phi$$

The rate of change of speed is the projection of the acceleration vector onto the tangent to the path, which can be obtained by dot-multiplying the acceleration with the unit tangent vector:

$$\begin{aligned}\dot{v} = a_t &= \mathbf{e}_t \bullet \mathbf{a} \\ &= (0.6)(50\ \text{m/s}^2) + (0.8)(120\ \text{m/s}^2) \\ &= 126\ \text{m/s}^2\end{aligned}$$

Circular Path

When the path is circular, r is constant, and Equations (3.13) and (3.14) reduce to

$$\begin{aligned}\mathbf{v} &= r\dot{\phi}\mathbf{e}_\phi \\ \mathbf{a} &= -r\dot{\phi}^2\mathbf{e}_r + r\ddot{\phi}\mathbf{e}_\phi\end{aligned}$$

Comparing these with Equations (3.2) and (3.6) (with $\rho = r$),

$$\mathbf{v} = v\mathbf{e}_t$$

$$\mathbf{a} = \dot{v}\mathbf{e}_t + \frac{v^2}{r}\mathbf{e}_n$$

we see that, for circular path motion, $\mathbf{e}_t = \mathbf{e}_\phi$, $\mathbf{e}_n = -\mathbf{e}_r$, and

$$v = r\dot{\phi} \tag{3.15}$$

$$a_n = r\dot{\phi}^2 \tag{3.16}$$

Example 3.7

A satellite is to be placed in a circular orbit over the equator at such an altitude that it makes one revolution around the earth per sidereal day (23.9345 hours). The gravitational acceleration is $(3.99 \times 10^{14}\ \text{m}^3/\text{s}^2)/r^2$, where r is the distance from the center of the earth. What is the altitude at which the satellite must be placed to achieve this period of orbit?

Solution

The angular speed of the line from the center of the earth to the satellite is

$$\dot{\phi} = \frac{2\pi\ \text{rad}}{(23.9345\ \text{h})(3600\ \text{s/h})} = 7.292 \times 10^{-5}\ \text{rad/s}$$

The acceleration has no tangential component, but the radial component in terms of the orbit radius and the angular speed will be $a_n = r(7.292 \times 10^{-5}\text{s}^{-1})^2$.

This acceleration is imparted by the earth's gravitational attraction, so that $r(7.292 \times 10^{-5}\text{s}^{-1})^2 = (3.99 \times 10^{14}\ \text{m}^3/\text{s}^2)/r^2$. This equation is readily solved for r, resulting in

$$r = \sqrt[3]{\frac{3.99\times10^{14}\,\text{m}^3/\text{s}^2}{(7.292\times10^{-5}\text{s}^{-1})^2}} = 42.2\times10^6\ \text{m}$$

The altitude will then be the difference between this value and the size of earth's radius, which is about 6.4×10^6 m: altitude = 35.8×10^6 m.

RIGID BODY KINEMATICS

The analysis of numerous mechanical systems rests on the assumption that the bodies making up the system are *rigid*. If the forces involved and the materials and geometry of the bodies are such that there is little deformation, the resulting predictions can be expected to be quite accurate.

The Constraint of Rigidity

If a body is **rigid**, the distance between each pair of points remains constant as the body moves. This constraint may be expressed in terms of a position vector $\mathbf{r}_{PQ}$ from a point P of the body to a point Q of the body, as indicated in Figure 3.4. If the magnitude of $\mathbf{r}_{PQ}$ is constant, then

$$\frac{d}{dt}|\mathbf{r}_{PQ}|^2 = \frac{d}{dt}(\mathbf{r}_{PQ}\bullet\mathbf{r}_{PQ}) = 2\mathbf{r}_{PQ}\bullet\frac{d\mathbf{r}_{PQ}}{dt} = 0 \qquad \text{(i)}$$

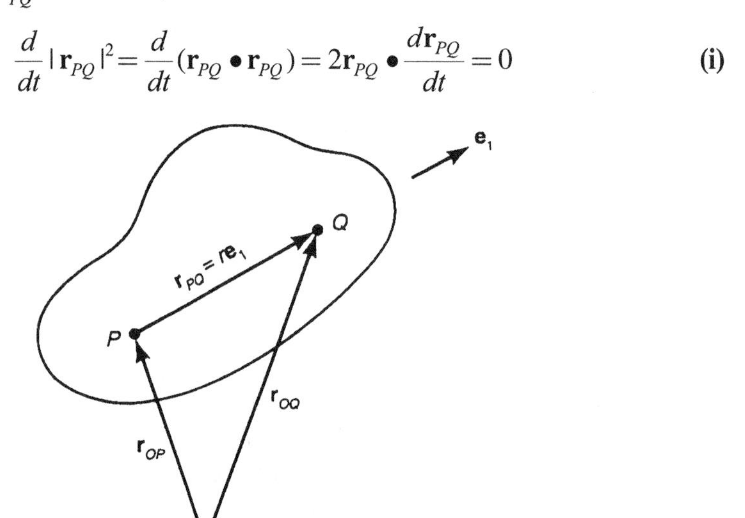

Figure 3.4

which indicates that $\dot{\mathbf{r}}_{PQ}$ is perpendicular to $\mathbf{r}_{PQ}$. Now, with a selected, fixed point designated as O, and vectors $\mathbf{r}_{OP}$ and $\mathbf{r}_{OQ}$ defined as indicated in Figure 3.4, differentiation of the vector relationship $\mathbf{r}_{PQ} = \mathbf{r}_{OQ} - \mathbf{r}_{OP}$ leads to the relationship

$$\frac{d\mathbf{r}_{PQ}}{dt} = \mathbf{v}_Q - \mathbf{v}_P \qquad \text{(ii)}$$

in which $\mathbf{v}_P$ and $\mathbf{v}_Q$ designate the velocities of P and Q, respectively. Finally, if we define $\mathbf{e}_1$ to be the unit vector in the direction of $\mathbf{r}_{PQ}$, so that

$$\mathbf{r}_{PQ} = r\mathbf{e}_1 \qquad \text{(iii)}$$

then substitution of (ii) and (iii) into (i) leads to

$$2r\mathbf{e}_1 \bullet (\mathbf{v}_Q - \mathbf{v}_P) = 0$$

or

$$\mathbf{e}_1 \bullet \mathbf{v}_Q = \mathbf{e}_1 \bullet \mathbf{v}_P \qquad \textbf{(3.17)}$$

This shows that *the projections of the velocities of any two points of a rigid body onto the line connecting the two points must be equal*. This is intuitively plausible; otherwise the distance between the points would be changing. This frequently provides the most direct way of evaluating the velocities of various points within a mechanism.

Example **3.8**

As the crank OQ in Exhibit 7 rotates clockwise at 200 rad/s, the piston P moves vertically. What will be the velocity of the piston at the instant when the angle θ is 50 degrees?

Solution

Since point Q must follow a circular path, its speed may be determined from Equation (3.15): $v_Q = (0.075\text{ m})(200\text{ s}^{-1}) = 15$ m/s, with the direction of $\mathbf{v}_Q$ as indicated in the figure. Because the cylinder wall constrains the piston, its velocity is vertical. The connecting rod PQ is rigid, so the velocities of the points P and Q must

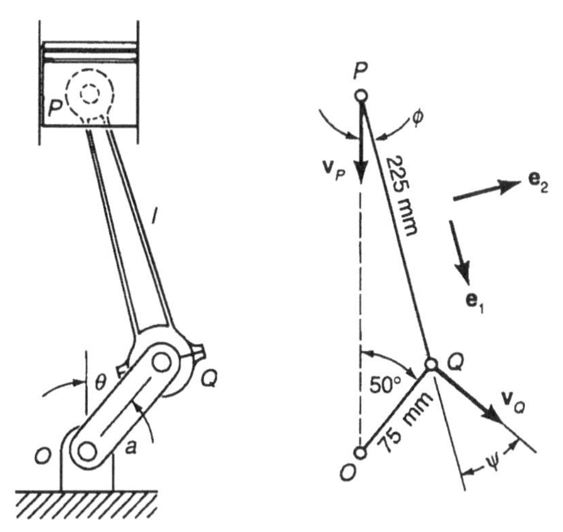

Exhibit 7

satisfy $v_P \cos\phi = v_Q \cos\psi$. The trigonometric rule of sines, applied to the triangle OPQ, gives

$$\sin\phi = \frac{a}{l}\sin\theta = \frac{75}{225}\sin 50^\circ$$

which yields $\phi = 14.8°$. The other required angle is then $\psi = 90° - \theta - \phi = 25.2°$. Once these angles are determined, the constraint equation yields the speed of the piston:

$$v_P = \frac{\cos\psi}{\cos\phi} v_Q = 14.04\text{ m/s}$$

The Angular Velocity Vector

If a rigid body is in *plane motion*, that is, if the velocities of all points of the body lie in a fixed plane, then its orientation may be specified by the angle θ between two fixed lines, one of which passes through the body, as indicated in Figure 3.5. The rate of change of this angle is central to the analysis of the velocities of various points of the body.

To determine this relationship, consider Figure 3.6, which shows a position vector from the point P to point Q, both fixed in the moving body. Two configurations are shown, one at time t and another after an arbitrary change during a time increment Δt. $\mathbf{e}_1$ is defined to be the unit vector in the direction of $\mathbf{r}_{PQ}(t)$, and $\mathbf{e}_2$ is

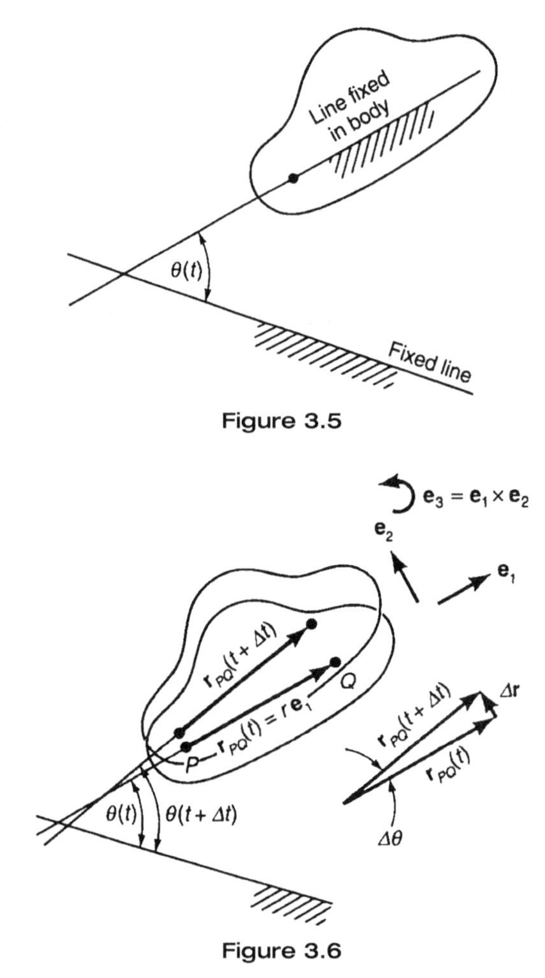

Figure 3.5

Figure 3.6

defined to be the unit vector of P and Q, 90° counterclockwise from $\mathbf{e}_1$. Both $\mathbf{e}_1$ and $\mathbf{e}_2$ are further assumed to lie in the plane of motion; this assumption is convenient but not limiting. The vector diagram in Figure 3.6 shows the change in $\mathbf{r}_{PQ}$ to be given by the approximation $\Delta\mathbf{r} \approx r\Delta\mathbf{e}_2$. Dividing both sides by the time increment Δt and letting this increment approach zero leads to the relation

$$\frac{d\mathbf{r}_{PQ}}{dt} = r\dot{\theta}\mathbf{e}_2$$

The scalar $\dot{\theta}$ will be denoted also by ω. Note from the definition of θ that a positive value of ω indicates a counterclockwise rotation, whereas a negative value of ω indicates a clockwise rotation. Another useful form of this relation may be written in terms of the **angular velocity vector,**

$$\omega = \omega\,\mathbf{e}_3$$

where $\mathbf{e}_3$ is defined to be $\mathbf{e}_1 \times \mathbf{e}_2$, oriented perpendicular to the plane of Figure 3.6. With this definition,

$$\frac{d\mathbf{r}_{PQ}}{dt} = \boldsymbol{\omega} \times \mathbf{r}_{PQ}$$

Note that this relation is valid for *any* two points in the body. (A pair of points different from those shown in Figure 3.6 might give rise to a different angle, but as the body moves, *changes* in this angle would equal *changes* in θ.)

It can be shown that for the most general motion of a rigid body (not restricted to planar motion) there also exists a unique angular velocity vector for which the same relation holds. However, in nonplanar motion, the angular velocity ω is not straightforwardly related to the rate of change of an angle, and its calculation requires a more extensive analysis than in the case of planar motion.

When $\dot{\mathbf{r}}_{PQ}$ is replaced with $\mathbf{v}_Q - \mathbf{v}_P$ according to Equation (3.ii) of the previous section, the important velocity relationship

$$\mathbf{v}_Q = \mathbf{v}_P + \omega \times \mathbf{r}_{PQ} \tag{3.20}$$

is obtained, which, for planar motion, becomes

$$\mathbf{v}_Q = \mathbf{v}_P + r\omega\mathbf{e}_2 \tag{3.21}$$

In the special case in which $\omega = 0$, this indicates that all points have the same velocity, a motion called **translation**. In the special case in which $\mathbf{v}_P = \mathbf{0}$, the motion is simply rotation about a fixed axis through P. Thus, in the general case, the two terms on the right of Equation (3.20) can be seen to express a superposition of a translation and a rotation about P. But since P can be selected *arbitrarily*, there are as many combinations of a translation and a corresponding "center of rotation" as the analyst wishes to consider!

In all of the these cases, the angular velocity is a property of the *body's* motion, and Equation (3.21) relates the velocities of *any* two points of a body experiencing planar motion. Dot-multiplication of each member of Equation (3.21) with $\mathbf{e}_2$ leads to the following means of evaluating the angular velocity of a plane motion in terms of the velocities of two points:

$$\omega = \frac{\mathbf{e}_2 \bullet \mathbf{v}_Q - \mathbf{e}_2 \bullet \mathbf{v}_P}{r} \tag{3.22}$$

That is, ω will be the difference between the magnitudes of the projections of the velocities of P and Q onto the perpendicular to the line connecting P and Q, divided by the distance between P and Q.

Example 3.9

What will be the angular velocity of the connecting rod in Example 3.8, at the instant when the angle θ is 50 degrees?

Solution

Referring to Exhibit 7 for the definition of $\mathbf{e}_2$, we see that

$$\omega = \frac{v_Q \sin\psi + v_Q \sin\phi}{l}$$

$$= \frac{(15\text{ m/s})\sin 25.2° + (14.04\text{ m/s})\sin 14.8°}{0.225\text{ m}} = 44.3\text{ rad/s}$$

The positive value indicates that the rotation is counterclockwise at this instant.

Instantaneous Center of Zero Velocity

For planar motion with $\omega \neq 0$, there always exists a point C' of the body (or an imagined extension of the body) that has zero velocity. If point P of Equation (3.21) is selected to be this special point, the equation reduces to $\mathbf{v}_Q = \mathbf{v}_{C'} + \omega \times \mathbf{r}_{C'Q} = r\omega\mathbf{e}_2$ where r is now the distance from C' to Q and $\mathbf{e}_2$ is perpendicular to the line connecting C' and Q. This latter property can be used to locate C' if the directions of the velocities of two points of the body are known.

Example 3.10

What is the location of the instantaneous center C' of the connecting rod in Examples 3.10 and 3.11? Use this to verify the previously determined values of the angular velocity of the connecting rod and the velocity of point P.

Solution

The velocity of any point of the connecting rod must be perpendicular to the line from C' to that point. Hence C' must lie at the point of intersection of the horizontal line through P and the line through Q perpendicular to $\mathbf{v}_Q$ (i.e., on the line through O and Q), as shown in Exhibit 3. The pertinent distances can be found as follows:

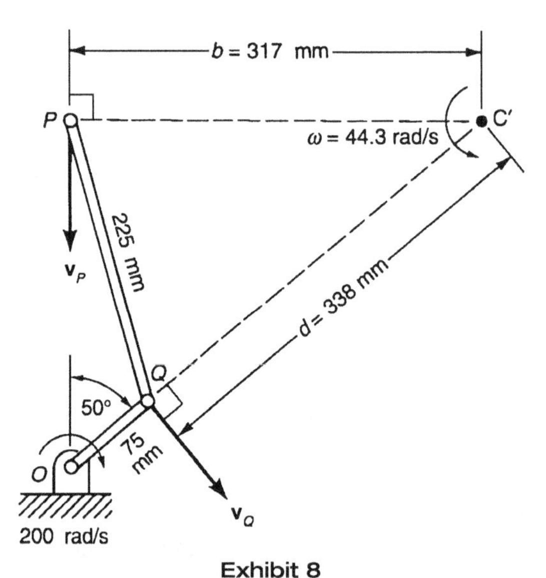

Exhibit 8

$$\begin{aligned} OP &= (75\text{ mm})\cos 50° + (225\text{ mm})\cos 14.8° = 266\text{ mm} \\ PC' &= OP\tan 50° = 317\text{ mm} \\ QC' &= OP\sec 50° - 75\text{ mm} = 338\text{ mm} \end{aligned}$$

The angular velocity of the connecting rod is then

$$\omega = \frac{v_Q}{QC'} = \frac{15\text{ m/s}}{0.339\text{ m}} = 44.3\text{ rad/s}$$

and the velocity of P is then

$$v_P = PC'\omega = (0.317 \text{ m})(44.3 \text{ s}^{-1}) = 14.04 \text{ m/s}$$

in agreement with values the previously obtained.

Example 3.11

Using the properties of the instantaneous center, determine the velocity of the point P on the rim of the rolling wheel in Example 3.5.

Solution

Since the wheel rolls without slipping, the point of the wheel in contact with the flat surface has zero velocity and is therefore its instantaneous center. The angular speed of the wheel is $\dot{\theta}$, and the distance from C' to P is readily determined from Exhibit 9:

$$r = 2b\sin\frac{\theta}{2}$$

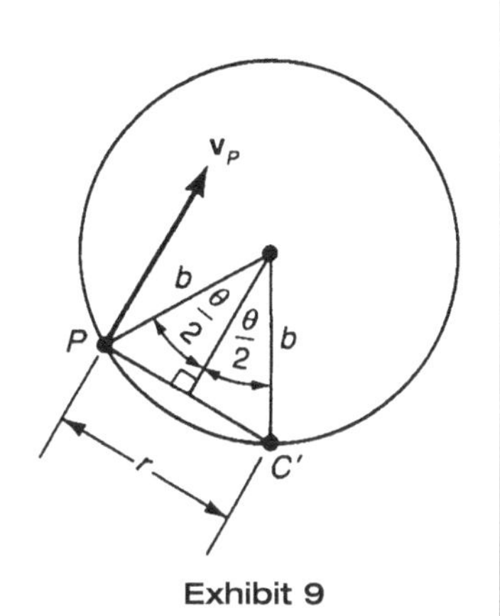

Exhibit 9

The velocity of point P then has the magnitude

$$v_P = r\omega = 2b\sin\frac{\theta}{2}\dot{\theta}$$

and the direction shown in Exhibit 9. This direction should be evident by inspection once it is realized that a positive $\dot{\theta}$ corresponds to clockwise rotation. The reader may find it instructive to recall the conventions for the choice of $\mathbf{e}_2$ and positive ω used in the derivation leading to Equation (3.21) and verify the agreement. Note the simplicity of this analysis as compared with the one expressing the position of P in a rectangular Cartesian coordinate system.

Accelerations in Rigid Bodies

Formally differentiating Equation (3.20) and substituting for $\dot{\mathbf{r}}_{PQ}$ using Equation (3.19) leads to

$$\mathbf{a}_Q = \mathbf{a}_P + (\alpha \times \mathbf{r}_{PQ}) + \omega \times (\omega \times \mathbf{r}_{PQ}) \quad \textbf{(3.23)}$$

in which the vector $\alpha = d\omega/dt$ is called the **angular acceleration** of the body. For planar motion, $\alpha = \alpha\mathbf{e}_3 = \dot{\omega}\mathbf{e}_3$ and $\omega \times (\omega \times \mathbf{r}) = -\omega^2\mathbf{r}$ so that

$$\mathbf{a}_Q = \mathbf{a}_P + r\alpha\mathbf{e}_2 - r\omega^2\mathbf{e}_1 \quad \textbf{(3.24)}$$

where $\mathbf{e}_1$ and $\mathbf{e}_2$ are defined as indicated in Figure 3.6.

Equivalent relationships, analogous to Equation (3.17) and Equation (3.22) for velocity, can be obtained by dot-multiplying this equation by $\mathbf{e}_1$ and by $\mathbf{e}_2$:

$$\mathbf{e}_1 \bullet \mathbf{a}_Q = \mathbf{e}_1 \bullet \mathbf{a}_P - r\omega^2 \quad \textbf{(3.25)}$$

$$\alpha = \frac{\mathbf{e}_2 \bullet \mathbf{a}_Q - \mathbf{e}_2 \bullet \mathbf{a}_P}{r} \quad \textbf{(3.26)}$$

Example 3.12

If the speed of the crank in Examples 3.6 through 3.8 is constant, what are the acceleration $\mathbf{a}_P$ of the piston and the angular acceleration α of the connecting rod at the instant when the angle θ is 50 degrees (Exhibit 10)?

Solution

When the crank speed is constant, the acceleration of Q is entirely centripetal, of magnitude

$$a_Q = r\omega^2$$

$$a_Q = (0.075 \text{ m})(200 \text{ s}^{-1})^2 = 3000 \text{ m/s}^2$$

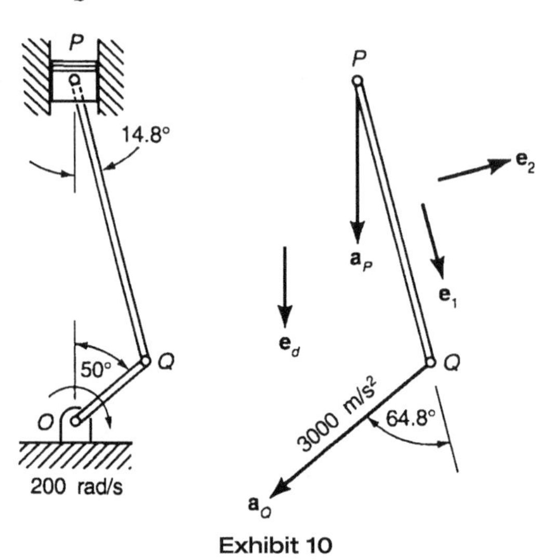

Exhibit 10

and directed toward the center of curvature O of the path of Q. The acceleration of P is vertically upward or downward. To determine the direction, we define a downward unit vector $\mathbf{e}_d$ and let $\mathbf{a}_p = a_p\mathbf{e}_d$ (see Exhibit 10). A positive value of a_P then indicates a downward acceleration and a negative value an upward acceleration. These expressions for a_Q and $\mathbf{a}_P$ are substituted into Equation (3.25), along with the previously determined angular velocity of the rod, giving

$$(3000 \text{ m/s}^2) \cos 64.8^\circ = a_P \cos 14.8^\circ - (0.225 \text{ m})(44.3 \text{ s}^{-1})^2$$

which yields

$$a_P = 1779 \text{ m/s}^2$$

The angular acceleration α of the rod can then be determined from Equation (3.26):

$$\alpha = \frac{(3000 \text{ m/s}^2)\cos 154.8^\circ - (1779 \text{ m/s}^2)\cos 104.8^\circ}{0.225 \text{ m}} = -10{,}050 \text{ rad/s}^2$$

The negative value indicates that the angular acceleration is clockwise; that is, the 44.3-rad/s counterclockwise angular velocity is rapidly decreasing at this instant.

NEWTON'S LAWS OF MOTION

Every element of a mechanical system must satisfy Newton's second law of motion; that is, the resultant force **f** acting on the element is related to the acceleration **a** of the element by

$$\mathbf{f} = m\mathbf{a} \tag{3.27}$$

in which m represents the mass of the element. Newton's third law requires that the force exerted on a body A by a body B is of equal magnitude and opposite direction to the force exerted on body B by body A. These laws and their logical consequences provide the basis for relating motions to the forces that cause them.

Applications to a Particle

A **particle** is an idealization of a material element in which its spatial extent is disregarded, so that the motion of all of its parts is completely characterized by the path of a geometric *point*. When the accelerations of various parts of a system differ significantly, the system is considered to be composed of a number of particles and analyzed as described in the next section.

Example 3.13

An 1800-kg aircraft in a loop maneuver follows a circular path of radius 3 km in a vertical plane. At a particular instant, its velocity is 210 m/s directed 25 degrees above the horizontal as shown in Exhibit 11. If the engine thrust is 16 kN greater than the aerodynamic drag force, what is the rate of change of the aircraft's speed, the magnitude of the aircraft's acceleration, and the aerodynamic lift force?

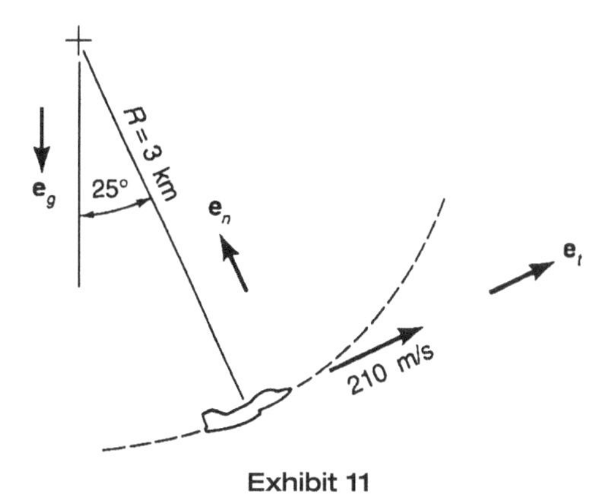

Exhibit 11

Solution

Since the dimensions of the aircraft are small compared with the radius of the path, all of its material elements can be considered to have essentially the same motion, so treating the aircraft as a particle as described above is reasonable.

The forces acting on the aircraft are shown on the free-body diagram, Exhibit 12. The thrust **T**, the drag **D**, and the lift **L** all result from aerodynamic pressure from the surrounding air and engine gas. The lift is defined to be the component of the total force that is perpendicular to the flight path, and arises primarily from the wings. The force of gravity, mg, is the only other force arising from a source external to the free body. The left-hand side of Equation (3.27) is the resultant of

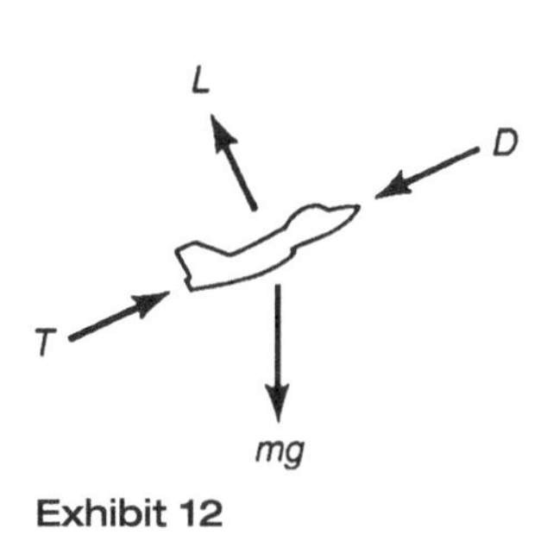

Exhibit 12

these forces, whereas the right-hand side is obtained from Equation (3.6). Thus, Newton's second law is written in this case as

$$(T-D)\mathbf{e}_t + L\mathbf{e}_n + mg\mathbf{e}_g = m\left(\dot{v}\mathbf{e}_t + \frac{v^2}{R}\mathbf{e}_n\right)$$

Two independent equations arise from this two-dimensional vector equation. Dot multiplication with $\mathbf{e}_t$ yields $(T - D) + mg\mathbf{e}_t \bullet \mathbf{e}_g = m\dot{v}$ because $\mathbf{e}_t$ and $\mathbf{e}_n$ are always perpendicular vectors and thus their dot product is 0. This equation, rearranged, leads to the rate of change of speed:

$$\dot{v} = \frac{T-D}{m} - g\sin 25° = \frac{16{,}000\ \text{N}}{1800\ \text{kg}} - (9.81\ \text{m/s}^2)\sin 25° = 4.74\ \text{m/s}^2$$

Dot multiplication with $\mathbf{e}_n$ yields

$$L + mg\mathbf{e}_n \bullet \mathbf{e}_g = \frac{mv^2}{R}$$

which then allows us to determine the magnitude of lift force,

$$L = m\left(g\cos 25° + \frac{v^2}{R}\right) = (1800\ \text{kg})\left[(9.81\ \text{m/s}^2)\cos 25° + \frac{(210\ \text{m/s})^2}{3000\ \text{m}}\right] = 42.5\ \text{kN}$$

The magnitude of the acceleration is then determined by combining the tangential and normal components found above:

$$|\mathbf{a}| = \sqrt{(4.74\ \text{m/s}^2)^2 + \left[\frac{(210\ \text{m/s})^2}{3000\ \text{m}}\right]^2} = 15.45\ \text{m/s}^2$$

Example 3.14

Two blocks are interconnected by an inextensible, massless line through the pulley arrangement shown in Exhibit 13. The inertia and friction of the pulleys are negligible. The coefficient of friction between the block of mass m_1 and the horizontal surface is μ. What is the acceleration of the block of mass m_2 as it moves downward?

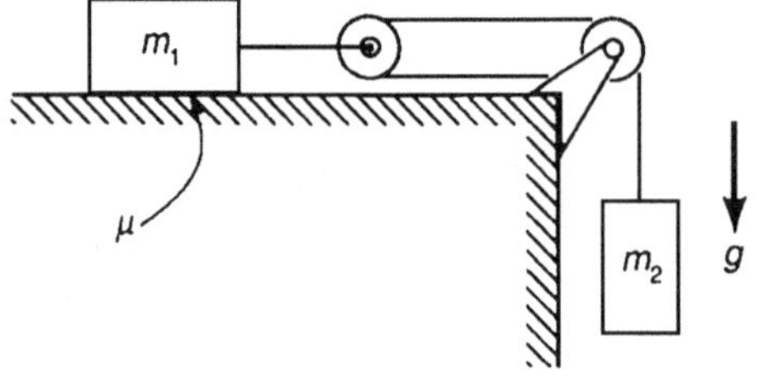

Exhibit 13

Solution

Since the motion of each block is a translation, the acceleration of each element in a block is the same; hence their spatial extensions may be ignored and each block may be treated as a particle.

Moment equilibrium of each pulley requires that the tension be the same in each part of the longer line around the pulleys. Denoting this tension by T, the two free-body diagrams in Exhibit 14 are used to write expressions for Newton's second law. Since the acceleration of the block on the left has no vertical component, $R_1 - m_1 g = 0$. Denoting its rightward acceleration by a_1, consideration of the horizontal forces leads to $2T - \mu R_1 = m_1 a_1$. Denoting the downward acceleration of the other block by a_2, application of Newton's second law to the other free body yields $m_2 g - T = m_2 a_2$. To relate the two accelerations, consider the pulley connected to the horizontal block. The instantaneous center of this pulley is at its rim and directly below its center. Thus, the speed of the pulley's upper rim is twice that of its center, so that the speed of the vertically moving block is twice that of the horizontally moving block. Since this is true at all times, we have $a_2 = 2a_1$.

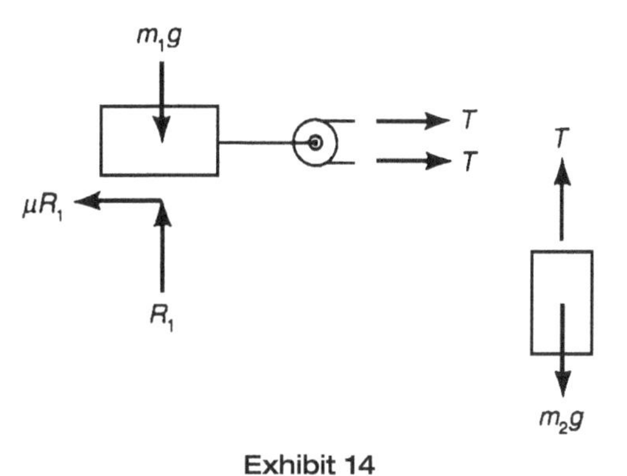

Exhibit 14

Eliminating R, T, and a_1 from these four equations now leads to

$$a_2 = \frac{4m_2 - 2\mu m_1}{4m_2 + m_1} g$$

Observe that if $m_2 > \mu m_1/2$, the acceleration is downward, whereas if $m_2 < \mu m_1/2$, the acceleration is upward, implying that the downward velocity will reach zero, after which time the friction force will no longer equal μR_1.

Systems of Particles

A mechanical **system** is any collection of material elements of fixed identity whose motion we may wish to consider. Such a system is treated as a collection of particles in which the individual particles must obey Newton's laws of motion.

It proves to be very useful to separate the forces acting on a system into those arising from sources outside the system and arising from the interaction between members of the system, as shown in Figure 3.7. That is, the resultant force on the ith particle is written as

$$\mathbf{f}_i = \mathbf{f}_{ie} + \sum_j \mathbf{f}_{i/j}$$

in which $\mathbf{f}_{ie}$ represents the resultant of all forces on the ith particle arising from sources external to the system, and $\mathbf{f}_{i/j}$ represents the force exerted on ith particle by the jth particle. With this notation, Newton's third law may be expressed as $\mathbf{f}_{j/i} = -\mathbf{f}_{i/j}$.

Now, each particle moves according to Newton's second law:

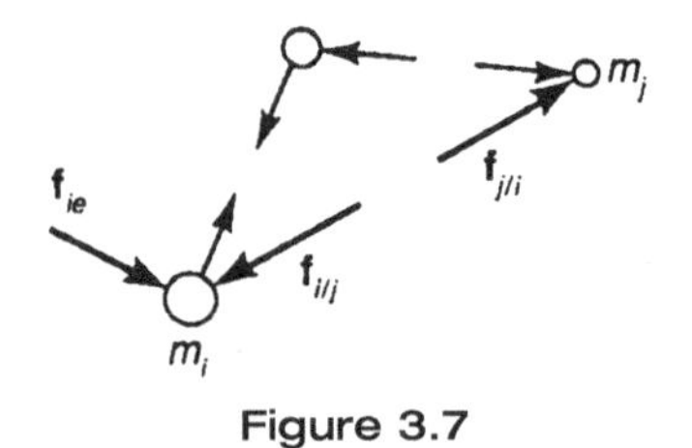

Figure 3.7

$$\mathbf{f}_{ie} + \sum_j \mathbf{f}_{i/j}$$

in which m_i denotes the mass of the ith particle and $\mathbf{a}_i$ its acceleration. There are as many such equations as there are particles in the system; if all such equations are added, the result is

$$\sum_i \mathbf{f}_{ie} + \sum_i \sum_j \mathbf{f}_{i/j} = \sum_i m_i \mathbf{a}_i$$

In view of Newton's third law, the internal forces can be grouped as pairs of oppositely directed forces of equal magnitude, and so their sum vanishes, leaving

$$\sum_i \mathbf{f}_{ie} = \sum_i m_i \mathbf{a}_i \quad \textbf{(3.28a)}$$

That is, the resultant of all *external* forces is equal to the sum of the products of the individual masses and their corresponding accelerations.

Linear Momentum and Center of Mass

The right-hand member of Equation (3.28a) can be expressed alternatively in terms of the **linear momentum** of the system, which is defined as

$$\mathbf{p} = \sum_i m_i \mathbf{v}_i \quad \textbf{(3.29)}$$

in which $\mathbf{v}_i$ denotes the velocity of the ith particle. Differentiation of this equation results in

$$\frac{d\mathbf{p}}{dt} = \sum_i m_i \mathbf{a}_i$$

which is the same expression appearing in Equation (3.28a). Hence an alternative to Equation (3.28a) is

$$\sum_i \mathbf{f}_{ie} = \frac{d\mathbf{p}}{dt} \quad \textbf{(3.28b)}$$

which states that the sum of the external forces is equal to the time rate of change of the linear momentum of the system.

The **center of mass** of the system is a point C located, relative to an arbitrarily selected reference point O, by the position vector $\mathbf{r}_C$, which satisfies the defining equation

$$m\mathbf{r}_C = \sum_i m_i \mathbf{r}_i \tag{3.30}$$

in which m denotes the total mass of the system and $\mathbf{r}_i$ is a position vector from O to the ith particle. Differentiation of this equation leads to

$$m\mathbf{v}_C = \sum_i m_i \mathbf{v}_i \tag{3.31}$$

which shows that the linear momentum is the product of the total mass and the velocity of the center of mass. Another differentiation yields

$$m\mathbf{a}_C = \sum_i m_i \mathbf{a}_i$$

which provides still another way of expressing Equation (3.28a):

$$\sum_i \mathbf{f}_{ie} = m\mathbf{a}_C \tag{3.28c}$$

This is sometimes called the **principle of motion of the mass center**. It indicates that the mass center responds to the resultant of external forces exactly as would a single particle having a mass equal to the total mass of the system.

Example 3.15

A motor inside the case shown in Exhibit 15 drives the eccentric rotor at a constant angular speed ω. The distance from the rotor bearing to its center of mass is e, the mass of the rotor is m_r, and the mass of the nonrotating housing is $m - m_r$. (That is, the total mass of the rotor and housing together is m.) The housing is free to translate horizontally, constrained by the rollers, and under the influence of a spring of stiffness k and a dashpot that transmits a force to the housing of magnitude c times the speed of the housing in the direction opposite to that of the velocity of the housing. Write the differential equation that governs the extension $x(t)$ of the spring from its relaxed position.

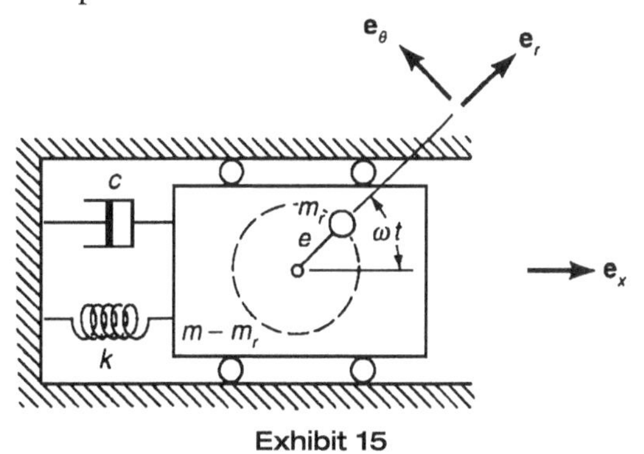

Exhibit 15

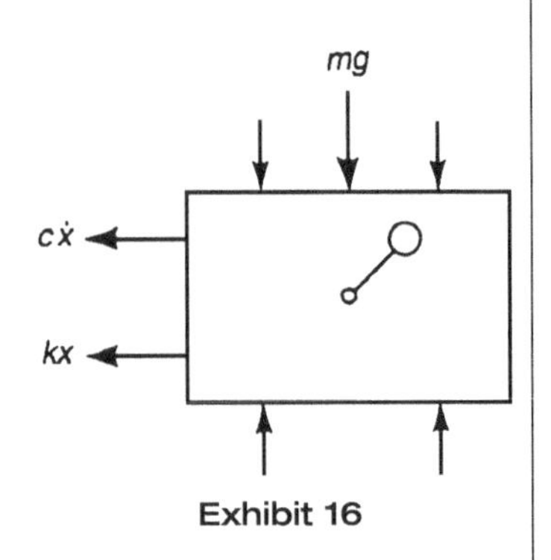

Exhibit 16

Solution

Consider the system consisting of the housing and rotor together. The free-body diagram (Exhibit 16) shows forces acting on this system from sources *external* to it. It does *not* include the torque necessary to maintain constant rotor speed nor the reaction at the bearing, these being internal, action-reaction pairs.

Note that when x is positive (the spring extended), the force exerted by the spring on the housing acts to the left, and when x is negative (the spring compressed), this force acts to the right. Both situations are depicted properly by the label kx on the arrow; that is, this indicates that the force equals $-kx\mathbf{e}_x$ in all cases. The same consideration applies to the arrow and label representing the force from the dashpot. The sum of all external forces then is

$$\sum_i \mathbf{f}_{ie} = -(kx + c\dot{x})\mathbf{e}_x + f_y\mathbf{e}_y$$

The acceleration of the housing is simply $\ddot{x}\mathbf{e}_x$, while that of the mass center of the rotor can be most readily determined by using Equation (3.24), letting P be the center of the bearing and noting that $\alpha = 0$. This gives the acceleration of the mass center of the rotor as

$$\mathbf{a} = \ddot{x}\mathbf{e}_x - e\omega^2\mathbf{e}_r$$

Substitution into Equation (3.28a) results in

$$-(kx + c\dot{x})\mathbf{e}_x + f_y\mathbf{e}_y = (m - m_r)\ddot{x}\mathbf{e}_x + m_r(\ddot{x}\mathbf{e}_x - e\omega^2\mathbf{e}_r)$$

The forces f_y are neither known nor of interest for our purpose; they may be eliminated from the equation by dot-multiplying each member with $\mathbf{e}_x$, which leads to

$$-(kx + c\dot{x}) = m\ddot{x} - m_r e\omega^2 \cos \omega t$$

A "standard" form of this equation is obtained by placing the dependent variable and its derivatives on one side and the known function of time on the other:

$$m\ddot{x} + c\dot{x} + kx = m_r e\omega^2 \cos \omega t$$

The same result can be obtained using either Equation (3.28b) or (3.28c).

Impulse and Momentum

The integral with respect to time of the resultant of external forces is called the **impulse** of this resultant force:

$$\mathbf{g} = \int_{t_1}^{t_2} \sum_i \mathbf{f}_{ie}\, dt$$

Integration of both members of Equation (3.28b) results in the following integrated form:

$$\mathbf{g} = \mathbf{p}(t_2) - \mathbf{p}(t_1) \qquad \textbf{(3.32)}$$

This states that impulse of the resultant external force is equal to the change in momentum of the system. Since it is a vector equation, we may obtain up to three independent relationships from it, one for each of three dimensions.

A special case occasionally arises, in which one or more components of the impulse are absent. The corresponding components of momentum then remain constant, and are said to be **conserved**.

Example 3.16

A rocket is simulated by a vehicle that is accelerated by the action of the passenger throwing rocks in the rearward direction as the vehicle moves along the roadway as shown in Exhibit 17. At a certain time, the mass of the vehicle, passenger, and supply of rocks is m, and all are moving at speed v. The passenger then launches a rock of mass m_1 with a rearward velocity of magnitude v_e *relative to the vehicle.* What is the increase Δv in the speed of the vehicle resulting from this action?

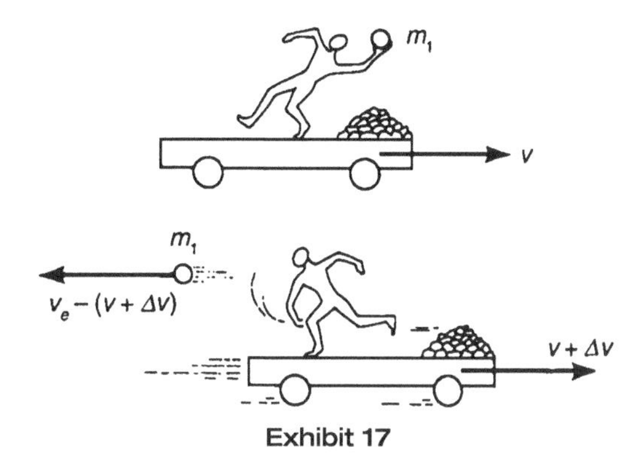

Exhibit 17

Solution

Assuming there is negligible friction at the wheels, the system consisting of passenger, vehicle, and rocks has no external forces acting on it in the direction of travel. The horizontal component of momentum is therefore conserved; that is, it is the same after the rock is launched as it was prior to the launching. After this is written in detail as

$$\mathrm{p}_{\text{final}} - \mathrm{p}_{\text{rock}} = \mathrm{p}_{\text{initial}}$$

$$(m - m_1)(v + \Delta v) - m_1[v_e - (v + \Delta v)] = mv$$

the equation can then be solved for the increase in vehicle speed:

$$\Delta v = \frac{m_1}{m} v_e$$

Moments of Force and Momentum

Equations (3.28) and (3.29) are valid regardless of the lines of action of the forces $\mathbf{f}_{ie}$. For example, the acceleration of the center of mass and the change in momentum of the system shown in Figure 3.8 will be the same for each of the different points of application of the force. However, there are characteristics of the motions induced by these forces that *do* depend on the lines of action, some of which are revealed by considering *moments* of forces.

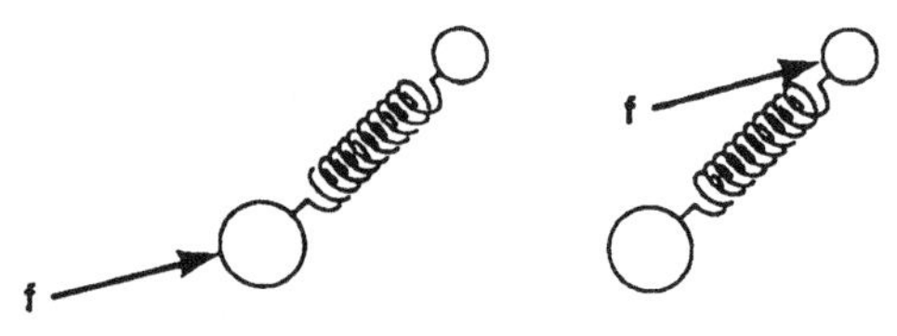

Figure 3.8

Consider again the external and internal forces acting on two typical particles of a system (Figure 3.9). Let the position vectors $\mathbf{r}_i$ and $\mathbf{r}_j$ locate the ith and jth particles, respectively, with respect to a selected point O. If the equation expressing Newton's second law for the ith particle is cross-multiplied by $\mathbf{r}_i$ and all such equations are added, the result is

$$\sum_i \mathbf{r}_i \times \mathbf{f}_{ie} + \sum_i \sum_j \mathbf{f}_{i/j} = \sum_i \mathbf{r}_i \times m_i \mathbf{a}_i$$

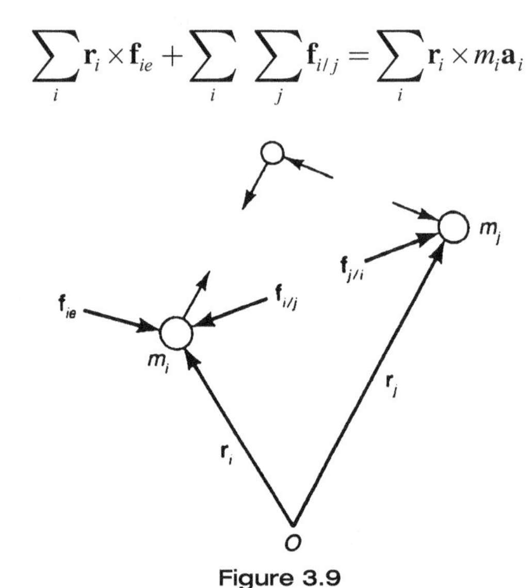

Figure 3.9

Now, if the forces $\mathbf{f}_{i/j}$ and $\mathbf{f}_{j/i} = -\mathbf{f}_{i/j}$ have a common line of action, then

$$\mathbf{r}_i \times \mathbf{f}_{i/j} + \mathbf{r}_j \times \mathbf{f}_{j/i} = 0$$

That is, the **moments** of the members of each action-reaction pair cancel one another, leaving the **moment equation** for the system:

$$\mathbf{M}_O = \sum_i \mathbf{r}_i \times m_i \mathbf{a}_i \tag{3.33a}$$

in which the moment of the external forces is evaluated as in the previous chapter:

$$\mathbf{M}_O = \sum_i \mathbf{r}_i \times \mathbf{f}_{ie}$$

Example 3.17

The pendulum in Exhibit 18 consists of a stiff rod of negligible mass with two masses attached, and it swings in the vertical plane about the frictionless hinge at O under the influence of gravity. What will be the angular acceleration of the pendulum in terms of angular displacement θ and the other parameters indicated in the sketch?

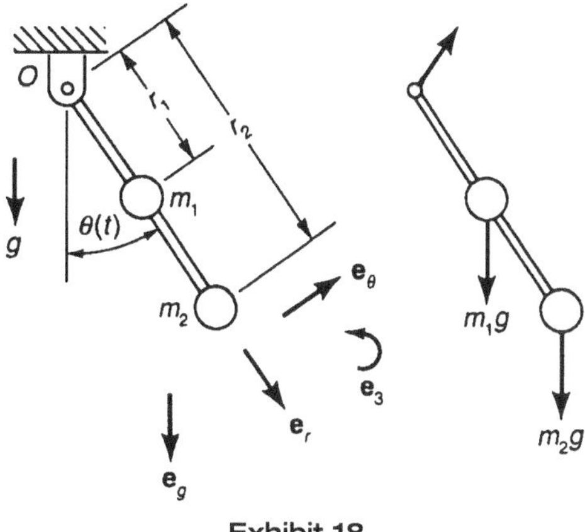

Exhibit 18

Solution

The free-body diagram shows the forces external to the system consisting of the rod together with the two particles. Since the reaction at the support is unknown and is of no interest for our purpose, a good strategy would be to consider moments about this point. Referring to the free-body diagram, we evaluate the resultant moment as usual:

$$\mathbf{M}_O = (r_i\mathbf{e}_r)\times(m_1 g\mathbf{e}_g) + (r_2\mathbf{e}_r)\times(m_2 g\mathbf{e}_g) = -(m_1r_1 + m_2r_2)g\sin\theta\,\mathbf{e}_3$$

Since each particle follows a circular path with center at O, their accelerations may be expressed as $\mathbf{a}_i = r_i\dot{\theta}^2\mathbf{e}_r + r_i\ddot{\theta}\mathbf{e}_\theta$. The right-hand member of the moment law, Equation (3.33a), is then evaluated in this case as

$$\begin{aligned}\sum \mathbf{r}_i \times m_i\mathbf{a}_i &= (r_1\mathbf{e}_r)\times m_1r_1(-\dot{\theta}^2\mathbf{e}_r + \ddot{\theta}\mathbf{e}_\theta) + (r_2\mathbf{e}_r)\times m_2r_2(-\dot{\theta}^2\mathbf{e}_r + \ddot{\theta}\mathbf{e}_\theta)\\ &= (m_1r_1^2 + m_2r_2^2)\ddot{\theta}\mathbf{e}_3\end{aligned}$$

Substitution into the moment law, Equation (3.33a), results in

$$-(m_1r_1 + m_2r_2)g\sin\theta\,\mathbf{e}_3 = (m_1r_1^2 + m_2r_2^2)\ddot{\theta}\,\mathbf{e}_3$$

or

$$\ddot{\theta} = -\frac{m_1r_1 + m_2r_2}{m_1r_1^2 + m_2r_2^2}g\sin\theta$$

The **moment of momentum** or **angular momentum about point** O is defined as

$$\mathbf{H}_O = \sum_i \mathbf{r}_i \times m_i\mathbf{v}_i \tag{3.34}$$

Now, if O is fixed in the inertial frame, then $\mathbf{v}_i = \mathbf{r}_i$, and it follows that

$$\frac{d\mathbf{H}_O}{dt} = \sum_i(\dot{\mathbf{r}}_i \times m_i\mathbf{v}_i + \mathbf{r}_i \times m_i\dot{\mathbf{v}}_i) = \sum_i \mathbf{r}_i \times m_i\mathbf{a}_i$$

This provides an alternative way of writing the moment law, as

$$\mathbf{M}_O = \frac{d\mathbf{H}_O}{dt} \tag{3.33b}$$

In the preceding example, the angular momentum about O is

$$\mathbf{H}_O = (r_1\mathbf{e}_r)\times(m_1 r_1\dot{\theta}\mathbf{e}_\theta) + (r_2\mathbf{e}_r)\times(m_2 r_2\dot{\theta}\mathbf{e}_\theta) = \left(m_1 r_1^2 + m_2 r_2^2\right)\dot{\theta}\mathbf{e}_3$$

Differentiating this expression and substituting for the right-hand member of Equation (3.33b) leads to the result achieved using Equation (3.33a).

As with forces and linear momentum, there are situations in which the moment of external forces vanishes. Then, Equation (3.33b) implies that the angular momentum about O remains constant, or is *conserved.*

Example **3.18**

Suppose the pendulum of Example 3.17 is suspended at rest when it is struck by a small projectile, which becomes imbedded in the lower ball (Exhibit 19). What angular velocity ω is imparted to the pendulum?

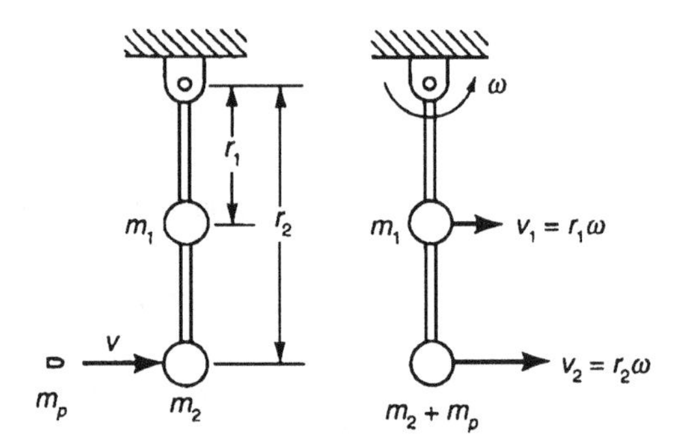

Exhibit 19

Solution

During the collision, which can induce a large reaction at the support as well as between the projectile and ball, the moment of forces external to the system—the pendulum and the projectile—will be zero. Hence, the angular momentum of this system prior to impact will equal that immediately after impact:

$$r_2 m_p v\mathbf{e}_3 = r_1 m_1 v_1\mathbf{e}_3 + r_2(m_2 + m_p)v_2\mathbf{e}_3 = [m_1 r_1^2 + (m_2 + m_p)r_2^2]\omega\mathbf{e}_3$$

Hence,

$$\omega = \frac{r_2 m_p v}{m_1 r_1^2 + (m_2 + m_p)r_2^2}$$

WORK AND KINETIC ENERGY

The integration in Equation (3.9) has extensive implications that will be examined in this section.

A Single Particle

If the form of the tangential acceleration indicated in Equation (3.9) is merged with Equation (3.6), the result can be used to express Newton's second law as

$$\mathbf{f} = m\left(v\frac{dv}{ds}\mathbf{e}_t + \frac{v^2}{\rho}\mathbf{e}_n\right) \qquad \textbf{(i)}$$

Now, if each member is dot-multiplied by an increment of change of position, $d\mathbf{r} = ds\mathbf{e}_t$, and the resulting scalars are integrated, they become

$$\int_{\mathbf{r}_1}^{\mathbf{r}_2} \mathbf{f} \bullet d\mathbf{r} = \int_{s_1}^{s_2} f_t\, ds = W_{1-2}$$

and

$$m\int_{v_1}^{v_2} v\, dv = \tfrac{1}{2}mv_2^{\,2} - \tfrac{1}{2}mv_1^{\,2}$$

The integral W_{1-2} is called the **work** done on the particle by the force **f** as the particle moves from position 1 to position 2. The scalar $T = ½\, mv^2$ is called the **kinetic energy** of the particle. Since (i) holds throughout any interval, a consequence of Newton's second law is the **work-kinetic energy relationship**:

$$W_{1-2} = T_2 - T_1 \qquad \textbf{(3.35)}$$

Work is considered positive when the kinetic energy of the system has been increased. When enough information is available to permit evaluation of the work integral, this provides a useful way of predicting the change in the speed of the particle.

Example 3.19

A 3.5-Mg airplane is to be launched from the deck of an aircraft carrier with the aid of a steam-powered catapult. The force that the catapult exerts on the aircraft varies with the distance s along the deck as shown in Exhibit 20. If other forces are negligible, what value of the constant f_0 is necessary for the catapult to accelerate the aircraft from rest to a speed of 160 km/h at the end of the 30-m travel?

$$f(s) = \frac{f_0}{1 + \dfrac{s}{30\text{m}}}$$

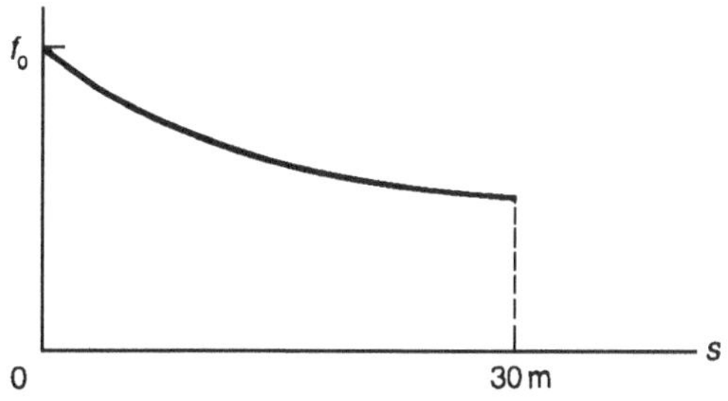

Exhibit 20

Solution

Letting d stand for the 30-m travel, the work done on the aircraft will be

$$W = \int_0^d \frac{f_0 ds}{1+\frac{s}{d}} = (f_0 d)\ln\left(1+\frac{s}{d}\right)\Bigg|_0^d = (f_0 d)\ln 2 = (20.8\,\text{m})f_0$$

This will equal the change in kinetic energy, which is initially zero.

$$\Delta T = (0.5)mv^2 - 0$$

$$\Delta T = \frac{1}{2}(3500\text{ kg})\left[\left(160\frac{\text{km}}{\text{h}}\right)\frac{1000\text{ m/km}}{3600\text{ s/h}}\right]^2 = 3.46\times10^6\,\text{J}$$

The work-kinetic energy relationship

$$W = \Delta T$$

$$(20.8\text{ m})f_0 = 3.46\times10^6\text{ N}\bullet\text{m}$$

implies that the constant f_0 must have the value

$$f_0 = 166\text{ kN}$$

Work of a Constant Force

A commonly encountered force of constant magnitude and direction is that of gravity near the earth's surface. When a constant force acts on a particle as it moves, the work done by the force can be evaluated as indicated in Figure 3.10. The increment of work as the particle undergoes an increment $d\mathbf{r}$ of displacement can be expressed as

$$dW = \mathbf{f}_0 \bullet d\mathbf{r} = |\mathbf{f}_0||d\mathbf{r}|\cos\measuredangle_{\mathbf{f}_0}^{d\mathbf{r}} = f_0 dq$$

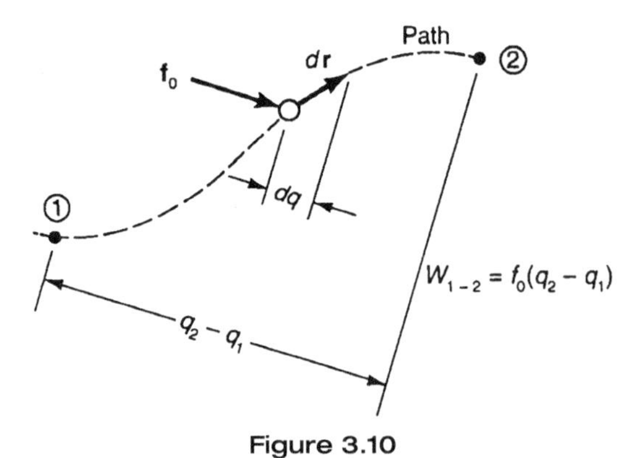

Figure 3.10

in which dq is the component of the displacement increment that is parallel to the force. Since the force is constant,

$$W_{1-2} = f_0\int_1^2 dq = f_0(q_2 - q_1) \qquad \textbf{(3.36)}$$

Thus, any movement of the particle that is perpendicular to the direction of the force has no effect on the work. In other words, the work is the same as would have been done if the particle had moved rectilinearly through a distance of $(q_2 - q_1)$ in a direction parallel to the force.

Example 3.20

How fast must the toy race car be traveling at the bottom of the hill to be able to coast to the top of the hill (see Exhibit 21)?

Solution

As the car moves up the hill, the work done by the force of gravity will be $W_g = -mgh$. The work done by friction forces may be negligible if the wheels are well made. If this is the case, the work done by all forces is approximately that due to

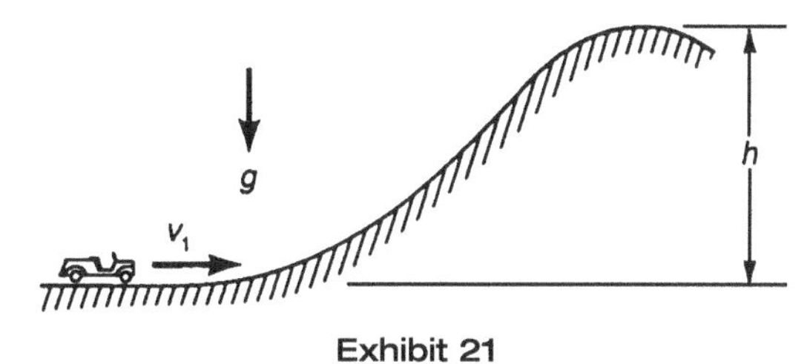

Exhibit 21

gravity. The speed of the car as it nears the hilltop can approach zero, so the work-kinetic energy equation may be written as

$$-mgh = T_2 - T_1 = 0 - \frac{1}{2}mv_1^2$$

which implies a minimum required speed of

$$v_1 = \sqrt{2gh}$$

With friction, the required speed will be somewhat greater.

Distance-Dependent Central Force

A force that remains directed toward or away from a fixed point is called a **central force**. Examples of forces for which the magnitude depends only on the distance from the particle to a fixed point are the force of gravitational attraction and the force from an elastic, tension-compression member with one end pinned to a fixed support. Figure 3.11 shows a particle P moving with such a central force acting on it; the dependence on distance is expressed by the function $f(r)$, with the convention that a positive value of f indicates an attractive force and a negative value of f a repulsive force.

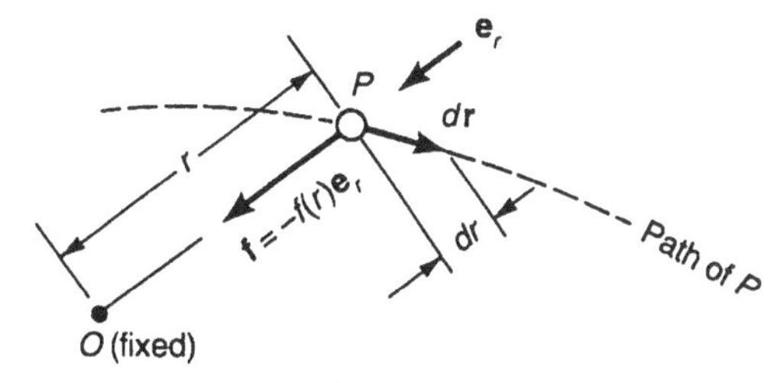

Figure 3.11

Now, the increment of work may be written as

$$dW = \mathbf{f} \bullet d\mathbf{r} = -f(r)\mathbf{e}_r \bullet d\mathbf{r}$$

Referring to the figure, we see that $\mathbf{e}_r \bullet d\mathbf{r} = |d\mathbf{r}| \cos \measuredangle_{\mathbf{er}}^{d\mathbf{r}}$ is equal to the change dr in radial distance r. Thus,

$$W_{1-2} = -\int_{r_1}^{r_2} f(r)\, dr \qquad \textbf{(3.37)}$$

Similar to the case of the constant force, the work done by a central force through an arbitrary motion is the same as would be done for a rectilinear motion, but in the radial direction.

Example 3.21

The elastic spring in Exhibit 22 has a linear force-displacement characteristic; that is, it exerts a force equal to the stiffness k times the amount it is stretched from its relaxed length l_0. As the particle moves from position 1 to position 2, what is the work done by the spring force on the particle?

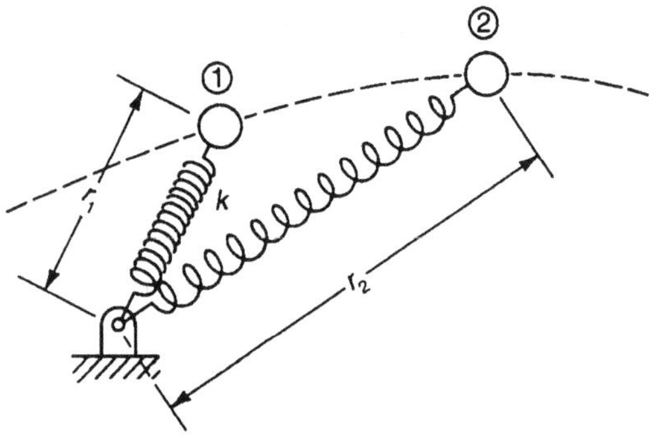

Exhibit 22

Solution

This is a case of a central distance-dependent force with $f(r) = -k(r - l_0)$. Equation (3.37) then becomes

$$W_{1-2} = -k\int_{r_1}^{r_2} (r - l_0)\, dr$$

A more convenient form results if we introduce the amount of spring extension as $\delta = (r - l_0)$. The integral then becomes

$$W_{1-2} = -k\int_{\delta_1}^{\delta_2} \delta\, d\delta = -\frac{k}{2}\left(\delta_2^2 - \delta_1^2\right)$$

Example 3.22

A 0.6-kg puck slides on a horizontal surface without friction under the influence of the tension in a light cord that passes through a small hole at O (see Exhibit 23). A spring under the surface imparts a tension in the cord that is proportional to the distance from the hole to the puck; its stiffness is $k = 30$ N/m. At a certain instant, the puck is 200 mm from the hole and moving at 2 m/s in the direction indicated in the top view. If the spring is in its relaxed position when the puck is at the hole, what is the maximum distance from the hole reached by the puck?

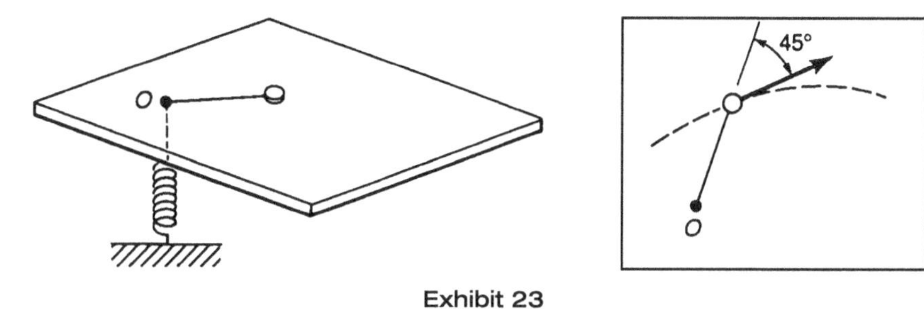

Exhibit 23

Solution

With the initial and maximum distances denoted by r_1 and r_2, respectively, the work done on the puck by the force from the cord from the initial position to that of maximum distance will be

$$W_{1-2} = -\frac{k}{2}\left(r_2^2 - r_1^2\right)$$

Since this is the only force that does work, this value must equal the change in kinetic energy:

$$-\frac{k}{2}\left(r_2^2 - r_1^2\right) = \frac{m}{2}\left(v_2^2 - v_1^2\right)$$

The moment about O of the force from the cord is zero, so the angular momentum about the hole is conserved. Because there is no radial component of velocity at the maximum distance, the angular momentum there is simply r_2mv_2. Hence,

$$r_1\ mv_1\ \sin\ 45° = r_2mv_2$$

These two equations contain the unknowns r_2 and v_2. Isolating v_2 from the latter, substituting this expression into the energy equation, and rearranging leads to the equation

$$\left(\frac{r_2}{r_1}\right)^4 - \left(1 + \frac{mv_1^2}{kr_1^2}\right)\left(\frac{r_2}{r_1}\right)^2 + \frac{mv_1^2}{kr_1^2}\sin^2 45° = 0$$

When the given values are substituted, the quadratic formula yields

$$\left(\frac{r_2}{r_1}\right) = 1.618$$

as the largest root so that the maximum distance reached is r_2 = 1.618 (200 mm) = 324 mm.

Example 3.23

A torpedo expulsion device operates by means of gas expanding within a tube that holds the torpedo. When test-fired with the tube firmly anchored, a 550-kg torpedo leaves the tube at 20 m/s. In operation, a 30-Mg submarine is traveling at 5 m/s when it expels a 550-kg torpedo in the forward direction. What are the speeds of the submarine and torpedo immediately after expulsion?

Solution

Considering the two-body system consisting of the submarine and torpedo, let us assume that the external forces remain in balance during expulsion. Then $W_e = 0$. Assuming also that the gas pressure depends only on the position of the torpedo relative to the submarine, the work W_{12} done by the internal forces will be the same during actual operation as during the test-firing. The work-kinetic energy relationship for the test-firing yields $W_{12} = ½ (550 \text{ kg})(20 \text{ m/s})^2 = 110 \text{ kJ}$, and for the operating condition the relationship is

$$110 \text{ kJ} = \frac{1}{2}(30{,}000 \text{ kg})\left[v_1^2 - (5 \text{ m/s})^2\right] + \tfrac{1}{2}(550 \text{ kg})\left[v_2^2 - (5 \text{ m/s})^2\right]$$

Also, if the external forces are in balance, momentum will be conserved:

$$(30{,}550 \text{ kg})(5 \text{ m/s}) = (30{,}000 \text{ kg})v_1 + (550 \text{ kg})v_2$$

These two relationships give the desired speeds as

$$v_1 = 4.6 \text{ m/s},\ v_2 = 24.8 \text{ m/s}$$

The 19.8-m/s boost in speed given the torpedo is slightly less than when it is fired from the firmly anchored tube. However, the speed of the torpedo relative to the submarine is

$$v_2 - v_1 = 20.2 \text{ m/s}$$

or slightly higher than in the fixed-tube test.

Two special cases of the work done by internal forces are of interest. The simpler is that in which the particles are constrained so that the distances between all pairs remain fixed, that is, the case of a rigid body. In this case, all the $dr_{i/j}$ are zero and the work-kinetic energy equation, Equation (3.38), reduces to $W_e = \Delta T$.

Another special case occurs when the force T_{ij} depends only on the distance $r_{i/j}$. That is, the force is not a function of relative velocity or previous history of deformation. Then the work integral $-\int T_{ij} dr_{i/j}$ is a function only of the distance between the particles and does not depend on the manner in which the particles move to reach a particular configuration. This would be the case, for example, with elastic spring interconnections or gravitational interactions. In this case, we can define the potential functions as

$$V_{ij}(r_{i/j}) = \int_{(r_{i/j})_0}^{r_{i/j}} T_{ij}(\rho_{i/j})d\rho_{i/j}$$

and if their sum is denoted by

$$V = \sum_{i-j} V_{ij}$$

the work-energy integral becomes

$$W_e = \Delta T + \Delta V$$

That is, when the internal forces are all conservative, the work done by the external forces is equal to the change in total mechanical energy within the system.

KINETICS OF RIGID BODIES

If a system of particles is structurally constrained so that the distance between every pair of particles remains constant as the system moves, it forms a rigid body. Thus, the laws of kinetics in the previous section are applicable, along with the kinematics relationships reviewed earlier. Of the kinematics relationships, Equation (3.28c) will be useful in the form given, whereas the moment equation, Equation (3.33a), must be specialized to relate accelerations to angular velocity and angular acceleration.

In the general (three-dimensional) case, both the moment equation and kinematics relationships become considerably more complicated than they are for planar motion, and will be outside the scope of this review.

Moment Relationships for Planar Motion

Figure 3.12 shows the plane of the motion of a rigid body, with a point P (to be selected by the analyst for moment reference) along with an element of mass dm, which in the following is analogous to the mass m_i in the earlier analysis of a set of particles. The summation appearing in Equation (3.33a) will be written as an integral in this case, because the body is viewed as having continuously distributed mass. The element of mass is located relative to P by the position vector $\mathbf{r} = r\mathbf{e}_1 + z\mathbf{e}_3$. Its acceleration is related to that of point P through Equation (3.24), and the moment equation, Equation (3.33a), may be written as

$$\begin{aligned}\mathbf{M}_P &= \int_m (r\mathbf{e}_1 + z\mathbf{e}_3)\times(\mathbf{a}_P + r\alpha\mathbf{e}_2 - r\omega^2\mathbf{e}_1)\,dm \\ &= \left(\int_m \mathbf{r}\,dm\right)\times\mathbf{a}_P + \left(\int_m r^2\,dm\right)\alpha\mathbf{e}_3 - \alpha\int_m zr\mathbf{e}_1\,dm - \omega^2\int_m zr\mathbf{e}_2\,dm\end{aligned}$$

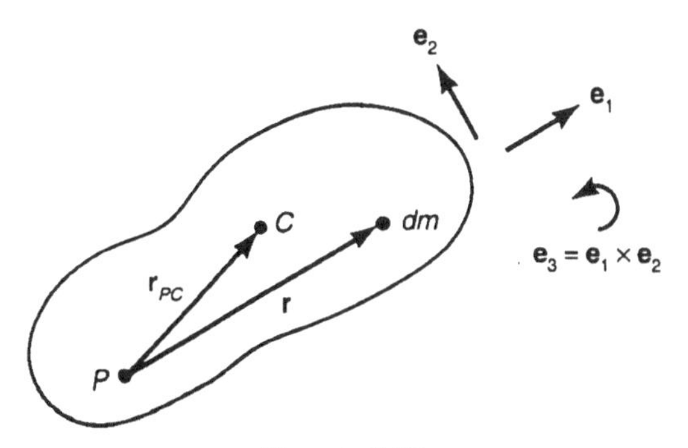

Figure 3.12

If the body's mass is distributed symmetrically with respect to the plane of motion through P, the last two integrals in the last line will vanish; without such symmetry, these terms indicate the possibility of components of moment in the plane of motion. Thus, even for *plane* motion, the distribution of mass may imply forces *perpendicular* to the plane of motion. These will not be pursued in detail here, but the reader must be aware of this possibility. The other two integrals will be of concern. The first is exactly the expression one would write to determine the location of the center of mass from point P:

$$\int_m \mathbf{r}\, dm = m\mathbf{r}_{PC}$$

The second integral is called the **moment of inertia** of the body about the axis through P and perpendicular to the plane of motion:

$$\int_m r^2 dm = I_P$$

The moment about this axis is thus related to accelerations through

$$M_{P3} = \mathbf{e}_3 \bullet \mathbf{M}_P = I_P\alpha + \mathbf{e}_3 \bullet (m\mathbf{r}_{PC} \times \mathbf{a}_P) \qquad \textbf{(3.40a)}$$

By using Equation (3.24) to relate the acceleration of P to that of the mass center C, it is possible to express this moment law in the alternative form

$$M_{P3} = I_C\alpha + \mathbf{e}_3 \bullet (m\mathbf{r}_{PC} \times \mathbf{a}_C) \qquad \textbf{(3.40b)}$$

where I_C is the moment of inertia of the body about an axis through C perpendicular to the plane of motion. The two moments of inertia are related through the **parallel axis formula**

$$I_P = I_C + md^2$$

in which d is the distance between P and C.

Two special cases warrant attention. If P is chosen to be the mass center C, then $\mathbf{r}_{PC} = 0$, and the relationship is

$$M_{C3} = I_C\alpha$$

If the body is hinged about a fixed support and P is selected to be on the axis of the hinge, then

$\mathbf{a}_P = 0$, and the relationship is

$$M_{P3} = I_P\alpha$$

These last two relationships indicate the moment of inertia of the body is the property that provides resistance to changes in the angular velocity, much as mass provides resistance to changes in the velocity of a particle. For bodies of simple geometry, the integrals have been evaluated in terms of mass and the geometry, and results can be found in tabulated summaries. More complicated bodies can require tedious work to estimate the moment of inertia, or there are experiments based on the implications of Equation (3.40) that can be used to determine it. It is common to specify the moment of inertia by giving the mass of the body and its **radius of gyration**, k_P, defined by $I_P = mk_P^2$.

Example 3.24

A 23-kg rotor has a 127 mm radius of gyration about its axis of rotation. What average torque about its fixed axis of rotation is required to bring the rotor from rest to a speed of 200 rpm in 6 seconds?

Solution

The moment of inertia of the rotor is

$$I = (23\text{ kg})(0.127\text{ m})^2 = 0.371\text{ kg} \bullet \text{m}^2$$

and the average angular acceleration is

$$\alpha = \frac{(200\text{ rpm})}{6\text{ s}}\frac{(2\pi\text{ rad/r})}{(60\text{ s/min})} = 3.49\text{ rad/s}^2$$

For fixed-axis rotation,

$$M = I\alpha = (0.371\text{ kg} \bullet \text{m}^2)(3.49\text{ s}^{-2}) = 1.29\text{ N} \bullet \text{m}$$

Example 3.25

The car with rear-wheel drive in Exhibit 24 has sufficient power to cause the drive wheels to slip as it accelerates. The coefficient of friction between the drive wheels and roadway is μ. What is the acceleration in terms of g, μ, and the dimensions shown?

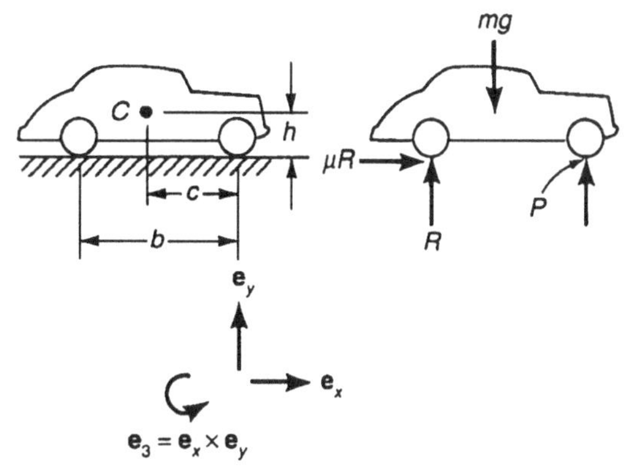

Exhibit 24

Solution

Assuming the car does not rotate, every point will have the same acceleration, $\mathbf{a} = a_x\mathbf{e}_x$. The free-body diagram shows the forces external to the car, with the label on the horizontal force at the drive wheels accounting for the fact that the friction limit has been reached there. Equation (3.28c) implies that $\mu R = ma_x$. Since the reaction R in this equation is unknown, another relationship must be introduced. Of several that could be written (for example, forces in another direction, moments about a selected point), it would be best if no additional unknowns are introduced. Thus, to avoid bringing the unknown reaction at the front wheels into the analysis, consider moments about point P, which are related by the moment law, Equation (3.40b):

$$-bR + c\,mg = I(0) + \mathbf{e}_3 \bullet [m(-c\mathbf{e}_x + h\mathbf{e}_y) \times a_x\mathbf{e}_x] = -mha_x$$

Eliminating R between this equation and the friction equation leads to

$$a_x = \frac{\mu c}{b - \mu h} g$$

Example 3.26

The uniform slender rod in Exhibit 25 slides along the wall and floor under the effects of gravity. If friction is negligible, what is the angular acceleration in terms of g, l, and the angle θ?

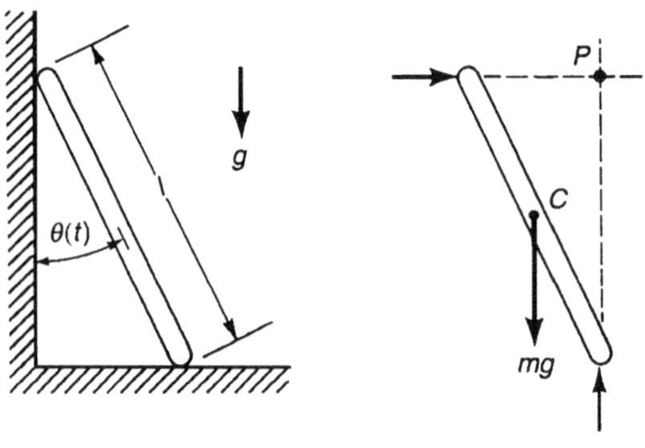

Exhibit 25

Solution

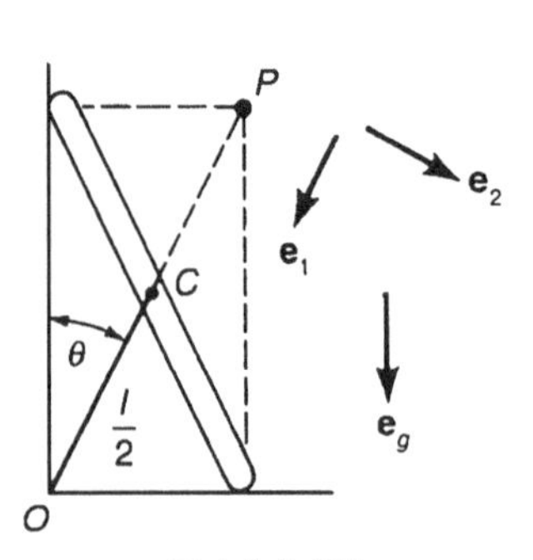

Exhibit 26

As indicated on the free-body diagram, the reactions from the wall and floor are horizontal and vertical since there is no friction. Because neither of these is known, a good strategy would be to avoid dealing with them; to this end, consider moments about point P, which will be related by Equation (3.40b). To evaluate the acceleration of the mass center C, observe that it follows a circular path with radius $l/2$ and center at O (see Exhibit 26). Thus, the acceleration of C is

$$\mathbf{a}_C = \frac{l}{2}(\dot{\theta}^2 \mathbf{e}_1 + \ddot{\theta}\mathbf{e}_2)$$

Referring to the free-body diagram, we can write the moment of forces about P as

$$\mathbf{M}_P = \frac{l}{2}\mathbf{e}_1 \times mg\mathbf{e}_g = \frac{1}{2} mgl \sin\theta \mathbf{e}_3$$

Any of a number of references gives the moment of inertia of a slender, uniform rod about an axis through its center as

$$I_C = \frac{1}{12} ml^2$$

Substituting the above into the moment law, Equation (3.40b), yields

$$\frac{1}{2} mgl \sin\theta = \frac{1}{12} ml^2 \ddot{\theta} + \mathbf{e}_3 \bullet \left[m\frac{l}{2}\mathbf{e}_1 \times \frac{l}{2}(\dot{\theta}^2 \mathbf{e}_1 + \ddot{\theta}\mathbf{e}_2) \right] = \frac{ml^2}{3}\ddot{\theta}$$

which leads to the desired angular acceleration:

$$\ddot{\theta} = \frac{3g}{2l}\sin\theta$$

Example 3.27

The uniform, slender beam of length l is suspended by the two wires in the configuration shown in Exhibit 27 when the wire on the left is cut. Immediately after the wire is severed (that is, while the velocities of all points are still zero), what is the tension in the remaining wire?

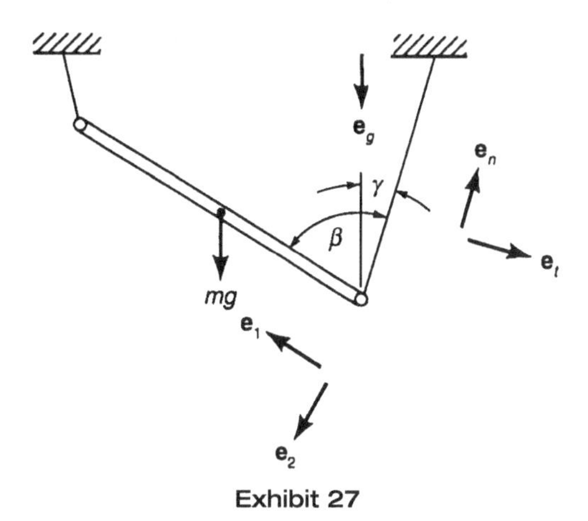

Exhibit 27

Solution

The free-body diagram (Exhibit 28) shows the desired force and the only other force acting on the beam, that of gravity. Summing moments about the center of mass gives a relationship between the desired tension and the angular acceleration of the bar:

$$\frac{l}{2}T\sin\beta = \frac{ml^2}{12}\alpha \tag{i}$$

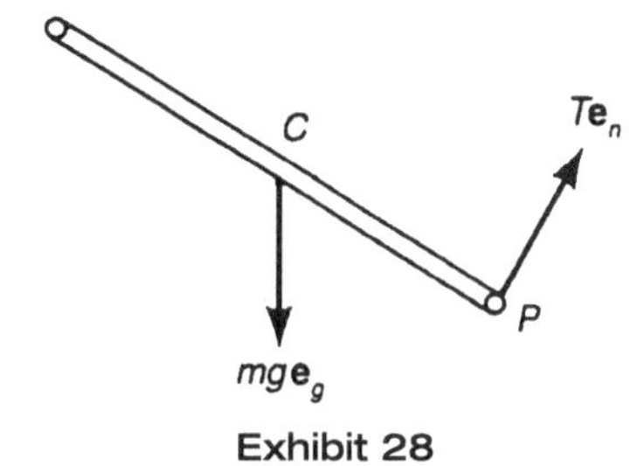

Exhibit 28

We also know that the sum of all forces is related to the acceleration of the mass center by

$$T\mathbf{e}_n + mg\mathbf{e}_g = m\mathbf{a}_C \tag{ii}$$

Since the end P is constrained by the wire to follow a circular path, its acceleration may be expressed by

$$\mathbf{a}_P = a_t\mathbf{e}_t + \frac{v_P^2}{R}\mathbf{e}_n \tag{iii}$$

in which a_t is another unknown quantity. Finally, this acceleration is related to that of the center of mass by

$$\mathbf{a}_C = \mathbf{a}_P + \frac{l}{2}\alpha\mathbf{e}_2 - \frac{l}{2}\omega^2\mathbf{e}_1 \tag{iv}$$

Since velocities are still zero, the centripetal terms v_P^2/R and $\frac{l}{2}\omega^2$ are both zero. With this simplification, Equations (ii), (iii), and (iv) readily combine to give

$$T\mathbf{e}_n + mg\mathbf{e}_g = m\left(a_t\mathbf{e}_t + \frac{l}{2}\alpha\mathbf{e}_2\right)$$

To avoid dealing with the unknown a_t, we may dot-multiply each term in this equation by $\boldsymbol{e}_n$ with the result

$$T + mg\mathbf{e}_n \bullet \mathbf{e}_g = \frac{1}{2}ml\alpha\mathbf{e}_n \bullet \mathbf{e}_2$$

Referring to the specified geometry, the dot products are evaluated as

$$\mathbf{e}_n \bullet \mathbf{e}_g = \cos\,(180° - \gamma) = -\cos\,\gamma$$

$$\mathbf{e}_n \bullet \mathbf{e}_2 = \cos\,(90° + \beta) = -\sin\,\beta$$

and the equation can be written as

$$T + \frac{1}{2}ml\alpha\sin\beta = mg\cos\gamma \qquad \textbf{(v)}$$

Now α is readily eliminated by substituting the expression for α obtained from Equation (i) into Equation (v), leading to the desired value of the tension:

$$T = \frac{mg\,\cos\gamma}{1 + 3\sin^2\beta}$$

Work and Kinetic Energy

If a rigid body has a number of forces $f_1, f_2, \ldots, f_n$ applied at points $P_1, P_2, \ldots, P_n$, the time rate at which these forces do work on the body (that is, the power transmitted to the body) can be evaluated as

$$\frac{dW}{dt} = \mathbf{f}_1 \bullet \frac{d\mathbf{r}_1}{dt} + \mathbf{f}_2 \bullet \frac{d\mathbf{r}_2}{dt} + \cdots + \mathbf{f}_n \bullet \frac{d\mathbf{r}_n}{dt} = \sum_i \mathbf{f}_i \bullet \mathbf{v}_i$$

But $\mathbf{v}_i$, the velocity of point P_i, can be related to the velocity of a selected point P of the body:

$$\mathbf{v}_i = \mathbf{v}_P + \omega \times \mathbf{r}_{Pi}$$

so that the power can also be expressed as

$$\frac{dW}{dt} = \sum_i \mathbf{f}_i \bullet (\mathbf{v}_P + \omega \times \mathbf{r}_{Pi}) = \left(\sum_i \mathbf{f}_i\right) \bullet \mathbf{v}_P + \left(\sum_i \mathbf{r}_{Pi} \times \mathbf{f}_i\right) \bullet \omega$$

$$= \mathbf{f} \bullet \mathbf{v}_P + \mathbf{M}_P \bullet \omega \qquad \textbf{(3.41)}$$

in which **f** is the resultant of all of the forces. P may be selected as any point of the body, and M_P is the resultant moment about P. For example, as a rotor turns about a fixed axis, there may be forces from the support bearings in addition to an accelerating torque about the axis of rotation. If the point P is selected to be on the axis of rotation, $\mathbf{v}_P$ will be zero, and the power transmitted to the rotor (which will, of course, induce a change in its kinetic energy) is simply the dot-product of the torque and the angular velocity. Negative or positive values are possible, depending on the angle between $\mathbf{M}_P$ and ω (that is, whether the moment component is in the same or the opposite direction as the rotation).

The kinetic energy of a rigid body is the sum of the kinetic energies of its individual elements, whose velocities can be related to the velocity of a selected point P and the angular velocity ω. Referring to Figure 3.12, we write this for plane motion as

$$T=\frac{1}{2}\int_m |\mathbf{v}|^2\, dm=\frac{1}{2}\int_m (\mathbf{v}_P+\omega\times\mathbf{r})\bullet(\mathbf{v}_P+\omega\times\mathbf{r})dm$$
$$=\frac{1}{2}\left(\int_m dm\right)v_P^2+\mathbf{v}_P\bullet\left[\omega\times\left(\int_m \mathbf{r}dm\right)\right]+\frac{1}{2}\int_m |\omega\times\mathbf{r}|^2 dm$$

The integral in the first term is simply the mass m of the body. The integral in the second term is related to the mass and position of the center of mass by

$$\int_m \mathbf{r}dm=m\mathbf{r}_{PC}$$

For plane motion, $\omega\times\mathbf{r}=\omega r\mathbf{e}_2$, so the last term becomes

$$\int_m |\omega\times\mathbf{r}|^2 dm=\omega^2\int_m r^2 dm=I_P\omega^2$$

The expression for kinetic energy for plane motion of a rigid body is then

$$T=\frac{1}{2}mv_P^2+m\mathbf{v}_P\bullet(\omega\times\mathbf{r}_{PC})+\frac{1}{2}I_P\omega^2 \tag{3.42}$$

Example 3.28

The wheel in Exhibit 29 is released from rest and rolls down the hill with sufficient friction to prevent slipping. Its mass is m, its radius is r, and its central radius of gyration is k. After the center of the wheel has dropped a vertical distance h, what is the speed of the center of mass of the wheel?

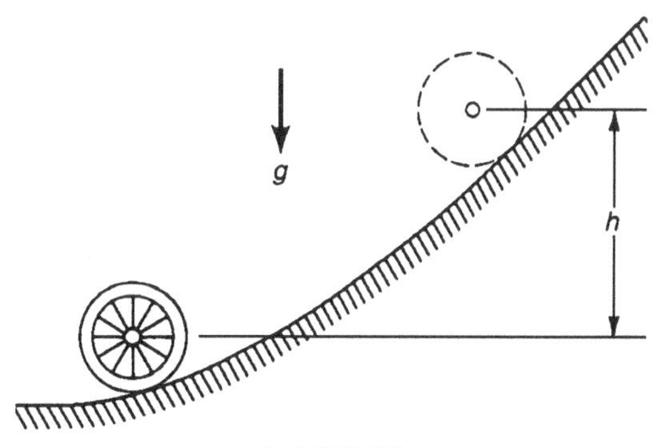

Exhibit 29

Solution

Since the velocity of the contact point is zero, the work of the force there is zero. The work of the force of gravity is then the total work done on the wheel and is simply $W = mgh$. Since the contact point is the instantaneous center, the speed of the center of the wheel is readily related to the angular velocity, $\omega = v_C/r$. The kinetic energy can be written from Equation (3.2), with P selected as any point on the wheel. If we choose P to be the center of mass,

$$T = \frac{1}{2}mv_C^2 + 0 + \frac{1}{2}mk^2\omega^2 = \frac{1}{2}\left(1 + \frac{k^2}{r^2}\right)mv_C^2$$

If, instead, we choose P to be the instantaneous center, the kinetic energy is

$$T = \frac{1}{2}m(0)^2 + 0 + \frac{1}{2}(mk^2 + mr^2)\omega^2 = \frac{1}{2}\left(1 + \frac{k^2}{r^2}\right)mv_C^2$$

Since the work must equal the change in kinetic energy,

$$mgh = \frac{1}{2}\left(1 + \frac{k^2}{r^2}\right)mv_C^2$$

and the speed of the center will be

$$v_C = \sqrt{\frac{2gh}{1 + \frac{k^2}{r^2}}}$$

SELECTED SYMBOLS AND ABBREVIATIONS

Symbol or Abbreviation	Description
$\mathbf{a}$	acceleration
$\mathbf{a}_t$	tangential component of acceleration
$\mathbf{e}_t$	unit vector tangent to path
$\mathbf{e}_n$	unit vector in principal normal direction
$\mathbf{e}_i$	unit vector in direction indicated by the specific value of i
$\mathbf{f}$	resultant force
g	gravitational field intensity
$\mathbf{g}$	impulse resultant force
$\mathbf{H}_o$	angular momentum about O
I_P	moment of inertia about P
κ	local curvature of path
k_P	radius of gyration about P
M_i	mass of ith particle
M	total mass
$\mathbf{M}_P$	moment of forces about P
N	coefficient of kinetic fraction
ρ	radius of curvature of path

(Continued)

(Continued)

Symbol or Abbreviation	Description
r	position vector
s	distance along path
t	time
$\boldsymbol{T}$	kinetic energy
v	velocity
$\boldsymbol{W}$	work
α	angular acceleration
μ	coefficient of sliding friction
ω	angular velocity

MATERIAL PROPERTIES

All engineering products are made of materials. Thus, engineers become directly involved with materials, whether they be design engineers, production engineers, or applications engineers. Their familiarity with a wide spectrum of materials becomes particularly important as they advance through management and into administration, where they must oversee the activities of additional engineers on their technical staffs.

The way that an engineering product performs in service is a consequence of the combination of the components of the product. Thus, a cellular phone must have the diodes, resistors, capacitors, and other components that function together to meet its design requirements. Likewise, a competitively produced car must possess a carefully designed engine with its numerous parts, as well as safety features and operating characteristics that meet customer approval. Materials are pertinent to each and every design consideration.

Just as it is to be expected that the internal circuitry of a four-function hand calculator will differ from the internal circuitry of its multifunctional scientific counterpart, the internal structure of a steel gear differs from that of the sheet steel to be used in an automotive fender. Their roles, and therefore their properties, are designed to be different.

The variations in the internal structures of materials that lead to property differences include variations in atomic coordination and electronic energies, differences in internal geometries (microstructures), and the incorporation of larger structures, sometimes called macrostructures. Each of these is considered in the following sections, along with procedures for obtaining desired structures and properties.

ATOMIC ORDER IN ENGINEERING MATERIALS

Atoms, Ions, and Electrons

There is an order within atoms. Each atom has an integer number of protons. That number is called the **atomic number**. The natural elements possess, progressively, 1 to 92 of these protons, which carry a positive charge. A neutral atom has a number of electrons equal to the number of protons. Each electron is negative with a charge of 1.6×10^{-17} coulombs. Electrons are only allowed in given orbitals (called shells) that correspond to specific allowed energy levels.

With the exception of the principal isotope of hydrogen, each atom possesses neutrons. While these are charge-neutral, they add to the mass of each atom. The protons and neutrons reside in the nucleus of the atom. Figure 3.13 shows the Bohr model of a sodium atom. The lighter elements contain approximately equal numbers of protons and neutrons; however, in heavier elements the number of neutrons exceeds the number of protons. Furthermore, the number of neutrons per atom is not fixed. Thus, we encounter several **isotopes** for most atoms.

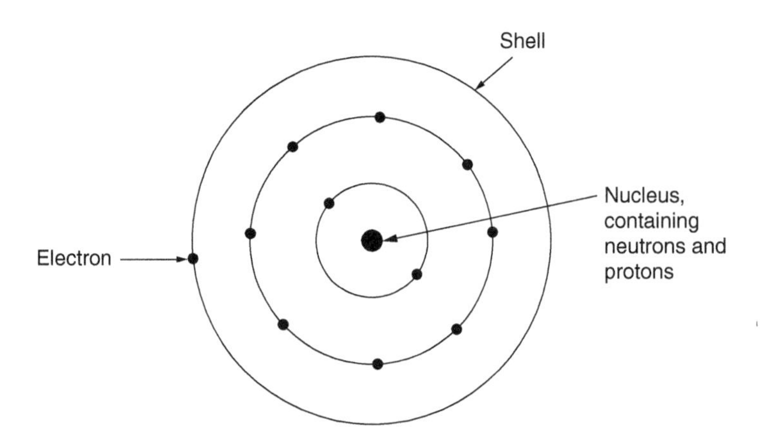

Figure 3.13 Bohr model of a sodium atom

Table 3.1 Data for selected atoms

Element	Protons Electrons Atomic No.	Neutrons (in natural isotopes)	Atomic Mass Unit	Grams per Avogadro's Number*
Hydrogen	1	0 or 1	1.008	1.008
Carbon	6	6 or 7	12.011	12.011
Oxygen	8	8, 9, or 10	15.995	15.995
Chlorine	17	18 or 20	35.453	35.453
Iron	26	28, 30, 31, or 32	55.847	55.847
Gold	79	118	196.97	196.97
Uranium	92	142, 143, or 146	238.03	238.03

*Avogadro's number = 6.022×10^{23}

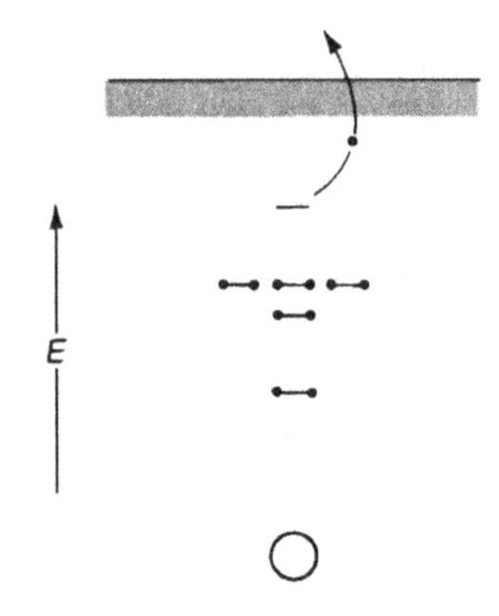

Figure 3.14 Ionization energy (schematic for sodium). Electrons reside at specific energy **states** (levels). Energy must be supplied to remove an electron from an atom, producing a positive ion. Two electrons (of opposite magnetic spins) may reside in each state.

Since the mass of an electron is appreciably less than 1% of that of protons and neutrons, the mass of an atom is directly related to the combined number of the latter two. By definition, an **atomic mass unit** (amu) is 1/12 of the mass of a carbon isotope that has six protons and six neutrons, C^{12}. The **atomic mass** of an element is equal to the number of these atomic mass units. (Selected values are listed in Table 3.1.) Thus while there are integer numbers of neutrons, protons, and electrons in each atom, the mass generally is not an integer, because more than one isotope is typically present.

A limited number of electrons may be accepted or released by an atom, thus introducing a charge on the atom (due the difference in the number of protons and electrons). A charged atom is an **ion**. Negative ions that have accepted extra electrons are called **anions**. Positive ions that have released electrons are called **cations**. Because they are charged, ions respond to electric fields. These fields may involve macroscopic dimensions (in electroplating baths); or they may involve interatomic distances (in molecules). Unlike charges attract; like charges repel.

Energy is required to remove an electron from a neutral atom. Figure 3.14 shows this schematically for a sodium atom. Conversely, fixed quantities of energy are released when electrons are captured by a positive ion. These energy levels (**states**) associated with an atom are fixed. Furthermore each state may accept only two electrons, and these must have opposite magnetic characteristics. Electronic, magnetic, and optical properties of materials must be interpreted accordingly.

Molecules

Atoms can join to one another; this is called **bonding**. Strong attractive forces can develop between atoms by three mechanisms; (1) coulombic attraction between oppositely charged ions, forming ionic bonds; (2) sharing of electrons to fill outer shells, creating covalent bonds; and (3) formation of ion cores surrounded by valence electrons that have been excited above the Fermi level and have become free electrons, forming metallic bonds. More detailed examples of these three primary types of bonds follow.

Ionic bonds form between metallic and nonmetallic atoms. The metallic atoms release their valence electrons to become cations (which are positively charged) and the nonmetallic atoms accept them to become anions. Ionic bonds are nondirectional. For example, a sodium ion that has lost an electron (Na^+) associates with as many negatively charged chlorine ions (Cl^-) as space will allow. And each Cl^- ion will become *coordinated* with as many Na^+ ions as necessary to balance the charge. The resulting structure will continue to grow in three dimensions until all available ions are positioned. Energy is released with each added ion.

Covalent bonds form between atoms that share valence electrons in order to fill their outer shells. In the simplest case, two hydrogen atoms release energy as they combine to produce a hydrogen molecule:

$$2H \rightarrow H2 \quad \text{or} \quad 2H \rightarrow H\text{—}H \qquad \textbf{(3.43)}$$

The bond between the two involves a pair of shared electrons. This mechanism is common among many atomic pairs. In this case only one pair of atoms is involved. The covalent bonds of molecules are stereospecific; that is, they are between specific atoms and are therefore directional bonds.

In polymers a string or network of thousands or millions of atoms is bonded together. Examples include polyethylene, which has the structure shown in Figure 3.15(a), in which there is a backbone of carbon atoms that are covalently bonded; that is, they share pairs of valence electrons. Since each carbon atom has four valence electrons, it can form four covalent bonds, thus adding two hydrogens at the side of the chain in addition to the two bonds along the chain. Polyvinyl chloride, Figure 3.15(b), is related but has one of the four hydrogen atoms replaced by a chlorine atom.

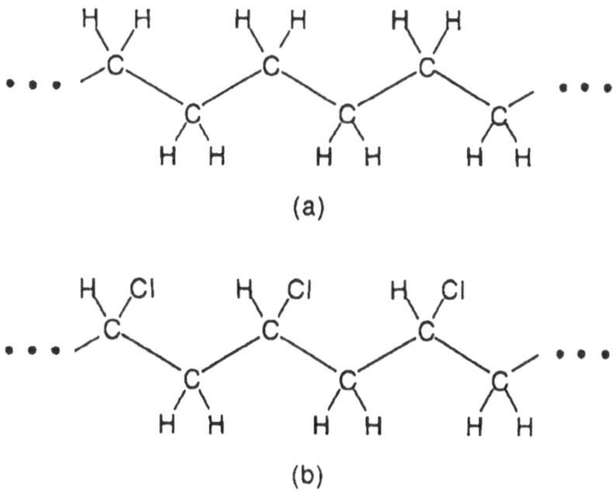

(a)

(b)

Figure 3.15 Covalent bonds: (a) polyethylene; (b) polyvinyl chloride

Metallic bonds form between the elements on the left side of the periodic table known as metals. Metallic bonds can form between atoms of the same element, to form a pure metal such as gold (Au), or between different metal atoms to form an alloy, such as brass, a mixture of copper (Cu) and zinc (Zn). The basis for the metallic bond is the formation of ion cores created when the metal's valence electrons are no longer associated with a specific atom. These electrons (called free electrons) move freely around surrounding metal ion cores, shielding them from each other. These freely moving electrons are often referred to as a *sea of electrons*. Metallic bonds are nondirectional.

The type and strength of bonding between atoms and molecules determines many properties of materials. As a general rule (and with other things being equal), the stronger the atomic bonding, the higher the melting temperature, hardness, and elastic modulus of materials. Ionically bonded materials are usually electrical and thermal insulators, while materials with metallic bonds have high electrical and thermal conductivities.

Crystallinity

The repetition of atomic coordinations in three dimensions produces a periodic structure that is called a **crystal**. The basic building block of a crystal is called a unit cell. Figure 3.16 illustrates the unit cell structure of iron. Each atom in this metal has eight nearest neighbors, which are symmetrically coordinated to give a **body-centered cubic** (BCC) crystal. About 30% of the metals have this structure. The atoms of aluminum and copper, among another 30% of the metals, become coordinated with 12 nearest neighbors with the result that **face-centered cubic** (FCC) crystal lattices are formed, as shown in Figure 3.17. A third group of metals form **hexagonal crystals** with **close packing** (HCP), as shown in Figure 3.18. As in the FCC crystals, each atom is coordinated with 12 nearest neighbors in HCP crystals.

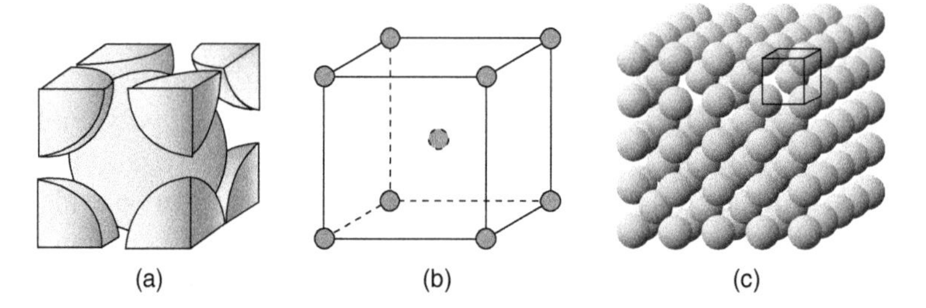

Figure 3.16 Body-centered cubic crystal structure: (a) a hard-sphere unit cell representation; (b) a reduced-sphere unit cell; (c) an aggregate of many atoms

It is possible to have a very high degree of perfection in crystals. For example, the repetition dimension (**lattice constant**) of the FCC lattice of pure copper is constant to the fifth significant figure (and to the sixth if the thermal expansion is factored in). This high degree of ordering provides a quantitative base for anticipating properties. Included are density calculations, certain thermal properties, and some of the effects of alloying.

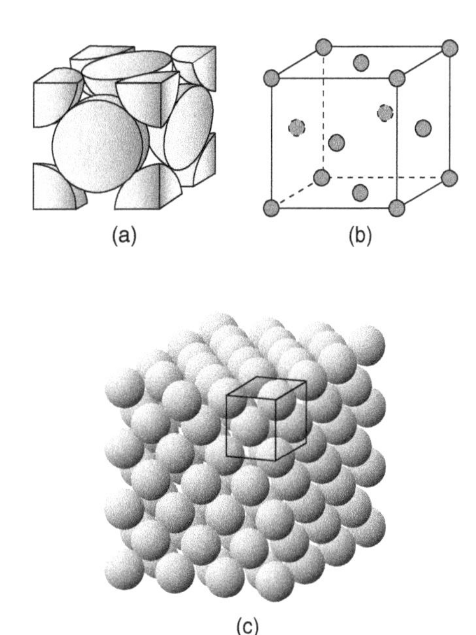

Figure 3.17 Face-centered cubic crystal structure: (a) a hard-sphere unit cell representation; (b) a reduced-sphere unit cell; (c) an aggregate of many atoms

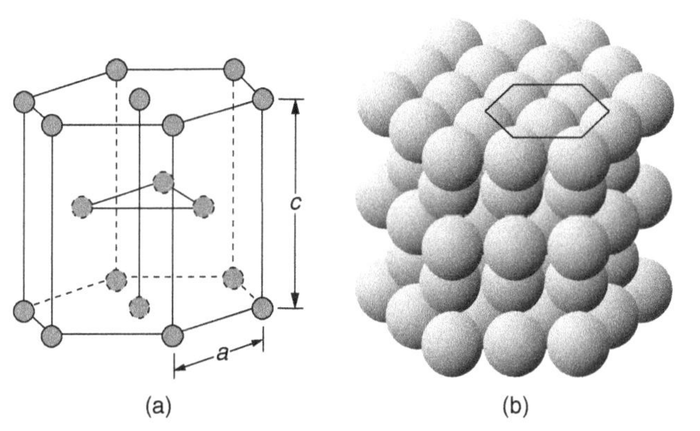

Figure 3.18 Hexagonal close-packed crystal structure: (a) a reduced-sphere unit cell (*a* and *c* represent the short and long edge lengths, respectively); (b) an aggregate of many atoms

Directions and Planes

Many properties are **anisotropic**; that is, they differ with direction and orientation. We can identify crystal directions by selecting any zero location (the origin) and determining the x, y, and z coordinates for any point along the direction ray. A corner of the unit cell is often used as the origin. The unit length is the edge length of the unit cell. The direction of easy magnetization in iron is parallel to one of the crystal axes. This is labeled the [100] direction, because that direction passes from the origin through a point that is one unit along the x-axis and zero units along the other two axes. Figure 3.19(a) shows a ray in the [120] direction and a ray in the $[1\bar{1}0]$ direction, where the overbar indicates a negative direction. Parallel directions carry the same label. We use square brackets, [ ], for closures for direction rays.

In cubic crystals, each of the four directions that are diagonal through the cube are identical (because the three axes are identical). We label these four directions as a *family* with pointed arrows for closures, <111>.

Figure 3.19(b) identifies the three shaded planes as (001), (210), and (111). Here the labeling procedure (known as indexing) is somewhat more complicated than for directions (but leads to simplified mathematics for complex calculations). As an example we will draw the (210) plane. We first invert the three indices, 1/**2**,

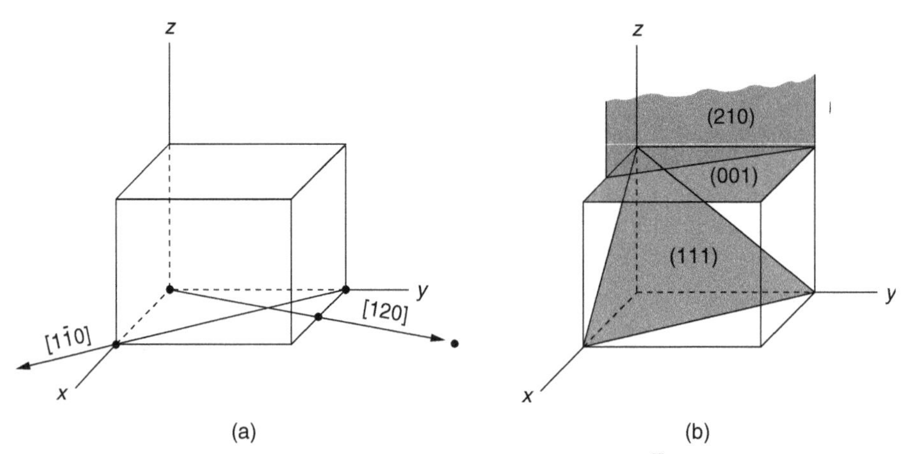

Figure 3.19 Crystal notation. (a) Directions, [120] and [1$\bar{1}$0]. The *x*, *y*, and *z* coordinates are used for crystal directions. Square brackets are used as closures. Negative coordinates are indicated with an overbar. (b) Planes, (001), (210), and (111). The reciprocals of the axial intercepts are used for crystal planes. Parentheses are used as closures. Any point may be selected as an arbitrary origin.

1/**1**, 1/**0**. These are the intercept dimensions of the plane across the three axes; specifically, 0.5 on the *x*-axis, 1.0 on the *y*-axis, and infinity along the *z*-axis. An adjacent parallel plane with intercepts of 1, 2, and ∞ carries the same (210) index. We use parentheses, ( ), as closures for the indices of individual planes, and braces, { }, as closures for a family of comparable planes. To index an unknown plane, the procedure is reversed: (1) Choose an origin that the plane does not pass through. (2) Determine the intercepts of the plane on the *x*, *y*, and *z* axes. (3) Take reciprocals of these intercepts. (4) Clear fractions and enclose in parentheses ( ).

Characteristics of Ordered Solids

There are several useful properties of unit cells that can be determined through geometric relations and 3-D visualization. These include:

- Number of atoms per unit cell
- Number of nearest-neighbor atoms (coordination number)
- Lattice parameter (spacing of atoms)
- Distance of nearest approach of atoms
- Atomic packing factor (the volume of atoms per unit volume of the solid)
- Density

Some of these relationships are given in Table 3.2 and in the NCEES *Fundamentals of Engineering Supplied-Reference Handbook* and therefore do not need to be memorized, but it is useful to see how these are determined. The following examples will illustrate these relationships.

Table 3.2 Characteristics of selected crystal structures

Unit Cell	Number of Atoms per Unit Cell	Coordination Number	Lattice Parameter	Packing Factor
BCC	2	8	$a = 4R/\sqrt{3}$	0.68
FCC	4	12	$a = 2R/\sqrt{2}$	0.72
HCP		12		0.72

Figure 3.16 showed a unit cell of α iron with an atom at its center. Inasmuch as each of the eight corner atoms is shared by the eight adjacent unit cells, we can note that there are (1 + 8/8 = 2) atoms per unit cell. From Table 3.1, each iron atom has a mass of 55.85 g per 0.6022×10^{24} atoms. Since iron forms a body-centered cubic crystal, its unit cell has a mass of 1.855×10^{-26} g. X-ray diffraction techniques give a lattice constant value of 2.866×10^{-18} m. As a result, the mass per unit volume, that is, the **density**, may be calculated to be nearly 7.88 g/cm^3. Careful density measurements give a value of slightly more than 7.87 g/cm^3 at ambient temperatures. The close agreement for the simple property of density implies that the concept of the crystal structure is valid.

The relationship between the size of the atoms (atomic radius, r) and lattice parameter (a) for BCC crystals is demonstrated in Figure 3.20. Note that the atoms touch along the body diagonal. By inspection this gives a known length of the body diagonal of $4r$.

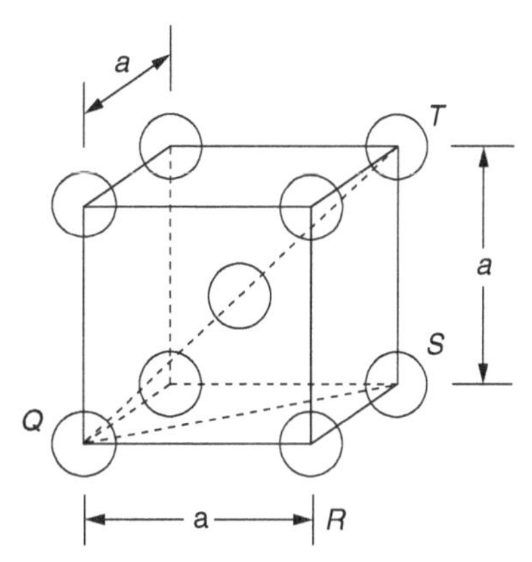

Figure 3.20 BCC unit cell

Using the triangle QRS

$$\| \overline{QS}^2 \| = a^2 + a^2 = 2a^2$$

and for triangle QST

$$\| \overline{QT}^2 \| = \| \overline{TS}^2 \| + \| \overline{QS}^2 \|$$

But $\overline{QT} = 4r$, r being the atomic radius. Also, $\overline{TS} = a$. Therefore,

$$(4r)^2 = a^2 + 2a^2$$

or

$$a = \frac{4r}{\sqrt{3}}$$

The relationship between r and a for FCC crystals can be similarly determined by noting that the atoms touch along the face diagonal as shown in Figure 3.21.

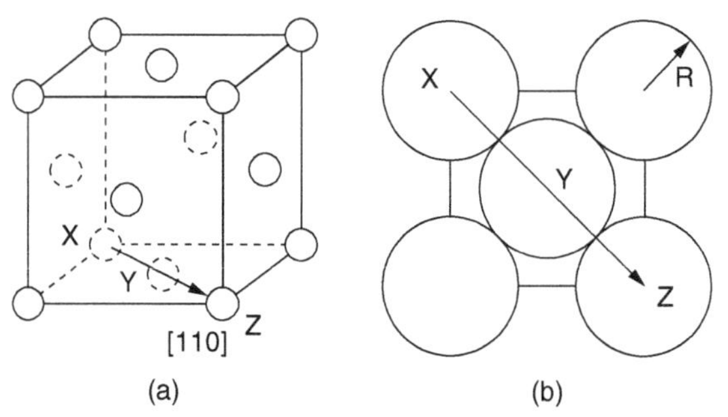

Figure 3.21 (a) Reduced-sphere FCC unit cell with the [110] direction indicated. (b) The bottom face-plane of the FCC unit cell in (a) on which is shown the atomic spacing in the [110] direction, through atoms labeled X, Y, and Z.

The plastic deformation of these solids is also related to crystal structure. To illustrate, in Figure 3.22 the {111} plane of an FCC structure is shown. The {111} planes of aluminum and other FCC metals are the most densely packed planes; each atom of those planes is surrounded by six other atoms in the same plane. There is no arrangement where more atoms could have been included. This is not true for other planes within the FCC crystal. Since there is a fixed number of atoms per unit volume, it is apparent that the interplanar spacings between parallel (111) planes through the centers of atoms must be greater than between planes of other orientations. It is thus not surprising that sliding (**slip**) occurs there at lower stresses than on other planes.

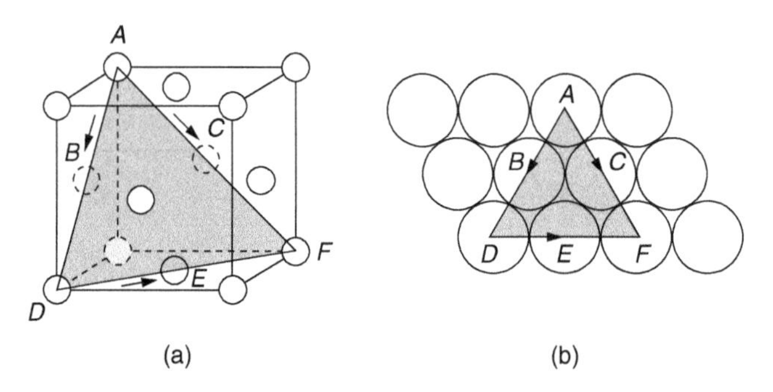

Figure 3.22 (a) A {111}<110> slip system shown within an FCC unit cell. (b) The (111) plane from (a) and three <110> slip directions (as indicated by arrows) within that plane comprise possible slip systems.

Not only is the required shear stress for slip low on the {111} planes, the <110> directions on those planes require less stress for slip than other directions. This is because the step distance between like crystal positions is the shortest, specifically, $2r$ where r is the atomic radius. We speak of the <110>{111} **slip system** for FCC metals. In a BCC metal, <111> {110} is the prominent slip system. Note especially that the {111} planes and the <110> directions are operative in FCC metals, whereas it is the {110} planes and the <111> directions in BCC metals. Hexagonal metals such as Mg, Zn, and Ti do not deform as readily as BCC and FCC metals because there are fewer combinations of directions and planes for slip by shear stresses.

Example 3.29

Carbon (12.011 amu) contains C^{12} and C^{13} isotopes with masses of 12.00000 amu and 13.00335 amu, respectively. What are the percentages of each?

Solution

$$12.011 = x\,(12.00000) + (1 - x)(13.00335)$$
$$1.00335\,x = 0.99335$$
$$x = 98.9\%$$

Carbon is 98.9% C^{12} and 1.1% C^{13}.

Example 3.30

Aluminum has a face-centered cubic unit cell, that is, an atom at each corner of the unit cell and an atom at the center of each face (see Figure 3.17). The Al–Al distance (= $2r$) is 0.2863 nm. Calculate the density of aluminum. (The mass of an aluminum atom is 26.98 amu.)

Solution

$$\text{Volume} = [2(0.2863 \times 10^{-9}\text{ m})/\sqrt{2}]^3 = 6.638 \times 10^{-29}\text{ m}^3$$
$$\text{Mass} = (8/8 + 6/2\text{ atoms})(26.98\text{ g}/6.022 \times 10^{23}\text{ atoms}) = 1.792 \times 10^{-22}\text{ g}$$
$$\text{Density} = (1.792 \times 10^{-22}\text{ g})/(6.638 \times 10^{-29}\text{ m}^3) = 2.700 \times 10^{6}\text{ g/m}^3 = 2.700\text{ g/cm}^3$$
$$\text{Actual density} = 2.699\text{ g/cm}^3$$

Example 3.31

What is the repeat distance along a <211> direction of a copper crystal that is face-centered cubic and has a unit cell dimension (lattice constant) of 0.3615 nm?

Solution

Select the center of any atom as the origin. Make a sketch of a cubic unit cell with that origin arbitrarily set at the lower left rear corner, as shown in Exhibit 30. One of the <211> directions, with coefficients of 2, 1, and 1, exits the first unit cell through the center of its front face, where another atom is centered (with no other intervening atoms).

$$d^2 = a^2 + \left(\frac{a}{2}\right)^2 + \left(\frac{a}{2}\right)^2 = \frac{6}{4}\,(0.3615\text{ nm})^2;\ d = 0.4427\text{ nm}$$

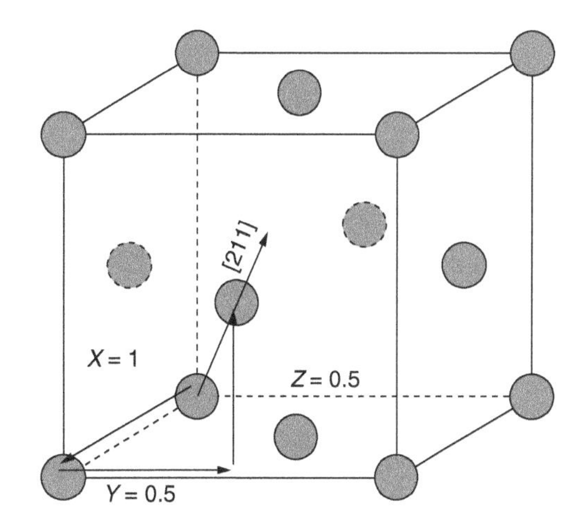

Exhibit 30 FCC unit cell

Example 3.32

Assuming spherical atoms, calculate the packing factor of a BCC metal.

Solution

The packing factor is the volume of atoms per unit volume of the solid. Based on Figure 3.16, there are (8/8 + 1) atoms/unit cell.

$$\text{Volume of 2 atoms in the BCC unit cell} = 2 \times (4\pi/3)r^3 = 8.38r^3$$

Since the cube diagonal is $4r$,

$$\text{Volume of unit cell} = a^3 = \left(\frac{4r}{\sqrt{3}}\right)^3 = 12.32r^3$$

$$\text{Packing factor} = \frac{8.38r^3}{12.32r^3} = 68\%$$

Example 3.33

How many atoms are there per mm^2 on one of the {110} planes of copper (FCC)?

Solution

From Example 3.31, $a_{Cu} = 0.3615$ nm. A {110} plane lies diagonally through the unit cell and is parallel to one of the axes. There are (4/4 + 2/2 atoms) in an area measuring a by $a\sqrt{2}$. The number of atoms is 2 atoms/[$(0.3615 \times 10^{-6}$ mm$)^2$ $\sqrt{2}$] = 10.8×10^{12}/mm2.

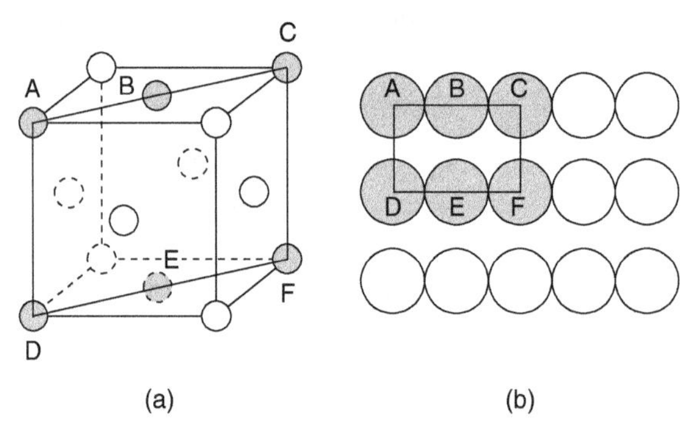

Exhibit 31 (a) Reduced-sphere FCC unit cell with (110) plane. (b) Atomic packing of an FCC (110) plane. Corresponding atom positions from (a) are indicated.

ATOMIC DISORDER IN SOLIDS

In the previous section, we paid attention to the orderly combinations that can exist in engineering materials. A variety of properties and behaviors are closely related to that ordering. Examples that were cited included density, slip systems, and molecular melting.

However, no solid has perfect order. There are always imperfections present, and these may be highly significant. A few missing potassium atoms in a compound such as KBr do not detectably affect the density; however, their absence introduces color. Likewise, the absence of a partial plane of atoms in a metal significantly modifies the shear stress required by a slip system for plastic deformation.

Also, a rubber is vulcanized by the joining (**crosslinking**) of adjacent molecules with only a minor compositional change (sulfur addition). As a final example, the thermal conductivity is doubled in diamond if the 1% of naturally present C^{13} has been removed.

Crystal defects can be characterized as point, line (one-dimensional), plane (two-dimensional), or volume (three-dimensional) defects.

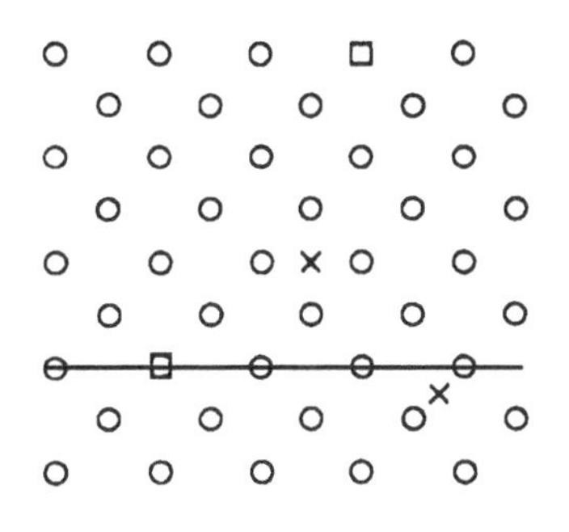

Figure 3.23 Point imperfections. These defects can originate from imperfect crystallization, or through a relocation of energized atoms.

Point Imperfections

Imperfections may be atomically local in nature. Missing atoms (**vacancies**), extra atoms (**interstitials**), **displaced** atoms, and impurity atoms are called **point imperfections** (Figure 3.23). Their existence facilitates the transport of atoms (**diffusion**), thus becoming important in materials processing. In service, the presence of point imperfections scatters internal waves and thus reduces energy transport. These include elastic waves for thermal conductivity, light waves for optical transparency, and electron waves for electrical conduction.

Lineal Imperfections

The principal defect of this type is a **dislocation**. It is most readily visualized (Figure 3.24) as a partial displacement along a slip system or an extra half-plane of atoms. Dislocations facilitate plastic deformation by slip; however, increased numbers of dislocations lead to dislocation tangles or *traffic snarls* and therefore to interference of slip. Thus, ductility decreases and strength increases. Dislocations may develop during initial crystal growth as well as from plastic deformation.

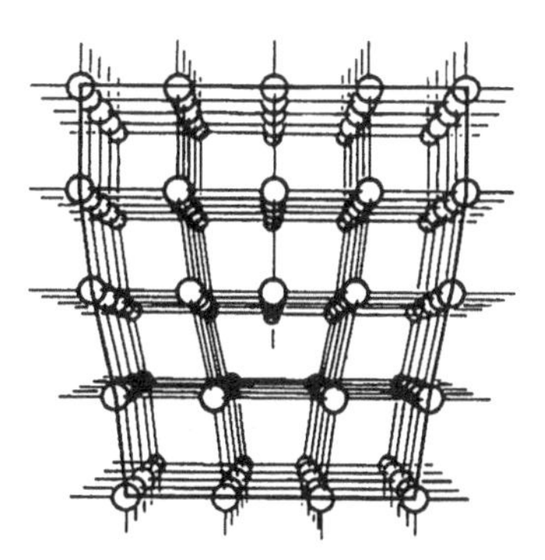

Figure 3.24 Dislocation, ⊥ (schematic). These lineal imperfections facilitate slip within crystals. Excessive numbers of defects, however, lead to their entanglement and a resistance to deformation. The resulting increase in strength is called **work hardening**.

Boundaries

No liquid or solid is infinite; each has a **surface**. The resulting two-dimensional boundary has a different structure and bonding than that encountered in the underlying material. Since atoms are absent from one side of the boundary, the atoms at these surfaces possess additional energy, and subsurface distortions are introduced.

Grain Boundaries

Boundaries also occur where two growing crystals meet. These are called grain boundaries. There are atoms on each side of the boundary; however, any misorientation between the two crystal grains leads to local inefficiencies in atomic packing. As a result some atom-to-atom distances are compressed; others are stretched. Both distortions increase the energy of the atoms along the grain boundaries. This **grain-boundary energy** introduces reactive sites for structural modification during processing and in service. The imperfect grain boundary is also an *avenue* for atomic diffusion within solid materials (Figure 3.25). Further, the mismatch at a grain boundary blocks slip that might otherwise occur, particularly at ambient temperatures where atom-by-atom mobility is limited.

Solutions

Both sugar and salt dissolve in water, each producing a **solution** (commonly called a syrup and a brine, respectively). There are many familiar solutions. Lower melting temperatures, increased conductivity, and altered viscosities commonly result.

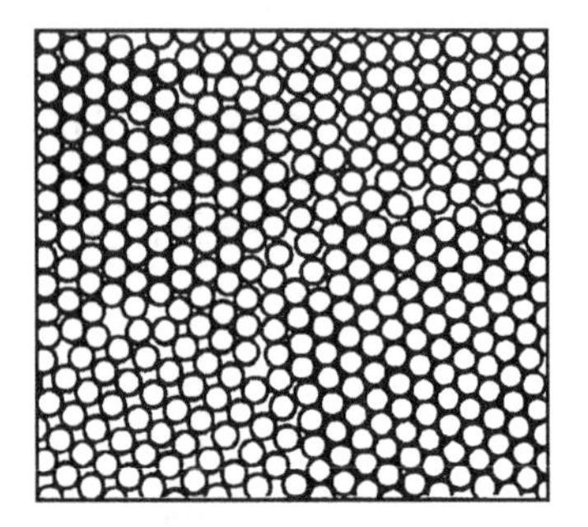

Figure 3.25 Grain boundaries (schematic). Most materials contain a multitude of grains, each of which is a separate crystal. The boundary between grains is a zone of mismatch. Atoms along grain boundaries possess added +energy because they are not as efficiently coordinated with their neighbors.

Solid Solutions

Impurities, both unwanted and intentional, may also dissolve into a solid. A crystal cannot be perfect when foreign atoms are present. Common brass (70Cu–30Zn) is a familiar example. Zinc atoms simply substitute for copper atoms to produce a **substitutional solid solution**. It has the face-centered cubic crystal structure of pure copper; however, with approximately one-third of the copper atoms replaced by zinc atoms, we can anticipate certain changes. First, the size of the unit cell is increased because zinc atoms are approximately 4% larger than copper atoms. (Fifteen percent mismatch is the practical limit for extensive substitutional solid solution.) Second, charge transport is greatly reduced because the electrons are scattered as they travel toward the positive electrode through locally varying electrical fields. Thus, the electrical and thermal **conductivities** are decreased. Also, atoms of a different size immobilize dislocation movements and in turn produce **solution hardening**.

There are some important situations in which small atoms can be positioned among larger atoms. The most widely encountered example is the solution of carbon in face-centered cubic iron at elevated temperatures. The result is an **interstitial solid solution**. Iron changes to body-centered cubic at ambient temperatures. The BCC iron cannot accommodate many carbon atoms in its interstices. This loss of interstitial solubility plays a major role in the various heat treatments of steel.

Amorphous Solids

Materials lose their crystallinity when they melt. The long-range order of the crystals is not maintained. As a liquid, the material is amorphous (literally, without form). Those materials that are closely packed—for example, metals—will expand on melting [Figure 3.26(a)]. Being thermally agitated, the atoms do not maintain close coordination with neighboring atoms (or molecules). A few materials, such as ice, which has stereospecific bonds, lose volume on melting and become more dense [Figure 3.26(b)].

In certain cases, crystallization may be avoided during cooling. An amorphous solid can result. Two common examples are window glass and the candy part of peanut brittle. In neither case is there time for the relatively complex crystalline structures to form. Very rapid cooling rates are required to avoid crystallization when the crystalline structures are less complex. For example, metals must be quenched a thousand degrees in milliseconds to avoid crystallization. The amorphous materials that result are considered to be **vitreous**, or glass-like.

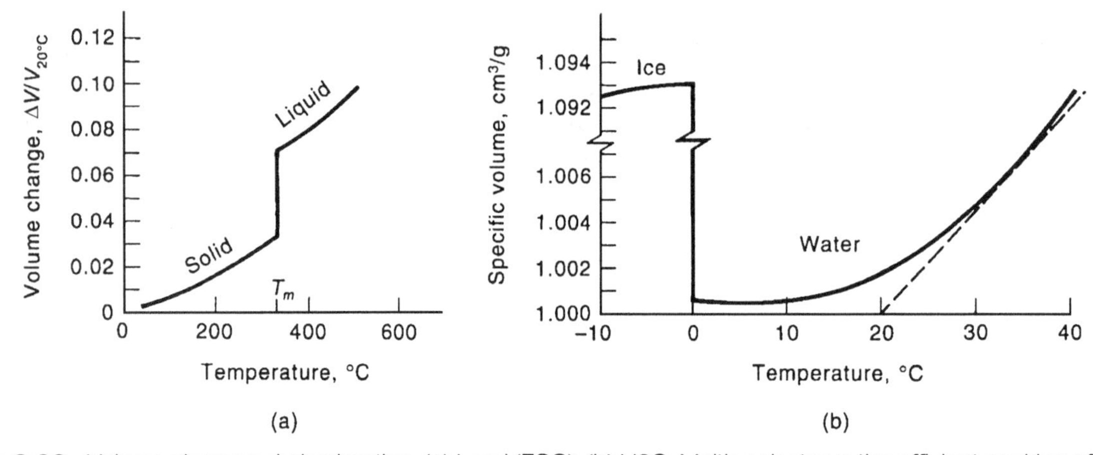

Figure 3.26 Volume changes during heating. (a) Lead (FCC); (b) H2O. Melting destroys the efficient packing of metallic atoms within solids, so most metals expand when melted. The crystalline structure of ice, silicon, and a number of related materials with stereospecific bonds have low, inefficient atomic packing within solids. Therefore, they lose volume when melted.

Example **3.34**

Sterling silver contains 92.5% silver and 7.5% copper by weight. What percentage of the atoms on a (111) plane are silver? Copper?

Solution

The alloy is a random solid solution; therefore the percentage of atoms on the (111) plane or any other plane will be the same as throughout the alloy. Change weight percent to atom percent. The atomic masses are 107.87 and 63.54 amu, respectively.

Basis: 1000 amu = 925 amu Ag + 75 amu Cu

Ag: 925 amu/(107.87 amu/Ag) = 8.58 Ag atoms

Cu: 75 amu/(63.54 amu/Cu) = 1.18 Cu atoms

Total: = 9.76 atoms

Ag atoms = 8.58/9.76 = 87.9%; Cu atoms = 12.1%

MICROSTRUCTURES OF SOLID MATERIALS

The atomic coordination within solids is on the nanometer scale. The resulting structures involve either crystalline solids, or amorphous solids such as the glasses. As discussed, certain properties arise from these atom-to-atom relationships. Other properties arise from longer-range structures, generally with micrometer to millimeter dimensions, called **microstructures**.

Atomic Movements in Materials

Our initial examination of crystals implied that an atom becomes permanently coordinated with adjacent atoms and remains fixed in position. This is not entirely true. In the first place, there is thermal vibration of the atoms within the crystal. Thus, while the lattice constant and the mean interatomic distances are fixed to several significant figures, the instantaneous interatomic distances vary. The amplitude of vibration increases with temperature. At the melting temperature, the crystal is literally *shaken apart*. As the melting point is approached, a measurable fraction of atoms jump out of their crystalline positions. They may return, or they may move to other sites, producing the vacancies and interstitials discussed in the previous section.

Diffusion

Within a single-component material, such as pure copper, there is equal probability that like numbers of copper atoms will jump in each of the coordinate directions. Thus, there is no net change.

Imagine, however, one location, x_1, in nickel containing 2000 atoms of copper for every mm^3 of nickel, whereas 1 mm to the right at x_2 there are 1000 atoms of copper for every mm^3. Although all copper atoms have the same probability for jumping in either direction, there is a net movement of copper atoms to the right simply because there are unequal numbers of copper atoms in the two locations. There is a copper **concentration gradient**, $\Delta C/\Delta x$. In this case

$$\Delta C/\Delta x = (C_2 - C_1 \text{ Cu/mm}^3)/(x^2 - x^1 \text{ mm}) = -(1000 \text{ Cu/mm}^3)/\text{mm} \quad \textbf{(3.44)}$$

The rate of diffusion, called the **flux**, J, is proportional to the concentration gradient

$$J = -D\frac{dC}{dx} \tag{3.45}$$

where D is the *diffusivity*, also called the **diffusion coefficient**. (Its units are m^2s^{-1} since the units for flux and concentration gradient are $m^{-2}s^{-1}$ and m^{-4}, respectively.)

The diffusion coefficient can be calculated by

$$D = D_o e^{-Q/(RT)}$$

where D_o (with units of m^2s^{-1}) is the proportionality constant, Q [with units of $J(mole)^{-1}$] is the activation energy, R is the gas constant, and T is absolute temperature.

Among the various factors that affect the diffusivity are (1) the size of the diffusing atom, (2) the crystal structure of the matrix, (3) bond strength, and (4) temperature. Comparisons are made in Table 3.3. The diffusivity of the C in Fe_{FCC} is higher than for the Fe in Fe_{FCC} because the diffusing carbon atom is smaller than the iron atom. The diffusivity of the C in Fe_{BCC} is higher than for the C in Fe_{FCC} because the latter contains more iron atoms per m^3. The FCC packing factor is higher than the BCC form. The diffusivity for Fe in Fe_{FCC} is lower than for Cu in Cu_{FCC} because the iron atoms are more strongly bonded than the copper atoms. (The melting temperatures provide evidence of this.) In each case, higher diffusion coefficients accompany higher temperatures.

The process engineer obtains many of the properties required of a material by heat-treating procedures. Diffusion plays the predominant role in achieving the required microstructures.

Table 3.3 Selected diffusion coefficients

Diffusing Atom	Host Structure	Diffusion Coefficient, D, m^2/s	
		500°C (930°F)	1000°C (1830°F)
Fe	FCC Fe	2×10^{-23}	2×10^{-16}
C	FCC Fe	5×10^{-15}	3×10^{-11}
C	BCC Fe	1×10^{-12}	2×10^{-9}
Cu	FCC Cu	1×10^{-18}	2×10^{-13}

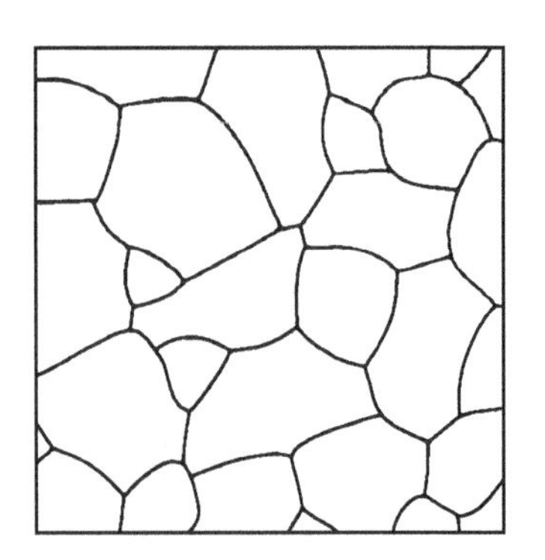

Figure 3.27 Grains (schematic). Each grain is a separate crystal. There is a surface of mismatch between grains because adjacent grains have unlike orientations. The **grain boundary area** is inversely related to grain size.

Structures of Single-Phase Solids

Many materials possess only one structure. Examples include copper wire, transparent polystyrene cups, and Al_2O_3 substrates for electronic circuits. The wire contains only face-centered-cubic crystals. The polystyrene is an amorphous solid with minimal crystallinity. The substrates have numerous crystals, all with the same crystalline structure. We speak of single phases because none of these materials contains a second structure.

Grains

Each of the individual crystals in a copper wire is called a **grain**. Recall from the previous section that adjacent crystals may be misoriented with respect to each other, and that there is a boundary between them. This is shown schematically in Figure 3.27.

The **grain size** is an important structural parameter, because the grain boundary area varies inversely with the grain size. Diffusion is faster along grain boundaries because there is less-perfect packing of the atoms and, consequently, a more open structure. At ambient temperatures, grain boundaries interfere with plastic deformation, thus increasing the strength. At elevated temperatures the grain boundaries contribute to creep and therefore are to be minimized. Grain boundaries also serve as locations that initiate structural changes within a solid.

Grain growth may occur in a single-phase material. The driving force is the fact that the atoms at the grain boundary possess extra energy. Grain growth reduces this excess energy by minimizing the boundary area. Higher temperatures increase the rate of grain growth because the atoms migrate faster. However, since the growth rate is inversely related to grain size, we see a decrease in the rate of growth with time.

The texture of a single-phase solid can also depend on **grain shape** and **orientation**. Even a noncrystalline material may possess a structure. For example, the molecular chains within a nylon fiber have been aligned during processing to provide greater tensile strength.

Phase Diagrams

Many materials possess more than one phase. A simple and obvious example is a cup containing both ice and water. While ice and water have the same composition, the structures of the two phases are different. There is a **phase boundary** between the two phases. A less obvious, but equally important material is the steel used as a bridge beam. The steel in the beam contains two phases: nearly pure iron (body-centered cubic), and an iron carbide, Fe_3C.

In these two examples, we have a mixture of two phases. Solutions are phases with more than one component. In the previous section, we encountered brass, a solid solution with copper and zinc as its components.

Although the steel beam just cited contains a mixture of two phases at ambient temperatures, it contained only one phase when it was red-hot during the shaping process. At that temperature, the iron was face-centered-cubic and was thereby able to dissolve all of the carbon into its interstices. No Fe_3C remained. The phases within a material can be displayed on a **phase diagram** as a function of temperature and composition.

Phase diagrams are useful to the engineer because they indicate the temperature and composition requirements for attaining the required internal structures and accompanying service properties. The phase diagram shows us (1) *what* phases to expect, (2) the *composition* of each phase that is present, and (3) the *quantity* of each phase within a mixture of phases.

What Phases?

Sterling silver contains 92.5Ag–7.5Cu (weight percent). Using the Ag–Cu phase diagram of Figure 3.28, we observe that this composition is liquid above 910°C; below 740°C, it contains a mixture of the α and β solid structures. The former is a solid solution of silver plus a limited amount of copper; the latter is a solid solution of copper plus a limited amount of silver. Between 740 and 810°C, only one phase is present, α. In that temperature range, all of the copper can be dissolved in the α solid solution. From 810 to 910°C, the alloy changes from no liquid to all liquid.

Phase Compositions?

Pure silver has a face-centered-cubic structure. Copper forms the same crystalline structure. Not surprisingly, copper atoms can be substituted for silver. But the solid solution in α is limited, because the silver atom is 13% larger in diameter than is copper. However, the solubility increases with temperature to 8.8 weight-percent copper at 780°C. This is shown by the boundary of the silver-rich shaded area of Figure 3.28. Within that shaded area, we have a single phase called α.

Conversely, the FCC structure of copper can dissolve silver atoms in solid solution. At room temperature the solubility is very small. Again, as the temperature is raised the solubility limit increases to 8 weight-percent silver (92 wt. % Cu) at 780°C. This copper-rich phase is called β.

Silver and copper are mutually soluble in a liquid solution. Above 1100°C there is no limit to the solubility. As the temperature is reduced below the melting point of copper (1084°C), the copper solubility limit decreases from 100% to 28%. Excess copper produces the β phase. Likewise, below the melting point of silver (962°C), the silver solubility limit decreases from 100% to 72%. Excess silver produces the α phase. The two solubility curves cross at approximately 72Ag–28Cu and 780°C. We call this low-melting liquid **eutectic**.

How Much of Each Phase in a Mixture?

At 800°C, a 72Ag–28Cu alloy is entirely liquid. At the same temperature, but with added copper, the solubility limit is reached at 32% Cu. Beyond that limit, still at 800°C, any additional copper precipitates as β. Halfway across the two-phase field of (L + β), there will be equal quantities of the two phases. Additional copper in the alloy increases the amount of β. Within this two-phase region the liquid remains saturated with 32% Cu, and the solid β is saturated with 8% Ag (thus, 92% Cu).

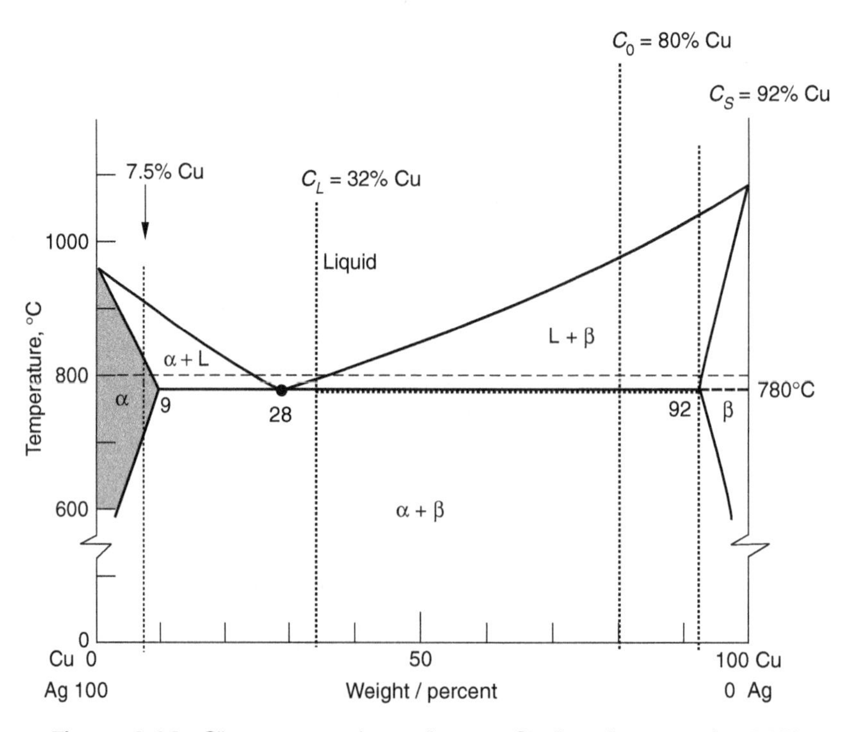

Figure 3.28 Silver-copper phase diagram. Sterling silver contains 7.5% copper; therefore, it has a single phase, α, in the 740–810°C range.

The *composition* of each phase is dictated by the solubility limits. (In a one-phase field, the solubility limits are not factors; the alloy composition and the phase composition are identical.) The *amount* of each phase in a two-phase field is determined by interpolation between the solubility limits using what is called the lever law. For example, at 800°C, an alloy consisting of 80% Cu and 20% Ag would consist of a liquid phase containing 32% Cu and a solid phase of 92% Cu. The weight fraction of liquid phase, W_L, and solid phase, W_s, is determined by $W_L + W_S = 1$.

$$W_L = \frac{C_s - C_O}{C_S - C_L} \quad \text{and} \quad W_S = \frac{C_O - C_L}{C_S - C_L}$$

Reaction Rates in Solids

A phase diagram is normally an equilibrium diagram; that is, all reactions have been completed. The time required to reach equilibrium generally increases with decreasing temperature. Thus, a material may not always possess the expected phases with predicted amounts or compositions. Even so, the phase diagram is valuable. For example, sterling silver (92.5Ag–7.5Cu) contains only one phase when equilibrated at 775°C (Figure 3.28). Slow cooling to room temperature precipitates β as a minor second phase, as would be expected when plenty of time is available. Rapid cooling, however, traps the copper atoms within the a solid solution. This situation is used to advantage, because the solid solution is stronger than the ($\alpha + \beta$) combination. Also, a single-phase alloy corrodes less readily. This explains why the *impure* sterling silver is commonly preferred over pure silver.

The selection of compositions and processing treatments is generally based on a knowledge of equilibrium diagrams plus a knowledge of how equilibrium is circumvented.

Microstructures of Multiphase Solids

The microstructure of a single-phase, crystalline solid includes the *size*, *shape*, and *orientation* of the grains. Variations in these properties are also found in multiphase solids. In addition, the microstructure of a multiphase solid may also vary in the *amount* of each phase and the *distribution* of the phases. In an equilibrated microstructure, the amounts of the phases may be predicted directly from the phase diagrams using the lever law. From Figure 3.29, a 1080 steel (primarily iron, with 0.80 percent carbon) will have twice as much Fe_3C (W_{Fe3C}) at room temperature as a 1040 steel (with 0.40 percent carbon). The Fe_3C is a hard phase. Therefore, with all other factors equal, we expect a 1080 steel to be harder than a 1040 steel, and it is.

$$W_{Fe_3C} = \frac{C_0 - C_\alpha}{C_{Fe_3C} - C_\alpha}$$

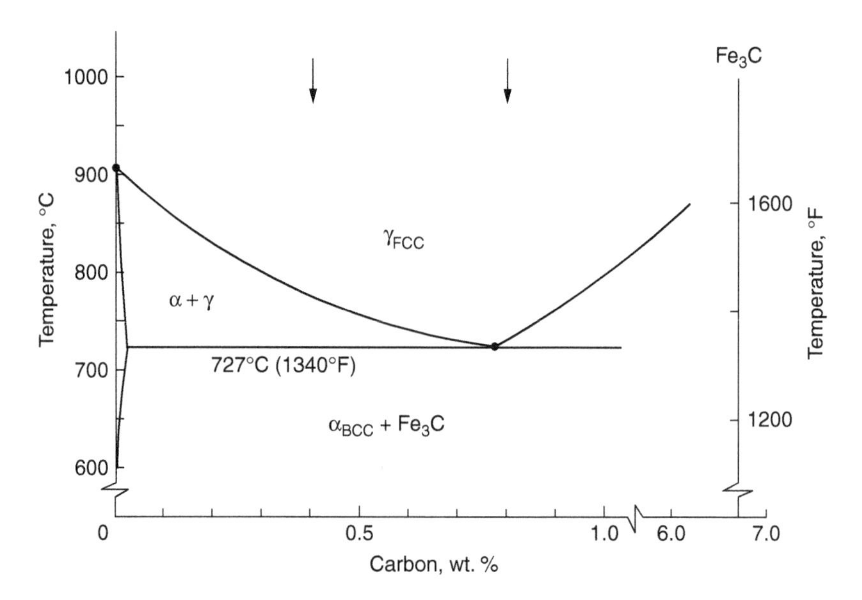

Figure 3.29 Portion of the Fe–Fe_3C phase diagram. Most steels are heat treated by initially forming austenite, γ, which is Fe_{FCC}. It changes to a mixture of ferrite, α, and carbide at lower temperatures.

The distribution of phases within microstructures is more difficult to quantify, so only descriptive examples will be cited. Similar to sterling silver, aluminum will dissolve several percent of copper in solid solution at 550°C. If it is cooled slowly, the copper precipitates as a minor, hard, brittle compound ($CuAl_2$) along grain boundaries of the aluminum. The alloy is weak and brittle and has little practical use. If the same alloy is cooled rapidly from 550°C, trapping the copper atoms within the aluminum grains as shown in Figure 3.30(a), the quenched solid solution is stronger and more ductile and has commercial uses. If the quenched alloy is reheated to 100°C, the $CuAl_2$ precipitates, as expected from the phase diagram shown in Figure 3.30(b) and (c). In this case, however, the precipitate is very finely dispersed within the grains of aluminum. The alloy retains its toughness, because the brittle $CuAl_2$ does not form a network for fracture paths. In addition, the strength is increased because the submicroscopic hard particles interfere with the deformation of the aluminum along slip planes. Alloys of this type are used in airplane construction because they are light, strong, and tough. Observe that all of the examples in this paragraph are for the same alloy. The properties have been varied through microstructural control.

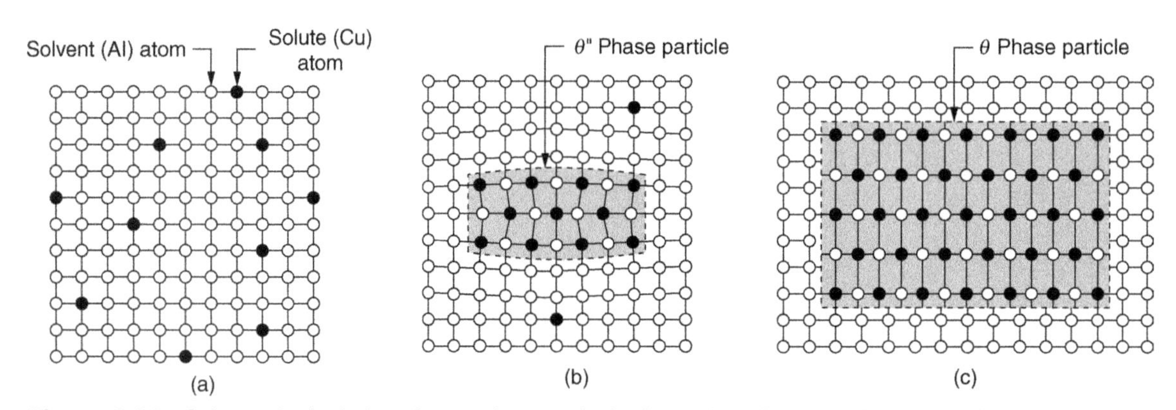

Figure 3.30 Schematic depiction of several stages in the formation of the equilibrium precipitate (*q*) phase. (a) A supersaturated a solid solution; (b) a transition, θ'', precipitate phase; (c) the equilibrium θ' phase, within the a-matrix phase. Actual phase particle sizes are much larger than shown here.

Two distinct microstructures may be produced in a majority of steels. In one, called **spheroidite**, the hard Fe_3C is present as rounded particles in a matrix of ductile ferrite (α). In the other, called **pearlite**, Fe_3C and ferrite form fine alternating layers, or lamellae. Spheroidite is softer but tougher; pearlite is harder and less ductile (Figure 3.31). The mechanical properties are controlled by the spacing of the Fe_3C, because the hard Fe_3C phase stops dislocation motion. The closer the spacing between Fe_3C particles, the greater the strength or hardness, but the lower the ductility.

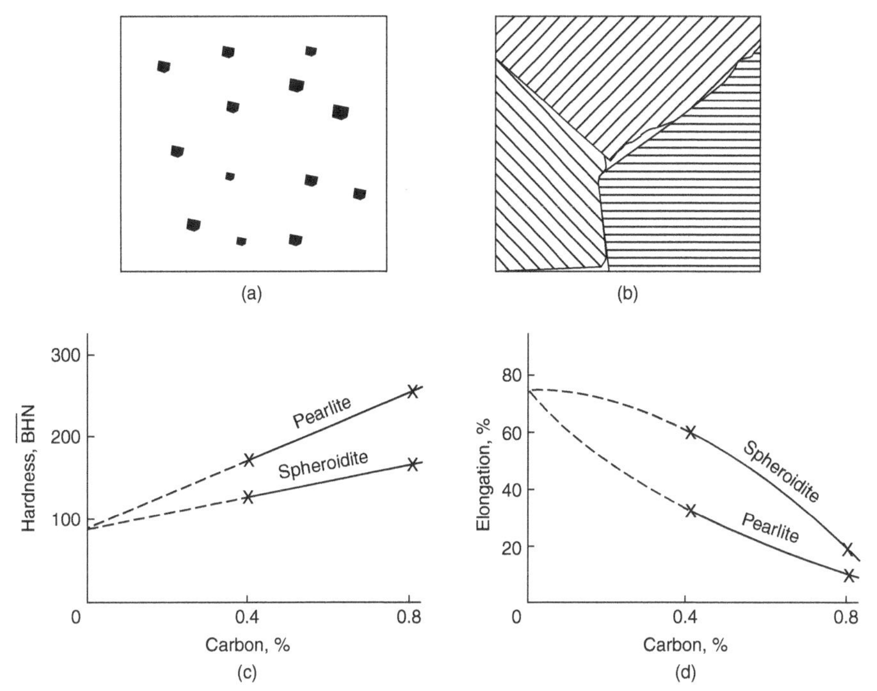

Figure 3.31 Phase distributions (schematic). The two samples are of the same steel, both containing ferrite (white) and carbide (black) but heat treated differently. Spheroidite is shown in (a), and pearlite in (b). The hardness of the two structures is shown in (c), and ductility is shown in (d).

Example **3.35**

At 780°C, an Ag–Cu liquid of eutectic composition (Figure 3.28) solidifies to solid α and solid β. (i) What are the compositions of the two solid phases just below the eutectic temperature? (ii) How much of each of these two phases exists per 100 grams of alloy?

Solution

(i) From Figure 3.28, the α phase is 91Ag–9Cu, and the β phase is 8Ag–92Cu.

(ii) Interpolation along the 780°C tie-line between the solubility limits yields

$$\beta\text{: (using Cu) } [(28 - 9)/(92 - 9)](100 \text{ g}) = 23 \text{ g}$$

$$\alpha\text{: (using Cu) } [(92 - 28)/(92 - 9)](100 \text{ g}) = 77 \text{ g}$$

or

$$\alpha\text{: (using Ag) } [(72 - 8)/(92 - 9)](100 \text{ g}) = 77 \text{ g}$$

Example 3.36

The phases and microstructures of an SAE 1040 steel may be related to the Fe-Fe_3C phase diagram (Figure 3.29). (i) Assume equilibrium. What are the phases and their weight percents at 728°C? At 726°C? (ii) Pearlite has the eutectoid composition of 0.77% C. What percent of the steel will *not* be contained in the pearlite?

Solution

(i) At 728°:

$$\alpha\ (0.02\%\ \text{C}): (0.77 - 0.40)/(0.77 - 0.02) = 49\%\ \alpha$$

$$\gamma\ (0.77\%\ \text{C}): (0.40 - 0.02)/(0.77 - 0.02) = 51\%\ \gamma$$

$$Fe_3C \text{ contains } 12/[12 + 3(55.85)] = 6.7\%\ \text{C}$$

At 726°:

$$\alpha\ (0.02\%\ \text{C}): (6.7 - 0.40)/(6.7 - 0.02) = 94\%\ \alpha$$

$$Fe_3C: (0.40 - 0.02)/(6.7 - 0.02) = 6\%\ Fe_3C$$

(ii) All of the pearlite, $\alpha + Fe_3C$, comes from the γ. This FCC phase changes to pearlite at 727°C. The α that was present at 728°C remains unchanged. Therefore, the answer is $P = 51\%$; and unchanged $\alpha = 49\%$.

MATERIALS PROCESSING

For many engineers, materials are first encountered in terms of handbook data or stockroom inventories. However, materials always have a prior history. They must be obtained from natural sources, then subjected to compositional modifications and complex shaping processes. Finally, specific microstructures are developed to achieve the properties that are necessary for extended service.

Extraction and Compositional Adjustments

Few materials are used in their natural form. Wood is cut and reshaped into lumber, chipboard, or plywood; most metals must be extracted from their ores; rubber latex is useless unless it is vulcanized.

Extraction from Ores

Most ores are oxides or sulfides that require chemical reduction. Commonly, oxide ores are chemically reduced with carbon- or CO-containing gas. For example,

$$Fe_2O_3 + 3\ CO \rightarrow 2\ Fe + 3\ CO_2 \tag{3.46}$$

Elevated temperatures are used to speed up the reactions and more completely reduce the metal. If the metal is melted, it is more readily separated from the accompanying gangue materials. The reduced product is a carbon-saturated metallic iron. As such, it has only limited applications. Normally, further processing is required.

Refining

Dissolved impurities must be removed. Even if the above ore were of the highest quality iron oxide, it would be necessary to refine it, because the metallic product is saturated with carbon. In practice, small but undesirable quantities of silicon and other species are also reduced and dissolved in the iron. They are removed by closely controlled **reoxidation** at chemically appropriate temperatures (followed by **deoxidation**). The product is a **steel**. Alloying additions are made as specified to create different types of steel.

Chemicals from which a variety of plastics are produced are refined from petroleum. The principal step of petroleum refinement involves selective distillation of liquid petroleum. Lightweight fractions are removed first. Controlled temperatures and pressures distill the molecular fractions that serve as precursors for polymers. Residual fractions are directed to other products.

Polymerization makes macromolecules out of the smaller molecules that are the product of distillation. **Addition polymerization** involving a C=C→C—C reaction is encountered in the polymerization of ethylene, $H_2C{=}CH_2$; vinyl chloride, $H_2C{=}CHCl$; and styrene, $H_2C{=}CH(C_6H_5)$.

$$C{=}C \rightarrow C{-}C{-} \tag{3.47}$$

Shaping Processes

The earliest cultural ages of human activities produced artifacts that had been shaped from stone, bronze, or iron. In modern technology, we speak of casting, deformation, cutting, and joining in addition to more specialized shaping procedures.

Casting

The concept of casting is straightforward. A liquid is solidified within a mold of the required shape. For metal casting, attention must be given to volume changes; in most cases there is shrinkage. In order to avoid porosity, provision must be made for feeding molten metal from a **riser**. **Segregation** may occur at the solidifying front because of compositional differences in the $(\alpha + L)$ range. (See the earlier discussion of phase diagrams and the discussion of annealing processes in a following subsection.)

A number of ceramic products are made by **slip casting**. The slip is a slurry of fine powders suspended in a fluid, usually water. The mold is typically of gypsum plaster with a porosity that absorbs water from the adjacent suspension. When the shell forming inside the mold is sufficiently thick, the remaining slip is drained. Subsequent processing steps are **drying** and **firing**. The latter high-temperature step bonds the powder into a coherent product.

The casting process is also used in forming polymeric products. Here the solidification is accomplished by polymerization. There is a chemical reaction between the small precursor molecules of the liquid to produce macromolecules and a resulting solid.

Deformation Processes

Deformation processes include forging, rolling, extrusion, and drawing, plus a number of variants. In each case, a force is applied, and a dimensional change results. **Forging** involves shape change by impact. **Rolling** may be used for sheet products as well as for products with constant cross sections, for example, structural beams. **Extrusion** is accomplished through open or closed dies. The former requires that the product be of uniform cross section, such as plastic pipe or siderails for aluminum ladders. Closed dies are molds into which the material is forced. These forming processes can be done at ambient temperature (cold working) or elevated temperatures ($T > 0.4T_m$ in Kelvin; hot working). T_m is the melting temperature.

The forming temperature has a strong effect on mechanical properties. Products formed at ambient temperature have high dislocation densities and hence greater strength and hardness but lower ductility than hot-worked products.

In general, ceramics do not lend themselves to the above deformation processes because they lack ductility. Major exceptions are the glasses, which deform not by crystalline slip but by viscous flow.

Cutting

Chiseling and sawing are cutting processes that predate history. Current technology includes **machining** in which a cutting tool and the product move with respect to each other. Depth and rates of cut are adjustable to meet requirements. **Grinding** is a variant of machining that is used for surface removal.

Joining

The process of **welding** produces a joint along which the abutting material has been melted and filled with matching metal. **Soldering** and **brazing** processes use fillers that have a lower melting point than the adjoining materials, which remain solid. There are glass solders as well as metallic ones. Adhesives have long been used for joining wood and plastic components, and many have now been developed to join metals and ceramics.

Annealing Processes

Annealing processes involve reheating a material sufficiently that internal adjustments may be made between atoms or between molecules. The temperature of annealing varies with (1) the material, (2) the amount of time available, and (3) the structural changes that are desired.

Homogenization

The dynamics of processing will produce segregation. For example, when an 80Cu–20Ni alloy starts to solidify at 1200°C, the first solid contains 30% Ni. When solidification is completed, the final liquid has only 12% Ni. Uniformity can be obtained if the alloy is reheated to a temperature at which the atoms can relocate by diffusion. There is a time-temperature relationship (log t vs. $1/T$). In this case an increase in temperature from 500 to 550°C reduces the necessary annealing time by a factor of eight.

Recrystallization

Networks of dislocations are introduced when most metals are plastically deformed at ambient temperatures (cold working). The result is a work hardening and loss of ductility. Whereas the resulting increase in strength is often desired, the property changes resulting from dislocations make further deformation processing more difficult. Annealing will remove the dislocations and restore the initial workable characteristics by forming new, strain-free crystals.

A one-hour heat treatment is a common shop practice because it allows for temperature equalization as well as scheduling requirements. For that time frame, it is necessary to heat a metal to approximately 40 percent of its melting temperature (on the absolute scale). Thus, copper that melts at 1085°C (1358 K) may be expected to recrystallize in the hour at 270°C (545 K). The recrystallization time is shorter at higher temperatures and longer at lower temperatures.

Grain Growth

Extended annealing, beyond that required for recrystallization, produces grain growth and therefore coarser grains. Normally, this is to be avoided. However, grain growth has merit in certain applications because grain boundaries hinder magnetic domain boundary movements, reducing creep. So, coarse grains are preferred in the sheet steel used in transformers, for example. At high temperatures, grain boundaries permit creep to occur under applied stresses, producing changes in dimensions.

Residual Stresses

Expansions and contractions occur within materials during heat-treating operations. These are isotropic for many materials; or they may vary with crystal orientation. Also, differential expansions exist between the two or more phases in a multiphase material. The latter, plus the presence of thermal gradients, introduce internal stresses, which can lead to delayed fracture if not removed.

It is generally desirable to eliminate these residual stresses by an annealing process called **stress relief**. The required temperature is less than that for recrystallization because atomic diffusion is generally not necessary; rather, adjustments are made through the local movement of dislocations. Stress relief is performed on metals before the final machining or grinding operation. Annealing is always performed on glass products, because any residual surface tension easily activates cracks in this nonductile material.

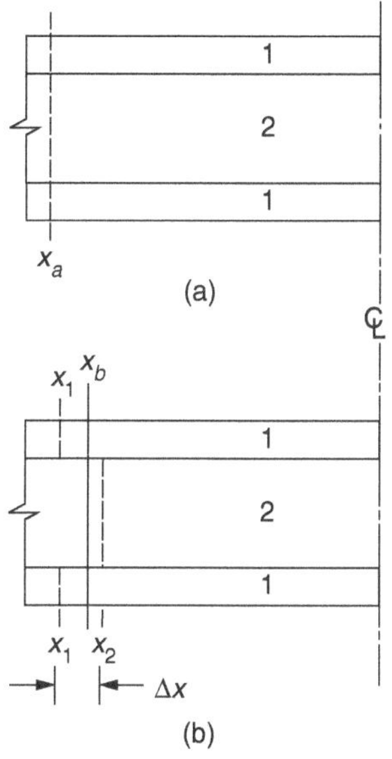

Figure 3.32 Induced stresses (sandwich glass). (a) The previously bonded composite containing glasses 1 and 2 is annealed. There are no internal stresses at x_a. (b) If the three layers were separate, the outer layers, which have a lower thermal coefficient, would contract less during cooling than the center glass (x_1 vs. x_2). Since the layers are bonded together, the restricted contraction of the center (x_a to x_b rather than x_a to x_2) induces compression within the surface layers (x_1 to x_b).

Induced Stresses

In apparent contradiction to the last statement, residual compressive stresses may be prescribed for certain glasses, since glass like most nonductile materials is strong in **compression** but weak in **tension**. As an example, a familiar dinnerware product is made from a *sandwich* glass sheet in which the *bread* layers have a lower thermal expansion than does the *meat* layer. The processing involves heating the dinnerware to relieve all stresses (annealing). As the dinnerware pieces are cooled, the center layer tries to contract more than the surface layers, placing the surfaces under compression (Figure 3.32). Any tension encountered in service must overcome the residual compression before a crack can propagate.

Time-Dependent Processes

We have seen in a previous section that sterling silver is solution treated to dissolve all of the copper within the silver-rich α phase. The single phase is preserved by rapidly cooling the alloy. This avoids the precipitation of the copper-rich β phase, as required for equilibrium. The cooling rate need not be drastic for sterling silver because it takes a minute or more to nucleate and grow the precipitate, β.

Even more time is available for the production of a **silicate glass**—a supercooled liquid. The necessary bond breaking and rearrangements are very slow. The available processing times can approach an hour or more. However, in order to produce a **metallic glass**, the cooling rate must approach 1000°C per *millisecond*. Otherwise, individual metal atoms rapidly order themselves into one of the crystalline patterns described earlier.

Martensitic Reactions

Most steel processing treatments initially heat the steel to provide a single-phase microstructure of γ or **austenite** (Fe_{FCC}). This is face-centered-cubic and dissolves all of the carbon that is present. Normal cooling produces **pearlite**, a lamellar microstructure containing layers of α or ferrite (Fe_{BCC}) and **carbide** (Fe_3C), as shown previously on the Fe–Fe_3C phase diagram (Figure 3.28).

$$Fe_{FCC} \xrightarrow{\text{cooling}} Fe_{BCC} + Fe_3C \qquad \textbf{(3.48)}$$

If **quenched**, a different structure forms:

$$Fe_{FCC} \xrightarrow{\text{quenching}} \text{Martensite} \qquad \textbf{(3.49)}$$

Martensite is a transition phase. It offers an interesting possibility for many applications because it is much harder than pearlite. Unfortunately, martensite is also very brittle, and its usefulness in steel is severely limited. Martensite will exist almost indefinitely at ambient temperatures, but reheating the steel provides an opportunity for the completion of the reaction of Equation (3.48).

$$\text{Martensite} \xrightarrow{\text{tempering}} Fe_3C \qquad \textbf{(3.50)}$$

The reheating process is called **tempering**. The product, which has a microstructure of finely dispersed carbide particles in a ferrite matrix, is both hard and tough. It is widely used and is called **tempered martensite**.

Hardenability

Tempered martensite is the preferred microstructure of many high-strength steels. Processing requires a sufficiently rapid quench to obtain the intermediate martensite, Equation (3.49), followed by tempering, Equation (3.50). This means severe quenching for products of Fe–C steels that are larger than needles or razor blades. Even then, martensite forms only at the quenched surface. The subsurface metal transforms directly to the ferrite and carbide, Equation (3.48).

The reaction rate of Equation (3.48) can be decreased by the presence of various alloying elements in the steel. Thus, gears and similar products commonly contain fractional percentages of nickel, chromium, molybdenum, or other metals. Quenching severity can be reduced, and larger components can be hardened throughout. These alloying elements delay the formation of carbide because, not only must small carbon atoms relocate, it is also necessary for the larger metal atoms to choose between residence in the ferrite or in the carbide.

Surface Modification

Products may be treated so that their surfaces are modified and therefore possess different properties than the original material. Chrome plating is a familiar example.

Carburizing

Strength and hardness of a steel increase with carbon content. Concurrently, ductility and toughness decrease. With these variables, the engineer must consider trade-offs when specifying steels for mechanical applications. An alternate possibility is to alter the surface zone. A common example is the choice of a low-carbon, tough steel that has had carbon diffused into the subsurface (<1 mm) to produce hard carbide particles. Wear resistance is developed without decreasing bulk toughness.

Nitriding

Results similar to carburizing are possible for a steel containing small amounts of aluminum. Nitrogen can be diffused through the surface to form a subscale containing particles of aluminum nitride. Since AlN has structure and properties that are related to diamond, wear resistance is increased for the steel.

Shot Peening

Superficial deformation occurs when a ductile material is impacted by sand or by hardened shot. A process employing shot peening places the surface zone in local compression and therefore lowers the probability for fracture initiation during tensile loading.

Example 3.37

Assume a single spherical shrinkage cavity forms inside a 2-kg lead casting as it is solidified at 327°C. What is the initial diameter of the cavity after solidification? (The greatest density of molten lead is 10.6 g/cm^3.)

Solution

Refer to Figure 3.26(a). Based on a unit volume at 20°C, lead shrinks from 1.07 to 1.035 during solidification.

The volume of molten lead is (2000 g)/(10.6 g/cm^3) = 188.7 cm^3.

Volume of solid lead at 327°C: (188.7 cm^3)(1.035/1.07) = 182.5 cm^3

Shrinkage: 188.7 cm^3 – 182.5 cm^3 = $\pi d^3/6$; d = 2.28 cm

Example 3.38

How much energy is involved in polymerizing one gram of ethylene (C_2H_4) into polyethylene, Equation (3.47). The double carbon bond possesses 162 kcal/mole, and the single bond has 88 kcal/mole.

Solution

As shown in Equation (3.47), one double carbon bond changes to two single bonds/mer:

$$(1\ \text{g})/(24 + 4\ \text{g/mole}) = 0.0357\ \text{moles}$$

The energy required to break 0.0357 moles of double bonds is

$$(1)(0.0357)(162\ \text{kcal}) = 5.79\ \text{kcal}$$

The energy released in joining twice as many single bonds is

$$(2)(0.0357)(-88\ \text{kcal}) = -6.29\ \text{kcal}$$

The net energy change is –6.29 + 5.79 kcal = 500 cal released/g.

Example 3.39

There are 36 equiaxed grains per mm^2 observed at a magnification of 100 in a selected area of copper. The copper is heated to double the average grain diameter. (i) How many grains exist per mm^2? (ii) What will be the percentage (increase, decrease) in grain boundary area?

Solution

(i) Doubling a lineal dimension decreases the number of grains by a factor of four, so there are 9 grains per mm^2.

(ii) Surface area is a function of the lineal dimension squared:

$$a_1/a_2 \propto d_1^2/d_2^2$$

Thus, a_1/a_2 = 0.25, or a 75% decrease in grain boundary area.

MATERIALS IN SERVICE

Products of engineering are made to be used. Conditions that are encountered in service most often vary tremendously from those present in the stockroom: static and dynamic loads, elevated and subambient temperatures, solar and nuclear radiation exposure, and many other reactions with the surrounding environment. All of these situations can lead to deterioration and even to failure. The design engineer should be able to anticipate the conditions of failure.

Mechanical Failure

Under ambient conditions, excessive loads can lead to bending or to cracking, the principal modes of mechanical failure. The former depends upon geometry and the stress level. **Stress** is defined as load divided by cross-sectional area, $\sigma = F/A$. Cracking (and succeeding fracture) includes those two considerations, plus the loading rate.

Yield Strength

When solids are stressed, strain occurs. Strain is the change in length divided by the original length, $\varepsilon = \Delta L/L_0$. The ratio of stress to strain is called the **elastic modulus**, $E = \sigma/\varepsilon$, and is initially constant in most solids as shown in Figure 3.33. The interatomic spacings are altered as the load is increased. Initial slip starts at a threshold level called the **yield strength**, σ_y, shown in Figure 3.34. Higher stresses will produce a permanent distortion, which will be called failure if the product was designed to maintain its initial shape. The toughness of a material is related to the energy or work required for fracture, a product of both strength and ductility. In a tensile test, the energy is the area under the stress-strain curve. Ductile fracture involves significant plastic deformation and hence absorbs much more energy than brittle fracture, which has little or no plastic deformation. A complete stress-strain curve is shown in Figure 3.35.

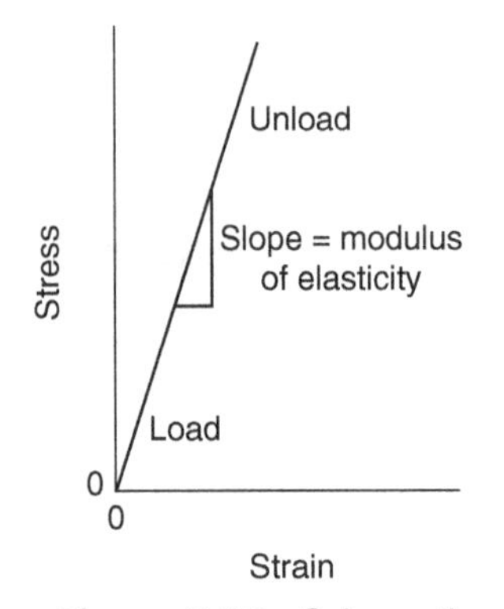

Figure 3.33 Schematic stress-strain diagram showing linear elastic deformation for loading and unloading cycles

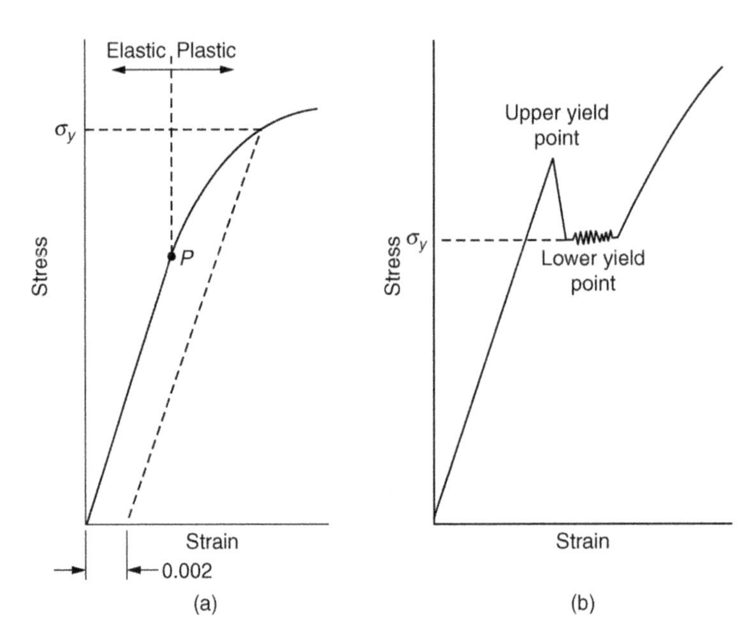

Figure 3.34 (a) Typical stress-strain behavior for a metal, showing elastic and plastic deformations, the proportional limit P, and the yield strength σ_y, as determined using the 0.002 strain offset method; (b) representative stress-strain behavior found for some steels, demonstrating the yield point phenomenon

Fracture

Breakage always starts at a location of **stress concentration**. This may be at a *flaw* of microscopic size, such as an abrasion scratch produced while cleaning eyeglasses, or it may be of larger dimensions, such as a hatchway on a ship.

With a crack, the **stress intensity factor**, K_I, is a function of the applied stress, σ, and of the square root of the crack length a:

$$K_I = y\sigma\sqrt{\pi a} \quad \textbf{(3.51)}$$

The proportionality constant, y, relates to cross-sectional dimensions and is generally near unity.

Fracture toughness is the **critical** stress intensity factor, K_{Ic}, to propagate fracture and is a property of the material. This corresponds to the yield strength, σ_y, being the critical stress to initiate slip. However, stronger materials generally have lower fracture toughness and *vice versa*.

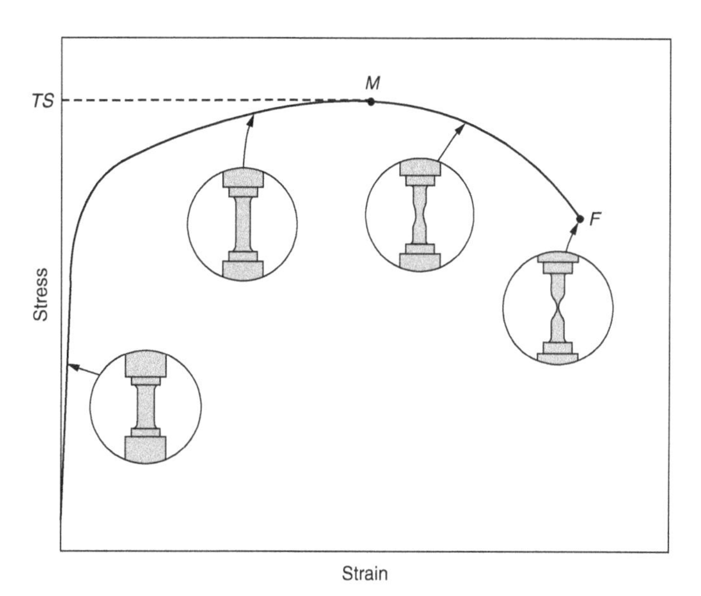

Figure 3.35 Typical engineering stress-strain behavior to fracture, point *F*. The tensile strength *TS* is indicated at point *M*. The circular insets represent the geometry of the deformed specimen at various points along the curve.

To illustrate the relationship between strength and toughness, consider a steel that has a yield strength, σ_y, of 1200 MPa, and a critical stress intensity factor, K_{Ic}, of 90 MPa • $m^{1/2}$.

In the presence of a 2-mm crack, a stress of 1135 MPa would be required to propagate a fracture, according to Equation (3.51). This is below the yield stress. If the value of K_{Ic} had been 100 MPa • $m^{1/2}$, fracture would not occur; rather, the metal would deform at 1200 MPa. A 2.2-mm crack would be required to initiate fracture without yielding.

Fatigue

Cyclic loading reduces permissible design stresses, as illustrated in Figure 3.36. Minute structural changes that introduce cracks occur during each stress cycle. The crack extends as the cycles accumulate, leading to eventual fracture. When this delayed behavior was first observed, it was assumed that the material got tired; hence the term *fatigue*. Steels and certain other materials possess an *endurance limit*, below which unlimited cycling can be tolerated.

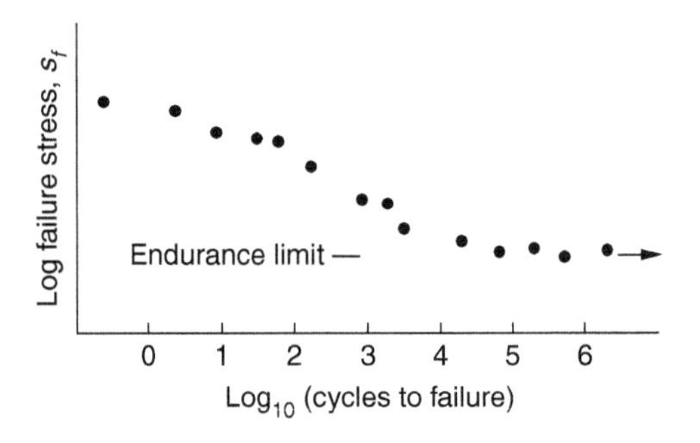

Figure 3.36 Cyclic fatigue. The stress for failure, *sf*, decreases as the number of loading cycles is increased. Most steels exhibit an endurance limit, a stress level for unlimited cycling. (A static tensile test is only one-fourth of a cycle.)

Example 3.40

(i) What is the maximum static force that can be supported without permanent deformation by a 2-mm-diameter wire that has a yield strength of 1225 MPa?

(ii) The elastic deformation at this threshold stress is 0.015 m/m. What is its elastic modulus?

Solution

(i) $\sigma_y = F/A$

$F = \pi(2 \text{ mm}/2)^2(1225 \text{ MPa}) = 3800 \text{ N}$

(ii) $E = \sigma/\varepsilon = 1225 \text{ MPa}/0.015 \text{ m/m} = 82{,}000 \text{ MPa}$

Example 3.41

The value of K_{Ic} for steel is 186 MPa•m$^{1/2}$. What is the maximum tolerable crack length, a, when the steel carries a nominal stress of 800 MPa? (Assume 1.1 as the proportionality constant.)

Solution

$$186 \text{ MPa}\cdot\text{m}^{1/2} = 800 \text{ MPa } (1.1)(\pi a)^{1/2}$$

$$a = 0.014 \text{ m} = 14 \text{ mm}$$

Thermal Failure

Melting is the most obvious change in a material at elevated temperatures. Overheating, short of melting, can also introduce microstructural changes. For example, the tempered martensite of a tool steel is processed so that it has a very fine dispersion of carbide particles in a tough ferrite matrix. It is both hard and tough and serves well for machining purposes. However, an excessive cutting speed raises the temperature of the cutting edge, causing the carbide particles to grow and softening the steel; if heating continues, failure eventually occurs by melting at the cutting edge. *High-speed* tools incorporate alloy additions, such as vanadium and chromium, that form carbides that are more stable than iron carbide. Thus, they can tolerate the higher temperatures that accompany faster cutting speeds.

Creep

As the name implies, **creep** describes a slow (<0.001%/hr) dimensional change within a material. It becomes important in long-term service of months or years. Slow viscous flow is commonly encountered in plastic materials. Refractories (temperature-resistant ceramics) will slowly slump when small amounts of liquid accumulate.

In metals, creep occurs when atoms become sufficiently mobile to migrate from compressive regions of the microstructure into tensile regions. Grain boundary areas are heavily involved. For this reason, coarser grained metals are advantageous for high-temperature applications. Three stages of creep are identified in Figure 3.37. Following the initial elastic strain, Stage 1 of creep is fairly rapid as stress variations are equalized. In Stage 2, the creep rate, dL/dt, is essentially constant. Design considerations are focused on this stage. Stage 3, which accelerates when the cross-sectional area starts to be reduced, leads to eventual rupture.

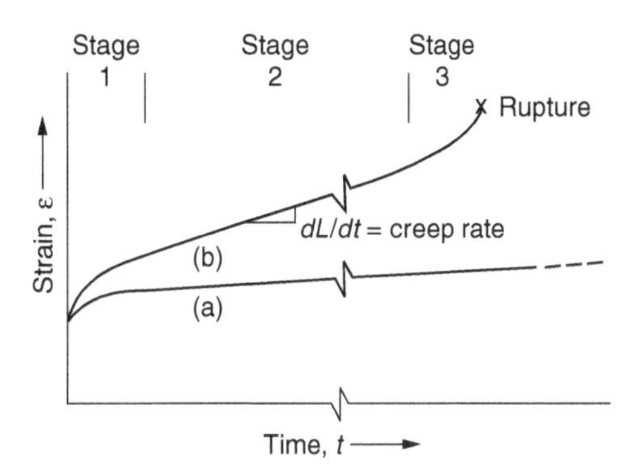

Figure 3.37 Creep. (*a*) Low stresses and/or low temperatures; (*b*) high stresses, high temperatures. The initial strain is elastic, followed by rapid strain adjustments (Stage 1). Design calculations are commonly based on the steady-state strain rate (Stage 2). Strain accelerates (Stage 3) when the area reduction becomes significant. Rupture occurs at *x*.

Spalling

Spalling is thermal cracking. It is the result of stress caused by differential volume changes during processing or service. As discussed earlier, stresses can be introduced into a material (1) by thermal gradients, (2) by anisotropic volume changes, or (3) by differences in expansion coefficients in multiphase materials. Cyclic heating and cooling lead to **thermal fatigue** when the differential stresses produce localized cracking.

The spalling resistance index (SRI) of a material is increased by higher thermal conductivities, k, and greater strengths, S; it is reduced with greater values of the thermal expansion coefficients, α, and higher elastic moduli. In functional form,

$$\mathrm{SRI} = f(kS/\alpha E) \tag{3.52}$$

Low-Temperature Embrittlement

Many materials display an abrupt drop in ductility and toughness as the temperature is lowered. In glass and other amorphous materials, this change is at the temperature below which atoms or molecules cannot relocate in response to the applied stresses. This is the **glass-transition temperature,** T_g. Metals are crystalline and do not have a glass transition. However, steels and a number of other metals have a **ductility-transition temperature** below which fracture is nonductile. The impact energy required for fracture can drop by an order of magnitude at this transition temperature. Thus it becomes a very significant consideration in design for structural applications. The **ductile-to-brittle transition temperature** (DBTT) is often measured using the Charpy Impact test, with samples soaked at different temperatures immediately prior to testing. Representative curves and the influence of carbon content on the DBTT in steel are shown in Figure 3.38.

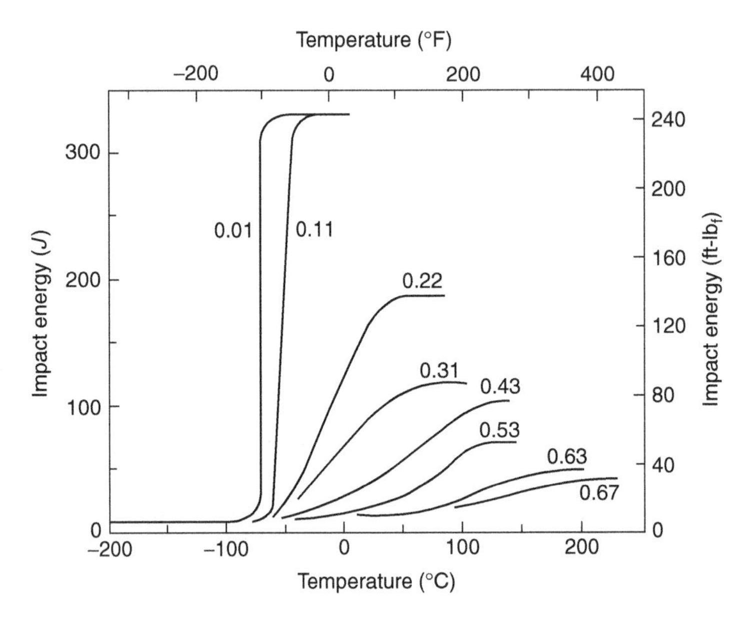

Figure 3.38 Influence of carbon content on the Charpy V-notch energy-versus-temperature behavior for steel

Radiation Damage

Unlike heat, which energizes all of the atoms and molecules within a structure, radiation introduces energy at *pinpoints* called **thermal spikes**. Individual bonds are broken, specific atoms are dislodged, molecules are ruptured, and electrons are energized. Each of these actions disorders the structure and alters the properties of the material. As expected, slip and deformation are resisted. Therefore, while strength and hardness increase, ductility and toughness decrease. Electrical and thermal conductivities drop within metals, because there is more scattering of the electrons as they move along the voltage or thermal gradient. These property changes, among others, are considered to be damaging changes, especially when they contradict carefully considered design requirements.

Damage Recovery

Partial correction of radiation damage is possible by annealing. Reheating a material allows internal readjustments, since atoms that have been displaced are able to relocate into a more ordered structure. It is similar to recrystallization after work hardening, where dislocations involving lineal imperfections composed of many atoms must be removed, except that radiation damage involves *pinpoint* imperfections, which means that it can be removed at somewhat lower temperatures than those required for recrystallization.

Chemical Alteration

Oxidation

Materials can be damaged by reacting chemically with their environments. All metals except gold oxidize in ambient air. Admittedly, some oxidize very slowly at ordinary temperatures. Others, such as aluminum, form a protective oxide surface that inhibits further oxidation. However, all metals—including gold—will oxidize significantly at elevated temperatures or in chemical environments that consume the protective oxidation.

Several actions are required for oxide scale to accumulate on a metal surface. Using iron as an example, the iron atom must be ionized to Fe^{2+} before it or the electrons move through the scale to produce oxygen ions, O^{2-}, at the outer surface. There, FeO or Fe_3O_4 accumulates. As the scale thickens, the oxidation rate decreases. However, exceptions exist. For example, the volume of MgO is less than the volume of the original magnesium metal. Therefore, the scale cracks and admits oxygen directly to the underlying metal. Also, an Al_2O_3 scale is insulating so the ionization steps are precluded.

Moisture

Moisture can produce chemical **hydration**. As examples, MgO reacts with water to produce $Mg(OH)_2$, and Fe_2O_3 can be hydrated to form $Fe(OH)_3$. Water can be absorbed into materials. Small H_2O molecules are able to diffuse among certain large polymeric molecules. Consequently, polymers such as the aramids, which we normally consider to be very strong, are weakened—a fact that the design engineer must consider in specifications.

Corrosion

Metallic corrosion is familiar to every reader who owns a car, since rust—$Fe(OH)_3$—is the most obvious product of corrosion. Oxidation produces positive ions and electrons:

$$M \rightarrow M^{n+} + ne^- \qquad \textbf{(3.53)}$$

The reaction stops unless the electrons are removed. Oxygen accompanied by water (Figure 3.39) is a common consumer of the electrons:

$$O_2 + 2\,H_2O + 4e^- \rightarrow 4\,(OH)^- \qquad \textbf{(3.54)}$$

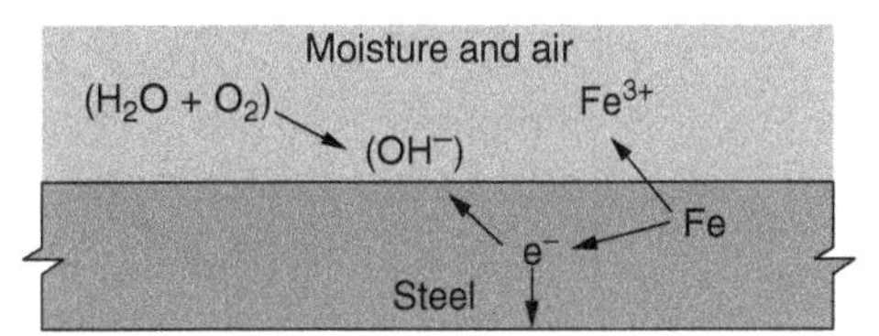

Figure 3.39 Rust formation. Electrons are removed from iron atoms and react with water and oxygen to produce $Fe(OH)_3$—rust

Alternatively, if ions of a metal with a low oxidation potential are present, they can be reduced, consuming electrons from the preceding corrosion reaction, Equation (3.53). Copper is cited as a common example:

$$Cu^{2+} + 2e^- \rightarrow Cu^0 \quad \textbf{(3.55)}$$

Electroplating uses this reaction advantageously to deposit metals from a solution by the addition of electrons. The relative reactivity of metals with respect to standard electrodes is represented by the electromotive force (emf) series given in Table 3.4. However, in real environments such as sea water the galvanic series, as shown in Table 3.5, is more commonly used to determine the likelihood of corrosion.

Table 3.4 The standard emf series

	Electrode Reaction	Standard Electrode Potential V^0(V)
↑	$Au^{3+} + 3^{e-} \rightarrow Au$	+1.420
	$O_2 + 4H^- + 4e^- \rightarrow 2H_2O$	+1.229
	$Pl^2+ + 2e^- \rightarrow Pl$	+1.200
	$Ag^+ + e^- \rightarrow Ag$	+0.800
Increasingly	$Fe^{3+} + e^- \rightarrow Fe^{2+}$	+0.771
inert (cathodic)	$O_2 + 2H_2O + 4e^- \rightarrow 4(OH^-)$	+0.401
	$Cu^{2+} + 2e^- \rightarrow Cu$	+0.340
	$2H^+ + 2e^- \rightarrow H_2$	0.000
	$Pb^{2+} + 2e^- \rightarrow Pb$	–0.126
	$Sn^{2+} + 2e^- \rightarrow Sn$	–0.136
	$Ni^2+ + 2e^- \rightarrow Ni$	–0.250
	$Co^{2+} + 2e^- \rightarrow Co$	–0.277
	$Cd^{2+} + 2e^- \rightarrow Cd$	–0.403
	$Fe^{2+} + 2e^- \rightarrow Fe$	–0.440
Increasingly	$Cr^{3+} + 3e^- \rightarrow Cr$	–0.744
inert (cathodic)	$Zn^{2+} + 2e^- \rightarrow Zn$	–0.763
	$Al^{2+} + 3e^- \rightarrow Al$	–1.662
	$Mg^{2+} + 2e^- \rightarrow Mg$	–2.363
	$Na^+ + e^- \rightarrow Na$	–2.714
↓	$K^+ + e^- \rightarrow K$	–2.924

Corrosion Control

Corrosion is minimized by a variety of means. Some involve the avoidance of one or more of the above reactions. Feed water for steam-power boilers is deaerated; surfaces are painted to limit the access of water and air; junctions between unlike metals are electrically insulated. Other control procedures induce a reverse reaction: sacrificial anodes such as magnesium are attached to the side of a ship; iron sheet is galvanized with zinc. Each corrodes preferentially to the underlying steel and forces the iron to assume the role of Equation (3.55). Corrosion may also be restricted by an impressed voltage. Natural gas lines utilize this procedure by connecting a negative dc voltage to the pipe.

COMPOSITES

Composites are not new. Straw in brick and steel reinforcing rods in concrete have been used for a long time. But there is current interest in the development of new composites by designing materials appropriately. It is possible to benefit from the positive features of each of the contributors in order to optimize the properties of the composite.

The internal structures of composites may be viewed as enlarged poly-component micro-structures, which were previously discussed. Attention is given to size, shape, amounts, and distribution of the contributing materials. However, there is commonly a significant difference in processing composites. Typically, the internal structure of a composite is a function of mechanical processing steps—mixing, emplacement, surface deposition, and so on—rather than thermal processing. These processing differences suggest a different approach in examining property-structure relationships.

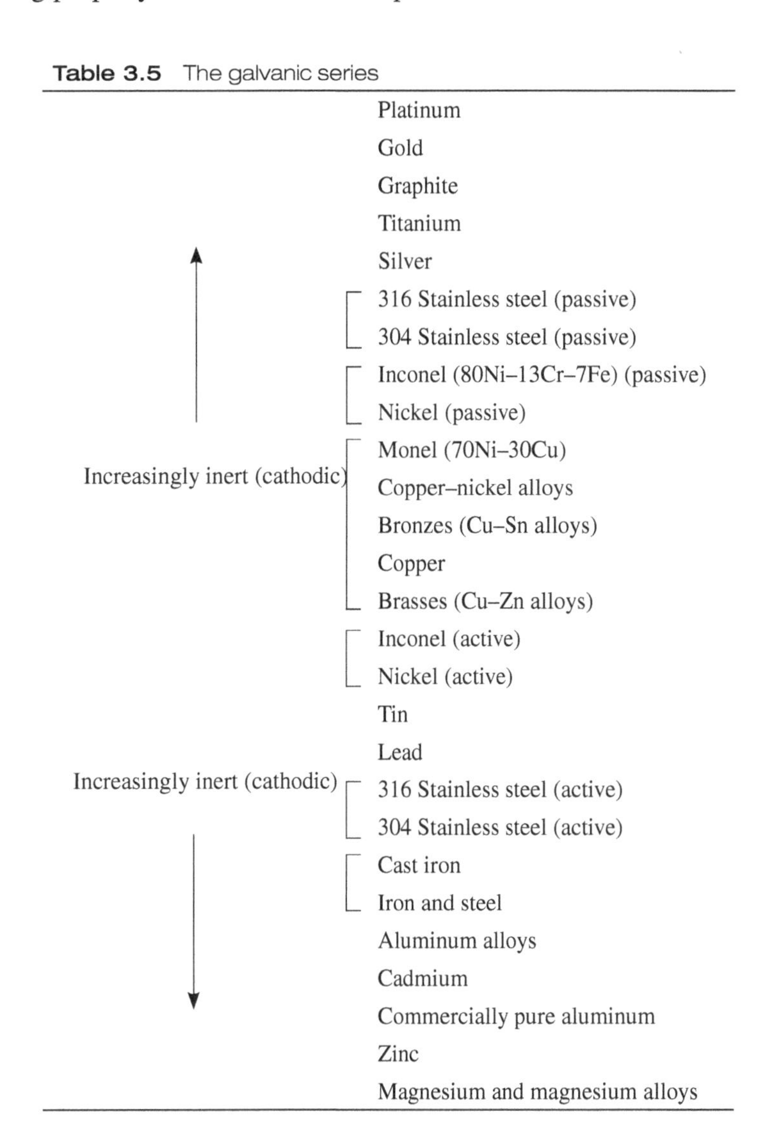

Table 3.5 The galvanic series

	Platinum
	Gold
	Graphite
	Titanium
↑	Silver
	316 Stainless steel (passive)
	304 Stainless steel (passive)
	Inconel (80Ni–13Cr–7Fe) (passive)
	Nickel (passive)
	Monel (70Ni–30Cu)
Increasingly inert (cathodic)	Copper–nickel alloys
	Bronzes (Cu–Sn alloys)
	Copper
	Brasses (Cu–Zn alloys)
	Inconel (active)
	Nickel (active)
	Tin
	Lead
Increasingly inert (cathodic)	316 Stainless steel (active)
	304 Stainless steel (active)
	Cast iron
	Iron and steel
	Aluminum alloys
↓	Cadmium
	Commercially pure aluminum
	Zinc
	Magnesium and magnesium alloys

Reinforced Materials

In familiar composites, such as reinforced concrete, the steel carries the tensile load. Also, there must be bonding between the reinforcement and the matrix. In a *glass* fishing rod, the glass fibers are all oriented longitudinally within the polymer matrix. If it is fractured, the break exhibits a splintered appearance with a noticeable amount of fiber pull-out.

Steel reinforcing bars (rebar) are used to reinforce concrete to meet strength and cost factors. Positioning is dictated by stress calculations. Rebar surfaces are commonly merloned (ridged circumferentially) for better anchorage.

Glass is widely used in fiber-reinforced plastics (FRP). The positioning of the fibers varies with the product. Continuous fibers are used in such structures as fuel storage tanks or rocket casings. However, chopped fibers are required when FRP are processed within molds. A matrix-to-glass bond must be achieved through chemical coatings at the interfaces.

Silicon carbide (SiC) and alumina (Al_2O_3) fibers are used increasingly in high-temperature composites with either metallic or ceramic matrices.

Inert fillers such as silica flour and wood flour serve multiple purposes in many polymers. They add strength, rigidity, and hardness to the product. In addition, they generally are less expensive than the matrix polymer.

Reinforcing bars or fibers are expected to carry the bulk of the tensile load. To be effective, the reinforcement must have a higher elastic modulus, E_r, than does the matrix because the reinforcement and the matrix undergo the same tensile strain when loaded ($\varepsilon_r = \varepsilon_m$). Therefore, $(E/\sigma)_r = (E/\sigma)_m$; and

$$\sigma_r / \sigma_m = E_r / E_m \tag{3.56}$$

Mixture Rules

We commonly study and analyze properties of composites in terms of the properties of the contributing materials. Mixture rules can then be formulated in which the properties are a function of the amounts and geometric distributions of each of the contributors. The simplest mixture rules are based on volume fraction, *f*. For example, the density of the mixture, ρ_m, is the volume-weighted average:

$$\rho_m = f_1\rho_1 + f_2\rho_2 + \ldots \tag{3.57a}$$

or simply

$$\rho_m = \sum f_i \rho_i \tag{3.57b}$$

Likewise, for *heat capacity*, *c*:

$$c_m = \sum f_i c_i \tag{3.58}$$

Directional Properties

Many composites are anisotropic. Laminates of two or more layers are bidirectional, as is a matted FRP that uses chopped fibers. The previously cited *glass* fishing rod has a structure with uniaxial anisotropy.

When considering conductivity, elastic moduli, strength, or other properties that are directional, attention must be given to the anisotropy of the composite. Consider the electrical resistance across a laminate that contains alternate layers of two materials with different resistances, R_1 and R_2. In a direction perpendicular to the laminate, the plies are in series. Thus the relationship is

$$R_{\perp} = L_1 R_1 + L_2 R_2 + \cdots \quad \textbf{(3.59)}$$

where L_i is the thickness of each ply. For unit dimensions, volume fractions, f_i, and the resistivities, ρ_i, are applicable:

$$\rho_{\perp} = f_1 \rho_1 + f_2 \rho_2 + \cdots \quad \textbf{(3.60)}$$

In contrast, for directions parallel to the laminate, resistivities follow the physics analog for parallel circuits:

$$\frac{1}{\rho_{\parallel}} = \frac{f_1}{\rho_1} + \frac{f_2}{\rho_2} + \cdots \quad \textbf{(3.61)}$$

The mixture rules for conductivities, either thermal, k, or electrical, σ, are inverted to those for resistivities:

$$\frac{1}{k_{\perp}} = \frac{f_1}{k_1} + \frac{f_2}{k_2} + \cdots \quad \textbf{(3.62)}$$

and

$$k_{\parallel} = f_1 k_1 + f_2 k_2 + \cdots \quad \textbf{(3.63)}$$

More elaborate mixture rules must be used when the structure of the composite is geometrically more complex, such as the use of particulate fillers or chopped fibers. In general, however, the property of the composite falls between the calculated values using the parallel and series versions of the preceding equations.

Preparation

Composites receive their name from the fact that two or more distinct starting materials are combined into a unified material. The application of a protective coating to metal or wood is one form of composite processing. The process is not as simple as it first appears because priming treatments are commonly required after the surface has been initially cleared of contaminants and moisture; otherwise, peeling and other forms of deterioration may develop. Electroplating, Equation (3.55), commonly requires several intermediate surface preparations so that the final plated layer meets life requirements.

Wood paneling combines wood and polymeric materials into large sheets. **Plywood** not only has dimensional merit, it transforms the longitudinal anisotropy into a more desirable two-dimensional material. Related products include chipboard and similar composites. Composite panels include those with veneer surfaces where appearance and technical properties are valued.

The concept of uniformly mixing particulate **fillers** into a composite is simple. The resulting product is isotropic.

Several considerations are required for the use of **fibers** in a composite. Must the fiber be continuous? What are the directions of loading? What is the shear strength between the fiber and the matrix? Chopped fibers provide more reinforcement than do particles, and at the same time permit molding operations. The **aspect ratio** (L/D) must be relatively low for die molding. Higher ratios, and therefore more reinforcement, are used in sheet molding. (Sheet molded products are used where strength is not critical in the third dimension.) **Continuous fibers** maximize the mechanical properties of FRP composites. Their uses, however, are generally limited to products that permit parallel layments.

Example 3.42

A rod contains 40 volume percent longitudinally aligned glass fibers within a plastic matrix. The glass has an elastic modulus of 70,000 MPa; the plastic, 3500 MPa. What fraction of a 700-N tensile load is carried by the glass?

Solution

Based on Equation (3.56):

$$(F_{gl}/0.4A)/(F_p/0.6A) = (70{,}000 \text{ MPa}/3500 \text{ MPa}) = 20$$

$$F_{gl} = 20(0.4/0.6)\,F_p = 13.3\,F_p$$

$$F_{gl}/(F_{gl} + F_p) = (13.3\,F_p)/(13.3\,F_p + F_p) = 93\%$$

Example 3.43

An electric highline cable contains one cold-drawn steel wire and six annealed aluminum wires, all with a 2-mm diameter. (The steel provides the required strength; the aluminum, the conductivity.) Using the following data, calculate (i) the resistivity and (ii) the elastic modulus of the composite wire.

$$\text{Steel: } \rho = 17 \times 10^{-6}\ \Omega \bullet \text{cm},\ E = 205{,}000 \text{ MPa}$$

$$\text{Aluminum: } \rho = 3 \times 10^{-6}\ \Omega \bullet \text{cm},\ E = 70{,}000 \text{ MPa}$$

Solution

(i) From Equation (3.61):

$$1/\rho_{\parallel} = (1/7)/(17 \times 10^{-6}\ \Omega \bullet \text{cm}) + (6/7)/(3 \times 10^{-6}\ \Omega \bullet \text{cm})$$
$$= 0.294 \times 10^{6}\ \Omega^{-1} \bullet \text{cm}^{-1}$$

$$\rho_{\parallel} = 3.4 \times 10^{-6}\ \Omega \bullet \text{cm}$$

(ii) We must write a mixture rule for the elastic modulus of a composite in parallel. Let A be the area of one wire.

Since $\varepsilon_c = \varepsilon_{Al} = \varepsilon_{St}$, $[(F/A)/E]_C = [(F/A)/E]_{Al} = [(F/A)/E]_{St}$

Also, $F_C = F_{Al} + F_{St} = F_C(f_{Al}A_C/A_C)(E_{St}/E_C) + F_C(f_{St}A_C/A_C)(E_{St}/E_C)$

Canceling,

$$E_C = f_{Al}E_{Al} + f_{St}E_{St} = (6/7)(70{,}000 \text{ MPa}) + (1/7)(205{,}000 \text{ MPa}) = 89{,}000 \text{ MPa}$$

The apparent modulus will be lower because there will also be cable extension by the straightening of the cable wire.

SELECTED SYMBOLS AND ABBREVIATIONS

Symbol or Abbreviation	Definition
A	area
a	unit cell length
a	crack length
amu	atomic mass unit
D	diffusion coefficient
D_o	proportionality constant
E	elastic modulus
ε	engineering strain
F	force
f	volume fraction
K_I	stress intensity factor
P	density
Q	activation energy
R	gas constant
r	atomic radius
σ	stress
σ_y	yield strength
SRI	Spalling Resistance Index
Tg	glass-transition temperature
Tm	melting temperature
V^0	standard electrode potential
W_L	weight fraction of liquid phase
W_S	weight fraction of solid phase
y	proportionality constant

ELECTRICAL QUANTITIES AND DEFINITIONS

Electrostatics

The subject of electric fields is sometimes omitted in a formal course in electrical engineering (the knowledge is frequently assumed from a prerequisite physics course). Thus, one needs to review electric fields and flux densities attributable to electrical charges. These fields, forces, and flux densities require three-dimensional vector notation. However, most problems will reduce in complexity to two dimensions; this reduction will permit quick graphical solutions.

Whereas the forces exerted between charges depend on whether the charges are in motion, one may start with stationary electric charges that produce electric fields. These fields may be defined in terms of the forces they produce on one another.

The smallest amount of charge that can exist is the charge of one electron, which is 1.602×10^{-19} coulombs (C). One coulomb of charge is thus equivalent to 6.24×10^{18} electrons. This is the amount of charge that is necessary to develop a force of one newton in an electric field of one volt per meter. An electric field, **E** (boldface for vector quantity), is the amount of force (F) that would be exerted on a *positive charge* (assuming it is concentrated at a point) *of one coulomb* if it were placed in that field:

$$\mathbf{F} = \mathrm{Q}\mathbf{E}$$

The electric field is not thought of as a point but rather as being distributed throughout a small region. This is a vector quantity; the direction of force on the one coulomb would be toward the point source of the field if that point were a negative charge. The units of measurement are newtons per unit of charge; alternately, it may be given in terms of volts/meter since

$$\frac{\text{force}}{\text{charge}} = \frac{\text{force} \times \text{distance}}{\text{charge} \times \text{distance}} = \frac{\text{energy}}{\text{charge} \times \text{distance}} = \frac{\text{voltage}}{\text{distance}}$$

For point charges, the force is directly proportional to the product of the two charges and inversely proportional to the square of the distance between them (similar to the laws of gravity).

$$\mathbf{F}_2 = \frac{Q_1 Q_2}{4\pi\varepsilon r^2}\mathbf{a}_{r12}, \quad \text{where} \tag{3.64a}$$

$\mathbf{F}_2$ = the force on charge 2 due to charge 1

Q_i = the ith point charge

r = the distance between charges 1 and 2

$\mathbf{a}_{r12}$ = a unit vector directed from 1 to 2

ε = the permittivity (or dielectric constant) of the medium

The constant of proportionality depends on the medium between the two charges. This constant is $1/(4\pi\varepsilon)$ or approximately 9×10^9 for free space. If the medium is not free space, the 9×10^9 is merely divided by the relative permittivity to give the simplified *in line* form as

$$F = (9 \times 10^9/\varepsilon_{rel})Q_1Q_2/r^2 \text{ newtons} \tag{3.64b}$$

$$\varepsilon = \varepsilon_{rel}\varepsilon_0 \text{ F/m} \tag{3.65}$$

and ε_0 = permittivity of free space = 8.85×10^{-12} farads/meter. Permittivity of air is approximately the same as that of free space or a vacuum (that is, $\varepsilon_{rel} = 1.0006$), but if the medium happens to be caster oil, ε_{rel} is almost 5.

A convenient way of thinking of these fields is to imagine flux lines radiating either away from or toward a point source, as shown in Figure 3.40. If one imagines a positive point source charge at the center of a sphere and arrows pointed away from the center, these arrows would be the flux lines (the bigger the charge, the more arrows).

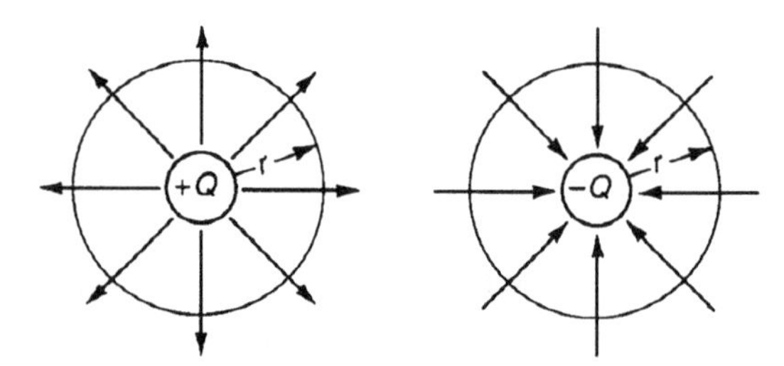

Figure 3.40 Electric flux lines around a charge

The flux density on the surface of the sphere (whose center is located at the point charge) would then be the number of arrows through a unit area on the sphere.

$$D = Q/A \tag{3.66a}$$

$$Q = \mathbf{D} \bullet \mathbf{A} \tag{3.66b}$$

where A is the area of the sphere ($4\pi r^2$) and Q is the quantity of charge in coulombs (assuming the area is normal to the flux). Equation (3.66b) is presented for those familiar with the dot and cross product vector notation (this notation guarantees the portion of the surface being considered is normal to the flux). If one were to divide the area of the sphere into very small areas, dA's, the sum of areas times the amount of flux through each area would be the amount of charge at the center of the sphere. More formally, as the size of each area approaches zero, and as one integrates over the entire area of the sphere, the total charge enclosed is Q; this is known as Gauss's Law, Equations (3.67a) and (3.67b). But, by using the dot-product notation, one is not limited to a spherical shape with the charge at the center; the formal law is then given as Equation (3.67b) or (3.67c):

$$Q = \sum D\ dA \text{ (for entire surface)} \tag{3.67a}$$

$$Q = \oint_s \mathrm{D} \bullet dA \tag{3.67b}$$

$$Q/\varepsilon = \oint_s \mathrm{E} \bullet dA \tag{3.67c}$$

The field strength, **E**, at the sphere's surface is then proportional to D,

$$\mathbf{E} = (D/\varepsilon)\mathbf{a}_r \tag{3.68}$$

where $\mathbf{a}_r$ is a unit radial vector direction. If there are a number of charges throughout a region, it is usually easier to use the flux density concept in solving problems. In this chapter, unit vectors will be denoted as **a** so that e can be reserved for other purposes.

Example 3.44

Assume three point charges, A, B, and C, as shown in Exhibit 32. Points A and B are 2 meters apart; Point C is on a perpendicular bisector between A and B and is 1 meter lower. Point A has 4×10^{-6} coulombs of negative charge, Point B has 10×10^{-6} coulombs of positive charge, and Point C has no charge yet. Determine the force and the field on B (in this case, due only to the charge on A) and the flux density at Point C.

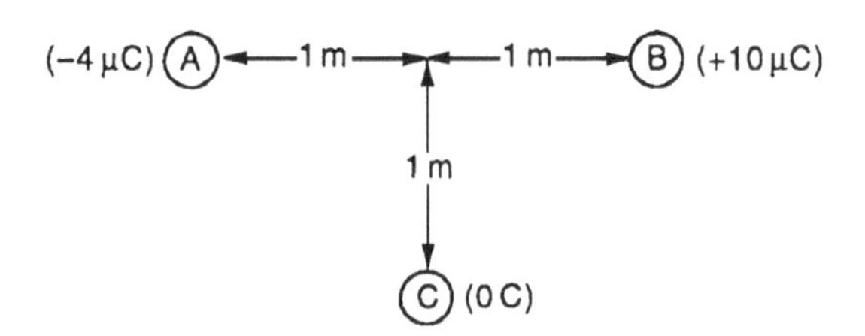

Exhibit 32 Point charge locations

Solution

The force on B due to A is directly proportional to the charges on A and B and inversely proportional to the square of the distance of separation. Using Equation (3.64),

$$\mathbf{F} = (9 \times 10^9)(-4 \times 10^{-6})(10 \times 10^{-6})/(2^2) = -9 \times 10^{-2} \text{ newtons}$$

with the direction of the force toward each other.

$$\mathbf{E} \text{ (at B)} = \mathbf{F}/Q_B = 9 \times 10^{-2}/10^{-5} = 9 \times 10^3 \text{ V/m}$$

(If the medium happened to be a special oil with relative permittivity 5.0, both **F** and **E** would only be one-fifth as large.) A more orderly solution (especially if several charges are involved—or none at C in this case) is to use the flux density relationship to find the individual *D*s, then convert to *E*s. All that is now necessary to find the net **E** is to use vector summation of the flux densities and divide by ε. To find the flux density at C attributable to charge A, imagine a sphere passing through C with its center at A; repeat for B.

$$D_{CA}\text{ (at C attributable to A)} = Q_A/(4\pi r^2)$$

$$\mathbf{E}_{CA} = (D/\varepsilon)\mathbf{a}_{rCA} = (Q_A/r^2)(1/4\pi\varepsilon)\mathbf{a}_{rCA}\text{ V/m}$$

Recall that $1/(4\pi\varepsilon)$ for free space is 9×10^9, then

$$\mathbf{E}_{CA} = \left[(4\times10^{-6})/\sqrt{2}^2\right](9\times10^9)\mathrm{a}_{,CA} = 1.8\times10^4\,\mathbf{a}_{,rCA}\text{ V/m}$$

$$\mathbf{E}_{CB} = \left[(10\times10^{-6})/\sqrt{2}^2\right](9\times10^9)\mathrm{a}_{,CA} = 4.5\times10^4\,\mathbf{a}_{,rCA}\text{ V/m}$$

To find the E_C net one may use the more formal procedure of finding the rectangular components of each of the field vectors since only two dimensions are involved. Then, summing the horizontal and vertical components, the net field vector is the square root of the sum of the squares. If one assumes the reference vector, $\mathbf{a}_r$, has the horizontal component a and the vertical component jb (the j implies the 90° or vertical axis), then

$$\mathbf{E}_{CA} = |\mathbf{E}_{CA}|(\cos\theta + j\sin\theta) = 1.8\times10^4(\cos 135° + j\sin 135°)$$
$$= (-1.27 + j1.27)\times10^4\text{ V/m}$$
$$\mathbf{E}_{CB} = 4.5\times10^4(\cos 225° + j\sin 225°) = (-3.18 - j3.18)\times10^4\text{ V/m}$$
$$\mathbf{E}_{Cnet} = (-4.45 - j1.91)\times10^4 = 4.84\times10^4\angle 203.2°\text{ V/m}$$

where the angle θ is the angle between the vector **E** and a horizontal reference line. However, solving for the vector field at C may be done faster by using a graphical solution, assuming reasonable accuracy. The points A, B, and C are set up (to their own scale); then, working with Point C, one places the vectors (using any suitable scale) that represent the two other fields. The reference direction and magnitude is then pictured as shown in Exhibit 33.

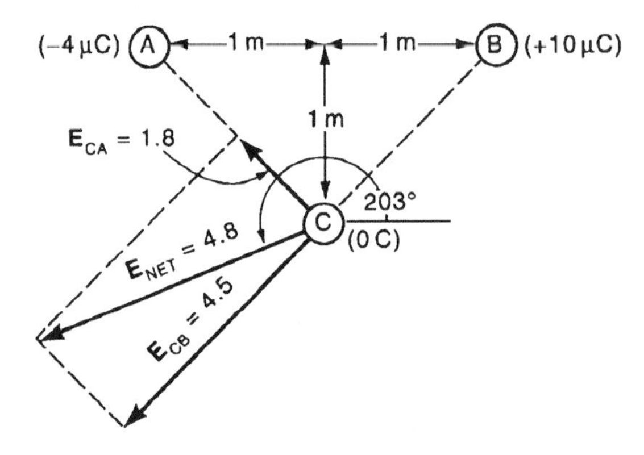

Exhibit 33 Sample problem vector fields (×104) at C

Now, assume the charge at C is $+5 \times 10^{-6}$ coulombs; find the net vector field at Point B. To find this field at Point B, merely remove the charge at B.

$$\mathbf{E}_{BA} = (Q_A/r^2)(9 \times 10^9)\mathbf{a}_{rBA} = 0.90 \times 10^4\ \mathbf{a}_{rBA}\ \text{V/m}$$
$$\mathbf{E}_{BC} = (Q_C/r^2)(9 \times 10^9)\mathbf{a}_{rBC} = 2.25 \times 10^4\ \mathbf{a}_{rBC}\ \text{V/m}$$

Again, solve graphically for the solution (Exhibit 34).

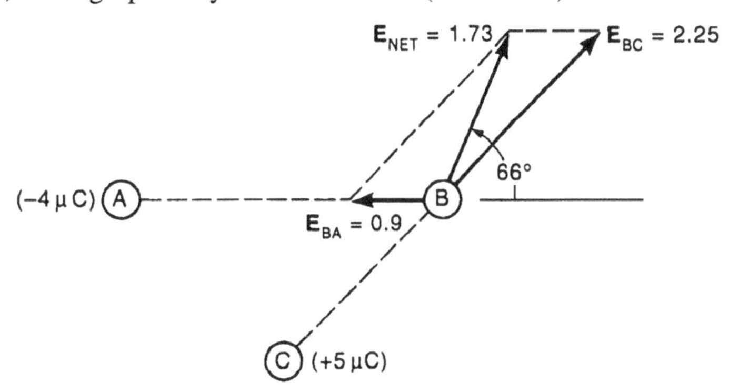

Exhibit 34 The electric field at B (×104)

The actual solution, using vector notation, is found to be **E** (at B) = 1.73×10^4 $\angle 66.5°$ volts/meter.

A more complex problem involving the x, y, and z axes is no more complicated except for the bookkeeping difficulty of solving a three-dimensional problem.

Before leaving the subject of static electric fields, the subject of energy storage in a capacitor (a pair of parallel plates separated by a distance, d) will be introduced. Capacitance is the ratio of the total stored charge uniformly distributed on the plates to the voltage difference between the plates. The separation distance, d, between plates will be considered to be small compared to the plate area (so that fringing may be neglected), then the field and flux lines will be perpendicular to the plates. The capacitance is

$$C = q_c(t)/v_c(t) \text{ farads (coulombs/volt)} \qquad \textbf{(3.69a)}$$

and for non-time-varying quantities

$$C = Q/(Ed) = AD/(Ed) = A(E\varepsilon)/(Ed) = A\varepsilon/d \qquad \textbf{(3.69b)}$$

Permittivity may now be defined in a slightly different manner, where ε equals the charge induced on one square meter of the capacitor plates by an electric field intensity of one volt/meter. Thus, for a free space separation of the plates, 8.85×10^{-12} coulombs is induced on one square meter of a conducting plate by an electric field having an intensity of one volt per meter.

Magnetic Quantities and Definitions

Magnetic effects are related to the motion of charges, or currents. From the previous section on electrostatics, a force of one newton is produced by a charge of one coulomb in an electric field intensity of one volt per meter. Electric current, on the other hand, can be thought of as moving charges. **Current** is the time rate of change of the electric charge passing through a surface area. This definition is expressed as

$$i = dq/dt \qquad \textbf{(3.70)}$$

The unit of current is the **ampere**, A, which equals one coulomb per second.

From the concept of moving charges, one can begin to understand magnetic fields. For permanent magnets (due to *static* magnetic fields) one recalls from physics that for some materials the molecular structure has the electron orbits of the atoms aligned. These tiny moving charges of electrons produce tiny currents. This alignment results in magnetic fields; actually ferromagnetic materials can be thought of as a large number of magnetic domains, with the domains being mostly aligned. However, when these (magnetic) domains are in disarray or randomly aligned, the material is unmagnetized.

Oersted, in 1819, observed that a magnetic flux existed about a wire carrying an electric current. (Flux lines can almost be visualized by observing the pattern of sprinkled iron filings on a piece of paper held over one pole of a magnet.) A few years after Oersted observed the effect of magnetic flux, Ampere found that wire coils carrying a current acted in the same manner as magnets. Simply stated, a coil of several turns of wire produced a stronger magnetic flux than only one turn of wire for the same current. And, if there were a ferromagnetic material to carry (or to provide a path for) the magnetic flux, the flux strength would be much greater.

Consider a toroidal ferromagnetic ring wrapped with a coil of several turns of wire, with the coil connected to a variable current source. Assume the current is zero and the material is not magnetized to begin with, then with an increase of current, an increase in flux results. One tends to think that the relationship would be linear; but the relationship is actually rather complex for a ferromagnetic material. As the current is further increased, the flux tends to level off. The leveling off is caused by most of the magnetic domains in the material aligning themselves in the same direction; consequently, increasing the current beyond the *knee of the bend* does not significantly increase the flux. If the current is decreased, most (but not all) of the magnetic domains return to random directions for zero current. The *going up* path is not necessarily the same as the *going down* path; these paths are known as the hysteresis curve for a particular material. When the current reaches zero on the *coming down* side, the material is left partially magnetized; this is called the residual magnetism. Again, assuming the material is unmagnetized to begin with, the flux is caused by the current through the number of turns of wire. This product of the current and the number of turns of wire, or amp-turns ($A \bullet t$), is frequently called the magneto-motive force (mmf) and indicated with a script F. The mmf is the source that causes the magnetic flux (ϕ). A plot of these two quantities is given in Figure 3.41(b); this plot is referred to as the hysteresis curve.

If the width of the hysteresis curve were small (that is, the residual magnetic flux is small), only the first quadrant of the hysteresis curve is needed and can be drawn as a single line; then, only one curve per kind of magnetic material is needed.

Before a numerical problem can be solved, one needs to quantify and further define several terms. Rather than using flux, ϕ (units of webers), it is more appropriate to use flux density, B (units of tesla or webers/unit area). Flux density is the amount of flux passing through a unit area (normal to the flux). Also, rather than using the straight magneto-motive force designation, a more useful one is mmf/length, called the magnetic field intensity, H. These standardized quantities will allow a plot of B vs. H; this plot is always given for solid magnetic materials (that is, before any air gap might be cut into them).

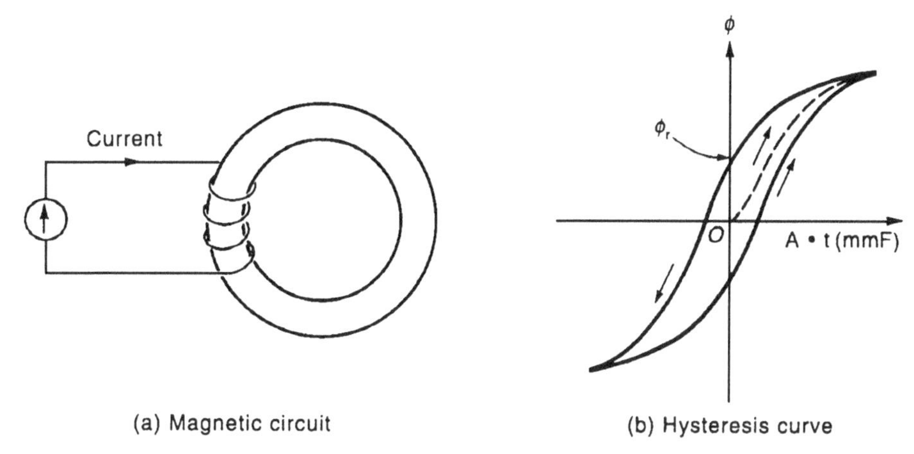

Figure 3.41 Magnetic characteristics

One needs to define further the direction of magnetic flux and fields; this is difficult to do without the concept of magnetic poles. To show this effect, consider an air gap cut into the ferromagnetic material. However, analogous to electrical charges being positive or negative, one can think of magnetic poles (they always occur in pairs) as north and south (recall, like poles repel and unlike ones attract). As an example, consider a freely suspended magnet (say, a magnetized needle in a compass) in the earth's magnetic field; one end of the magnet points toward the geographical north. By common usage, this end of the magnet is referred to as the *north-seeking pole*. For the toroidal ring with the air gap cut in it, if the positive direction of the current enters the top end of the coil, the direction of the flux in the ring will be clockwise, and the top of the air gap is the north pole, whereas the bottom is the south pole.

In the previous example, the magnetic flux in the air gap is concentrated and may be very high. The lines of magnetic flux produced in the material and the air gap are the same and continuous when there is a current in the wire. Without the ferromagnetic material in the path, the relationship between the current (causing the magnetic flux) and the resulting magnetic field would be linear but very weak. The linear constant of proportionality is called the permeability of free space, μ_0 (as in the air gap). This constant is

$$\mu_0 = 4\pi \times 10^{-7} \text{ F/m} \tag{3.71a}$$

For ferromagnetic materials, the relative permeability is nonlinear; for approximate calculations, it is sometimes linearized in the region of the curve before magnetic saturation is reached. The slope of the curve in the saturation region approaches that of free space or air, thus

$$\mu_r \mu_0 \rightarrow \mu_0 \tag{3.71b}$$

The relationship between B and H may then be expressed as the slope of a B vs. H curve:

$$\mu = \mu_r \mu_0 = B/H \tag{3.71c}$$

when B and H are normal to each other.

As stated previously, flux lines are continuous (that is, the lines of flux in the ring and air gap are the same), whereas the flux density in the ring may or may not be the same as that in the air gap—frequently they are considered the same by neglecting the fringing effect in the gap. Also, the cause of the flux (that is, the amp-turns) is thought of as being distributed along the whole ring, thus it is appropriate to use field intensity, H (amp-turns/meter or mmf/m). The length for this example is merely the mean circumference of the ring (the width of the air gap usually being negligible). Kirchhoff's voltage law in electrical series circuits states that the net voltage drop in a loop equals zero; one may use the same analogy for series magnetic circuits. That is, the net mmf in a magnetic series loop equals zero. Stated another way, the mmf source must equal the sum of the mmf drops (or losses) in the series circuit. Thus, summing these mmf drops in the iron and the air gap is all that is needed to find the mmf for the source.

Although the previous example involved a coil of wire, a more fundamental problem is one involving a long straight wire.

For a wire on the z-axis carrying a current,

$$\boldsymbol{H} = \frac{\boldsymbol{B}}{\mu} = \frac{\boldsymbol{I}\mathbf{a}_{\theta}}{2\pi r} \quad \textbf{(3.72a)}$$

where

$\boldsymbol{H}$ = the magnetic field strength (amps/meter)
$\boldsymbol{B}$ = the magnetic flux density (tesla)
$\mathbf{a}_{\theta}$ = the unit vector in positive θ direction in cylindrical coordinates
$\boldsymbol{I}$ = the current
μ = the permeability of the medium (for air, $\mu = \mu_0 = 4\pi \times 10^{-7}$ H/m)

The flux density at some radial distance, r, from the center line of the wire is given in Equation (3.72b).

$$B = (\mu i)(2\pi r) \quad \textbf{(3.72b)}$$

For the direction of the current shown in Figure 3.42, the flux density direction is shown with the head of an arrow as $(\odot)$ and the tail of an arrow as $(\otimes)$; this notation will be the same for current.

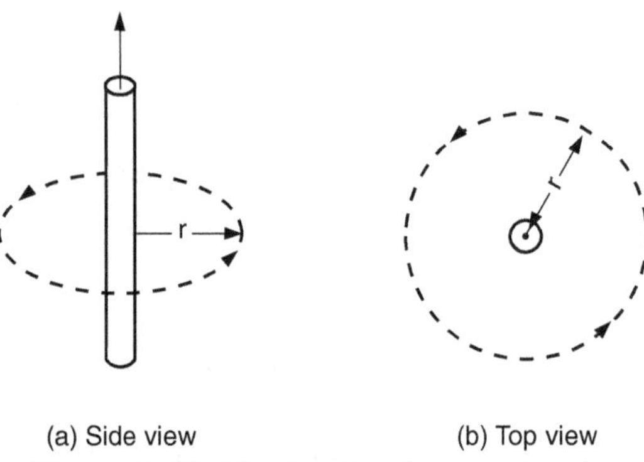

(a) Side view (b) Top view
Figure 3.42 The flux density around a wire

The direction of the flux may be remembered by using the right-hand rule: place the thumb in the direction of the current; the partially closed fingers will point in the direction of the magnetic flux. Now, if the long wire were formed into a circular coil of radius r, one could use calculus to consider a differential length of the wire and integrate around the closed loop to find the flux density within the loop. If there were several turns, N, for the loop, the equation for the flux density would be

$$B = (\mu N i)/r \quad \textbf{(3.73)}$$

The right-hand rule may also be used to find the direction that a current would flow if a voltage is induced (or generated) in a wire by the relative motion of a magnetic field and a wire (see Figure 3.43). Consider a straight wire being moved within a magnetic field (assume the wire and motion is normal to the magnetic field). A voltage induced in the wire will be proportional to the strength of the field, the length (l) of wire within the field, and the velocity, v, of the wire.

$$e \text{ (induced voltage)} = Blv \quad \textbf{(3.74a)}$$

For a coil of N turns enclosing flux ϕ

$$e = -N\,d\phi/dt \quad \textbf{(3.74b)}$$

where ϕ = the flux (webers) enclosed by the N conductor turns and

$$\phi = \int_A B \cdot dA$$

The direction of the flux is from the north pole face to the south pole face. It will help to think of the wire within the field as being a voltage generator (the induced voltage); then if there is a closed path (perhaps through an external resistive load) so that a current could flow, the polarity of the *generator* must be plus where the current (thumb direction) leaves the magnetic field and minus where it enters the field.

In this section, vector notation involving three-dimensional space has been kept to a minimum; the FE examination will probably not include this added complication for magnetic fields. Should such a problem occur, one could still attempt to tailor the problem using vector components. As an example, assume one is asked to find an induced voltage for the motion of a wire in a magnetic field as depicted in Figure 3.43 using Equation (3.74a). However, assume also that the wire velocity direction is not normal to the magnetic field but at an angle θ from the horizontal. The solution and the equation would be the same, except that one must use the vertical component of the velocity rather than the magnitude of the entire velocity.

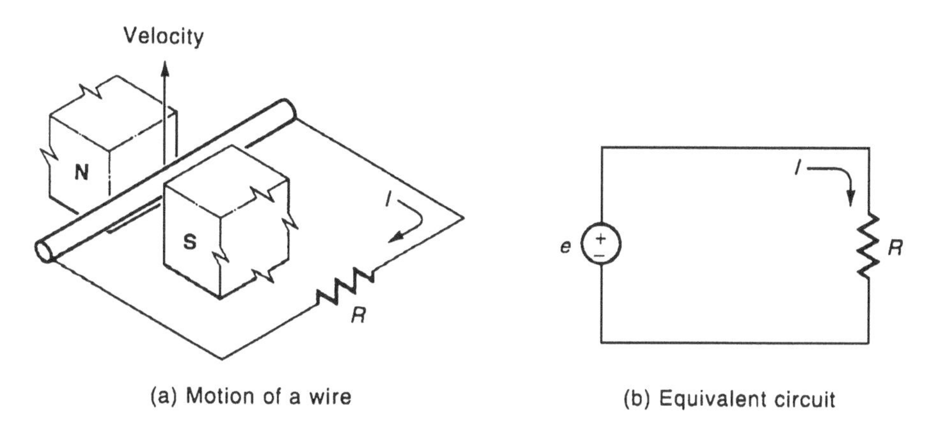

Figure 3.43 Induced voltage

Power, Energy, and Measurements

Before the discussion of power, energy, and measurements is undertaken, three items need to be reviewed. The first is notation, the second is the calculation of resistance, and the third is a more complete definition of voltage.

Notation

In this section, the notation for time-dependent quantities will normally be given by lowercase, italic letters (for example, for voltage, current, and such, $v(t)$, $i(t)$, where the (t) may or may not be included). A constant value, such as a battery voltage, will be given in uppercase, italic letters (for example, V and I); the effective or rms (root mean square) values and other quantities that have a nonvarying value (such as the average power) will also be expressed as an uppercase letter. Where confusion is possible, a subscript is normally used. Further discussion of notation will be presented where appropriate.

Resistance

Most reviews assume that one is familiar with simple wire resistance. Although the actual resistance of a wire conductor is usually considered to be negligible in circuit analysis, it is easily calculated. The parameters are shown in Figure 3.44.

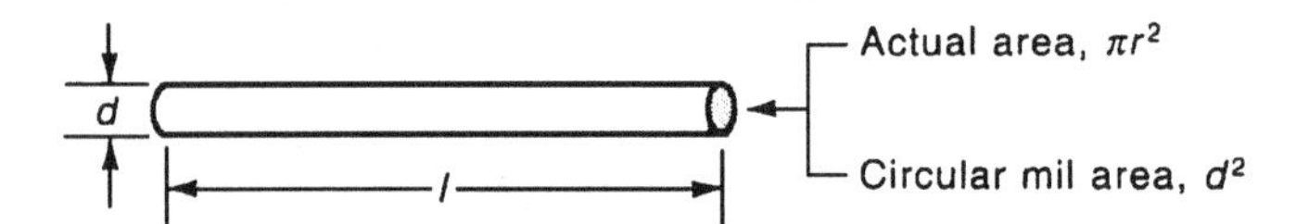

Figure 3.44 Wire resistance parameters

The resistance is proportional to its length and inversely proportional to its cross-sectional area; the constant of proportionality is the resistivity, ρ:

$$R = \frac{\rho L}{A} \qquad \textbf{(3.75)}$$

The resistivity of a particular conductor is normally given for a standard temperature of 20°C. However, the units depend on the length and area. For the MKS system of units, the value of ρ for copper is $1.7 \times 10^{-8}\ \Omega \bullet \text{m}$ (for aluminum, it is almost twice this value).

Example 3.45

As an example, for a 5-meter length of 12-gauge (approximate diameter of 2 mm) copper wire at room temperature (near 20°C) find the resistance, R.

Solution

$$R = (1.7 \times 10^{-8})(5/[\pi(0.001)^2]) = 0.027\ \Omega$$

On the other hand, for another temperature, the resistivity is modified by the temperature coefficient, α, to be

$$\rho = \rho_0[1 + \alpha(\mathrm{T} - \mathrm{T}_0)] \qquad \textbf{(3.76a)}$$

or

$$R = R_0[1 + \alpha(\mathrm{T} - \mathrm{T}_0)] \qquad \textbf{(3.76b)}$$

where $(\mathrm{T} - \mathrm{T}_0)$ is the change of temperature from 20°C.

Generally, these two equations will solve most resistive type problems. However, here is one note of caution: unfortunately, the inch, foot, and circular mil units for area are still in use. These units can be confusing since the area of the wire may be given in circular mils; this area omits π in the true area, πr^2, computation and, instead, includes it in the resistivity constant. In this case, the resistivity for copper wire is 10.4 Ω circ-mils/foot. Actually, the computation is made easier; assume a length of 5 feet for a wire diameter of 0.03 inches (or 30 mils):

$$R = (10.4)(5)/(30)^2 = 0.0578\ \Omega$$

The key to this type of problem is to check carefully by dimensional analysis to determine what units are being used. Wire tables, listing these parameters, should be furnished with the EIT examination when they are needed.

Voltage

Whereas a simple definition of **voltage** is the potential difference that will cause a current of one ampere to flow through a resistance of one ohm, a more formal definition is *a charge of one coulomb receives or delivers an energy of one joule moving through a voltage of one volt.* The instantaneous voltage is defined by Equation (3.77).

$$v = dw/dq \tag{3.77}$$

In other words, if a unit quantity of electricity (coulomb) gives up energy equal to one joule as it proceeds from one point to another, the difference in potential is one volt. The joule is sometimes called the coulomb-volt. Likewise, the energy acquired by an electron when it is raised through a difference of potential of one volt is called an electron-volt. The definition of voltage leads directly into the subject of power and energy.

Power and Energy

Power (instantaneous, i.e., lowercase italic notation) is the rate of change of energy, dw/dt, or

$$p(t) = dw/dt = (dw/dq)(dq/dt) = v(t)(i) \tag{3.78}$$

The total energy is then given in Equation (3.79). (The function of time, t, is implied and will be dropped for simplicity.)

$$w = \int_0^T p\,dt = \int_0^T vi\,dt \tag{3.79}$$

Energy Storage

Energy may be stored in some electrical circuit elements. For electrical circuits, there are three basic elements: resistance, inductance, and capacitance. For the resistor, electrical energy cannot be stored since it is turned into heat; however, both the inductor and capacitor are capable of energy storage (see Figure 3.45).

The inductor, L (measured in units of henries, H), usually a coil of wire that can produce a magnetic field, stores energy in the magnetic field; the capacitor stores energy within its electrical field (i.e., within its dielectric medium). The circuit symbols and voltage-current relationships are given in Figure 3.46. Equations (3.80) and (3.81) give the energy stored as

$$v_L = L(di_L/dt) \tag{3.80a}$$

$$i_C = C(dv_C/dt) \tag{3.80b}$$

$$w_L = (1/2)Li_L^2 \tag{3.81a}$$

$$w_C = (1/2)Cv_C^2 \tag{3.81b}$$

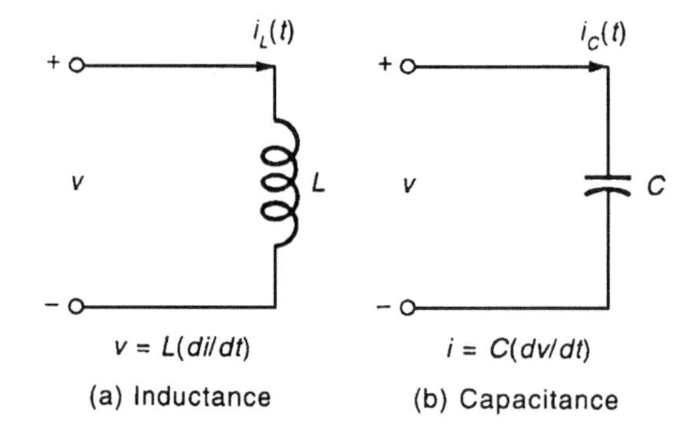

Figure 3.45 Energy storage elements

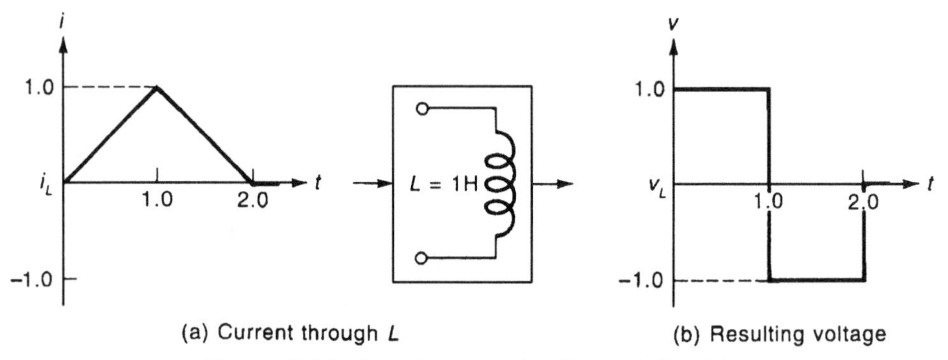

Figure 3.46 Inductor current-voltage relationship

The voltage across an inductor is a function of the rate of change of current. If there is no rate of change (or direct current), then the voltage must be zero, and the inductor acts like a short circuit or zero resistance. Actually, there is always some wire resistance, but usually this is negligible. For current to flow in a capacitor, there must be a rate of change of voltage. If there is no rate of change, there is no current, and the capacitor acts like an open circuit (this will be discussed in more detail later). As an example, assume the current (from a variable current source) through an inductor is as given in Figure 3.46(a); then the voltage across the inductor must be as shown in Figure 3.46(b).

As shown in Figure 3.47, the power *taken* (absorbed or stored) by the inductor is a function of time until $t = 1$ second and is positive; for the current with a negative slope, power is *given up* (or returned to the circuit source) from $t = 1$ second until $t = 2$ seconds. On the other hand, the energy stored (also a function of time) follows the square law given directly in Equation (3.81a). The resulting curves are found by splitting Equation (3.79):

$$w = \int_0^2 p\,dt = \int_0^1 vi\,dt + \int_1^2 vi\,dt$$

Average Voltage and Current

All DC quantities, being constant values, are represented by uppercase letters. However, for changing quantities one must be careful with notation. The formal definition involves the time period of interest, T. The average voltage (or current) is given by

$$V_{avg} = (1/T)\int_0^T v\,dt \tag{3.82}$$

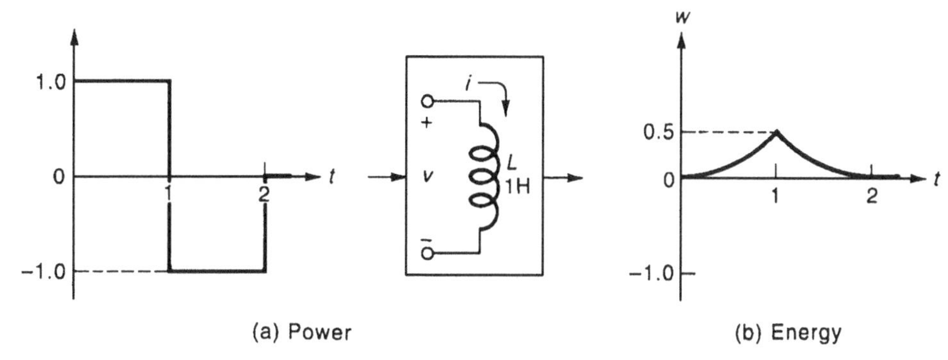

Figure 3.47 Power and energy plot for the current/voltage curves of Figure 3.46

or the area under the voltage curve divided by the period of time. Of course, the same holds for current.

If the positive and negative areas balance over a specific period of time, then the net area is zero. It should then be obvious that the average voltage from a sine wave generator over a full period, T (or 2π), is zero:

$$V_{avg} = (1/T)\int_0^T V_{max} \sin \omega t \, dt = 0$$

Over a *half period*, $(1/2)T$, the average voltage is

$$V_{avg} = (2/T)\int_0^{T/2} V_{max} \sin \omega t \, dt = 2V_{max}/\pi$$

(*Caution*: a half wave over a full period has an average value of V_{max}/π.)

Average Power

Since power is defined as the product of voltage and current, then for simple DC voltage or current sources, the average power is simply the product of the current and voltage. This is written as uppercase, italic P (without subscript) as $P = VI$. (Generally, use this equation only for DC quantities, as ac or other types of wave forms may have a phase shift. This will be discussed later.)

Effective Values

For most wave forms, other than straight DC, one must define an effective or rms value of current and voltage since an average value of current or voltage could be zero for a sinusoidal signal. Another way of stating this is the following: "What equivalent value of current will cause the same heating power in a resistor as would a DC current?" The effective (or rms) value of current or voltage is defined as

$$I_{eff} = I_{rms} = I = \sqrt{(1/T)\int_0^T i^2 \, dt} \tag{3.83}$$

If $i = I_{max} \sin \omega t$ or $i = I_{max} \cos \omega t$, then

$$I = 0.707I_{max} = I_{max}/\sqrt{2} \tag{3.84a}$$

$$V = 0.707V_{max} = V_{max}/\sqrt{2} \tag{3.84b}$$

Although Equation (3.84) and Figure 3.48 are for sinusoids, the effective or rms values may be found for any periodic wave form.

For sinusoids, the average power may involve phase angles (more on this later); however, one is safe in finding the power developed as heat in a resistor by using

$$P_{avg} = P = I^2 R = \left(\mathrm{I_m} / \sqrt{2}\right)^2 \ R \text{ watts} \qquad \textbf{(3.85)}$$

For a short example, assume a known sinusoidal current of i = 14.1 cos ωt flows through a resistor of 25 ohms. The effective, or rms, current is easily computed from Equation (3.83) or (3.84) and the power by Equation (3.85):

$$P = \left(14.1/\sqrt{2}\right)^2 (25) = 2500 \text{ watts}$$

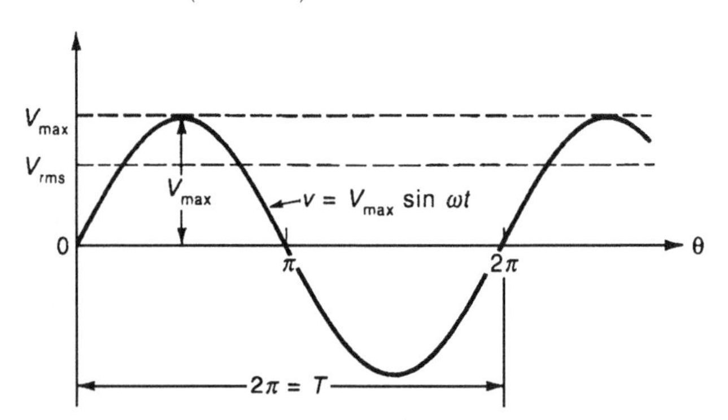

Figure 3.48 Effective or rms value for a sinusoid

If the sinusoid is replaced with another kind of periodic wave form, Equation (3.83) would, of course, still be used. However, sometimes a table of *standard wave forms* with a list of effective or rms values is available.

DC CIRCUITS AND RESISTANCE

Ohm's and Kirchhoff's Laws for a Single Source Network

The solutions of all circuits problems whether they are dc, ac, single source, or multiple source, involve the use of these laws. The first is Kirchhoff's voltage law (KVL):

For any closed loop in a circuit, the voltage algebraically sums to zero. $\Sigma v = 0$	or	The voltage rises equal the voltage drops in a closed loop. $\Sigma V_{rises} = \Sigma V_{drops}$

Another way of stating this relationship is to say the sum of the voltage rises is equal to the sum of the voltage drops in any loop. Both of these statements are true of both dc and instantaneous values. However, one needs to be careful when dealing with ac analysis with possibly different phase angles; this relationship will be discussed later.

The second is Kirchhoff's current law (KCL); this law involves any junction, or node, in an electrical circuit:

For any node in an electrical circuit, the net current algebraically sums to zero. $\Sigma i = 0$	or	The total current entering a junction equals the total current leaving. $\Sigma I_{in} = \Sigma I_{out}$

Restated, the sum of the currents entering a junction must equal the sum of the currents leaving the junction. Again, this is true of both dc and instantaneous values; however, for ac circuit analysis care is required in its application.

Ohm's law for resistance ($R = v/i$) is well known and is the same for dc, instantaneous quantitities, and ac values. For a series circuit (where the current is the same in every element) the equivalent resistance merely becomes

$$R_T = R_1 + R_2 + \cdots + R_n = \sum R \quad \textbf{(3.86a)}$$

For parallel circuits (where the voltage across every element is the same) the equivalent resistance, R_T, is

$$1/R_T = 1/R_1 + 1/R_2 + \cdots + 1/R_n = \sum G \quad \textbf{(3.86b)}$$

The conductance G is simply the inverse of the resistance (whose units are siemens or mhos, $1/\Omega$). From Equation (3.86b), one can quickly determine that the equivalent resistance for two resistors (or impedances) in parallel is $R_T = R_1R_2/(R_1 + R_2)$.

Example **3.46**

A simple example follows; for this type of circuit problem, it is usually quicker to convert the parallel resistors to an equivalent value. The simpler series circuit is easily solved: by finding the voltage drop across any element, the voltage across the equivalent resistance is found. Assume the question is to find the power dissipated in the 6-Ω resistor in Exhibit 35.

Solution

The equivalent parallel resistance is found to be 3/2 Ω. This value yields a total series resistance of 4 Ω as shown in Exhibit 35b. The series current is easily found to be 3 A; the voltage drop across the parallel resistance is $V_{eq} = I(R_T) = (3)(3/2) = 9/2$ volts. The current through the 6-Ω resistor is $I = V_{eq}/R_6 = (9/2)/6 = 3/4$ A. The power dissipated in the 6-Ω resistor is $P_{6\Omega} = I^2R = (3/4)^2 6 = 27/8 = 3.375$ watts.

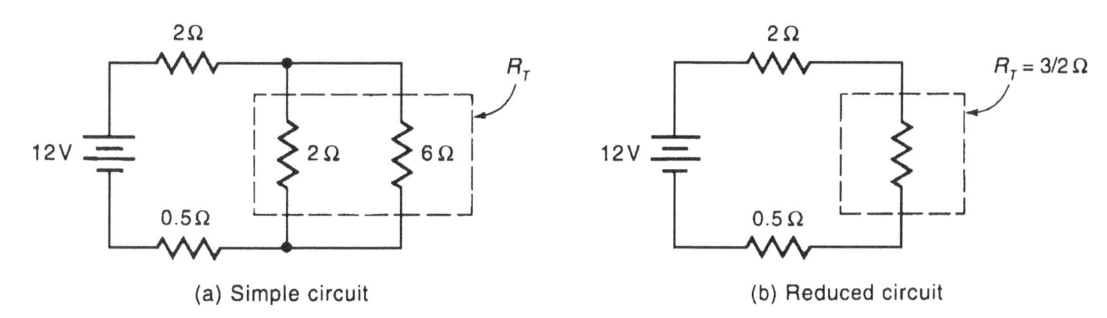

Exhibit 35 A simple voltage source circuit

Here, the power from the source equals the sum of the power dissipated in each of the resistors. It is instructive to compute the sum of the power lost; this is

$$P_{source} = V_{source} I = (12)(3) = 36 \text{ watts}$$

$$\sum P_{lost} = I^2\left(\sum R_{ser}\right) = 9(2 + 3/2 + 1/2) = 36 \text{ watts}$$

For this problem, a somewhat more formal method is to use the voltage dividing equation after R_T is found (all elements are now in series). The voltage across R_T may be calculated to be $V_T = (R_T / \Sigma R)V_{bat}$; then, the current through R_6 is determined as before.

Example 3.47

Consider another example with a current source: here, the series portion of the circuit should be reduced to one resistor; with all of the resistors now in parallel, the voltage across this group is easily found. Now the question is to find the amount of power dissipated in the 2-ohm resistor in Exhibit 36.

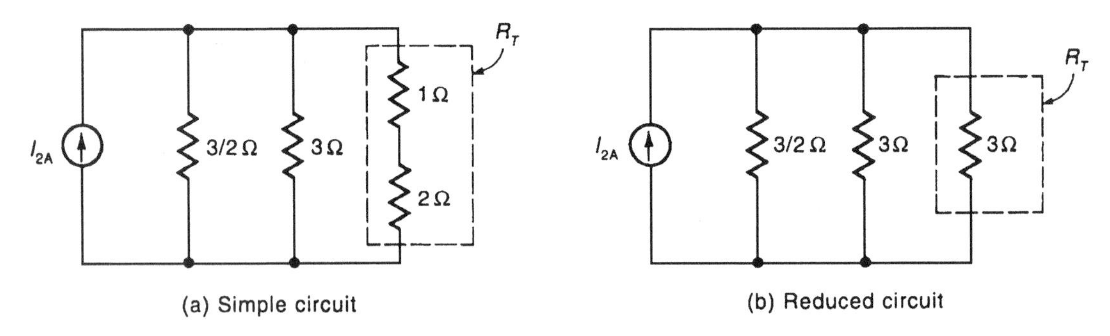

Exhibit 36 A simple single current source circuit

Solution

After finding the one series resistor, R_s, that is equivalent to the two resistances in series, all resistances are now in parallel. Then using Equation (3.86b), or merely taking two resistors at a time, one easily computes the total equivalent one resistance to be 3/4 Ω. The voltage across the parallel circuit is $RI = (3/4)(2) =$ 3/2 volts. The current through the series equivalent branch is $I_{se} = V_{par} / R_{se} = (3/2)/3$ = 1/2 amperes; the power dissipated in the 2-Ω resistor is $P = I^2R = (1/2)^2 2 = 0.5$ watts.

Again, a quick check is to find if the power taken from the source ($P_s = VI$ = (3/2)2 = 3 watts) and the power dissipated in the resistors are equal:

$$\sum P_{lost} = V^2(1/R_{3/2} + 1/R_3 + 1/R_{se}) = (3/2)^2(2/3 + 1/3 + 1/3) = 3 \text{ watts}$$

For this problem, as before, there is a somewhat more formal method of finding the current through R_{se} directly; $I_{se} = (G_{se}/\Sigma G)I_{tot}$. Knowing the branch current, the power is easily found.

Multiple Source Networks and Theorems

Although the term *multiple source circuits* implies multiple power output, this implication may be incorrect as the voltage or current source may actually absorb power. If the current is flowing into a voltage *source*—from plus to minus—the *source* is actually a *load* taking power (that is, a battery being charged). Therefore, it is very important to determine the direction of current (or voltage polarity as the case may be). The procedure is to assume current directions and carefully label these on a circuit diagram; then let the mathematics determine the actual direction. One way to do this is to assume a current through each element (of course series elements will have the same current) of a circuit. This is in contrast to assuming loop or mesh currents; however, both methods amount to the same thing. A short example (see Figure 3.49) will show the method. (To keep the circuit diagrams uniform and to later allow for time varying voltage sources, a circle is used to indicate the voltage source rather than showing a battery.)

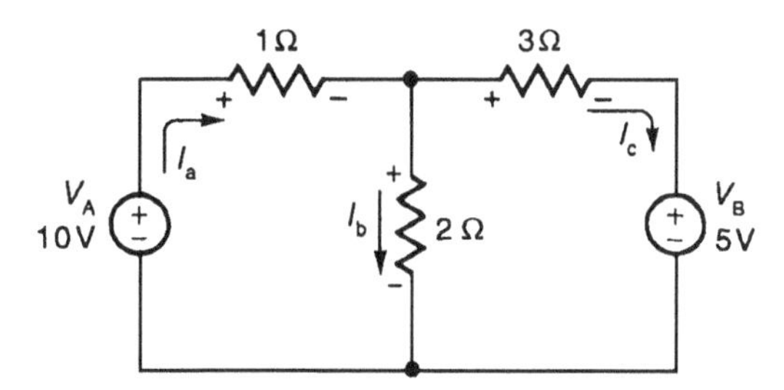

Figure 3.49 A two-loop network

After labeling the assumed current directions, place a plus sign where the current enters the resistor and a negative sign where it leaves. Then, by Kirchhoff's voltage law, sum the voltages around each loop:

$$V_A - R_1 I_a - R_2 I_b = 0 \qquad \textbf{(i)}$$
$$V_B + R_3 I_c - R_2 I_b = 0 \qquad \textbf{(ii)}$$

Since three variables are present, another equation is needed; Kirchhoff's current law will produce the third equation (actually J – 1 node junction equations are needed; here J = 2). Summing currents at one junction (say, the upper middle node) gives $I_a = I_b + I_c$.

Rewriting the equations for loops one and two (here, replacing all I_a), yields

$$V_A = R_1(I_b + I_c) + R_2 I_b, \qquad 10 = 3I_b + I_c \qquad \textbf{(i)}$$
$$V_B = R_2 I_b - R_3 I_c, \qquad 5 = 2I_b - 3I_c \qquad \textbf{(ii)}$$

Solving for the currents, the results are

$$I_a = 3.64 \text{ A}, \quad I_b = 3.18 \text{ A}, \quad I_c = 0.455 \text{ A}$$

Since all of the currents are positive, the assumed directions of the currents are correct; therefore, the voltage *source* B is really a load (perhaps a battery being charged). As a check, the sum of the voltages around the outside loop should equal zero (starting at the upper left-hand corner): Does 1(3.64) + 3(0.455) + 5 – 10 = 0? Or, 3.64 + 1.365 + 5 – 10 = 0.005 (acceptable).

Another way of analyzing the circuit is to use the *mesh* technique of analysis. Notice in the previous two-loop problem, the current through the middle resistance that was identified as I_b is nothing more than $I_a - I_c$. If we now consider going around the entire left loop as I_a and around the entire right loop as I_c, the currents mesh at the middle resistance. Because voltages are summed around each loop, the net voltage across the middle resistor (when going around the left loop) is $RI_a - RI_c$; the loop currents are in opposite directions. The same kind of equation is written for the right-hand loop. Thus one current variable (I_b) is eliminated. There are still other ways of analyzing the circuit, as will be pointed out later.

Example 3.48

Another example that involves mixed sources of current and voltages is shown in Exhibit 37. Find the voltage across the current source for this circuit configuration.

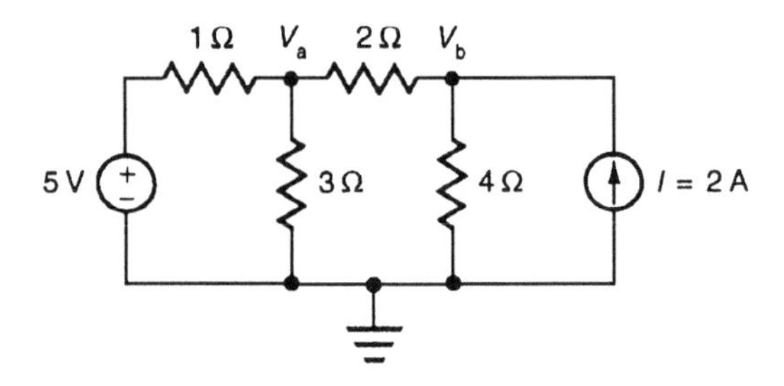

Exhibit 37 Mixed source circuit

Solution

Here, the node-voltage method of analysis will be used. This method finds voltages with respect to some common reference point called a ground node (see Exhibit 37). By wisely selecting this common point to include as many junctions as possible, the number of simultaneous equations may be reduced. The node voltage will be considered as a voltage rise from the common reference node (that is, current directions may then be assumed away from the node while writing the equations for that particular node).

$$\text{At } V_a, \sum I \qquad 0 = (V_a - 5)/1 + (V_a - V_b)/2 + (V_a - 0)/3$$

$$\text{At } V_b, \sum I \qquad 0 = (V_b - V_a)/2 + (V_b - 0)/4 + (-2)$$

Simplifying (multiplying through by 6 and 4, respectively) and collecting terms,

$$30 = 11V_a - 3V_b$$
$$8 = -2V_a + 3V_b$$

Solving these equations for the node voltages gives $V_a = 4.22$ V and $V_b = 5.48$ V. From these voltages, any desired currents or powers may be found. To check the solution, find the sum of the power for the current and voltage sources and compare this result with the sum of the power dissipated by the resistors ($P_{sources} = 14.9$ watts $= \Sigma P_{resistors}$). Either of the apparent sources could actually be a load by taking power from the other source. If the voltage source is a battery and it is found that the actual current flows through the battery from plus to minus, then the battery is being charged and acts as a load.

Network Reduction, Thevenin and Norton Equivalents

Most circuits may be greatly reduced in complexity and thus yield a simpler solution. This is especially so if a variable load is involved and requires a number of repeated calculations. Consider the circuit shown in Figure 3.50; many calculations would be involved if one were asked to compute the power dissipated in R_L for several different values of R_L. If one could reduce the network (all of that circuitry to the left of × – ×) to one simple voltage source in series with one resistance [see Figure 3.50(b)], then calculating the current to R_L would be simple and easy.

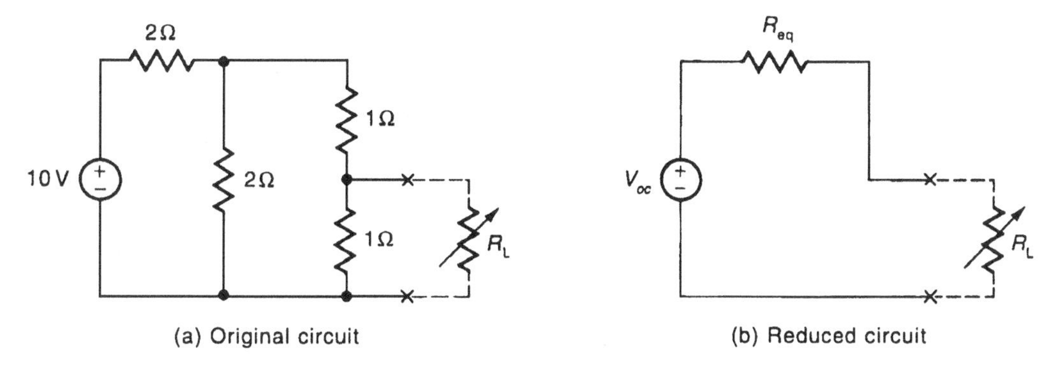

Figure 3.50 Circuit for Thevenin's reduction

Reducing the network is done with Thevenin's theorem. One can find the Thevenin voltage by merely opening the network at the point of interest (that is, temporarily removing R_L) and measuring the open circuit voltage at × – ×; the open circuit voltage is the Thevenin voltage. Next, place a short circuit (an ammeter, assumed to have zero resistance) across this section and measure the short circuit current. The Thevenin resistance is given as

$$R_{eq} = \frac{V_{OC}}{I_{SC}}$$

To find a Thevenin equivalent circuit on an examination, one cannot measure the various voltages or currents but must calculate them. Calculating the open circuit voltage does not present a problem, but more needs to be said about calculating the resistance.

There are at least two ways to calculate the Thevenin resistance. One is to calculate the short circuit current and proceed as before. The other method is to replace all voltage sources with their internal resistance (usually zero) and any current source with its internal resistance (usually infinity) and then calculate the resistance one would see looking back into the circuit (to the left of × – × in Figure 3.50). However, caution is needed if any dependent sources are present. For the values given in Figure 3.50, the calculated open circuit voltage is 5/3 volts, and the short circuit current is 5/2 amperes; thus the Thevenin equivalent circuit is as given in Figure 3.51(a).

The current source that is equivalent to the Thevenin circuit is Norton's equivalent circuit. Norton's circuit is the current source equivalency for Thevenin's voltage source [see Figure 3.51(b)]. The value of the resistance is the same except that it is usually stated in terms of conductance ($1/R$) and the units are mho (old) or siemens (new). The current source may be found directly from Thevenin's circuit by finding the short circuit current directly from Thevenin's circuit. If one is asked for a Norton equivalent circuit, one can first find Thevenin's circuit then convert to Norton's circuit.

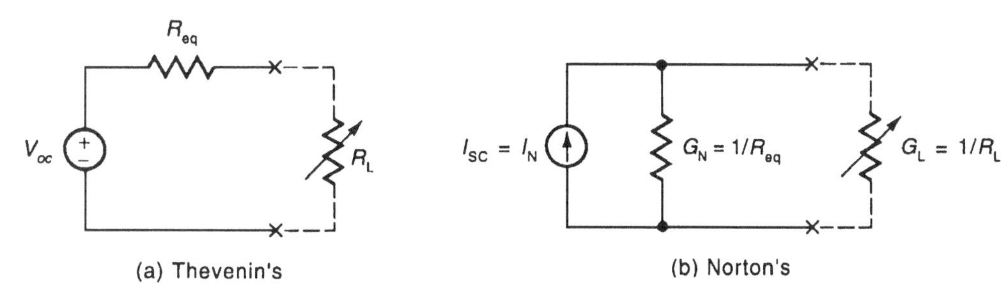

Figure 3.51 Equivalent circuits

Assume one wishes to determine the maximum power that could be dissipated in a variable load resistor that is nested in a multiple source circuit. First, it is necessary to know the size of the resistor. The **maximum power transfer theorem** (for dc circuits) states that maximum power is extracted from a circuit when the circuit is converted to Thevenin's equivalent circuit and the load resistance is equal to Thevenin's resistance. A consequence of this theorem is that half of the power is dissipated in the load and the other half in Thevenin's equivalent resistance. Consider the circuit of Figure 3.52; all that is necessary is to isolate the load resistor from being *buried* in the circuit. This is done by rearranging the circuit as shown in Figure 3.52(b). One way is to move the load resistor to the right side of the circuit and then find Thevenin's equivalent circuit for the portion on the left.

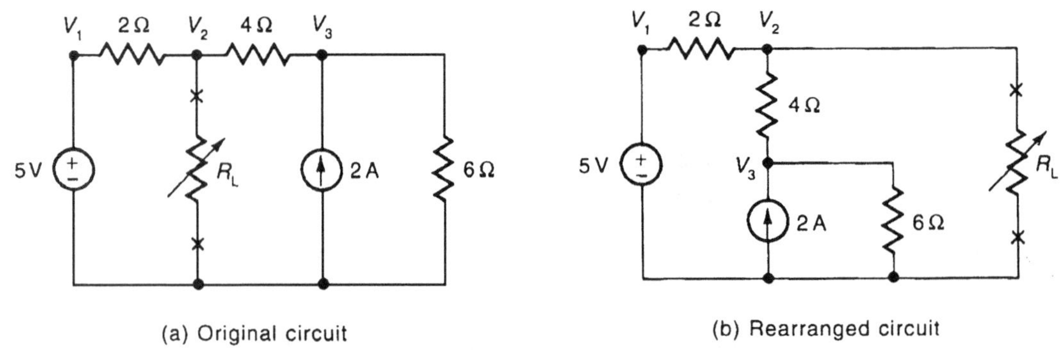

Figure 3.52 A maximum power transfer problem

For the rearranged circuit of Figure 3.52, the open circuit voltage (that is, with R_L temporarily disconnected) will be found by the node method of analysis:

$$\text{At } V_2, \sum I = (V_2 - V_1)/2 + (V_2 - V_3)/4 = 0$$

$$\text{At } V_3, \sum I = (V_3 - V_2)/4 - 2 + V_3/6 = 0$$

Solving (with $V_1 = 5$ volts) for $V_2 = V_T$ yields 6.116 volts. Thevenin's resistance can now be found by using the simpler method of merely replacing all independent voltage sources with zero and all independent current sources with an open circuit; then, looking back into the circuit × – ×, the equivalent resistance can now be found. This resistance yields $R_{oc} = 1.667\ \Omega$. From the maximum power relationship, $R_L = R_{oc}$. The power dissipated in $R_L = 5.61$ watts; see Figure 3.53.

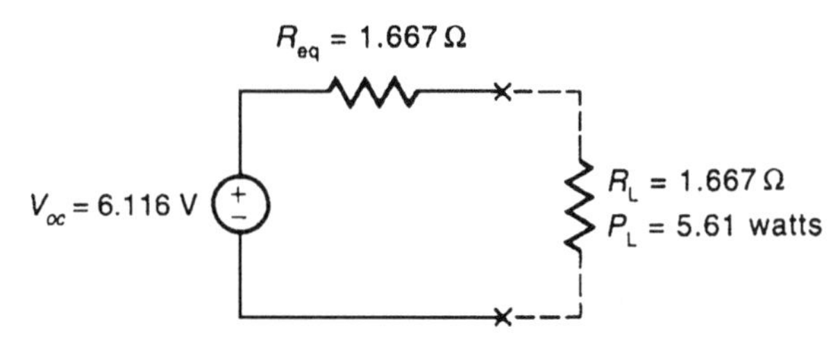

Figure 3.53 Final load resistance using Thevenin's circuit

Transient Response for a Single Energy Storage Element

If an energy storage element, such as a capacitor or inductor, is present in a circuit with one or more resistors, the problem solution will be a function of time. The voltage across a capacitor or current through an inductor cannot change instantaneously but takes time (unless infinite pulses are involved). It is assumed that the circuit is connected to either a voltage or current source or has some initial values.

Capacitive Circuits

If a capacitor (assume initially uncharged) is being charged from a voltage source through a resistor as shown in Figure 3.54(a), then it takes time for it to reach the charging voltage. The voltage vs. time solution is plotted in Figure 3.54(b). Here it is obvious that the amount of time for the capacitor voltage to reach its final value is infinite. On the other hand, if the capacitor already has an initial charge (or voltage) and is being discharged through a resistor, it again takes time for it to discharge [refer to Figure 3.55(b)]; this amount of time is also infinite.

Rather than use infinity as the changing time, it is more convenient to define a time constant, τ, to obtain a practical result. The time constant is some value of t that makes the exponent, x, of e^x, equal to -1. As an example, if $x = -t/RC$, then

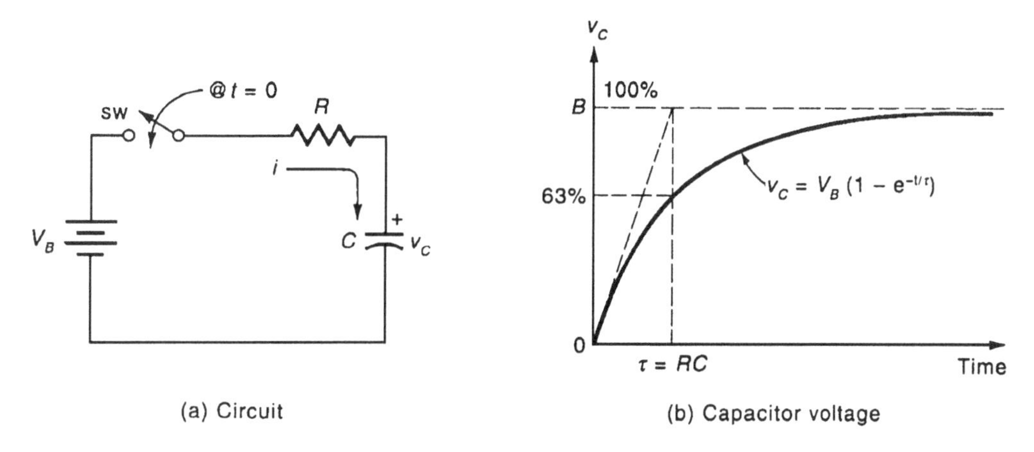

(a) Circuit (b) Capacitor voltage

Figure 3.54 The charging of a capacitor, τ = RC

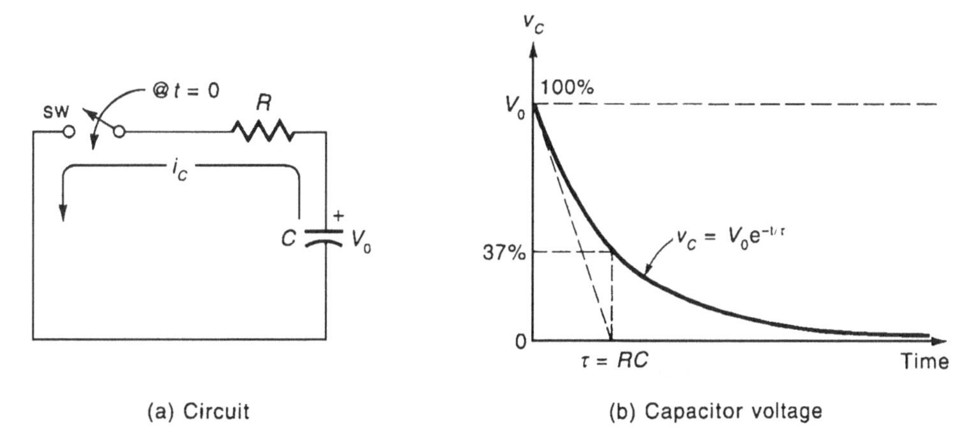

(a) Circuit (b) Capacitor voltage

Figure 3.55 The discharging of a capacitor, τ = RC

the time constant is equal to the value of RC. This value of e^{-1} is 0.3679, or approximately 37%; it is so often used in electrical transient analysis that it probably should be committed to memory. In Figure 3.54, V_B is the final value of the voltage, and at $t = \tau$ the parenthetical quantity is approximately 63% so the variable is within 37% of its final value. Also note that in Figure 3.55, the voltage decreases from the initial value, V_0, by 63% (or is within 37% of the final value) at $t = \tau$. If the initial slope of the curves in both Figures 3.54 and 3.55 were extrapolated forward, the intersection of the asymptotes with the horizontal axis would occur at $t = \tau$. To be realistic, *within* 37% is not good enough for engineering purposes, and a more practical approach is to consider the voltage at *almost* its final value in five time constants (well within 1%).

The equations in Figure 3.54 and Figure 3.55 are the solutions of the integral-differential equations that describe the circuit(s). Kirchhoff's voltage law states that sums of voltages around any closed loop must be equal to zero at *any* instant of time. In Figure 3.54, the switch closes at $t = 0$; the voltage around the loop (for $t >$ 0) is $V_B = Ri + v_c$, where v_c is q/C. Since current is defined as dq/dt, v_c is the integral of the present expression. If the initial voltage (or charge) on the capacitor was zero at the instant before the switch was closed, then $V_B = Ri + q/C = Ri + (I/C) \int_0^t i\,dt$.

The solution to this equation is

$$v_c(t) = v_c(0)e^{-t/RC} + V(I - e^{-t/RC}) \quad \textbf{(3.87a)}$$

and Equation (3.87a) frequently reduces to Equation (3.87b):

$$v_C = v_B(1 - e^{-t/\tau}) \quad \textbf{(3.87b)}$$

where $\tau = RC$. If t goes to infinity, then $v_C \to V_B$. If the initial value was not zero, as in Figure 3.55, it must also be considered. The differential equation used to obtain the solution for Figure 3.55(b) may be found by the node voltage method; use the lower junction of RC as the reference node, and the currents to the top voltage node are given by $i_C + i_R = 0 = C(dv/dt) + v/R$. Knowing the initial voltage, $V_C(0^-)$, the voltage across the capacitor just before the switch was closed, and that $V_C(0^-) = V_C(0) = V_C(0^+)$, the differential equation is easily solved. The solution, where $\tau = RC$, is $v_C = V_C(0)e^{-t/\tau}$.

A somewhat more comprehensive example involving a multiple time constant circuit is presented in Figure 3.56. The switch is assumed to be open for a long time (so long that any previous voltage across the capacitor has been reduced to zero because of R_3). The switch is then moved to the middle position for five seconds ($0 < t < 5$); it will then be switched to the lower position and remain there; this lower position will have a time, t', that starts at the time of switching to this new position. It is desired to know the current through the capacitor at $t = 9.8$ s (t' = 4.8 s). Although this problem may seem more comprehensive than most, the reader should follow the details of the solution to note simplification techniques. First, it should be obvious that the problem may be broken into two parts. The first part spans the first 5 seconds, and the second runs from $t = 5$ (or $t' > 0$) seconds. When the switch is in the middle position, one needs to make a Thevenin equivalent circuit (refer to Figure 3.57) to the left of $\times - \times$ and then solve for the capacitor voltage at $t = 5$s as done on the previous problem. This voltage at $t = 5$ s is easily computed from the capacitor charge equation as $v_C(t = 5) = 7.5(1 - e^{-5/3})$ = 6.083 volts.

One could almost obtain this value from a sketch, as at one time constant (3 seconds) the voltage would be 63% of 7.5 volts (4.72 volts), and at two time constants (6 seconds) the voltage would be 87% of 7.5 volts (6.49 volts). Thus at 5/3 time constant, a plot would yield about 6.1 volts. Of course the final voltage of Figure 3.57 is the initial voltage when the switch is moved from the middle to the bottom position in Figure 3.58. The voltage at $t' = 4.8$ ($t = 9.8$) seconds is 37% of the initial value, $0.37 \times 6.08 = 2.25$ volts. The current at this instant of time is $v_C/R_{Eq} = 1.87\ \mu A$.

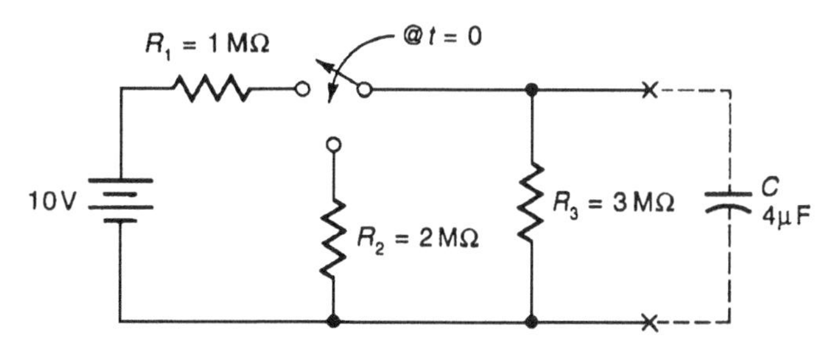

Figure 3.56 Multiple time constant circuit

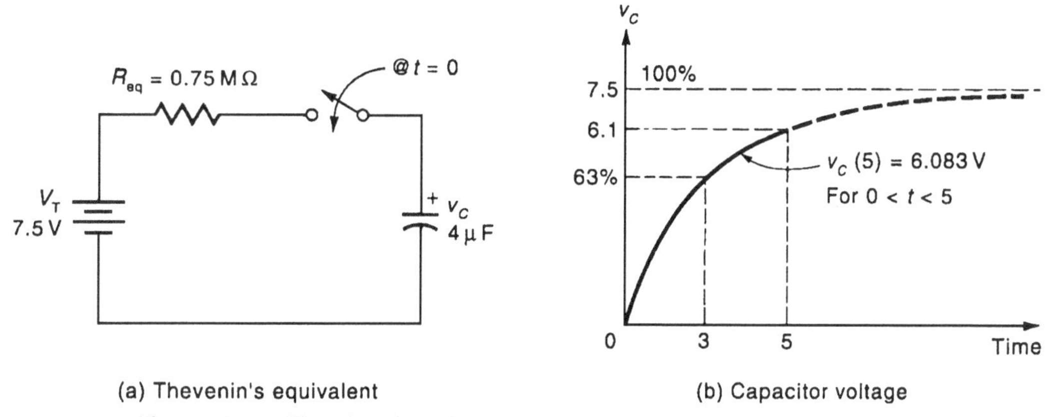

(a) Thevenin's equivalent (b) Capacitor voltage

Figure 3.57 The charging of a capacitor during $0 < t < 5$ seconds, for $\tau = 3$

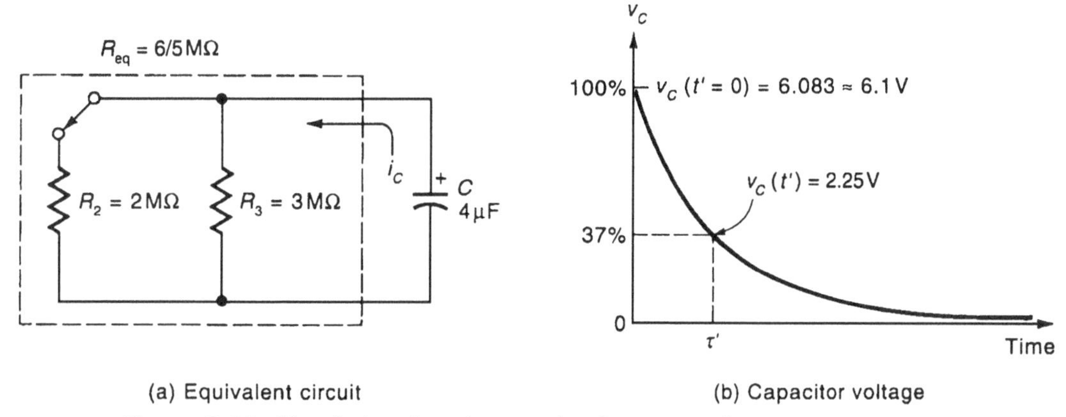

(a) Equivalent circuit (b) Capacitor voltage

Figure 3.58 The discharging of a capacitor for $t > 5$ (or $t' > 0$) seconds, $\tau' = 4.8$

Inductive Circuits

When an inductor (instead of a capacitor) is in a circuit, the current through the inductor takes time to change (like a voltage across a capacitor) and the circuit may be treated much like the previous voltage-capacitor relationship,

$$i(t) = i(0)e^{-Rt/L} + V/R(1 - e^{-Rt/L}) \tag{3.88}$$

An example *RL* circuit problem is presented in Figure 3.59. Here, the time constant, τ, is *L/R*. Assume zero initial current in the inductor. A voltage will be induced across an inductor equal to *L*(*di*/*dt*) because of the rate of change of magnetic flux through a coil or around a wire. The polarity of this voltage depends on whether the flux is increasing or decreasing. Unlike a capacitor, where the capacitor acts as an open circuit (that is, no current flow) after the *voltage* stabilizes, the inductor acts like a short circuit after the *current* stabilizes. Thus the final value of the current in Figure 3.59 is the applied voltage divided by the series resistance.

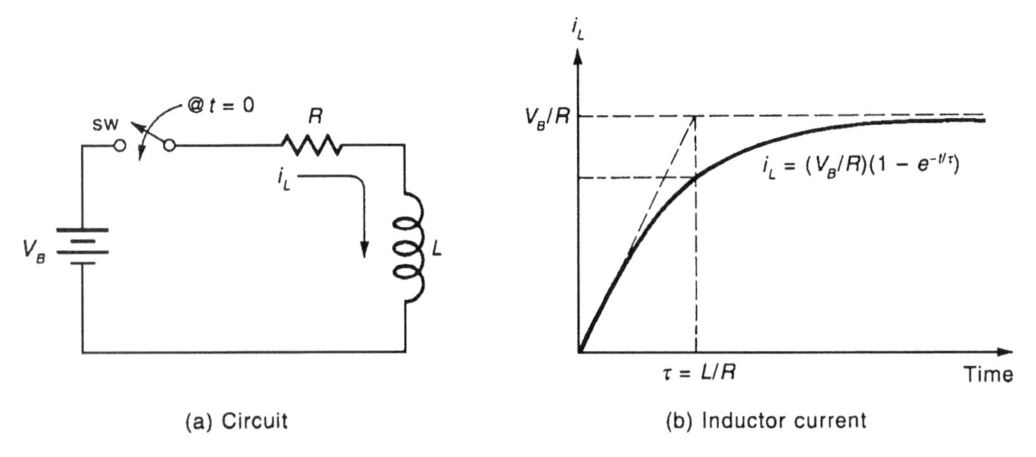

Figure 3.59 An RL circuit, τ = L/R

Example 3.49

An inductor has an initial current caused by the switch in Exhibit 38 being connected to a voltage source for a long period of time and then suddenly switched to the open position at $t = 0$. Find the current at $t = 0.1$ seconds and the voltage across the inductor at the same time.

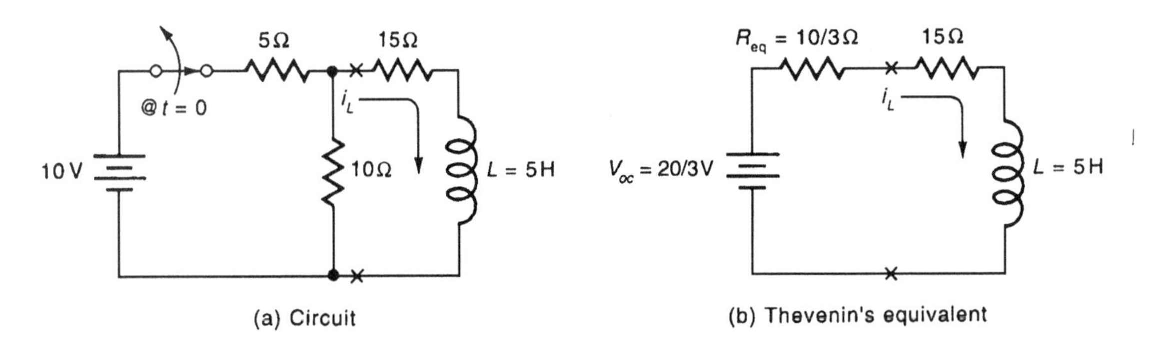

Exhibit 38 Inductor problem before the switch opens

Solution

Before the switch is open, the easy way to solve the problem is to make a Thevenin equivalent circuit of that portion of the circuit to the left of × – ×. While the switch is connected to the voltage source, the polarity of the voltage across the inductor is positive at the upper terminal of the inductor (while the voltage source is providing an increasing current in the inductor). Just before the switch opens the current in the inductor is $I(0^-)$. Because current cannot change instantaneously, $I(0^+)$ is the same current immediately after the switch is opened. However, the voltage polarity switches immediately as the energy stored in the inductor causes the current in the inductor is $I(0^-)$. Because current cannot change instantaneously, $I_0(0^+)$

is the same current immediately after the switch is opened. However, the voltage polarity switches immediately as the energy stored in the inductor causes the current to continue to flow in the same direction [that is, $L(di/dt)$ acts like a voltage source that is strong enough to keep the current flowing]. The current plot is given in Exhibit 39. The voltage is applied for a long period of time, so any rates of change, $L(di/dt)$, are completed; the inductor acts like a short circuit current and is easily found to be 0.3636. At $t(0)$, this current is the initial current $I(0) = I_0$ for the equation in Exhibit 39(c). The current at $t = 0.1$ second is found directly from the exponential equation where the time constant, $\tau = L/R$, has a resistance equal to the sum of the two resistors in series, $R = 25\ \Omega$, so $\tau = (5/25) = 0.2$ second. This current is $i = I_0 e_{-t/\tau} = 0.3636e^{-(0.1/0.2)} = 0.220$ A.

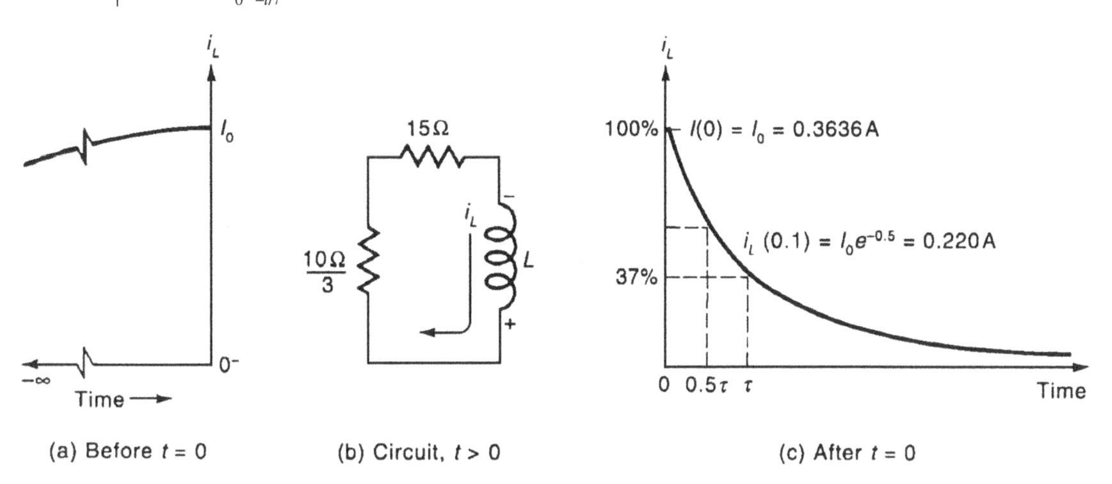

Exhibit 39 Current relationship in an inductor

The voltage across $L(di/dt)$ must be the same as the voltage drop across the resistors, since the loop is closed. Thus at $t = 0.2$ second, the voltage is $iR = 5.50$ volts; the polarity of this voltage is with the plus sign at the bottom of the inductance.

For an RLC circuit that involves two energy storage elements, L and C, the solution is complex. The governing equation is a second order differential equation whose solution may involve complex conjugate parameters.

AC CIRCUITS

In this review, the emphasis is on easily visualized graphical solutions rather than formal mathematical techniques. The review is based on assumptions that should be understood before the material is presented. The assumptions are the following:

- All sources are sinusoidal and are of the same frequency in any circuit unless specifically noted. For certain kinds of circuits, such as filters and resonant circuits, frequency changes will be made and will be considered in steady state and of the same frequency.
- Voltages and/or currents are considered at steady state. Any transients that may have resulted from the sudden closing of a switch are not considered. This assumption greatly simplifies the analysis; differential equation solutions will be bypassed and are replaced with phase-shifting operators. Although switch closing may be part of a circuit, it is assumed that one waits a short period of time (actually five time constants or more) for any transient or nonsinusoidal effects to die out before analyzing the circuit.

- Current or voltage sources are single phase unless otherwise noted (such as in three-phase circuit analysis).
- Circuits are linear. If a sinusoidal signal is applied to a circuit, then a current or voltage measured anyplace in the circuit will also be sinusoidal (with no harmonics). The only difference between an input and an output sinusoid will be a possible change in magnitude and phase.

An oversimplified pictorial (see Figure 3.60) will help one visualize a sinusoidal voltage source along with some of the above assumptions. The oversimplification is that of showing only one loop of wire in a magnetic field producing a pure sine wave. The frequency of the sine wave is directly proportional to shaft speed. And, the time for one revolution for a single pair of poles (called the period, T) is the inverse of the frequency. The wave is periodic and continuous and may be represented with a phasor.

Consider how one could draw a sine wave rather accurately by placing the tip of a pencil at the center of an X–Y plot (start with it laying horizontally to the right of the origin) and then allowing it to rotate counterclockwise through 360°. If one were to view the eraser from afar to the left and then to project horizontally the tip of the eraser onto the vertical axis of a time plot, the projection would trace out a pure sine wave (see Figure 3.61); the pencil could be thought of as a phasor, a two-dimensional vector. This phasor (length) represents the maximum value (height) of the sine wave (it is usually more convenient to substitute the rms value rather than the maximum value for the length). This maximum value will, at first, be used to demonstrate how the rotating phasor is visualized. When using these phasors, one's real concern is a time *snapshot* of the relative positions of the phasors. Also, rather than plotting the voltage vs. θ as in Figures 3.60 and 3.61, it is more convenient to plot it against ωt or $2\pi ft$. The angle the phasor passes through, with respect to some reference position or another phasor, represents a phase angle.

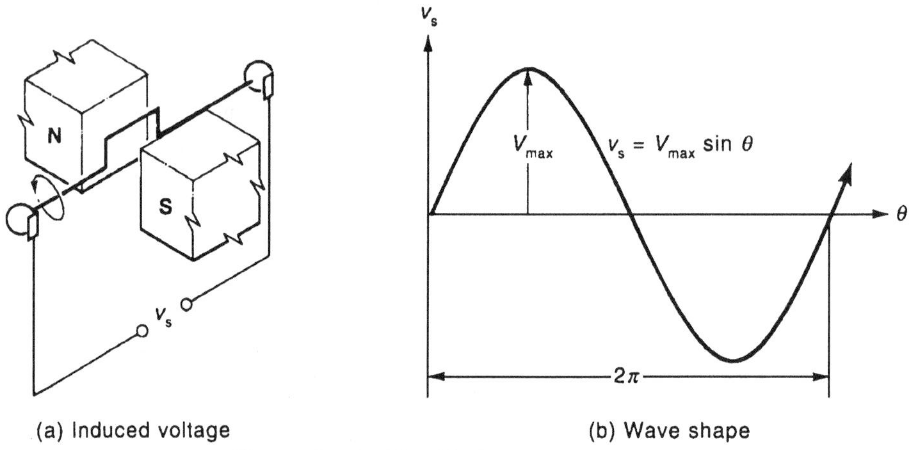

Figure 3.60 Simplified ac voltage source

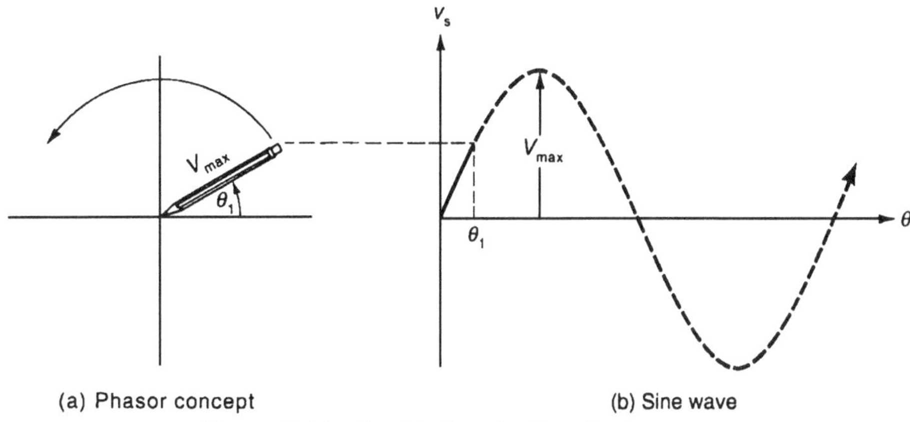

(a) Phasor concept (b) Sine wave

Figure 3.61 Graphical production of a sine wave

As an example of the simplicity of ac circuit analysis when using this phasor representation, consider summing two ac voltage sources graphically as in Figure 3.62(b). Here, one would sum the two voltages and continually draw the sum of the two waves; the result is

$$v_{total} = V_{max1} \sin \omega t + V_{max2} \sin(\omega t + \alpha) = V_{total} \sin(\omega t - \theta)$$

This graphic summing produces the desired total voltage and also displays the phase relationship with respect to either of the original sine waves. However, in Figure 3.62(c), the parallelogram formed from the two phasors yields the resulting phasor (the length is $v_{max\ total}$) much easier than summing the two sine waves. Furthermore, it gives the phase angle directly. The result is the same, in terms of relative magnitude and phase angle, for any instant of time. A formal mathematical description is

$$v_{total} = Im[V_{max\ total} e^{j(\omega t - \theta)}] \quad \textbf{(3.89)}$$

The notation *Im* implies the imaginary part of the expression that follows. A much simpler descriptive notation that most engineers use is to describe the representative phasor by a magnitude and its phase angle as $\mathbf{V}_{max\ total} \angle -\theta$. In electrical engineering, a positive rotation of the phasor is considered to be counterclockwise. Thus, in Figure 3.62, the parallelogram resultant phasor may be listed as being θ degrees behind (or lagging) the phasor V_1 or as ϕ degrees ahead (or leading) V_2.

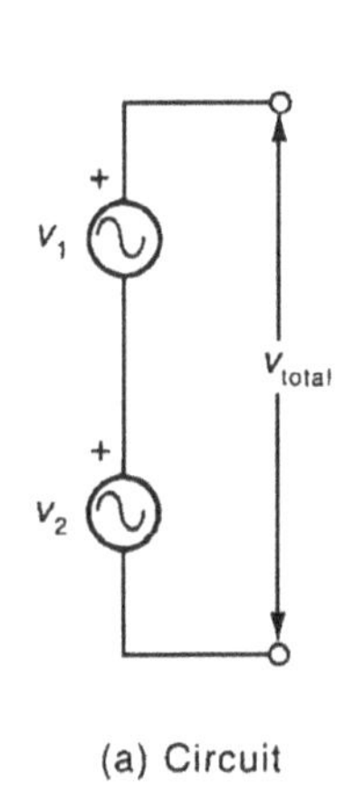

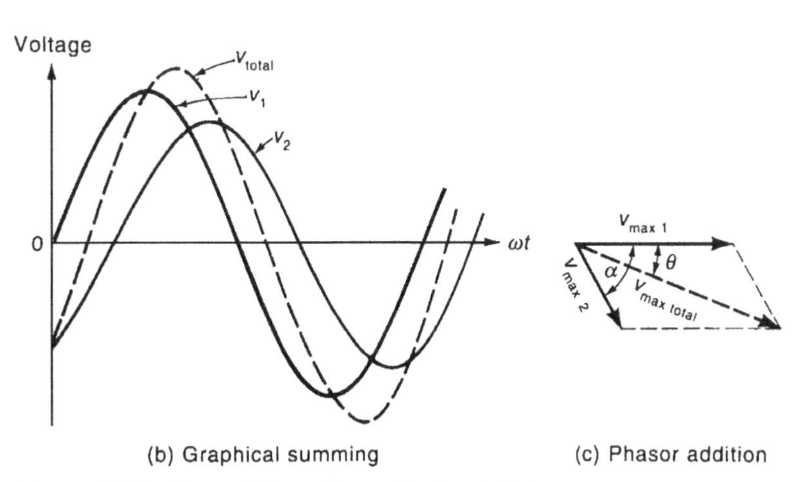

(a) Circuit (b) Graphical summing (c) Phasor addition

Figure 3.62 The addition of two voltage sources

Phasor Manipulation

Although adding two phasors is easily accomplished with the parallelogram, adding more than two might better be done by breaking each phasor into its x and y components and then summing the components. Multiplying phasors is done by multiplying the magnitudes and summing the angles, and division is done by dividing the magnitudes and subtracting the angles. As an example, two phasors **A** = $10\angle 0°$ and **B** = $5\angle 90°$ sum to **A+B** = $11.2\angle 26.6°$. And **AB** = $50\angle 90°$; or **A/B** = $2\angle -90°$. The concept of rotation is important. Assigning phasor **B** to be **B** $10\angle 180°$, then **B** = –**A**; one could say that **B** is the same as **A** if **A** were rotated through 180°. Here, multiplying phasor **A** by –1 is considered as operating on **A** to rotate it through 180°. One could then say that multiplying a phasor by the square root of –1 is also an operator that rotates the phasor through 90°. This imaginary value, the square root of –1, is referred to as j. The operator j then rotates the phasor by 90°. As an example, the only difference between a sine wave and a cosine wave (for the same amplitude) is that a sine wave is shifted through 90° to become a cosine wave,

$$\cos\theta = \sin(\theta + 90°) = j\sin\theta \tag{3.90}$$

Because of the effect of Equation (3.88), it will be more convenient to use the real (x-axis) and imaginary j (y-axis) notation when dealing with phasors.

For most ac voltages or currents, one is usually interested in rms values rather than peak or maximum quantities. It was previously shown that the rms [or in Equations (3.83) and (3.84) *what a meter would read*] value of a sine wave is its maximum value divided by the square root of two (or $V_{rms} = 0.707V_{max}$). Thus, one usually begins with rms magnitudes when using phasors. This makes voltage, current, and (especially) power calculations much simpler; the notation is even easier (the rms subscripts are dropped). In ac circuits there are really only three important devices: resistors (energy dissipating element), capacitors (energy storage element), and inductors (also an energy storage element).

Since all phasors are relative, plot phasor lengths as rms values rather than peak values.

Resistors

As in dc circuits, this resistive element dissipates power equal to the current squared times the resistance. This is not the only equation for power, but *it is the safest equation to use in ac circuits* unless one is very careful with notation and is experienced in ac circuit theory. Also, the voltage across a resistor is always in phase with the current through the resistor. The resistor stores no energy but dissipates the power in terms of heat.

Current and voltage are in phase for a resistor: $\mathbf{V}_R = R\mathbf{I}$.

By this notation, voltages and currents in bold indicate that they may be complex values.

CAPACITORS

The capacitor stores energy for half a cycle and gives it up on the other half, the net (real) energy is zero. However, it is sometimes convenient to refer to the product of voltage and current as imaginary power (using the symbol Q and calling it reactive power). The result of this energy interchange is that the current through the capacitor has a 90° phase difference for the two phasors; the current is ahead of the voltage (or leads the voltage). Whereas the resistance, R, impedes the flow of current, the impending quantity for a capacitor is the reactance, called X_c. Not only does X_c impede the current, but it also causes a phase shift of –90° so that the voltage across the capacitor is lagging the current by 90°. These descriptive words may be replaced by the symbol $-j$.

> The voltage drop across a capacitor is 90° behind the current.
>
> $$\mathbf{V}_C = -jX_C\mathbf{I}, \qquad \text{where } X_C = \frac{1}{2\pi fC}$$

The instantaneous voltage across a capacitor for a given current is

$$v = 1/C \int i \, dt \tag{3.91a}$$

and if the given current is $i = I_{\max} \sin \omega t$ then

$$v = 1/C \int I_{\max} \sin \omega t \, dt \tag{3.91b}$$

and the solution is

$$v = [-1/(\omega C)] I_{\max} \cos \omega t = [1/(\omega C)] I_{\max} (-j) \sin \omega t \tag{3.91c}$$

Using phasor notation, one can show that Equation (3.91c) is equivalent to

$$\mathbf{V}_C = -jX_C\mathbf{I}, \quad \text{where } X_C = 1/(\omega C) \tag{3.92}$$

It is obvious that a numerical value of X_C is a function of frequency. In fact, if the frequency is infinite, X_C is zero as in a short circuit; and if the frequency is zero, X_C is infinity as in an open circuit. This relationship between current and voltage is shown graphically in Figure 3.63.

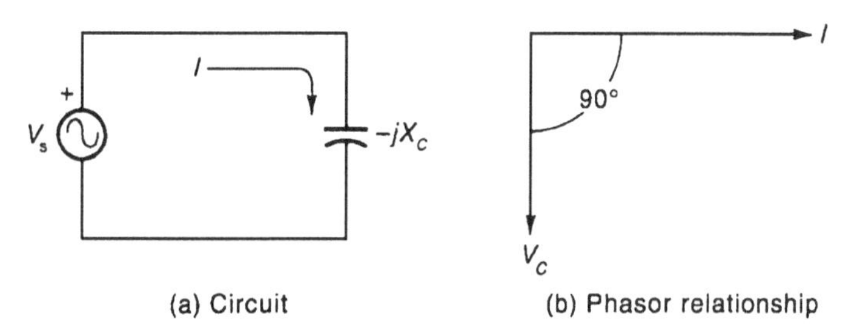

Figure 3.63 Voltage-current relationship for a capacitor. (The voltage and current are not to the same scale.)

Example 3.50

Consider a series R–C circuit whose measured current is 5 amperes (rms). Determine the voltage source required to produce this current. Assume that the voltage source is known to have a frequency of 60 Hz, with $R = 10$ ohms and $C = (0.1/377)$ farads. The factor $2\pi f$ for 60 Hz is 377.

Solution

The effect of the capacitor is to create a 90° phase shift between the current and voltage while the resistor produces no phase shift. The instantaneous Kirchhoff voltage equation is

$$v_{\text{total}} = Ri + v_C = Ri + (1/C)\int i\,dt \quad \textbf{(3.93a)}$$

Rather than solving the above equation, phasors will be used here in the form

$$\mathbf{V}_s = \mathbf{V}_R + \mathbf{V}_C = R\mathbf{I} - jX_C\mathbf{I} \quad \textbf{(3.93b)}$$

Now R and jX_C should be viewed as operators (they operate on the current to produce a voltage—and the j is to produce a 90 degree phase shift). Then X_C is calculated from $1/(2\pi fC)$ to be 10 ohms.

$$\mathbf{V}_s = 10\mathbf{I} - j10\mathbf{I} \quad \textbf{(3.93c)}$$

Remember, the $-j10$ operator multiplies the current phasor by ten and shifts it through –90°. With the two voltage phasors, one only needs the parallelogram construction to get the voltage source phasor along with its angular relationship to the current, as shown in Exhibit 40.

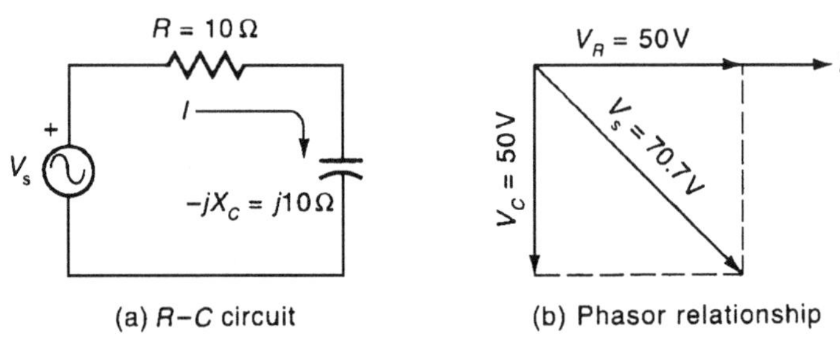

Exhibit 40 Voltage relationship for an *R–C* circuit

The voltage source is 70.7 volts with the voltage phasor lagging 45° behind the current. Or if the voltage source is considered to be the reference, one could say that the current leads the voltage by 45°. Observe that all of the (real) power from the source is dissipated by heat loss in the resistor (no real power is dissipated in the capacitor):

$$P_R = I^2R = 5^2(10) = 250 \text{ watts}$$

The power from the source, P_s, is the product of voltage and current times the power factor, which is defined as the cosine of the angle between the voltage and current. Thus

$$P_s = VI\cos 45° = 70.7 \times 5 \times 0.707 = 250 \text{ watts}$$

The reactive power, Q, is calculated as follows:

$$P_{\text{reactive}} = Q = VI\sin\theta = 70.7 \times 5 \times 0.707 = 250 \text{ VARs}$$

(*Note*: P_s and Q match only because θ is 45°.)

Inductors

Like the capacitor, the inductor stores energy for half a cycle and returns it to the circuit for the next half cycle. In circuit analysis, the inductor behaves in a similar manner as a capacitor except the sign is reversed; here the term is $+jX_L$ where X_L is directly proportional to the frequency and is equal to $\omega_L = 2\pi fL$. For the inductor, the voltage leads the current by 90°, or the current lags the voltage.

> A voltage drop across an inductor leads the current by 90°.
>
> $$\mathbf{V}_L = jX_L\mathbf{I}, \quad \text{where } X_L = 2\pi fL$$

Also, at zero frequency, the reactance is zero; and for an infinite frequency, the reactance is infinite. The phase relationship is shown in Figure 3.64.

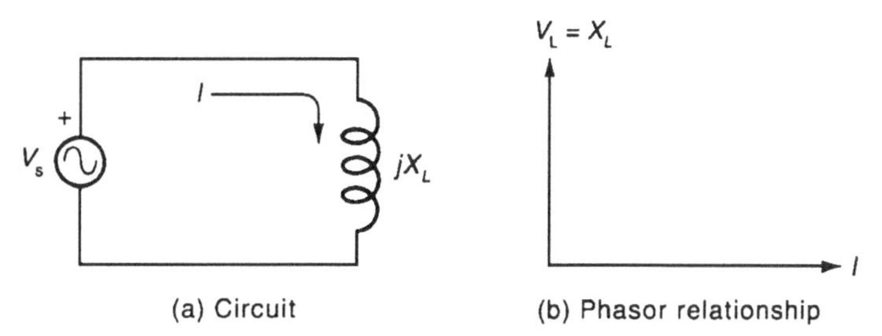

(a) Circuit (b) Phasor relationship

Figure 3.64 Voltage-current relationship for an inductor

Example **3.51**

Consider the circuit of Exhibit 41, where the current is, as before, 5 amperes, R is 10 Ω, and $L = 10/377$ henries.

Solution

Here X_L is ohms (Exhibit 41). The answers are unchanged from Example 3.50 except for the phase relationships.

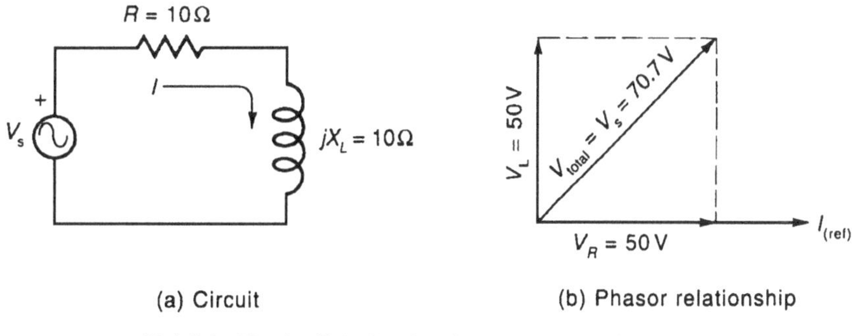

(a) Circuit (b) Phasor relationship

Exhibit 41 An *R-L* circuit voltage-current relationship

Impedance Relationship

In general, ac voltages and currents are related by a quantity called impedance (Z). It is usually complex, and it is sometimes referred to as Ohm's law for ac circuits. This association is easily explained by reference to either of the two previous example problems. For instance, in Exhibit 40 and Equation (3.93), the current was assumed to be known. If the current were unknown and the source voltage was given as 70.7 volts, then $\mathbf{V}_s = 70.7\angle 0°$ volts can be the known reference phasor. From Equation (3.93b), the unknown current is easily factored out of the expression to give $\mathbf{V}_s = (R - jX_C)\mathbf{I}$.

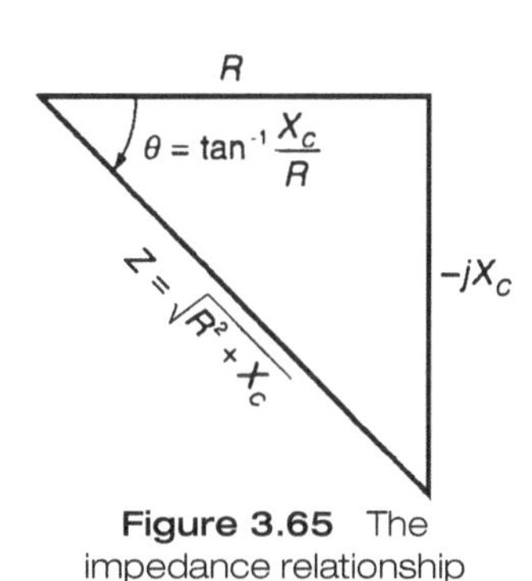

Figure 3.65 The impedance relationship

The parenthetical quantity—or the operators—for this problem, as in Equation (3.93c), is the complex impedance. It has a right-triangle relationship as in Figure 3.65 and is given by $\mathbf{Z} = R - jX_C = 10 - j10 = 10\sqrt{2}\angle -45°$.

The $10 - j10$ is known as the rectangular form, and the $10\sqrt{2}\angle -45^\circ$ is known as the polar form. The rectangular form is usually used when adding or subtracting is involved, and the polar form is used when one is multiplying or dividing. To finish the problem, the current is found as $\mathbf{I} = \mathbf{V}_s/\mathbf{Z} = (70.7\angle 0^\circ)/(10\sqrt{2}\angle -45^\circ) = \angle +45^\circ$ A.

Example 3.52

Another series circuit problem will expand on the complex impedance concept; see Exhibit 42. Here, all of the resistances and reactances and the source voltage are given for a particular frequency. It is desired to find the power dissipated in the 20-ohm load resistance and also to find the total power from the source voltage. Since the circuit is a series one, the current is common for all elements; therefore the current should be found. The voltage source is 100 volts.

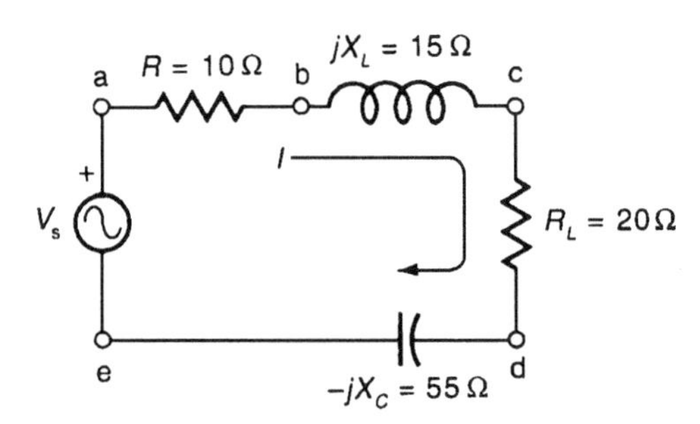

Exhibit 42 A series circuit with known parameters

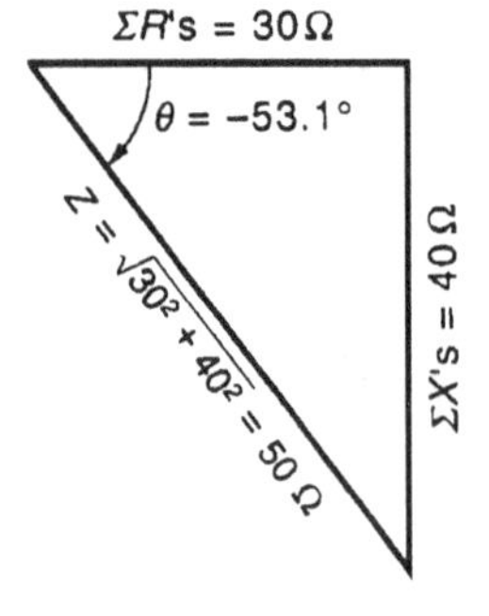

Exhibit 43 Impedance triangle

Solution

Find the total impedance to find the current (see Exhibit 43): $\mathbf{Z} = 10 + j15 + 20 - j55 = (10 + 20) + j(15 - 55) = 30 - j40 = 50 \angle -53.1^\circ\Omega$.

From this complex impedance, current and power can be found directly:

$$\mathbf{I} = \mathbf{V}/\mathbf{Z} = (100\angle 0^\circ)/(50\angle -53.1^\circ) = 2 \angle +53.1^\circ \text{ amperes}$$

$$P_{RL} = I^2R_L = 2^2(20) = 80 \text{ watts}$$

$$P_s = V_sI \cos\theta = 100 \times 2 \cos(-53.1^\circ) = 120 \text{ watts}$$

Check: $P_s = \Sigma I^2R = 2^2(10 + 20) = 120$ watts.

Example 3.53

Suppose, in Example 3.52, one is asked to find the magnitude and phase of the voltage across two points—say the voltage from point b to point d.

Solution

The phasor plot readily yields this answer. Since the current is common to all elements, the current should be made the reference phasor. Again, the currents and voltages are not necessarily plotted to the same scale. Rather than having all phasors emanating from the origin, it is convenient to plot the voltages in a cumulative fashion, from head to tail (see Exhibit 44).

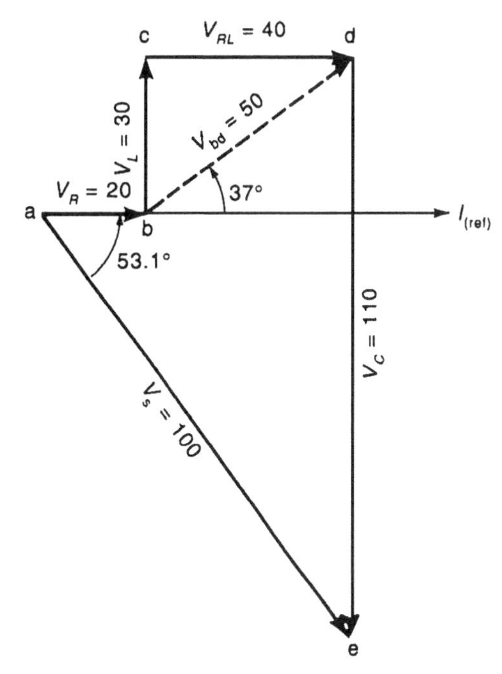

Exhibit 44 Plot of voltages vs. current

The voltage, V_{bd}, may be read directly from the plot to be 50 volts at an angle of 37° ahead of the current. An interesting aspect of this problem is brought out if the frequency of the source is doubled. Doubling the frequency causes no change in the resistive elements but dramatic changes in the reactive ones. For the inductor, X_L is 15 × 2 = 30 ohms and for the capacitor X_C = 0.5 × 55 = 27.5 ohms. The impedance then becomes $\mathbf{Z} = 10 + j30 + 20 - j27.5 = 30 + j2.5 = 30.1\angle 4.76°$ ohms. The current is $\mathbf{I} = (100\angle 0°)/30.1\angle 4.76° = 3.32\angle -4.76°$ A, and the power dissipated in the 20-ohm resistor is

$$P_{RL} = I^2R_L = 3.32^2 \times 20 = 220 \text{ watts}$$

At this point, several practical observations are in order. For approximate answers, if either the real or imaginary part is more than ten times the other, the hypotenuse is almost equal to the longest leg, and the angle is within a few degrees of being zero. Using this approximation for impedance, the approximate current is 3.3 amperes and P_{RL} is 222 watts. Another observation is that by changing the frequency, the power output has gone from only 80 watts to well over 200. There is a specific frequency to get maximum current and power; this is called the resonant frequency.

Resonant Frequency

For series circuits, the resonant frequency occurs when all of the reactive components cancel so that $X_L = X_C$, $\omega L = 1/\omega C$. For this particular frequency, the subscript notation may be given as *res*, *0*, or *n*.

Series Resonant Frequency

$$\omega_0 = 2\pi f_0 = \frac{1}{\sqrt{LC}}$$

The series resonant frequency concept leads directly to the maximum power transfer theorem, which should now be obvious. Maximum power occurs when the reactive components in an ac series circuit cancel.

Parallel ac circuits are no more difficult than series ones but may require slightly more bookkeeping; also, since many reciprocals are involved, additional definitions may be used to identify these reciprocals. As an example, the reciprocal of impedance, Z, is admittance, Y; the units of Y are siemens (old notation mhos) abbreviated S. The admittance may be further expanded to have a real and imaginary part. The real part is called the conductance, G, and the imaginary part is called the susceptance, B; both parts have the units of siemens. However, for simple problems, these definitions are optional; consider the example shown in Figure 3.66.

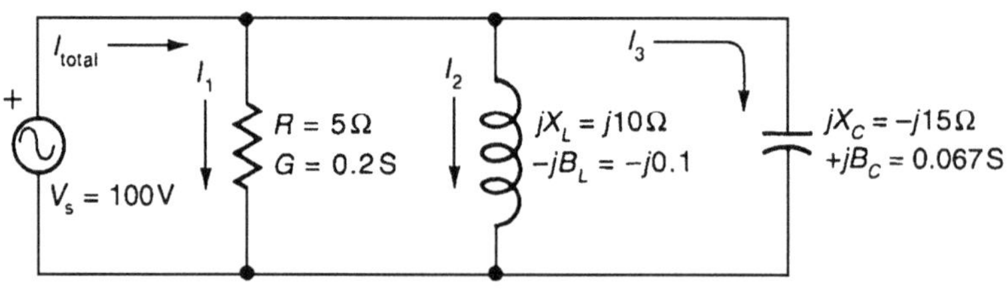

Figure 3.66 A parallel circuit

Example **3.54**

Solution

The equivalent impedance for this problem may be found by either dividing the source voltage by the total current (see Exhibit 45) or by taking the reciprocal of the total admittance.

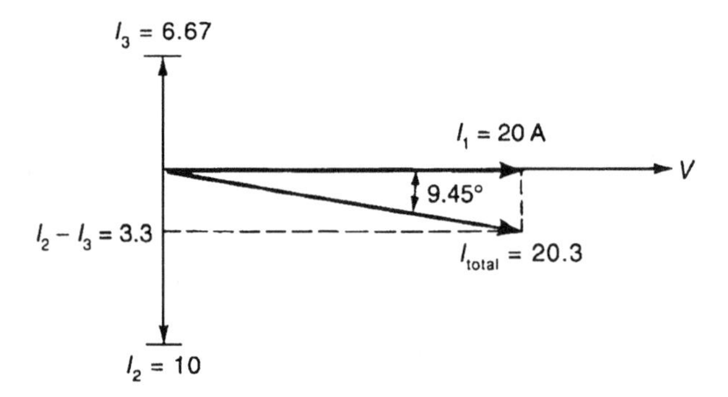

Exhibit 45 The phasor relationship for Figure 3.66

$$\mathbf{I}_{total} = \mathbf{I}_1 + \mathbf{I}_2 + \mathbf{I}_3$$

where

$$\mathbf{I}_1 = \mathbf{V}/R = (100\angle 0°)/5 = 20\angle 0° \text{ A}$$
$$\mathbf{I}_2 = \mathbf{V}/(jX_L) = (100\angle 0°)/(10\angle 90°) = 10\angle -90° \text{ A}$$
$$\mathbf{I}_3 = \mathbf{V}/(-jX_C) = (100\angle 0°)/(15\angle -90°) = 6.67\angle 90° \text{ A}$$
$$\mathbf{I}_{total} = 20 - j10 + j6.67 = 20 - j3.33 = 20.28\angle -9.46°$$

$$\mathbf{Z} = \mathbf{V}/\mathbf{I} = (100\angle 0°)/(20.28\angle -9.46°) = 4.93\angle 9.46° \text{ ohms}$$

However, using admittances, Y may be found directly:

$$G = 0.20 \text{ S}, \quad -jB_L = -j0.10 \text{ S}, \quad jB_C = j0.066 \text{ S}$$
$$\mathbf{Y} = G - jX_L + jX_C = 0.20 - j0.10 + j0.066 = 0.20 - j0.03 = 0.203\angle -9.45° \text{ S}$$
$$\mathbf{Z} = 1/\mathbf{Y} = 1/(0.203\angle -9.45°) = 4.93\angle 9.45° \text{ ohms}$$

Example 3.55

A final example of a combined series and parallel circuit is given in Exhibit 46. For this kind of problem, the reader is urged to make two separate plots to find the individual branch currents, then add the two current phasors on a third plot to find the total current. Find the power taken from the source.

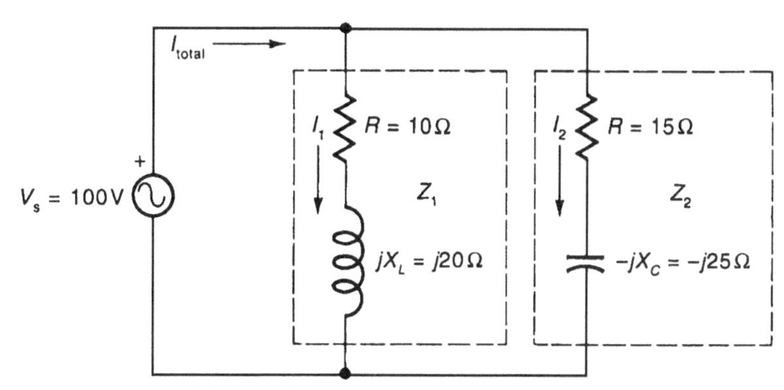

Exhibit 46 Combined series and parallel circuit

Exhibit 47 Phasor diagram of currents

Solution

First, find the currents:

$\mathbf{I}_1 = \mathbf{V}/\mathbf{Z}_1 = (100\angle 0°)/(10 + j20) = 4.47\angle{-63.4°} = 2.00 - j4.00$
$\mathbf{I}_2 = \mathbf{V}/\mathbf{Z}_2 = (100\angle 0°)/(15 - j25) = 3.43\angle 59.0° = 1.76 + j2.94$
$\mathbf{I}_{total} = \mathbf{I}_1 + \mathbf{I}_2 = (2.00 + 1.76) + j(-4.00 + 2.94) = 3.76 - j1.06 = 3.91\angle{-15.7°}$

See Exhibit 47 for the phasor diagram. The power from the source is now found to be

$$P_s = \mathbf{VI}_{total}\cos\theta = 100 \times 3.91 \times \cos(-15.7) = 376 \text{ watts}$$

(*Check*: $P_{total} = (\mathbf{I}_1)^2R_1 + (\mathbf{I}_2)^2R_2 = 200 + 176 = 376$ watts.)

Although calculations are straightforward, the reader is urged to sketch the phasor diagram. Again, for a multiple-choice problem, the graphical solution may be sufficiently accurate for one to select the correct answer.

Quality Factor

For parallel resonance as for series resonance circuits, Z still equals R at resonant frequency. For the series circuit the inductive and capacitive reactances cancel; and for parallel circuits, the inductive and capacitive susceptances (in siemens) cancel. In describing the behavior of these resonant circuits another descriptive quantity is frequently used—especially so for parallel circuits—which is called the *quality factor*, Q, of the circuit. For a series circuit Q is a measure of the energy stored in an inductor compared to the energy dissipated in the resistance; this ratio may also be given as

$$Q_{series} = \omega_o L/R = 1/(\omega_o CR) \quad \textbf{(3.94)}$$

where Q is dimensionless and may be quite high for a sharp resonant peak as a narrow bandwidth circuit. On the other hand, for a parallel circuit, Q may be defined (see Figure 3.66) as

$$Q_{parallel} = 1/Q_{series} = \omega_o RC = R/(\omega_o L) \quad \textbf{(3.95)}$$

and the bandwidth, BW, may easily be found as

$$BW = \omega_o / Q_{parallel} \text{ (rad/s)} \quad \textbf{(3.96)}$$

Transfer Function

The transfer function is the ratio of desired response of a system to the input (or excitation) when all the initial conditions are zeros.

Example 3.56

Find the transfer function $T_{(s)} = \dfrac{V_2(s)}{V_1(s)}$ of the circuit shown in Exhibit 48.

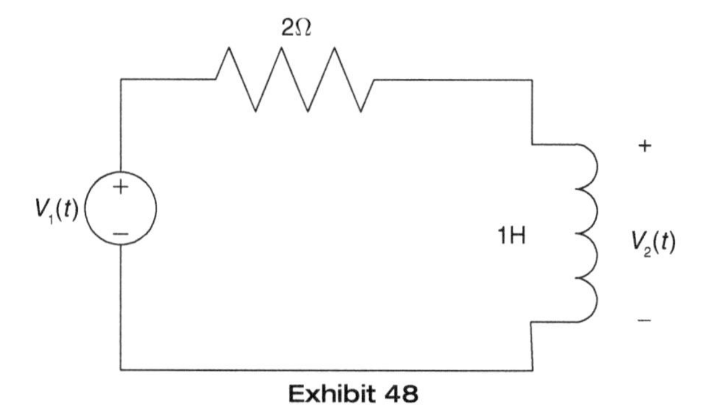

Exhibit 48

Solution

The s-domain circuit is drawn, as shown in Exhibit 49, by replacing the inductor with s •, the source by $V_1(s)$, and maintaining the resistor as such. Using voltage division, $T(s) = \dfrac{V_2(s)}{V_1(s)} = \dfrac{s}{s+2}$.

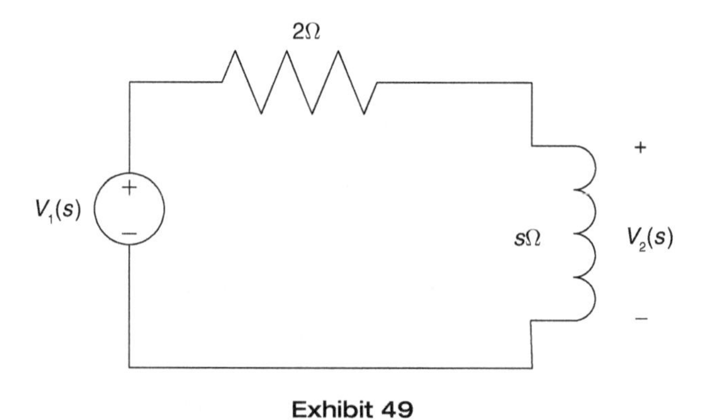

Exhibit 49

For a transfer function H(s), a plot of |H(jω)| versus ω and a plot of the phase angle of H(jω) versus ω are called the frequency response plots. For convenience, a plot of 20 log |H(jω)| dB (decibel) versus log ω, and a plot of the phase angle of H(jω) versus log ω are usually drawn. These are called Bode plots.

ELECTRIC MACHINES

AC Machines

The speed of rotation of a synchronous motor and that of a rotating magnetic field, in either a synchronous motor or an induction motor, are decided by the equation

$$n_s = \frac{120f}{p} \tag{3.97}$$

where f = line frequency in Hz, p = the number of poles, and the speed is in rpm. For three-phase induction motors, the actual speed of rotation will always be less than the speed of the rotating field. This difference is expressed by a factor called slip, as given by the equation

$$slip = \frac{n_s - n}{n_s} \tag{3.98}$$

DC Machines

For direct current machines, the voltage induced in the armature, E_a, is directly proportional to the product of speed, n, and the magnetic flux, ϕ, generated by the field. The magnetic flux is directly proportional to the field current, I_f. The relation between E_a and I_f is determined experimentally.

Example 3.57

Exhibit 50 shows the equivalent circuit of a separately excited DC generator where R_a is the armature resistance, R_L is the load resistance, and E_a is the induced EMF of the armature. It is determined experimentally that the induced EMF is 100 V for a field current of 1 A on no load at 1000 rpm. Find the load voltage, V_L, at 800 rpm if the field current is 1 A, R_a is 1 Ω, and R_L is 10 Ω.

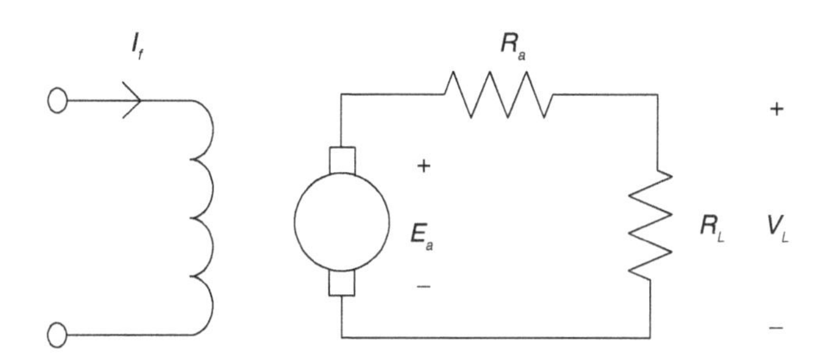

Exhibit 50

Solution

Since the induced EMF $E_a = K_a n\phi$, and flux $\phi = K_f I_f$, where K_a and K_f are constants, at any given I_f, E_a is proportional to speed n. As the field current is fixed, E_a is directly proportional to speed. Then, at 800 rpm, $E_a = 100\frac{800}{1000} = 80$ V.

Using KVL, $80 = I_a(1 + 10)$; $I_a = 7.27$ A; load voltage $V_L = 10I_a = 72.7$ V.

ELECTRONIC CIRCUITS

A thorough review of electronics is not undertaken here. But two topics—the solid state diode and the operational amplifier—have been selected as deserving attention. The complexity of certain aspects of both solid state theory and integrated circuit theory is well beyond the scope of this review, so only the idealized devices will be dealt with.

Diodes

The solid state theory of diode operation depends on the particular kind of diode being considered. For example, for the junction diode, knowledge of solid state theory and the behavior of majority and minority carriers in the presence of an electric field is desirable. But when an idealized diode is treated, the full theory is not essential. For a p–n junction diode, a very simplistic explanation of the operation at the junction between a *p* and an *n* type semiconductor is that in each type of material there are many free charges available; these are holes (p, positive) and electrons (n, negative), respectively. If the diode is biased in favor of forcing positive charges near the boundary (the positive voltage being connected to the p material and the negative terminal to the n material), a rapid recombining of the charges takes place [see Figure 3.67(a)]; this is the direction of easy current flow. On the other hand, if the diode is reverse biased [Figure 3.67(b)], the free charges are attracted toward their bias polarities, leaving a dearth of charges at the junction for very little recombination and almost no current flow. When the bias is in

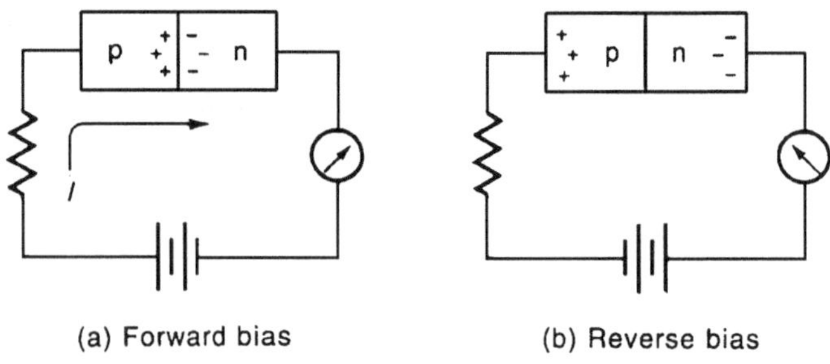

(a) Forward bias (b) Reverse bias

Figure 3.67 Charges inside a junction diode

the forward direction, the voltage necessary to cause the charges to recombine is fairly small but still could be significant (see Figure 3.68). The voltage is near half of one volt (approximately 0.4 V for germanium and near 0.7 for silicon) and is nonlinear. In the reverse direction, the current is essentially zero until breakdown voltage is reached; this breakdown is referred to as Zener voltage, which ranges from a few volts to several hundred volts. When designing circuits that use these diodes, care must be taken to work well within the reverse peak breakdown voltage. If one ignores the nonlinearity by assuming that the half volt is negligible for the forward direction, then the symbol shown in Figure 3.68(b) represents the ideal diode (the arrowhead side is the anode, and the other side is the cathode). These diodes, of course, have many applications; in this review, the applications will be limited to the ideal case in pure rectifying circuits.

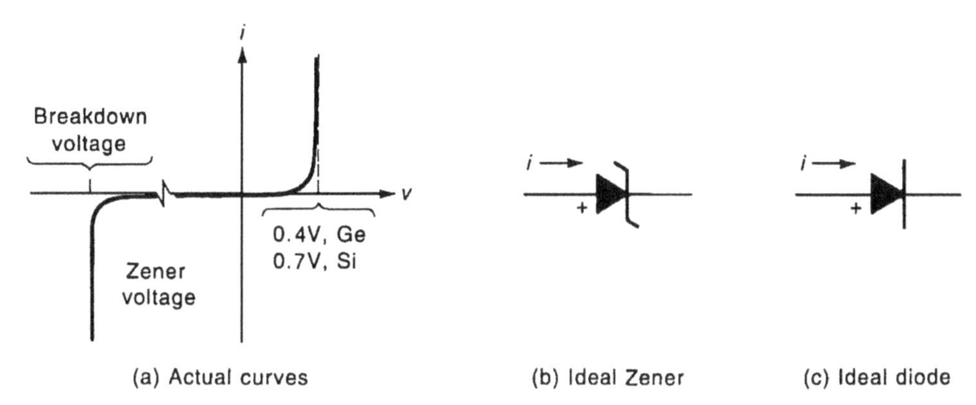

Figure 3.68 Junction diode voltage–current characteristics

Rectifying Circuits

Most diode rectifying circuits are used to convert ac voltages to *pulsed dc* voltages (see Figure 3.69). The diode is used to convert an ac voltage to a half-wave sinusoid so that the meter measurement found is the average current. Rectification is not limited to sinusoid; the following table (see Table 3.6) shows a sample of various kinds of signals and kinds of rectification, either full- or half-wave, relationships.

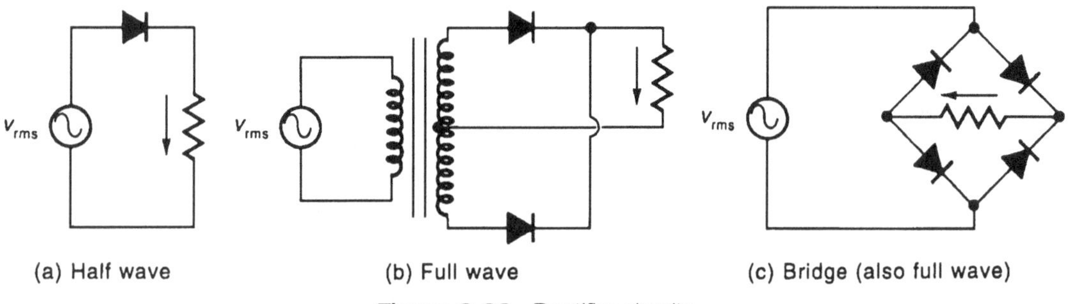

Figure 3.69 Rectifier circuits

Table 3.6 Rectified values for three signals

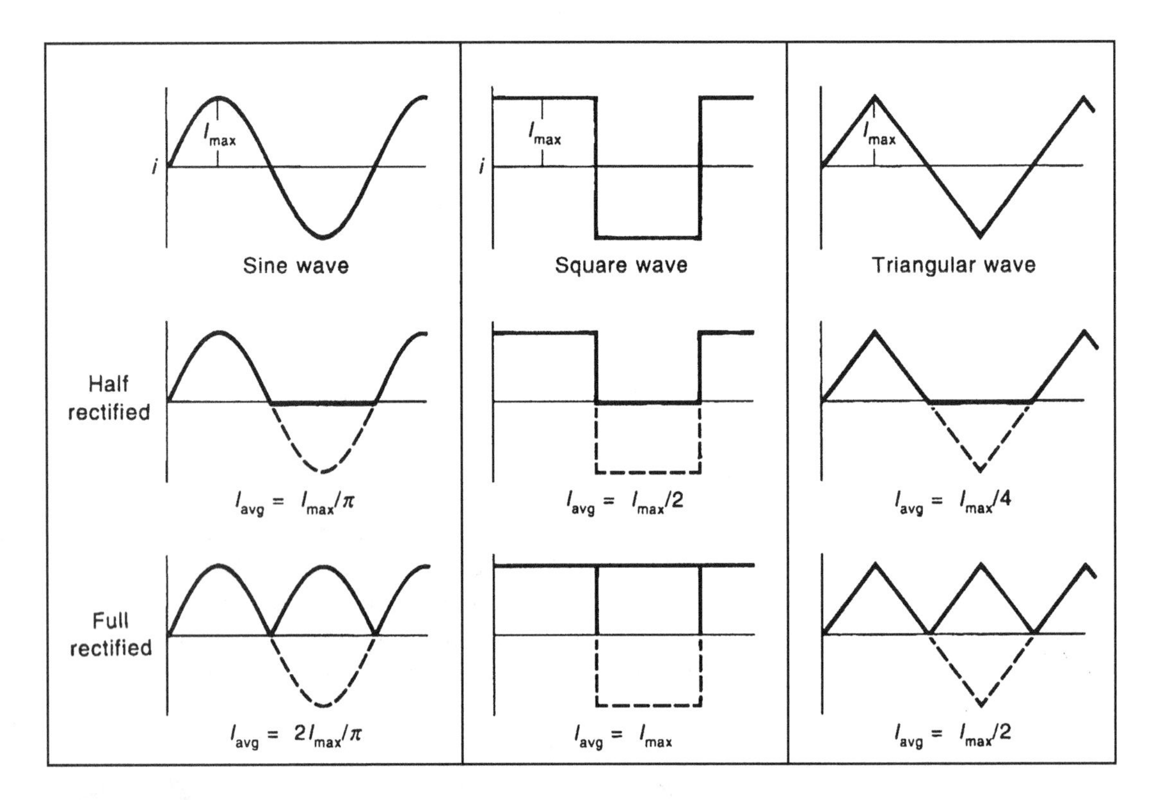

Example 3.58

A triangular wave form with a peak value of 50 volts is rectified by a half-wave rectifier and goes to a load resistor of 100 ohms (see Exhibit 51); the ammeter in the circuit is a dc type that measures average current. What is this current?

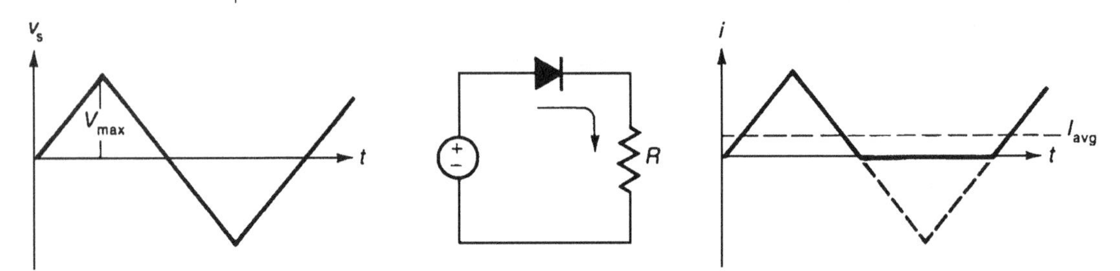

Exhibit 51 Half-wave rectifying circuit

Solution

The current is v/R and is found from the equation for average values as

$$I_{avg} = 1/R\left(1/T\int_0^T v\,dt\right) = 1/R\left[1/T\int_0^{T/2} v\,dt + 1/T\int_{T/2}^{T} 0\,dt\right] \qquad \textbf{(3.99)}$$

The integral is the area of the triangle over a half period divided by a full period, T.

$$I_{avg} = [(V_{max}/R)(0.5)(0.5T)]/T = (50/100)(1/4) = 0.125\text{ A}$$

Or, the current can be found by reading directly from Table 3.6; the average reading is one-quarter of the peak (or maximum) value of the current.

Operational Amplifiers

The operational amplifier is a high gain differential amplifier circuit (see the following list) that has been highly developed over the years. Since the cost has gone from a few hundred dollars (old vacuum tube era) to a few cents for highly developed integrated circuits, the applications for this device cover almost all areas of engineering instrumentation. This discussion focuses on an ideal operational amplifier that is treated externally as a black box. This operational amplifier (referred to as an *op-amp*) differs from a normal amplifier. For example, a home hi-fi audio amplifier's frequency response is considered good if it amplifies voltages over a frequency range from approximately 20 Hz to 20,000 Hz. An op-amp has much higher amplification. It also has several other significant characteristics:

- The amplification is usually of the order 100,000 or more. It is based on the input voltage being the difference between two very small voltages, $v_d = v_p - v_n = v_1 - v_2$. These small voltages are designated as positive and negative; however, the actual polarity of the applied voltages could be either. This implies that the output voltage is zero if the two small input voltages are equal.
- The amplification is flat over a frequency range from zero Hz (that is, *down to dc*) to some very high frequency (high compared to the highest frequency being amplified). Before the low-drift transistor, great care and clever circuit design were required to obtain a *dc amplification*.
- The currents to the actual input pins (both positive and negative) are very small and may usually be neglected; this means that the input resistance (or impedance) is very high. Also, the output resistance (or impedance) is very low; again, "low" means relatively small compared with the external circuitry values.
- The device is linear over a known range, which means that the superposition theorem applies over a range of voltages (for example, if the input voltage were doubled, the output voltage would double).

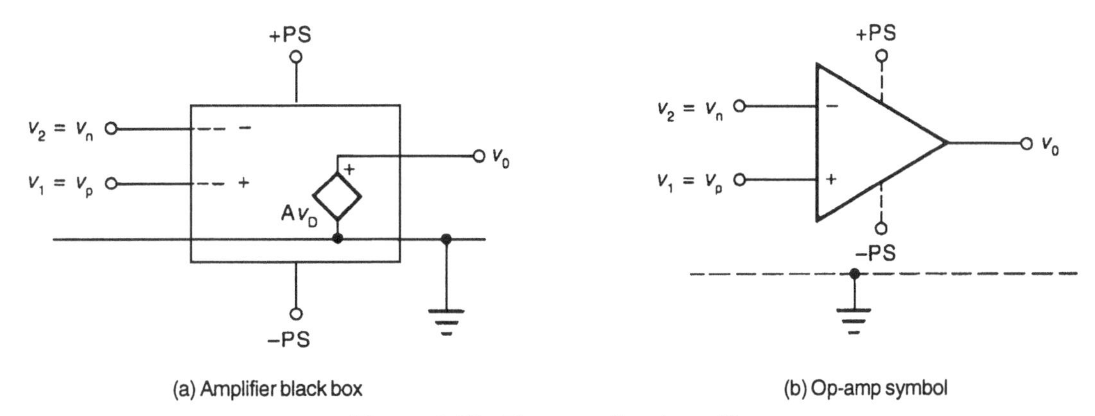

Figure 3.70 The operational amplifier

One may visualize an op-amp as the equivalent circuit shown in Figure 3.70(a). In this figure, the positive and negative power supply (+PS, –PS) connections are shown, but, as in Figure 3.70(b), they may not be shown. Since input voltages may be positive or negative, the power supply voltages are both relative to some common point, usually referred to as ground (signal ground). The power supply voltages must be larger than the largest expected output voltage within the limits of the op-amp. These constraints are normally assumed and are not usually stated or even shown on a diagram [see Figure 3.70(b)]; the triangular symbol may have the vertical side shown as slightly rounded in some diagrams.

The op-amp is used in circuits designed for use in either the inverting or the noninverting mode. Consider the inverting mode circuit (see Figure 3.71); here, since the two voltage input pins are assumed to go to an open circuit, the input currents for both the plus and minus pins are zero, and the voltage difference between these two pins is essentially zero. Since the gain is so high and output voltages are in the several-volts range, the input voltages are in the microvolt range. This makes the circuit analysis especially easy because if one knows one input voltage, the other is essentially the same (actually only a few microvolts different). So, if one input happens to be grounded [as in Figure 3.71(a)], the other input is almost zero. The equations for the output voltage given in Figure 3.71(a) are easily obtained by realizing that the current into the op-amp itself is zero. By summing currents at Node 1, i_a must equal $-i_f$ and the voltage at the junction is near zero volts. Since the node voltage is essentially zero, i_1 is v_a/R_1 and i_f is v_0/R_f. Thus, the circuit amplifies (by a factor of $-R_f/R_1$) as it inverts and has a relatively low current input at v_a (to the external circuit if R_1 is high) and whose output acts almost like an ideal voltage source. For more than one input, one simply sums the currents at Node 1 to yield the equation in Figure 3.71(b).

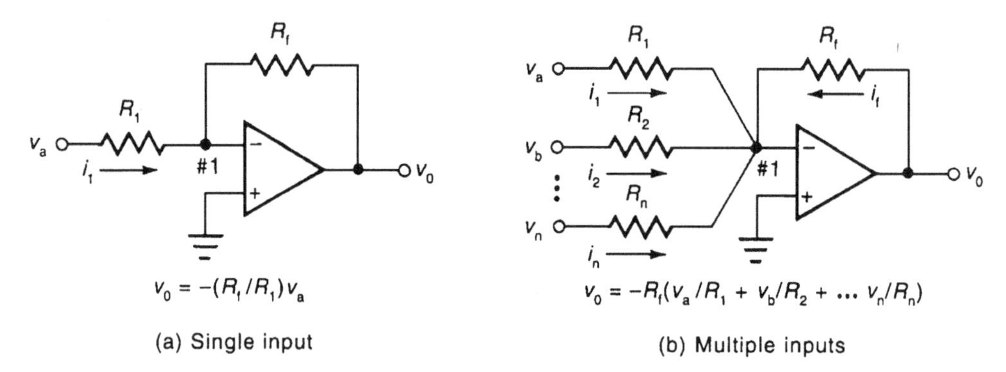

Figure 3.71 The op-amp in the inverting mode

As an example problem [for Figure 3.71(b)], four different transducers produce a possible maximum ac voltage of 0.1, 1.0, 5.0, and 10 volts, respectively. Each transducer is to be recorded on a one channel recorder whose desirable input signal level is one volt but whose input impedance will not allow a direct connection to the transducers. Since only one input at a time may be recorded, it is necessary only to determine each resistance ratio for a summing op-amp circuit. Summing is not required, but multiplying by a constant is. The ratios are easily found; in changing the level of the first voltage from 0.1 to 1, the ratio is ten. The common feedback resistor is arbitrarily chosen to be 100k ohms (the typical range of values runs from a few kilo-ohms to several megohms), then R_1 must be 10k ohms. Calculate the numerical values for the other resistor ratios shown in the circuit of Figure 3.72. When using the op-amp as a summer (all inputs connected at the same time), the instantaneous algebraic sum of the input voltages multiplied by their resistor ratio values will be summed together. They should not exceed the specified linear output voltage.

The other standard circuit configuration for the op-amp is the noninverting mode. For this, the negative input pin is normally fed from the output [perhaps through a resistive network as in Figure 3.73(b)], and the plus pin terminal is connected as an input. For the direct-feedback design [see Figure 3.73(a)], the output voltage—being also the minus terminal input—must be within a few microvolts of the plus terminal. Thus, this circuit has an output that is essentially the same as the input, and this circuit has a gain of unity with no change in polarity. The circuit is usually referred to as a buffer or voltage follower; the advantages are that it takes almost no power from an input source (that is, it does not "load" the input), and the output is almost as though it were an ideal voltage source. For the circuit shown in Figure 3.73(b), the minus input voltage must follow a certain percentage of the output and is considered a noninverting amplifier whose gain is v0 /vin = Gain = 1 + R2/R1. As an example, if a nonverting op-amp with a gain of three is desired, the resistor ratio must be two. If R2 is 10 kΩ, then R1 must be 5 kΩ.

The key to solving these operational amplifier circuits is that the current input to the op-amp itself is considered to be zero. And, if either the plus or minus input voltage is known (or is a ratio of some other voltage), the other input voltage is considered to be the same. The application of Kirchhoff's laws to the rest of the external circuit will usually yield the correct answer.

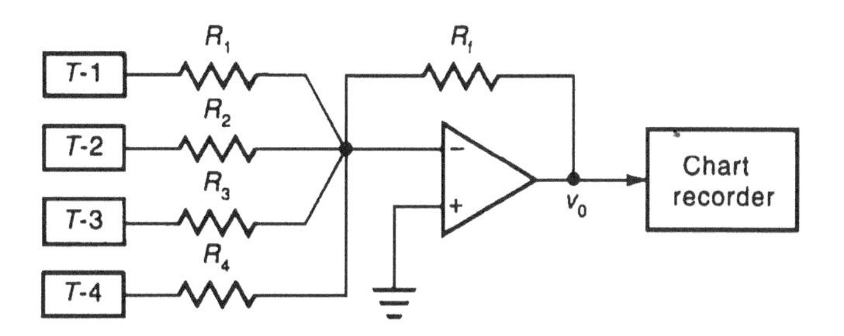

Figure 3.72 An op-amp circuit for matching voltage levels of various transducers. (*Caution*: only one switch to be closed at any one time.)

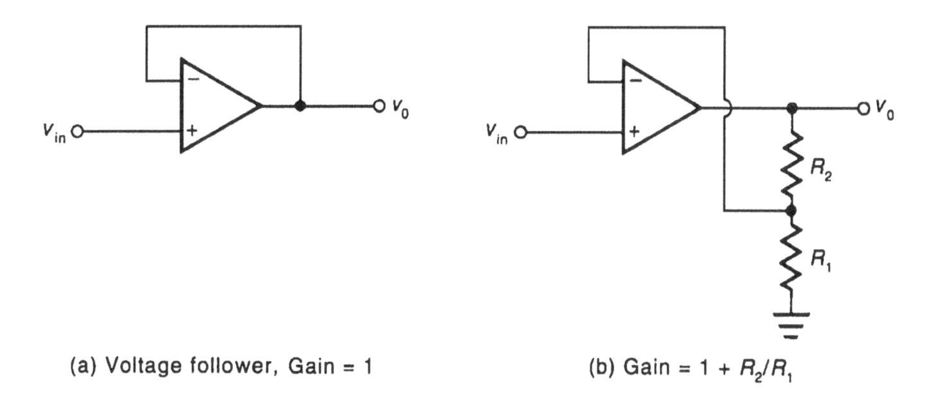

Figure 3.73 Noninverting op-amp circuits

TRANSFORMERS

The transformer is principally used in ac circuits to convert voltages from one level to another through the medium of magnetic fields. There are many other uses for the transformer, but they are specialized (such as pulse transformers). For ac circuits (implying sinusoidal wave forms), the ideal transformer is considered as lossless with 100% efficiency; actually, the efficiency of a typical transformer is greater than 90%. For these ideal devices, the product of the input volt-amps equals the product of output volt-amps (for larger power transformers, $kVA_{in} = kVA_{out}$). The product is the apparent power (VA) rather than real power (watts).

The nameplate rating of the transformer is important. The manufacturer gives the normal operating conditions; these nameplate ratings include the frequency, the voltage and the voltage ratio, and the kVA rating. The voltage ratios are the same as the turns ratio, a, and the current ratios are inversely related to the turns ratio. Whether a voltage is stepped up or down depends on which side one considers as primary and as secondary. For this discussion, assume the left side is primary (1) and the right side is secondary (2).

Example 3.59

For example, consider a transformer with a nameplate rating of 5 k*VA*, 60 Hz, and 880:220 V. The primary side might come from an 880-volt source, and the secondary would be at 220 volts.

Solution

The voltage/current rating is always given in rms values, and the turns ratio, of course, is 4:1. The current (or load) on the secondary side could be as high as I_2 = VA/V_2 = 5000/220 = 22.7 amperes, whereas the primary side would be I_1 = 5000/880 = 5.68 amperes, or 1/4 of 22.7 amperes. Note (from Exhibit 52) that a resistive load on the secondary side is $R_L = V/I = 220/22.7 = 9.69$ ohms, whereas it would be 880/5.68 = 155 ohms on the primary side. This equivalent resistance on the left side is a^2 times R_L:

$$R'_L = a^2 R_L = (4)^2 \times 9.69 = 155\ \Omega$$

Here, $kVA_{in} = kVA_{out}$ and $kW_{in} = kW_{out}$ because the load is a pure resistive one (the power factor of the load is unity). If the load were $5 + j5$ (or $Z_L = \sqrt{2} \times 5\angle 45°$), the current would be

$$I_2 = V_2/Z_L = 220/\left(\sqrt{2} \times 5\angle 45°\right) = 31.1\angle{-45°}\ \text{A}$$

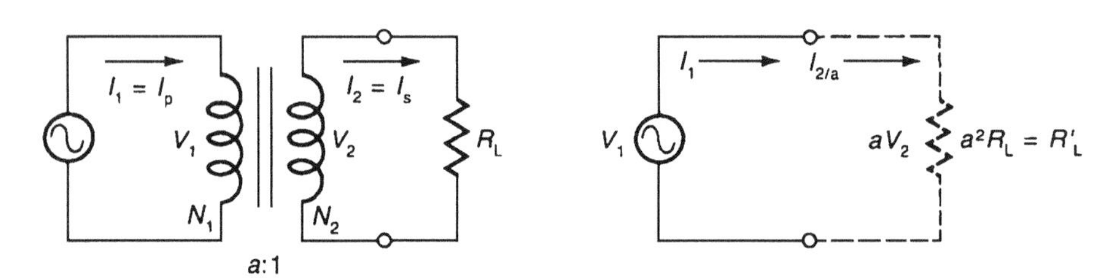

(a) Circuit with load (b) Circuit with equivalent load

Exhibit 52 A typical two-winding, loaded transformer

The current on the left side is $I_1 = I_2/a = 7.78$ amperes. These exceed the nameplate values, and a larger transformer would have to be selected. The equivalent impedance, Z'_L, of the load if reflected to the left side is

$$Z'_L = a^2 Z_L = (4)^2 5 + j(4)^2 5 = 80 + j80\ \Omega$$

Caution is needed here because the power-in/power-out relationship is misleading, $P = VI\cos\theta = 220\,(31.1)\,(0.707) = 4837$ W, or less than 5 kW; the transformer rating is 40 kVA, not necessarily 40 kW.

This chapter has reviewed selected electrical fundamentals. The depth of coverage has been limited. The reader who has more study time should select a text written about electrical engineering for all engineers. Of the many books available, several stand out. Any edition of the following should be available at a library or bookstore:

- Carlson and Gisser, *Electrical Engineering Concepts and Applications,* Addison-Wesley.
- Clement and Johnson, *Electrical Engineering Science*, McGraw-Hill.
- Smith, *Circuits, Devices and Systems*, Wiley.

SELECTED SYMBOLS AND ABBREVIATIONS

Symbol or Abbreviation	Definition
A	area
A	amperes
A • t	amp-turns
ac	alternating current
B	magnetic flux density; susceptance
C	coulomb
C	capacitance
dc	direct current
E	electric field
e	induced voltage
ε	permittivity
F	farad
F	force
G	conductance
H	henry
H	magnetic field intensity
I_c	core loss
I_e	exciting current
I_f	field current
I_i	current
I_m	magnetizing current
K	dielectric constant
KCL	Kirchhoff's Current Law
KVL	Kirchhoff's Voltage Law
L	inductance
l	length
ma	milliamp
mv	millivolts
mmf	magneto-motive force
N	number of turns
N_{ag}	newton air gap
n	speed
n_s	synchronous speed
P_i, p	power

(Continued)

(Continued)

Symbol or Abbreviation	Definition
ϕ	magnetic flux
Q	point change, reactive power
R	resistance
R_{eq}	Thevenin's resistance
rms	root mean square
S	siemens
s	slip
T	tesla
T	period
T_d	developed torque
θ	angle
V, v	voltage
v	velocity
V_i, V_{oc}	terminal voltage, Thevenin's voltage
W	work
w	total energy
Y	admittance
Z, z	impedance
N_p	number of pole pairs

REFERENCE

Van Vlack, L. *Elements of Materials Science and Engineering*, 6th ed. Addison-Wesley, 1989

PROBLEMS

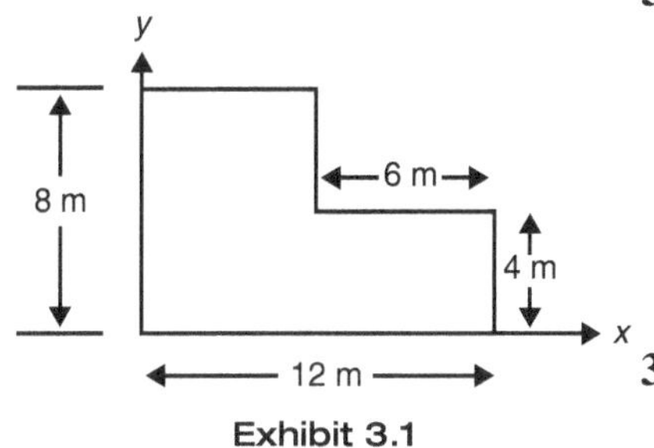

Exhibit 3.1

3.1 The uniform density flat plate shown in Exhibit 3.1 has a mass of 480 kg. The density of the body is most nearly:
a. 10 kg/m2
b. 4.6 kg/m2
c. 6.7 kg/m2
d. 11.8 kg/m2

3.2 Assuming the density of the plate in Exhibit 3.1 is 9 kg/m^2, the mass moment of inertia of the body is most nearly:
a. 10,400 kg-m^2
b. 18,300 kg-m^2
c. 9,600 kg-m^2
d. 7,400 kg-m^2

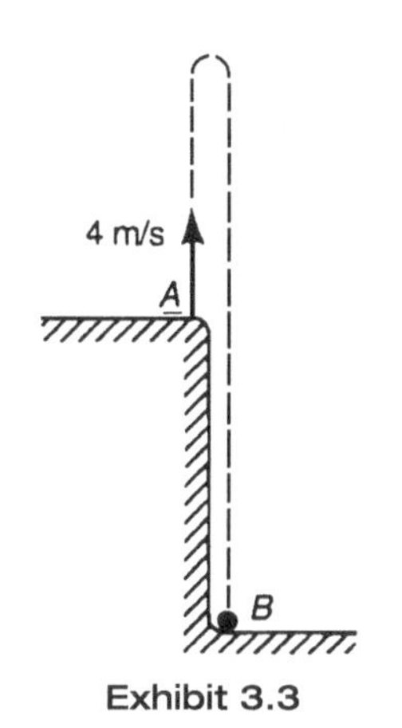

Exhibit 3.3

3.3 A particle is thrown vertically upward from the edge *A* of the ditch shown in Exhibit 3.3. If the initial velocity is 4 m/s, and the particle is known to hit the bottom, *B*, of the ditch exactly 6 seconds after it was released at *A*, determine the depth of this ditch. Neglect air resistance.
a. 24.0 m
b. 152.6 m
c. 200 m
d. 176.6 m

3.4 The slider *P* in Exhibit 3.4 is driven by a complex mechanism in such a way that (i) it remains on a straight path throughout; (ii) at the instant t = 0, the slider is located at A, (iii) at any general instant of time, the velocity of *P* is given by $v = (3t2 - t + 2)$ m/s. Determine the distance of *P* from point *O* when t = 2 s.
a. 26 m
b. 10 m
c. 6 m
d. 12 m

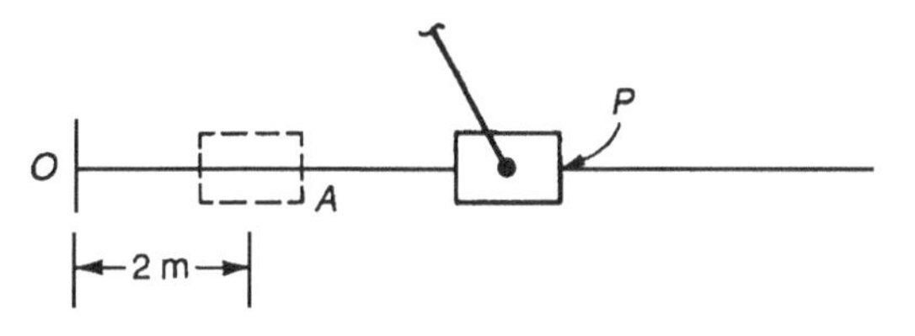

Exhibit 3.4

3.5 A particle in rectilinear motion starts from rest and maintains the acceleration profile shown in Exhibit 3.5. The displacement of the particle in the first 8 seconds is:

a. 4 m
b. 28 m
c. 24 m
d. 20 m

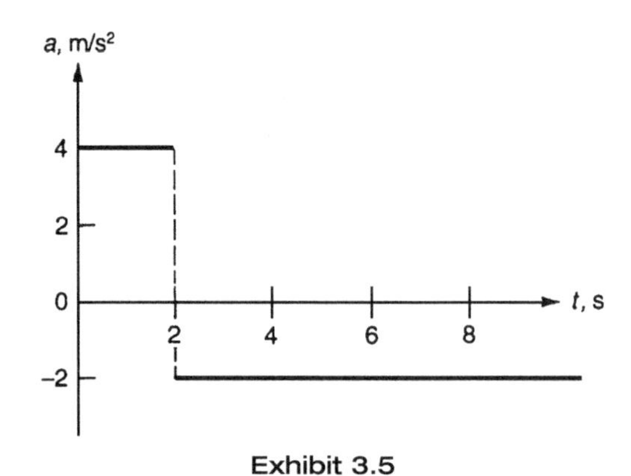

Exhibit 3.5

3.6 A ball is thrown by a player (Exhibit 3.6) from a position 2 m above the ground surface with a velocity of 40 m/s inclined at 60° to the horizontal. Determine the maximum height, *H*, the ball will attain.

a. 63.2 m
b. 61.2 m
c. 30.6 m
d. 31 m

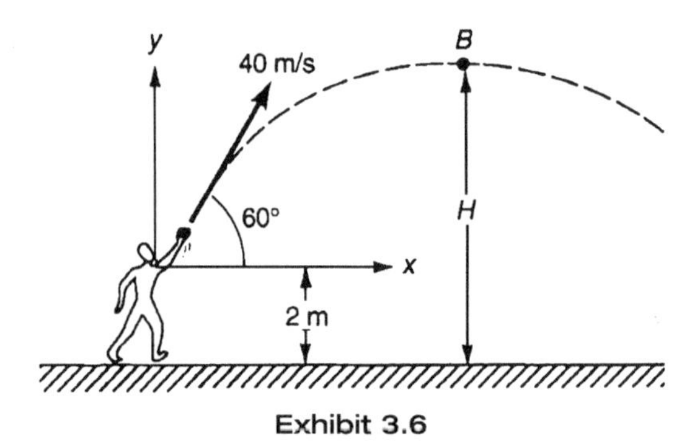

Exhibit 3.6

3.7 A golf ball (Exhibit 3.7) is struck horizontally from point A of an elevated fairway. Determine the initial speed that must be imparted to the ball if the ball is to strike the base of the flag stick on the green 140 meters away. Neglect air friction.

a. 34.3 m/s
b. 103 m/s
c. 90 m/s
d. 19.2 m/s

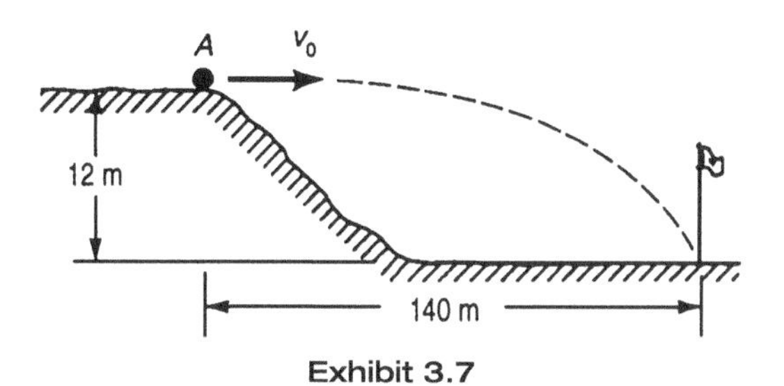

Exhibit 3.7

3.8 In Exhibit 3.8, the rod R rotates about a fixed axis at O. A small collar B is forced down the rod (toward O) at a constant speed of 3 m/s relative to the rod. If the value of θ at any given instant is $\theta = (t^2 + t - 2)$ rad, find the magnitude of the acceleration of B at time $t = 1$ second, when B is known the be 1 meter away from O.

a. 8.0 m/s^2
b. 20.2 m/s^2
c. 18.4 m/s^2
d. 3.0 m/s^2

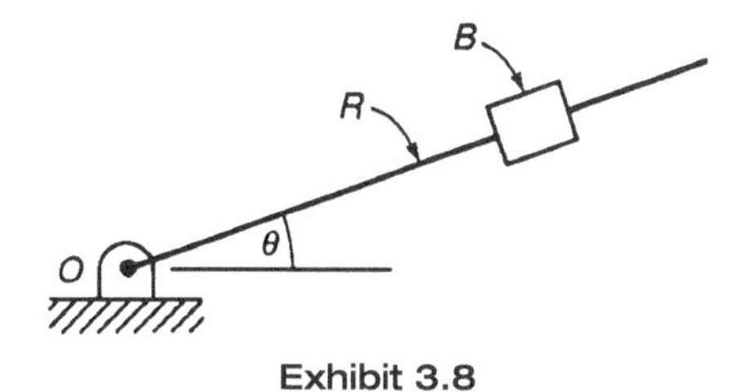

Exhibit 3.8

3.9 A rocket (Exhibit 3.9) is fired vertically upward from a launching pad at *B*, and its flight is tracked by radar from point *A*. Find the magnitude of the velocity of the rocket when $\theta = 45°$ if $\dot{\theta} = 0.1$ rad/s.

a. 36 m/s
b. 180 m/s
c. 90 m/s
d. 360 m/s

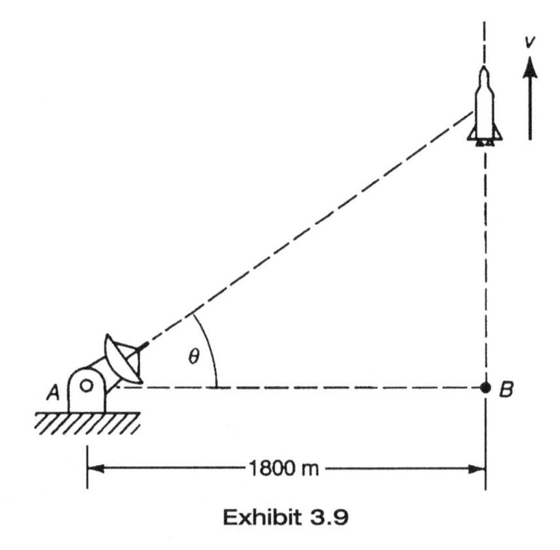

Exhibit 3.9

3.10 A particle is given an initial velocity of 50 m/s at an angle of 30° with the horizontal as shown in Exhibit 3.10. What is the radius of curvature of its path at the highest point, *C*?

a. 19.5 m
b. 255 m
c. 221 m
d. 191 m

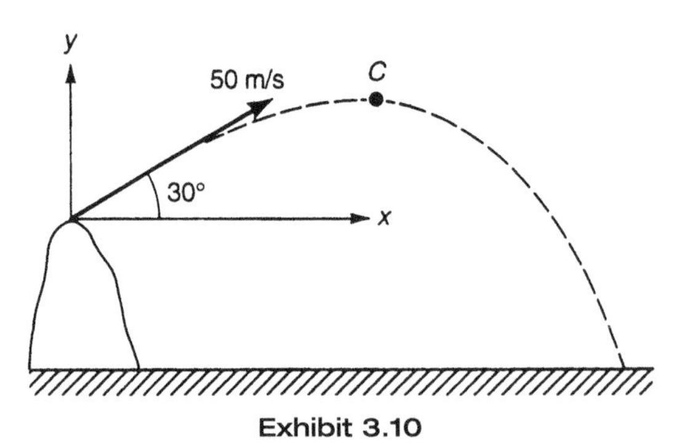

Exhibit 3.10

3.11 An automobile moves along a curved path that can be approximated by a circular arc of radius 110 meters. The driver keeps his foot on the accelerator pedal in such a way that the speed increases at the constant rate of 3 m/s^2. What is the total acceleration of the vehicle at the instant when its speed is 20 m/s?

a. 22.0 m/s^2
b. 3.6 m/s^2
c. 3.0 m/s^2
d. 4.7 m/s^2

3.12 A pilot testing an airplane at 800 kph wishes to subject the aircraft to a normal acceleration of 5 *gs* in order to fulfill the requirements of an onboard experiment. Find the radius of the circular path that would allow the pilot to do this.

a. 502 m
b. 3308 m
c. 1007 m
d. 1453 m

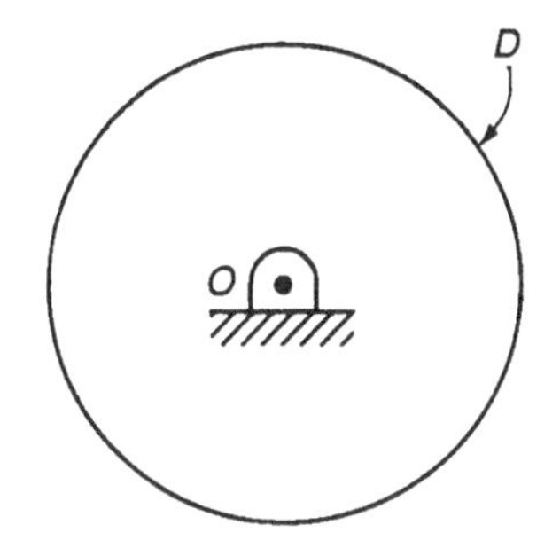

Exhibit 3.13

3.13 At the instant $t = 0$, the disk D in Exhibit 3.13 is spinning about a fixed axis through O at an angular speed of 300 rpm. Bearing friction and other effects are known to slow the disk at a rate that is k times its instantaneous angular speed, where k is a constant with the value $k = 1.2\ s^{-1}$. Determine when (from $t = 0$) the disk's spin rate is cut in half.

a. 6.5 s
b. 13.1 s
c. 0.8 s
d. 0.6 s

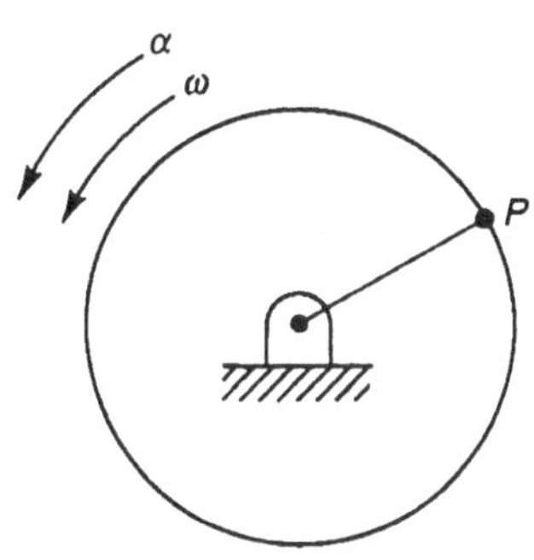

Exhibit 3.14

3.14 In Exhibit 3.14, a flywheel 2 m in radius is brought uniformly from rest up to an angular speed of 300 rpm in 30 s. Find the speed of a point P on the periphery 5 seconds after the wheel started from rest.

a. 10.5 m/s
b. 5.2 m/s
c. 62.8 m/s
d. 100.0 m/s

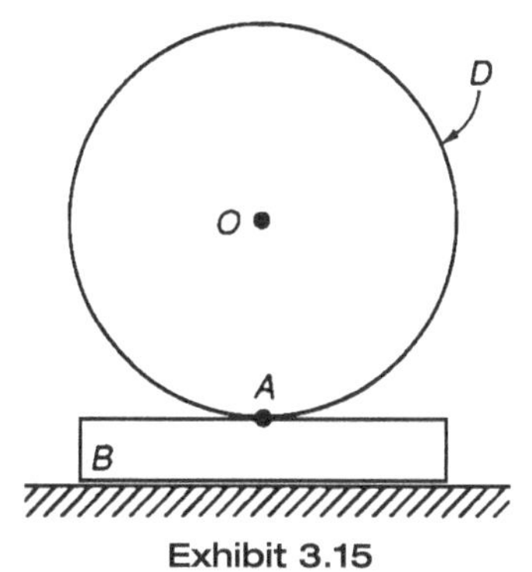

Exhibit 3.15

3.15 The block B (Exhibit 3.15) slides along a straight path on a horizontal floor with a constant velocity of 2 m/s to the right. At the same time, the disk, D, of 3-m diameter rolls without slip on the block. If the velocity of the center, O, of the disk is directed to the left and remains constant at 1 m/s, determine the angular velocity of the disk.

a. 0.3 rad/s counterclockwise
b. 2.0 rad/s counterclockwise
c. 0.7 rad/s counterclockwise
d. 0.7 rad/s clockwise

3.16 In Exhibit 3.16, the disk, D, rolls without slipping on a horizontal floor with a constant clockwise angular velocity of 3 rad/s. The rod, R, is hinged to D at A, and the end, B, of the rod touches the floor at all times. Determine the angular velocity of R when the line OA joining the center of the disk to the hinge at A is horizontal as shown.

a. 0.6 rad/s counterclockwise
b. 0.6 rad/s clockwise
c. 3.0 rad/s counterclockwise
d. 3.0 rad/s clockwise

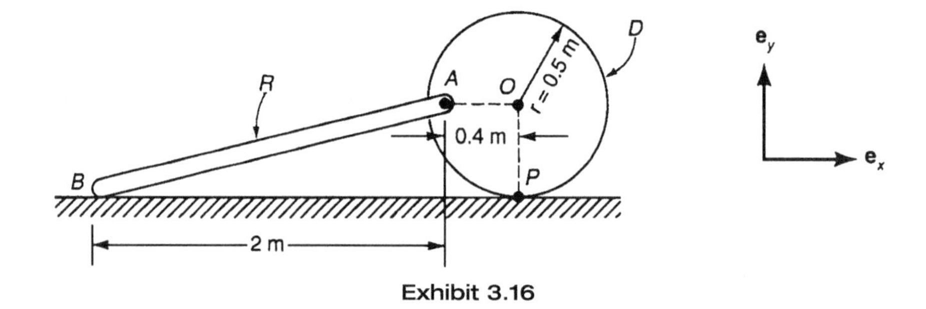

Exhibit 3.16

3.17 The fire truck in Exhibit 3.17 is moving forward along a straight path at the constant speed of 50 km/hr. At the same time, its 2-meter ladder OA is being raised so that the angle θ is given as a function of time by $\theta = (0.5t^2 - t)$ rad, where t is in seconds. The magnitude of the acceleration of the tip of the ladder when $t = 2$ seconds is:

a. 0
b. 4.0 m/s2
c. 2.0 m/s^2
d. 2.8 m/s^2

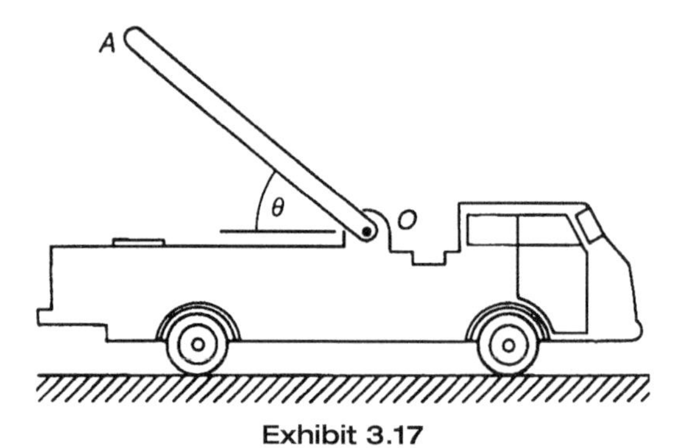

Exhibit 3.17

3.18 In Exhibit 3.18, the block B is constrained to move along a horizontal rectilinear path with a constant acceleration of 2 m/s^2 to the right. The slender rod, R, of length 2 m is pinned to B at O and can swing freely in the vertical plane. At the instant when $\theta = 0°$ (rod is vertical), the angular velocity of the rod is zero but its angular acceleration is 2.5 m/s^2 clockwise. Find the acceleration of the midpoint G of the rod at this instant ($\theta = 0°$).

a. 3.0 m/s2 ←
b. 0.5 m/s2 →
c. 2.5 m/s2 ←
d. 2.5 m/s2 →

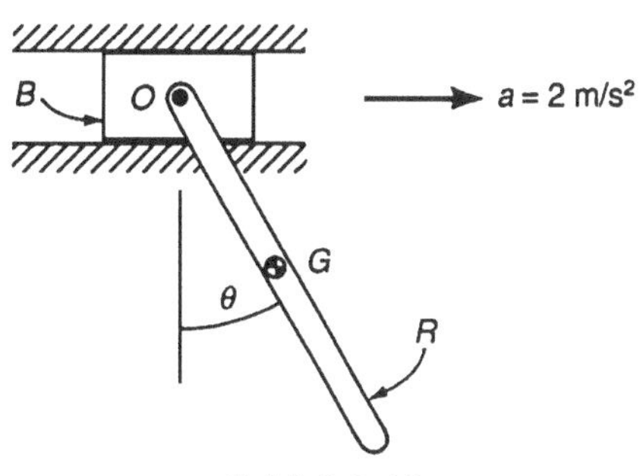

Exhibit 3.18

3.19 The block, B, in Exhibit 3.19, contains a square-cut circular groove. A particle, P, moves in this groove in the clockwise direction and maintains a constant speed of 6 m/s relative to the block. At the same time, the block slides to the right on a straight path at the constant speed of 10 m/s. Find the magnitude of the absolute velocity of P at the instant when $\theta = 30°$.

a. 8.7 m/s
b. 16 m/s
c. 4 m/s
d. 14 m/s

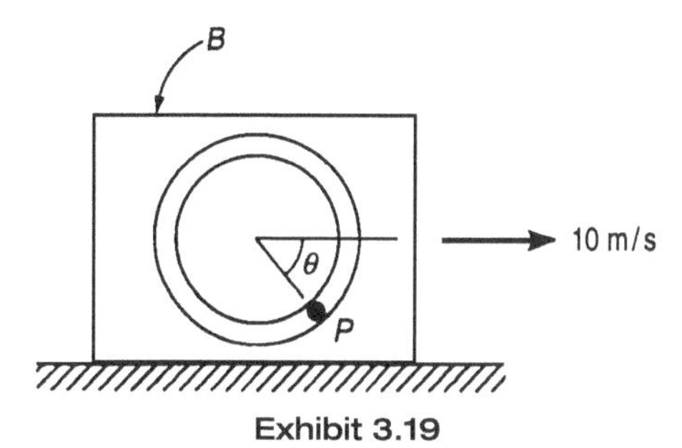

Exhibit 3.19

3.20 In Exhibit 3.20, a pin moves with a constant speed of 2 m/s along a slot in a disk that is rotating with a constant clockwise angular velocity of 5 rad/s. Calculate the absolute acceleration of this pin when it reaches the position C (directly above O). The unit vectors $\mathbf{e}_x$ and $\mathbf{e}_y$ are fixed to the disk.

a. $17.5\mathbf{e}_y$ m/s^2
b. $-17.5\mathbf{e}_y$ m/s^2
c. $-2.5\mathbf{e}_y$ m/s^2
d. $-22.5\mathbf{e}_y$ m/s^2

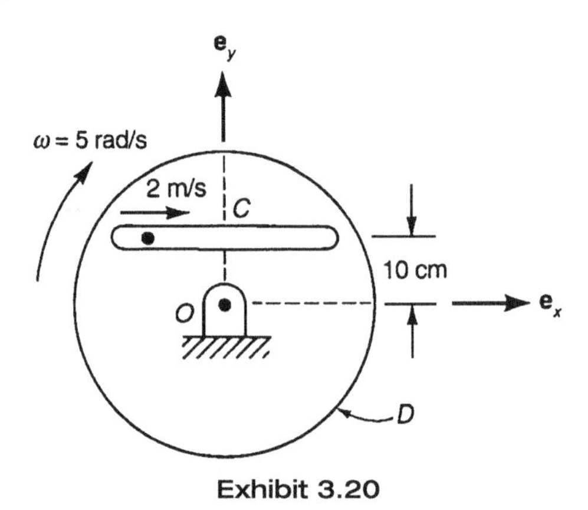

Exhibit 3.20

3.21 In Exhibit 3.21, a particle P of mass 5 kg is launched vertically upward from the ground with an initial velocity of 10 m/s. A constant upward thrust $T = 100$ newtons is applied continuously to P, and a downward resistive force $R = 2z$ newtons also acts on the particle, where z is the height of the particle above the ground. Determine the maximum height attained by P.

a. 6.0 m
b. 45.5 m
c. 15.8 m
d. 55.5 m

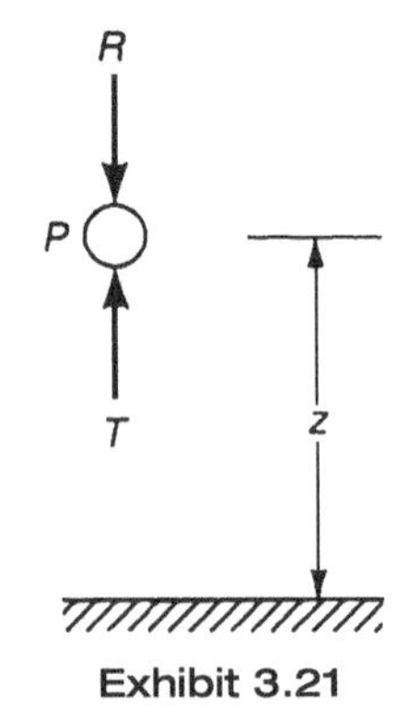

Exhibit 3.21

3.22 Determine the force P required to give the block shown in Exhibit 3.22 an acceleration of 2 m/s^2 up the incline. The coefficient of kinetic friction between the block and the incline is 0.2.

a. 39.2 N
b. 21.9 N
c. 44.6 N
d. 49.8 N

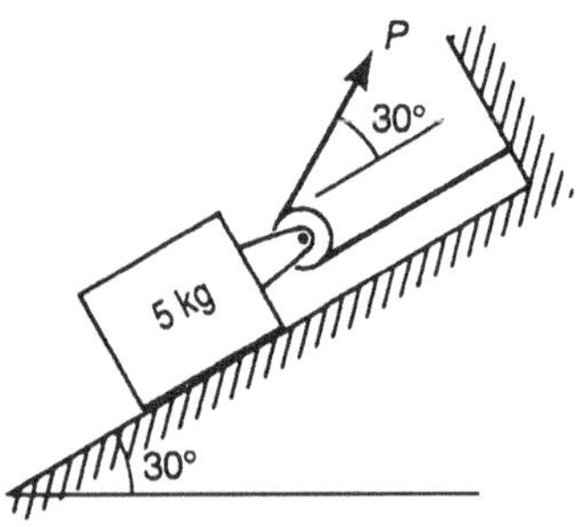

P
30°
5 kg
30°

Exhibit 3.22

3.23 In Exhibit 3.23 the rod R rotates in the vertical plane about a fixed axis through the point O with a constant counterclockwise angular velocity of 5 rad/s. A collar B of mass 2 kg slides down the rod (toward O) so that the distance between B and O decreases at the constant rate of 1 m/s. At the instant when $\theta = 30°$ and $r = 400$ mm, determine the magnitude of the applied force P. The coefficient of kinetic friction between B and R is 0.1.

a. 9.9 N
b. 11.9 N
c. 10.5 N
d. 0.3 N

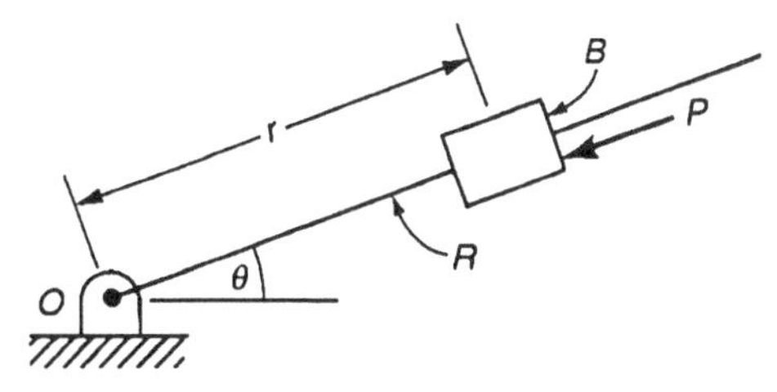

Exhibit 3.23

3.24 The 3-kg collar in Exhibit 3.24 slides down the smooth circular rod. In the position shown, its velocity is 1.5 m/s. Find the normal force (contact force) the rod exerts on the collar.

a. 12.2 N ↖
b. 24.4 N ↘
c. 19.2 N ↖
d. 12.2 N ↘

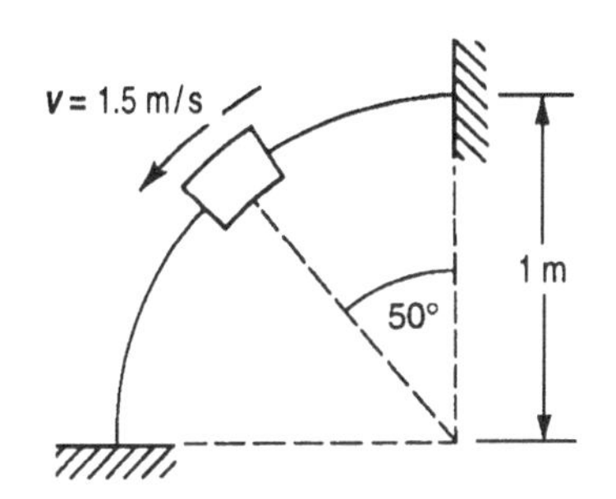

Exhibit 3.24

3.25 Forklift vehicles, Exhibit 3.25, tend to roll over if they are driven too fast while turning. For a vehicle of mass m with a mass center that describes a circle of radius R, find the relationship between the forward speed u and the vehicle dimensions and path radius at the onset of tipping.

a. u = (Rgb/H)0.5
b. u = (RgH/b)0.5
c. u = (gh)0.5
d. u = (bg)0.5

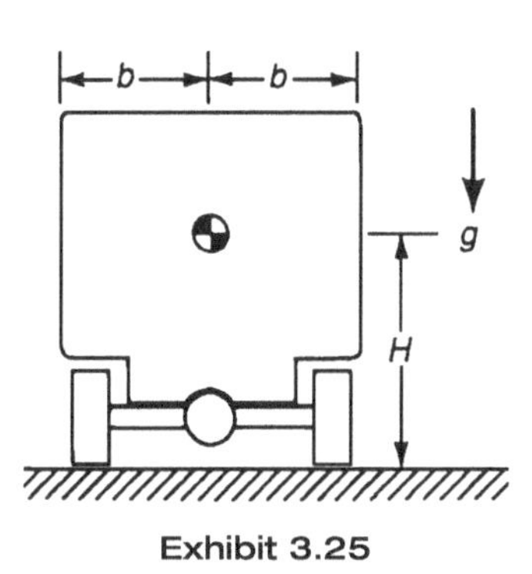

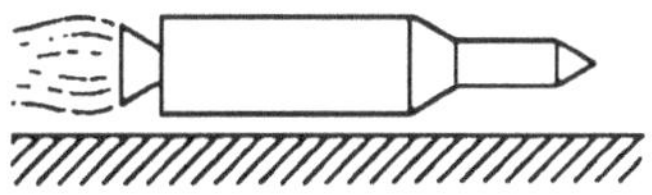

Exhibit 3.25

3.26 A toy rocket of mass 1 kg is placed on a horizontal surface, and the engine is ignited (Exhibit 3.26). The engine delivers a force equal to $(0.25 + 0.5t)$ N, where t is time in seconds, and the coefficient of friction between the rocket and the surface is 0.01. Determine the velocity of the rocket 7 seconds after ignition.

a. 14.0 m/s
b. 3.7 m/s
c. 13.3 m/s
d. 26.3 m/s

Exhibit 3.26

3.27 A 2000-kg pickup truck is traveling backward down a 10° incline at 80 km/hr when the driver notices through his rearview mirror an object on the roadway. He applies the brakes, and this results in a constant braking (retarding) force of 4000 N. How long does it take the truck to stop?

a. 11.1 s
b. 74.9 s
c. 2.3 s
d. 13.0 s

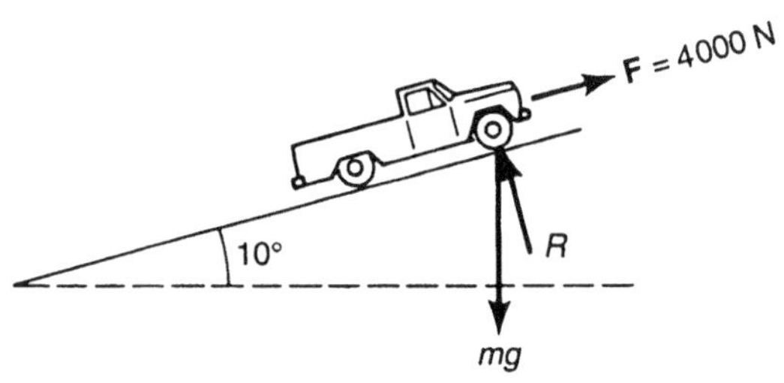

Exhibit 3.27

3.28 In Exhibit 3.28, a particle C of mass 2 kg is sliding down a smooth incline with a velocity of 3 m/s when a horizontal force $P = 15$ N is applied to it. What is the distance traveled by C between the instant when P is first applied and the instant when the velocity of C becomes zero?

a. 1.7 m
b. 5.6 m
c. 2.8 m
d. 0.9 m

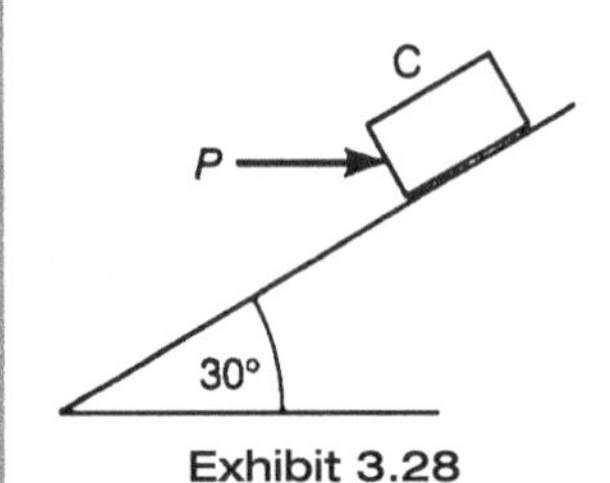

Exhibit 3.28

3.29 A particle moves in a vertical plane along the path ABC shown in Exhibit 3.29. The portion AB of the path is a quarter-circle of radius r and is smooth. The portion BC is horizontal and has a coefficient of friction μ. If the particle has mass m and is released from rest at A, determine the horizontal distance H that the particle will travel along BC before coming to rest.

a. μr
b. $2r$
c. r/μ
d. $2r/\mu$

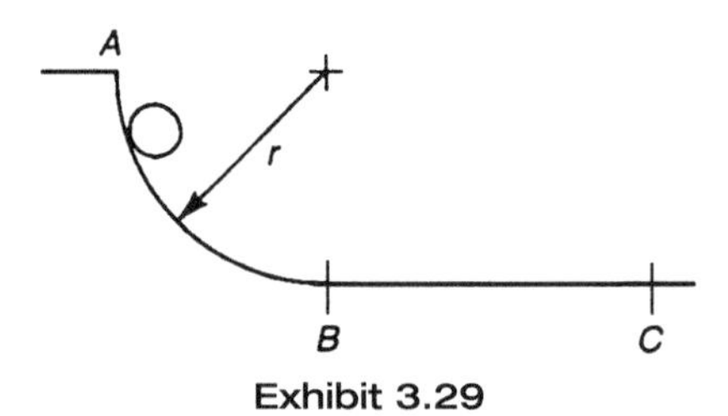

Exhibit 3.29

3.30 In Exhibit 3.30, a block of mass 2 kg is pressed against a linear spring of constant $k = 200$ N/m through a distance Δ on a horizontal surface. When the block is released at A, it travels along the straight horizontal path ADB and traverses point B with a velocity of 1 m/s. If the coefficient of kinetic friction between the block and the floor is 0.2, find Δ.

a. 0.22 m
b. 0.12 m
c. 0.26 m
d. 0.08 m

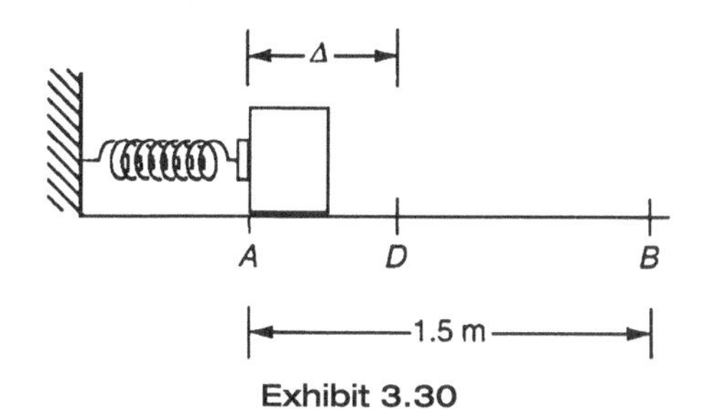

Exhibit 3.30

3.31 In Exhibit 3.31, a 6-kg block is released from rest on a smooth inclined plane as shown. If the spring constant $k = 1000$ N/m, determine how far the spring is compressed. Assume the acceleration of gravity, $g = 10$ m/s^2.

a. 0.40 m
b. 0.45 m
c. 0.83 m
d. 3.96 m

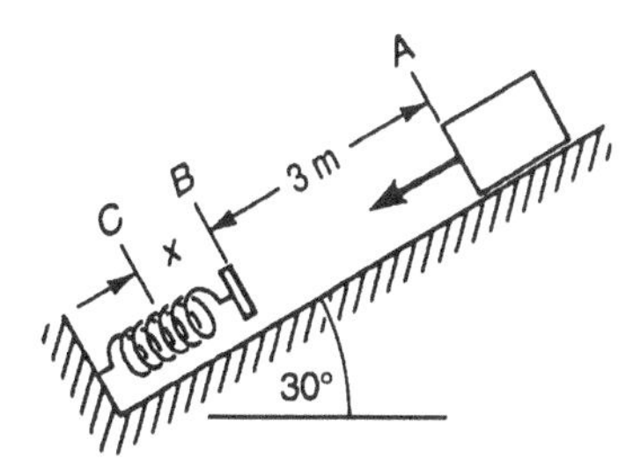

Exhibit 3.31

3.32 A train of joyride cars full of children in an amusement park is pulled by an engine along a straight-level track. It then begins to climb up a 5° slope. At a point *B*, 50 m up the grade when the velocity is 32 km/h, the last car uncouples without the driver noticing (Exhibit 3.32). If the total mass of the car with its passengers is 500 kg and the track resistance is 2% of the total vehicle weight, calculate the total distance up the grade where the car stops at point *C*.

a. 260 m
b. 37.6 m
c. 48.7 m
d. 87.6 m

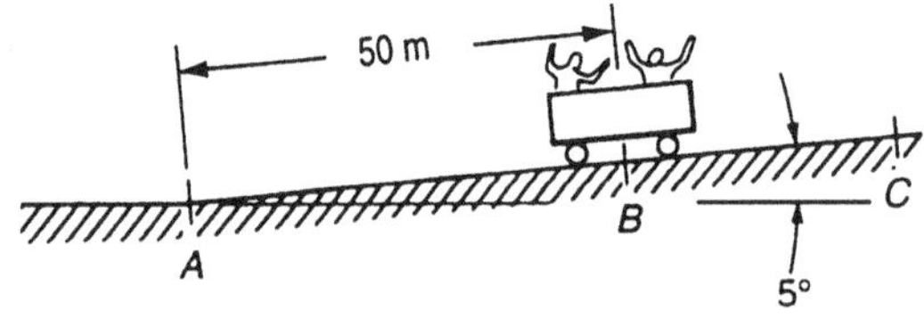

Exhibit 3.32

3.33 Two identical rods, each of mass 4 kg and length 3 m, are rigidly connected as shown in Exhibit 3.33. Determine the moment of inertia of the rigid assembly about an axis through the point *A* and perpendicular to the plane of the paper.

a. 19 kg-m^2
b. 23 kg-m^2
c. 18 kg-m^2
d. 15 kg-m^2

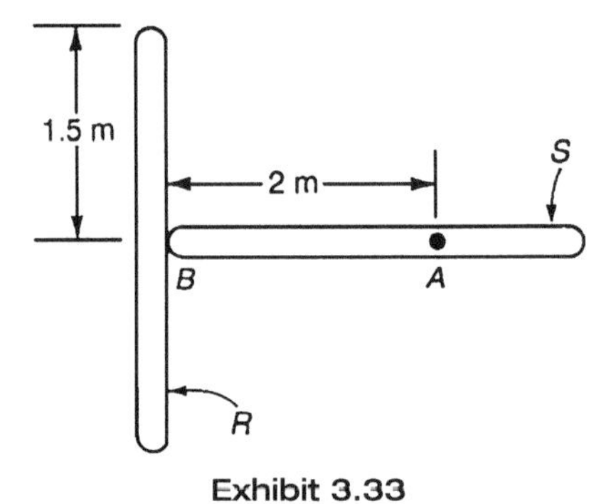

Exhibit 3.33

3.34 A torque motor, represented by the box in Exhibit 3.34, is to drive a thin steel disk of radius 2 m and mass 1.5 kg around its shaft axis. Ignoring the bearing friction about the shaft and the shaft mass, find the angular speed of the disk after applying a constant motor torque of 5 N-m for 5 seconds. The initial angular velocity of the shaft is 1 rad/s.

a. 8.3 rad/s
b. 7.3 rad/s
c. 5.2 rad/s
d. 9.3 rad/s

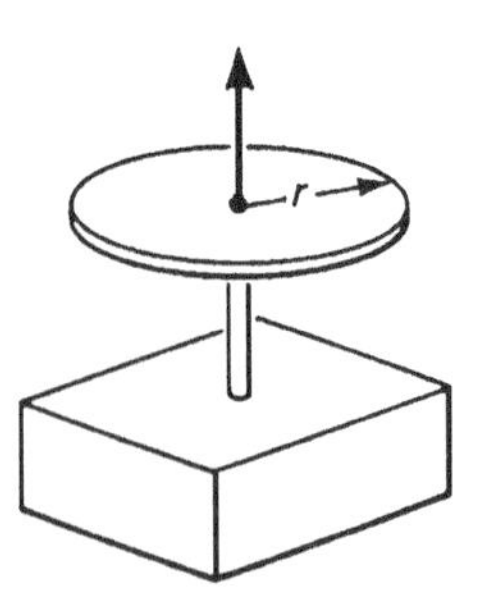

Exhibit 3.34

3.35 In Exhibit 3.35, the uniform slender rod R is hinged to a block B that can slide horizontally. Determine the horizontal acceleration α that must be given to B in order to keep the angle θ constant at 10°, balancing the rod in a tilted position.

a. 1.73 m/s^2
b. 0
c. 9.81 m/s^2
d. 9.66 m/s^2

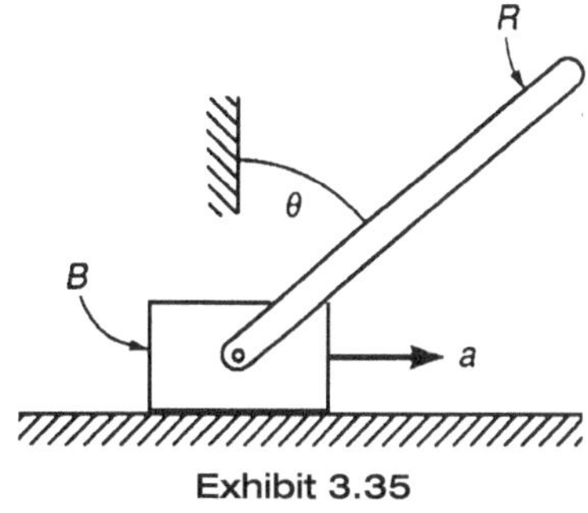

Exhibit 3.35

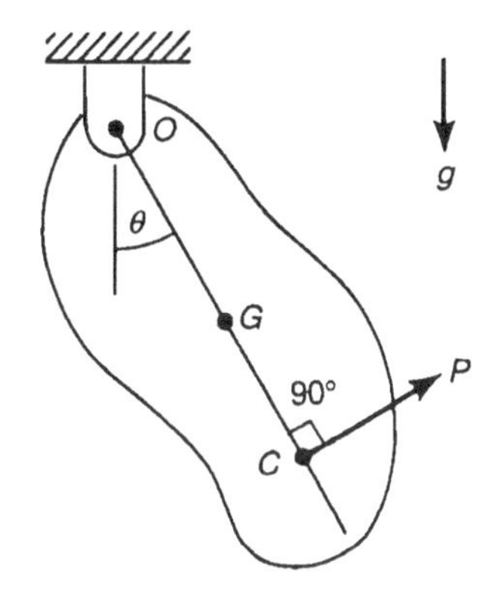

Exhibit 3.36

3.36 In Exhibit 3.36, a force P of constant magnitude is applied to the physical pendulum at point C and remains perpendicular to OC at all times. The pendulum moves in the vertical plane and has mass 3 kg; its mass center is located at G, and the distances are $OG = 1.5$ m, $OC = 2$ m. Also, $P = 10$ N and the radius of gyration of the pendulum about an axis through C and perpendicular to the plane of motion is 0.8 m. Determine the angular acceleration of the pendulum when $\theta = 30°$.

a. 5.31 rad/s^2 counterclockwise
b. 5.31 rad/s^2 clockwise
c. 0.26 rad/s^2 counterclockwise
d. 0.26 rad/s^2 clockwise

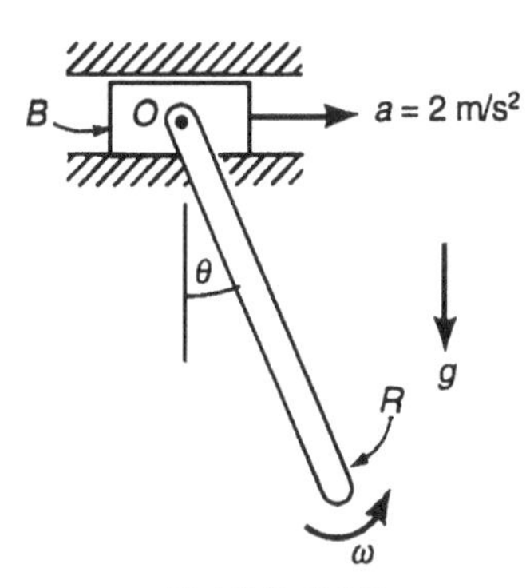

Exhibit 3.37

3.37 In Exhibit 3.37, the block B moves along a straight horizontal path with a constant acceleration of 2 m/s^2 to the right. The uniform slender rod R of mass 1 kg and length 2 m is connected to B through a frictionless hinge and swings freely about O as B moves. Determine the horizontal component of the reaction force at O on the rod when $\theta = 30°$ and $\omega = 2$ rad/s counterclockwise.

a. 4.31 N →
b. 4.31 N ←
c. 6.31 N →
d. 6.31 N ←

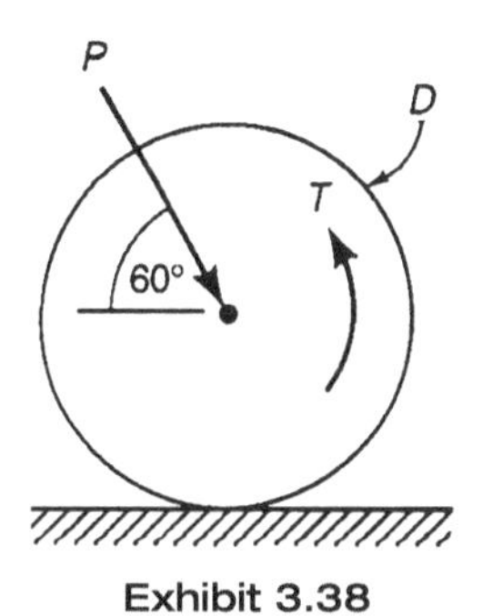

Exhibit 3.38

3.38 In Exhibit 3.38, a homogeneous cylinder rolls without slipping on a horizontal floor under the influence of a force $P = 6$ N and a torque $T = 0.5$ N-m. The cylinder has radius 1 m and mass 2 kg. If the cylinder started from rest, what is its angular velocity after 10 seconds?

a. 8.3 rad/s
b. 6.8 rad/s
c. 1.7 rad/s
d. 0.68 rad/s

3.39 A slender rod of length 2 m and mass 3 kg is released from rest in the horizontal position (Exhibit 3.39) and swings freely (no hinge friction). Find the angular velocity of the rod when it passes a vertical position.

a. 4.43 rad/s
b. 3.84 rad/s
c. 7.68 rad/s
d. 5.43 rad/s

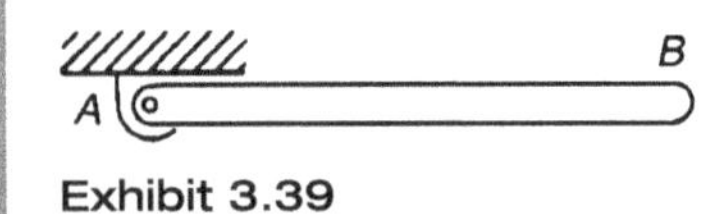

Exhibit 3.39

3.40 The uniform slender bar of mass 2 kg and length 3 m is released from rest in the near-vertical position as shown in Exhibit 3.40, where the torsional spring is undeformed. The rod is to rotate clockwise about *O* and come gently to rest in the horizontal position. Determine the stiffness *k* of the torsional spring that would make this possible. The hinge is smooth.

a. 47.8 N-m/rad
b. 37.5 N-m/rad
c. 0.7 N-m/rad
d. 23.8 N-m/rad

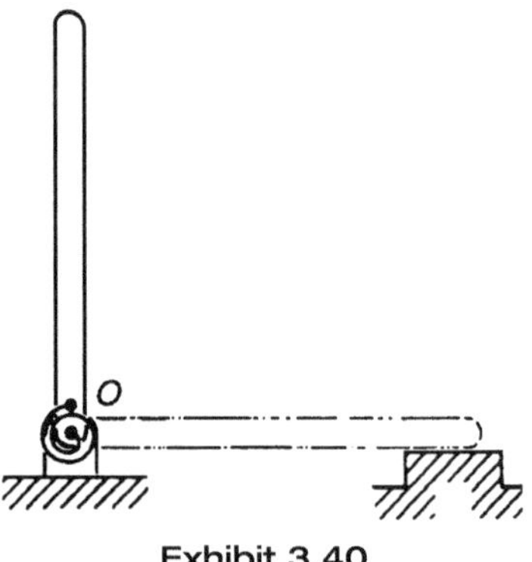

Exhibit 3.40

3.41 A solid homogeneous cylinder is released from rest in the position shown in Exhibit 3.41 and rolls without slip on a horizontal floor. The cylinder has a mass of 12 kg. The spring constant is 2 N/m, and the unstretched length of the spring is 3 m. What is the angular velocity of the cylinder when its center is directly below the point *O*?

a. 1.33 rad/s
b. 1.63 rad/s
c. 1.78 rad/s
d. 2.31 rad/s

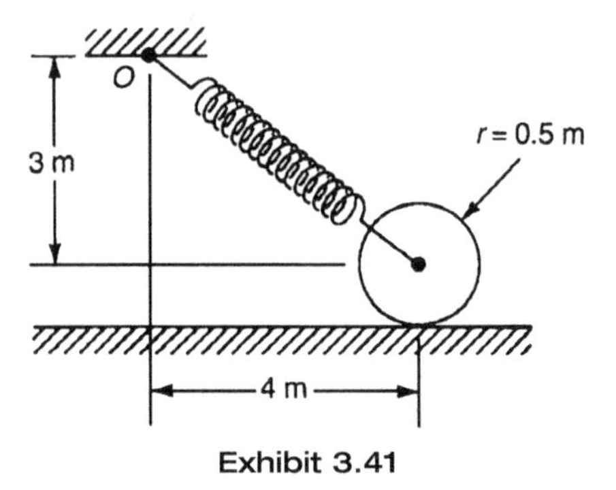

Exhibit 3.41

3.42 For a neutral atom:
a. the atomic mass equals the mass of the neutrons plus the mass of the protons
b. the atomic number equals the atomic mass
c. the number of protons equals the atomic number
d. the number of electrons equals the number of neutrons

3.43 All isotopes of a given element have:
a. the same number of protons
b. the same number of neutrons
c. equal numbers of protons and neutrons
d. the same number of atomic mass units

3.44 Which of the following statements is *FALSE*?
a. An anion has more electrons than protons.
b. Energy is released when water is solidified to ice.
c. Energy is required to remove an electron from a neutral atom.
d. Energy is released when a H_2 molecule is separated into two hydrogen atoms.

3.45 Select the correct statement.
a. Crystals possess long-range order.
b. Within a crystal, like ions attract and unlike ions repel.
c. A body-centered cubic metallic crystal (for example, iron) has nine atoms per unit cell.
d. A face-centered cubic metallic crystal (for example, copper) has 14 atoms per unit cell.

3.46 In a cubic crystal, a is the edge of a unit cell. The shortest repeat distance in the [111] direction of a body-centered cubic crystal is:
a. $a\sqrt{2}$
b. $2a$
c. $a\sqrt{3}/2$
d. $a\sqrt{3}/4$

3.47 All but which of the following data are required to calculate the density of aluminum in g/m^3?
a. Avogadro's number, which is 6.0×10^{23}
b. atomic number of Al, which is 13
c. crystal structure of Al, which is face-centered cubic
d. atomic mass of Al, which is 27 amu

3.48 The atomic packing factor of gold, an FCC metal, is:
a. $(4\pi r^3/3)/(4r/\sqrt{3})^3$
b. $4(4\pi r^3/3)/(4r/\sqrt{2})^3$
c. $4(4\pi r^3/3)/(r/\sqrt{2})^3$
d. $4(2r/\sqrt{2})^3/(4\pi r^3/3)$

3.49 Ethylene is C_2H_4. To meet bonding requirements, how many bonds are present?
a. 6 single
b. 4 single and 2 double
c. 1 double and 4 single
d. 12 single

3.50 Gold is FCC and has a density of 19.3 g/cm^3. Its atomic mass is 197 amu. Its atomic radius, *r*, may be calculated using which of the following?
a. $19.3\ \text{g/cm}^3 = (197)(6.02 \times 10^{23})/[(4r/)\sqrt{2^3}]$
b. $19.3\ \text{g/cm}^3 = 2\ (197/6.02 \times 10^{23})/[(4r/)\sqrt{2^3}]$
c. $19.3\ \text{g/cm}^3 = 4\ (197/6.02 \times 10^{23})/[(4r/)\sqrt{2^3}]$
d. $19.3\ \text{g/cm}^3 = 6\ (197)(6.02 \times 10^{23})/[(4r/)\sqrt{2^3}]$

3.51 Each of the following groups of plastics is thermoplastic *EXCEPT*:
a. polyvinyl chloride (PVC) and a polyvinyl acetate
b. phenolics, melamine, and epoxy
c. polyethylene, polypropylene, and polystyrene
d. acrylic (Lucite) and polyamide (nylon)

3.52 Styrene resembles vinyl chloride, C_2H_3C1, except that the chlorine is replaced by a benzene ring. The mass of each mer is:
a. 8(12) + 9(1) amu
b. 26 + 78 amu
c. 27 + 6(12) + 6(1) amu
d. 2(12) + 3(1) + 77 amu

3.53 The $<1\bar{1}0>$ family of directions in a cubic crystal include all but which of the following? (An overbar is a negative coefficient.)
a. [110]
b. $[0\bar{1}1]$
c. [101]
d. $[1\bar{1}1]$

3.54 The {112} family of planes in a cubic crystal includes all but which of the following directions?
a. (212)
b. (211)
c. $(1\bar{1}2)$
d. (121)

3.55 Crystal imperfections include all but which of the following?
a. Dislocations
b. Displaced atoms
c. Interstitials
d. Dispersions

3.56 A dislocation may be described as a:
a. displaced atom
b. shift in the lattice constant
c. slip plane
d. lineal imperfection

3.57 A grain within a microstructure is:
a. a particle the size of a grain of sand
b. the nucleus of solidification
c. a particle the size of a grain of rice
d. an individual crystal

3.58 Which of the following does *NOT* apply to a typical brass?
a. An alloy of copper and zinc
b. A single-phase alloy
c. An interstitial alloy of copper and zinc
d. A substitutional solid solution

3.59 Sterling silver, as normally sold:
a. is pure silver
b. is a supersaturated solid solution of 7.5% copper in silver
c. is 24-carat silver
d. has higher conductivity than pure silver

3.60 Atomic diffusion in solids matches all but which of the following generalities?
a. Diffusion is faster in FCC metals than in BCC metals.
b. Smaller atoms diffuse faster than do larger atoms.
c. Diffusion is faster at elevated temperatures.
d. Diffusion flux is proportional to the concentration gradient.

3.61 The proportionality constant for a particular gas diffusing through copper at 1000°C is 0.022 cm^2/s and the activation energy is 97 kJ/mole. Find the diffusion coefficient.
a. 1.6×10^{-6} cm^2/s
b. 1.9×10^{-6} cm^2/s
c. 2.3×10^{-6} cm^2/s
d. 4.4×10^{-6} cm^2/s

3.62 Grain growth involves all but which of the following?
a. Reduced growth rates with increased time
b. An increase in grain boundary area per unit volume
c. Atom movements across grain boundaries
d. A decrease in the number of grains per unit volume

3.63 Imperfections within metallic crystal structures may be any but which of the following?
a. Lattice vacancies and extra interstitial atoms
b. Displacements of atoms to interstitial site (Frenkel defects)
c. Lineal defects or slippage dislocations caused by shear
d. Ion pairs missing in ionic crystals (Shottky imperfections)

3.64 All but which of the following statements about solid solutions are correct?

a. In metallic solid solutions, larger solute atoms occupy the interstitial space among solvent atoms in the lattice sites.
b. Solid solutions may result from the substitution of one atomic species for another, provided radii and electronic structures are compatible.
c. Defect structures exist in solid solutions of ionic compounds when there are differences in the oxidation state of the solute and solvent ions, because vacancies are required to maintain an overall charge balance.
d. Order-to-disorder transitions that occur at increased temperatures in solid solutions result from thermal agitation that dislodges atoms from their preferred neighbors.

3.65 In ferrous oxide, $Fe_{1-x}O$, 2% of the cation sites are vacant. What is the Fe^{3+}/Fe^{2+} ratio?

a. 2/98
b. 0.04/0.94
c. 0.04/0.96
d. 0.06/0.94

3.66 A solid solution of MgO and FeO contains 25 atomic percent Mg^{2+} and 25 atomic percent Fe^{2+}. What is the weight fraction of MgO? (Mg: 24; Fe: 56; and O: 16 amu)

a. 40/(40 + 72)
b. 24/(24 + 56), or 4/80
c. 25/(25 + 25)
d. (25 + 25)/(50 + 50)

3.67 The boundary between two metal grains provides all but which of the following?

a. An impediment to dislocation movements
b. A basis for an increase in the elastic modulus
c. A site for the nucleation of a new phase
d. Interference to slip

3.68 If 5% copper is added to silver:

a. the hardness is decreased
b. the strength is decreased
c. the thermal conductivity is decreased
d. the electrical resistivity is decreased

3.69 All but which of the following statements about diffusion and grain growth are correct?

a. Atoms can diffuse both within grains and across grain (crystal) boundaries.
b. The activation energy for diffusion through solids is inversely proportional to the atomic packing factor of the lattice.
c. Grain growth results from local diffusion and minimizes total grain boundary area. Large grains grow at the expense of small ones, and grain boundaries move toward their centers of curvature.
d. Net diffusion requires an activation energy and is irreversible. Its rate increases exponentially with temperature. It follows the diffusion equation, in which flux equals the product of diffusivity and the concentration gradient.

3.70 Refer to the accompanying Mg-Zn phase diagram, Exhibit 3.70. Select an alloy of composition C (71Mg–29Zn) and raise it to 575°C so that only liquid is present. Change the composition to 60Mg–40Zn by adding zinc. When this new liquid is cooled, what will be the first solid to separate?

a. A solid intermetallic compound
b. A mixture of solid intermetallic compound and solid eutectic C (71Mg–29Zn)
c. A solid eutectic C (71Mg–29Zn)
d. A solid solution containing less than 1% intermetallic compound dissolved in Mg

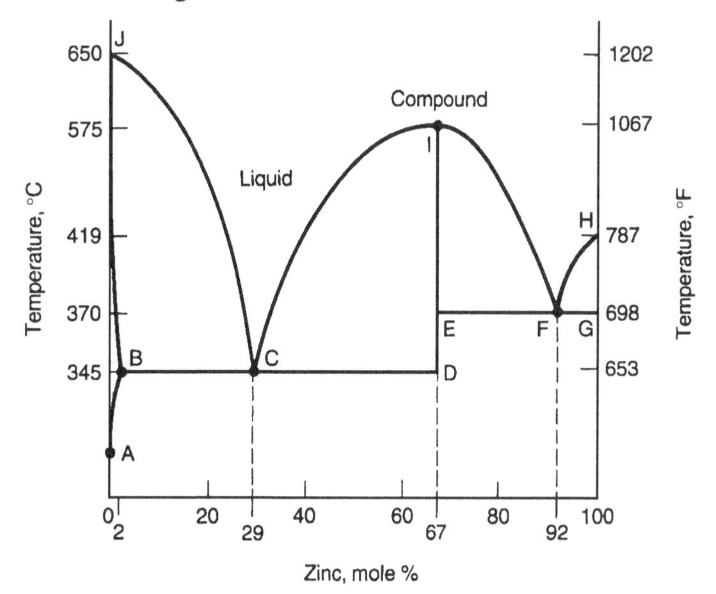

Exhibit 3.70 Magnesium-zinc phase diagram

3.71 Refer to the Mg-Zn phase diagram of Exhibit 3.70. Which of the following compounds is present?

a. Mg_3Zn_2
b. Mg_2Zn_3
c. MgZn
d. $MgZn_2$

3.72 Refer to Exhibit 3.73, a schematic sketch of the Fe-Fe_3C phase diagram. All but which of the following statements are *TRUE*?

a. A eutectoid reaction occurs at location C, 727°C (1340°F).
b. The eutectic composition is 99.2 weight percent Fe and 0.8 weight percent C.
c. A peritectic reaction occurs at K, 1500°C (2732°F).
d. A eutectic reaction occurs at G, 1130°C (2202°F).

3.73 Refer to Exhibit 3.73, the Fe-Fe_3C phase diagram. Pearlite contains ferrite (α) and carbide (Fe_3C). The weight fraction of carbide in pearlite is:

a. 0.8%
b. BC/CD
c. CD/BD
d. BC/BD

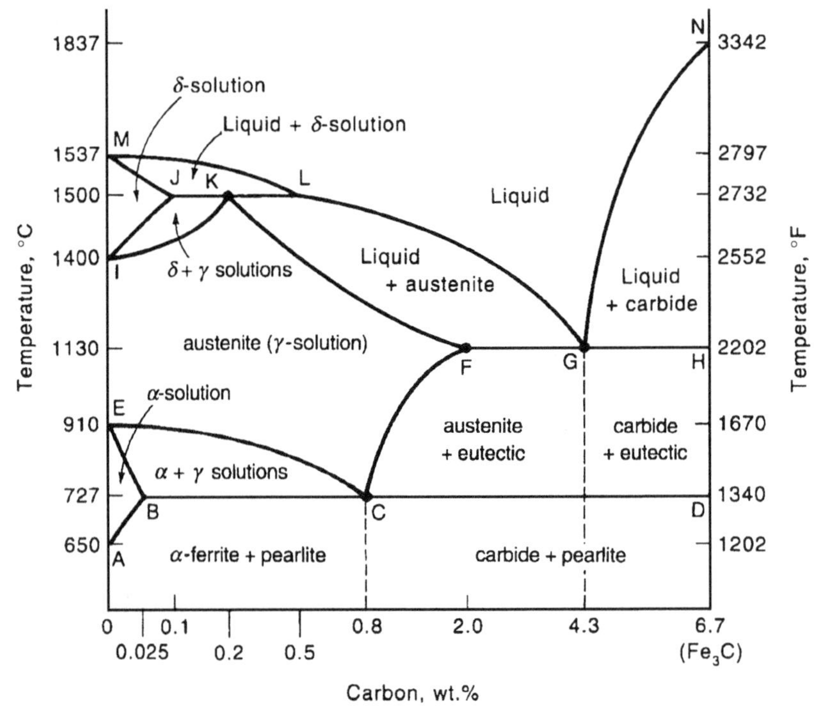

Exhibit 3.73 Iron-iron carbide phase diagram (schematic)

3.74 Consider the Ag-Cu phase diagram (Figure 3.28). Silver-copper alloys can contain *approximately* half liquid and half solid at all but which of the following situations?

a. 40Ag–60Cu at 781°C
b. 81Ag–19Cu at 781°C
c. 20Ag–80Cu at 910°C
d. 50Ag–50Cu at 779°C

3.75 Refer to the Ag-Cu phase diagram of Figure 3.28. Which of the following statements is *FALSE*?

a. The solubility limit of copper in the liquid at 900°C is approximately 62%.
b. The solubility limit of silver in b at 700°C is approximately 5%.
c. The solubility limit of silver in b at 900°C is approximately 7%.
d. The solubility limit of copper in b at 800°C is approximately 92%.

3.76 Other factors being equal, diffusion flux is facilitated by all but which of the following?
a. Smaller grain sizes
b. Smaller solute (diffusing) atoms
c. Lower concentration gradients
d. Lower-melting solvent (host structure)

3.77 In using copper as an example, all but which of the following statements are applicable for grain growth?
a. Atoms jump the boundary from large grains to small grains.
b. Grain size varies inversely with boundary area.
c. Grain growth occurs because the boundary atoms possess higher energy than interior atoms.
d. Grain growth occurs because larger grains have less boundary area.

3.78 A phase diagram can provide answers for all but which of the following questions?
a. What are the direction of the planes at a given temperature?
b. What phases are present at a given temperature?
c. What are the phase compositions at a given temperature?
d. How much of each phase is present at a given temperature?

3.79 In the Ag-Cu system (Figure 3.28), three equilibrated phases may be present at:
a. 930°C
b. 880°C
c. 830°C
d. 780°C

3.80 During heating, a 72Ag–28Cu alloy (Figure 3.28) may have any but which of the following equilibrium relationships?
a. ~25% β at 600°C, where β contains ~2% Ag
b. less than 50% β (95Cu–5Ag) at 700°C
c. ~74% α(93Ag–7Cu) at 750°C
d. α(26% Cu) and β(32% Cu) at 800°C

3.81 During cooling, a (20Ag–80Cu) alloy (Figure 3.28) has all but which of the following equilibrium situations?
a. The first solid forms at 980°C
b. The first solid contains 5% Ag
c. The second solid appears at 780°C
d. The last liquid contains 32% Cu

3.82 Add copper to 100 g of silver at 781°C (Figure 3.28). Assume equilibrium. Which statement is *TRUE*?
a. Liquid first appears with the addition of exactly 9 g of copper.
b. The last α disappears when approximately 39 g of copper has been added.
c. The solubility limit of copper in solid silver (α) is 28% copper.
d. Solid β first appears with the addition of 92 g of copper.

3.83 All but which of the following statements about strain hardening is correct?

a. Strain hardening is produced by cold working.
b. Strain hardening is relieved during annealing above the recrystallization temperature.
c. With more strain hardening, more time-temperature exposure is required for relief.
d. Strain hardening is relieved during recrystallization. Recrystallization produces less strained and more ordered structures.

3.84 Which process is used for the high-temperature shaping of many materials?

a. Reduction
b. Recrystallization
c. Polymerization
d. Extrusion

3.85 All but which of the following processes strengthens metals?

a. Precipitation processes that produce submicroscopic particles during a low-temperature heat treatment
b. Increasing the carbon content of low-carbon steels
c. Annealing above the recrystallization temperature
d. Mechanical deformation below the recrystallization temperature (cold working)

3.86 All but which of the following statements about the austentite-martensite-bainite transformations are correct?

a. Pearlite is a stable lamellar mixture consisting of BCC ferrite (α) plus carbide (Fe_3C). It forms through eutectoid decomposition during slow cooling of austenite. Most alloying elements in steel retard this transformation.
b. Martensite has a body-centered structure of iron that is tetragonal and is supersaturated with carbon. It forms by shear during the rapid quenching of austenite (FCC iron).
c. Tempering of martensite is accomplished by reheating martensite to precipitate fine particles of carbide within a ferrite matrix, thus producing a tough, strong structure.
d. Bainite and tempered martensite have distinctly different microstructures.

3.87 Steel can be strengthened by all but which of the following practices?

a. Annealing
b. Quenching and tempering
c. Age or precipitation hardening
d. Work hardening

3.88 Residual stresses can produce any but which of the following?

a. Warpage
b. Distortion in machined metal parts
c. Cracking of glass
d. Reduced melting temperatures

3.89 The reaction ($\gamma \rightarrow \alpha + Fe_3C$) is most rapid at:
a. the eutectoid temperature
b. 10°C above the eutectoid temperature
c. the eutectic temperature
d. 100°C below the eutectoid temperature

3.90 Grain growth, which reduces boundary area, may be expected to:
a. decrease the thermal conductivity of ceramics
b. increase the hardness of a solid
c. decrease the creep rate of a metal
d. increase the recrystallization rate

3.91 Rapid cooling can produce which one of the following in a material such as sterling silver?
a. Homogenization
b. Phase separation
c. Grain boundary contraction
d. Supersaturation

3.92 Martensite, which may be obtained in steel, is a:
a. supersaturated solid solution of carbon in iron
b. supercooled iron carbide, Fe_3C
c. undercooled FCC structure of austenite
d. superconductor with zero resistivity at low temperature

3.93 Alloying elements produce all but which one of the following effects in steels?
a. They alter the number of atoms in a unit cell of austenite.
b. They increase the depth of hardening in quenched steel.
c. They increase the hardness of ferrite in pearlite.
d. They retard the decomposition of austenite.

3.94 Hardenability may be defined as:
a. resistance to indentation
b. the hardness attained for a specified cooling rate
c. another measure of strength
d. rate of increased hardness

3.95 The linear portion of the stress-strain diagram of steel is known as the:
a. irreversible range
b. scant modulus
c. modulus of elasticity
d. elastic range

3.96 The ultimate (tensile) strength of a material is calculated from:
a. the applied force divided by the true area at fracture
b. the applied force times the true area at fracture
c. the tensile force at the initiation of slip
d. the applied force and the original area

3.97 All but which one of the following statements about slip are correct?

a. Slip occurs most readily along crystal planes that are least densely populated.
b. Slip, or shear along crystal planes, results in an irreversible plastic deformation or permanent set.
c. Ease of slippage is directly related to the number of low-energy slip planes within the lattice structure.
d. Slip is impeded by solution hardening, with odd-sized solute atoms serving as anchor points around which slippage does not occur.

3.98 When a metal is cold worked more severely, all but which one of the following generally occur?

a. The recrystallization temperature decreases.
b. The tensile strength increases.
c. Grains become equiaxed.
d. Slip and/or twinning occur.

3.99 All but which of the following statements about the rusting of iron are correct?

a. Contact with water or oxygen is required for rusting to occur.
b. Halides aggravate rusting, a process that involves electrochemical oxidation-reduction reactions.
c. Contact with a more electropositive metal restricts rusting.
d. Corrosion occurs in oxygen-rich areas.

3.100 All but which of the following statements about mechanical and thermal failure is *TRUE*?

a. Creep is time-dependent, plastic deformation that accelerates at increased temperatures. Stress rupture is the failure following creep.
b. Ductile fracture is characterized by significant amounts of energy absorption and plastic deformation (evidenced by elongation and reduction in cross-sectional area).
c. Fatigue failure from cyclic stresses is frequency-dependent.
d. Brittle fracture occurs with little plastic deformation and relatively low energy absorption.

3.101 The stress intensity factor is calculated from:

a. yield stress and crack depth
b. applied stress and crack depth
c. tensile stress and strain rate
d. crack depth and strain rate

3.102 Service failure from applied loads can occur in all but which of the following cases?

a. Cyclic loading, tension to compression
b. Glide normal to the slip plane
c. Cyclic loading, low tension to higher tension
d. Stage 2 creep

3.103 Brittle failure becomes more common when:
a. the endurance limit is increased
b. the glass-transition temperature is decreased
c. the critical stress intensity factor is increased
d. the ductility-transition temperature is increased

3.104 Where applicable, all but which of the following procedures may reduce corrosion?
a. Avoidance of bimetallic contacts
b. Sacrificial anodes
c. Aeration of feed water
d. Impressed voltages

3.105 A fiber-reinforced rod contains 50 volume percent glass fibers ($E = 70$ GPa, $\sigma_y = 700$ MPa) and 50 volume percent plastic ($E = 7$ GPa). The glass carries what part of a 5000-N tensile load?
a. [(7000 MPa)(0.5)]/[70,000 MPa)(0.5)] = 0.1; $F_{gl} = (0.1)(5000\text{ N}) = 500\text{ N}$
b. [(70,000)(0.5) + 7000(0.5)] = 5000/x; $x = 0.0002$
c. $[(F_{gl}/0.5A)/(F_p / 0.5A)] = (70/7) = 10;\quad F_{gl}/(F_p + F_{gl}) = 10F_p/(10F_p + F_p) = 0.91$
d. (70)/[70 + 2(7)] = 0.83

3.106 For a parallel plate capacitor separated by an air gap of 1 cm and with an applied dc voltage across the plates of 500 volts, determine the force on an electron mass of 18.2×10^{-31} kg inserted in the space. The mass of an electron is 9.1×10^{-31} kg.
a. 3.2×10^{-14} N
b. 1.6×10^{-14} N
c. 9.1×10^{-31} N
d. 1.6×10^{-19} N

3.107 Assume a point charge of 0.3×10^{-3} C at an origin. What is the magnitude of the electric field intensity at a point located 2 meters in the x-direction, 3 meters in the y-direction, and 4 meters in the z-direction from the origin?
a. 500 kV/m
b. 5 kV/m
c. 93 kV/m
d. 9.3 MV/m

3.108 An infinite sheet of charge with a positive charge density, σ, has an electric field of:

$$\mathbf{E} = [\sigma/(2\varepsilon_0)]_{ax} \text{ for } x > b$$

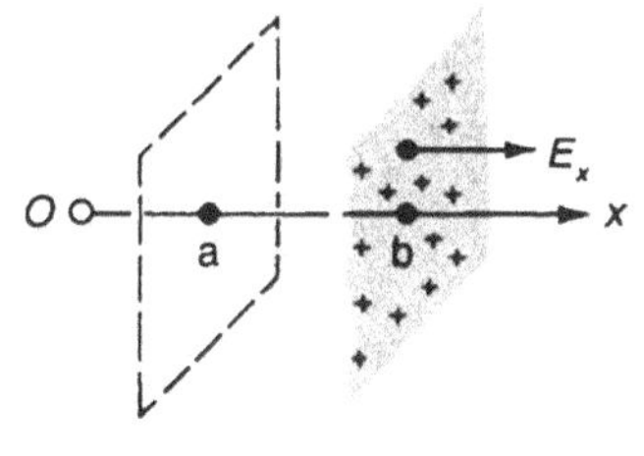

Exhibit 3.108 Charge placement

If a second sheet of charge with a charge density of $-\sigma$ is then placed (see Exhibit 3.108), what is the electric field for $a < x < b$?
a. $(\sigma/\varepsilon_0)_{ax}$
b. 0
c. $(-\sigma/\varepsilon_0)_{ax}$
d. $(\sigma/2\varepsilon_0)_{ax}$

3.109 Two equal charges of 10 μC are located one meter apart on a horizontal line, and another charge of 5 μC is placed one meter below the first charge (forming a right triangle). What is the magnitude of the force on the 5 μC charge?

a. 0.09×10^6 N
b. 12.6×10^4 N
c. 6.39×10^4 N
d. 63×10^{-2} N

3.110 For a coil of 100 turns wound around a toroidal core of iron with a relative permeability of 1000, find the current needed to produce a magnetic flux density of 0.5 tesla in the core. The dimensions of the core are given in Exhibit 3.110.

a. 390 A
b. 39 A
c. 1.2 A
d. 12.2 A

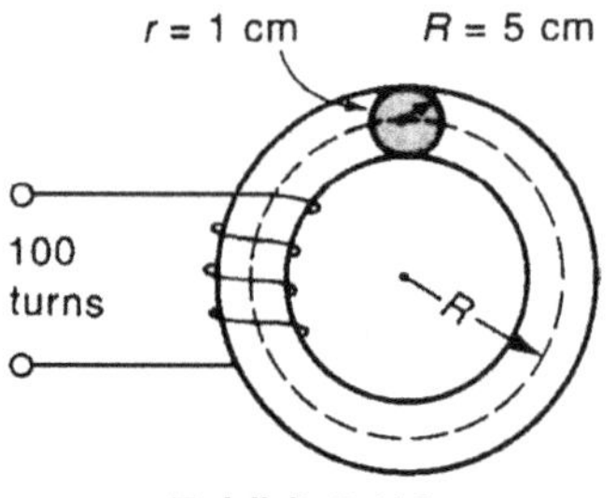

Exhibit 3.110

3.111 Two long straight wires, bundled together, have a magnetic flux density around them. One wire carries a current of 5 amperes, and the other carries a current of 1 ampere in the opposite direction. Determine the magnitude of the flux density at a point 0.2 meters away (i.e., normal to the wires).

a. $2\pi \times 10^{-6}$ T
b. $4\pi \times 10^{-6}$ T
c. 4×10^{-6} T
d. $16\pi \times 10^{-6}$ T

3.112 The root mean square of $i(t)$ for Exhibit 3.112 is:

a. 1.6 A
b. 3.0 A
c. 2.0 A
d. 0.0 A

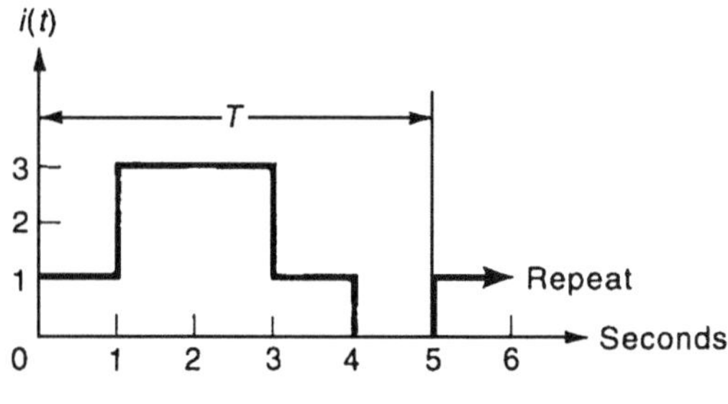

Exhibit 3.112 A current wave

3.113 A current phases through a 0.2-henry inductor. The current increases in a linear fashion from a value of zero at $t = 0$ to 20 A at $t = 20$ seconds. What is the amount of energy stored in the inductor at 10 seconds?
a. 0 J
b. 10 J
c. 40 J
d. 1 J

3.114 A sine wave of 10 volts (rms) is applied to a 10-ohm resistor through a half-wave rectifier. What is the average value of the current (i.e., what would a dc meter read)?
a. 0.32 A
b. 0.45 A
c. 1.0 A
d. 0.9 A

3.115 Two resistors of 2 ohms each are connected in parallel, another resistor of 1 ohm is connected in series with the parallel combination, and the resistive combination is connected to a 2-volt source. How much power is dissipated in either one of the parallel resistors?
a. 0.25 W
b. 0.5 W
c. 1.0 W
d. 1.5 W

3.116 A current source of 2 amperes is connected to four resistors, all in parallel. The resistors have values of 1, 2, 3, and 4 ohms, respectively. How much power is dissipated in the 2-ohm resistor?
a. 8.0 W
b. 4.0 W
c. 2.0 W
d. 0.46 W

3.117 Three resistors of 2 ohms each are connected in a "T" arrangement, and each side of the T is connected to its own battery voltage source. The battery on the left side is 1 volt (+ on top) and the battery on the right side is 2 volts (– on top). What is the power dissipated in the resistor in the middle leg?
a. 0 W
b. 4.5 W
c. 0.056 W
d. 0.89 W

3.118 For the circuit in Exhibit 3.118, the load resistor, R_L, might "see" a Thevenin equivalent circuit in its place. What are the values of the equivalent circuit?
a. $V_{oc} = 5$ V, $R_{eq} = 2\ \Omega$
b. $V_{oc} = 5$ V, $R_{eq} = 1\ \Omega$
c. $V_{oc} = 4$ V, $R_{eq} = 2\ \Omega$
d. $V_{oc} = 1$ V, $R_{eq} = 2\ \Omega$

Exhibit 3.118 Original circuit

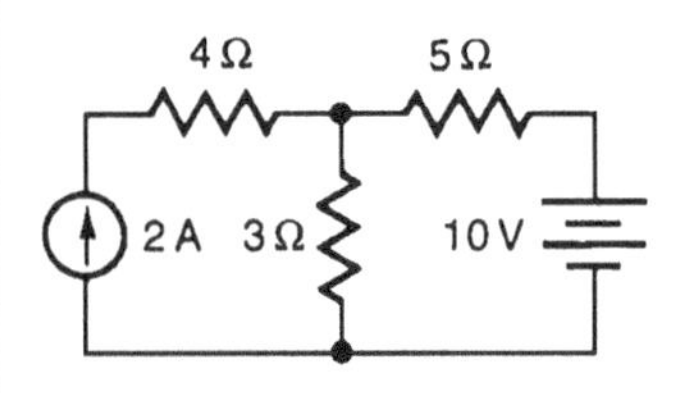

Exhibit 3.119 Circuit diagram

3.119 For the circuit shown in Exhibit 3.119, determine the magnitude of the voltage across the current source.

a. 16 V
b. 30 V
c. 20 V
d. 23 V

3.120 A charged 100-pF capacitor has initial voltage across it of 10 volts; a 100 kΩ resistor is suddenly connected across the capacitor (at $t = 0$) to discharge it. What is the magnitude of the instantaneous current at 20 microseconds?

a. 13.5 μA
b. 36.8 μA
c. 13.5 pA
d. 36.8 pA

3.121 Three 2-ohm resistors are arranged in "T" configuration. Connected to the left side is a 10-volt battery and switch; to the right is a 1-farad capacitor. The switch closes at $t = 0$. Find the time for the capacitor to reach 63% of its final voltage.

a. 0.2 s
b. 2.0 s
c. 0.3 s
d. 3.0 s

3.122 Three 2-ohm resistors are arranged in a "T" configuration. Connected to the left side is a 10-volt battery; to the right side is connected a switch and a 1-henry inductor. Assume the switch closes at $t = 0$. Find the amount of time for the inductor to reach 63% of its final current.

a. 0.2 s
b. 2.0 s
c. 0.3 s
d. 3.0 s

Exhibit 3.123 An ac circuit

3.123 For the circuit in the Exhibit 3.123, determine the power dissipated in the impedance, Z_L.

a. 1250 W
b. 625 W
c. 312 W
d. 1.7 kW

3.124 A resistance of 10 Ω, an inductor of 10/377 henries, and a capacitor of 20/377 farads are all connected in series to a 60-hertz, 100-volt source. Determine the magnitude of the current.

a. 3.3 A
b. 0.38 A
c. 10 A
d. $10\sqrt{2}$ A

3.125 If in Problem 3.124, all parameters were the same except that the frequency of the voltage source were doubled to 120 Hz, what would be the magnitude of the current?

a. 3.3 A
b. 0.38 A
c. 10 A
d. 5.5 A

3.126 A 10-ohm resistor and a capacitor with a capacitive reactance of $-j10\ \Omega$ are connected in parallel. The parallel combination is connected in series through an inductive reactance of $+j10\ \Omega$ to a 100-volt ac source. Determine the magnitude of the current from the source.

a. 6.7 A
b. 7.1 A
c. 10 A
d. $10\sqrt{2}$ A

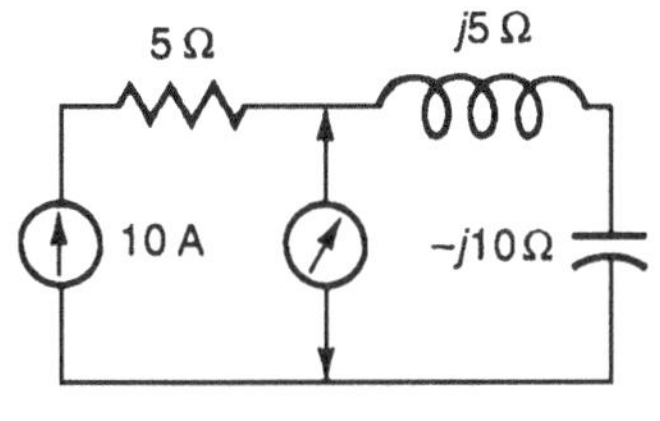

Exhibit 3.127 An ac circuit

3.127 An ac current source of 10 amperes is connected to a series circuit as shown in Exhibit 3.127. What would a voltmeter measure if the meter were connected across the inductor and capacitor combination?

a. 200 V
b. 150 V
c. 70 V
d. 50 V

3.128 For an *RLC* series circuit with $R = 10\ \Omega$, $L = 0.1$ henry, and $C = 0.1$ farad, all connected to an ac voltage source of 100 volts whose frequency could be varied, determine the frequency if maximum power is to be dissipated in *R*.

a. 10 Hz
b. 0 Hz
c. 1.6 Hz
d. 16 Hz

3.129 Find the transfer function $T(s) = \dfrac{V_2(s)}{V_1(s)}$ of the circuit shown in Exhibit 3.129.

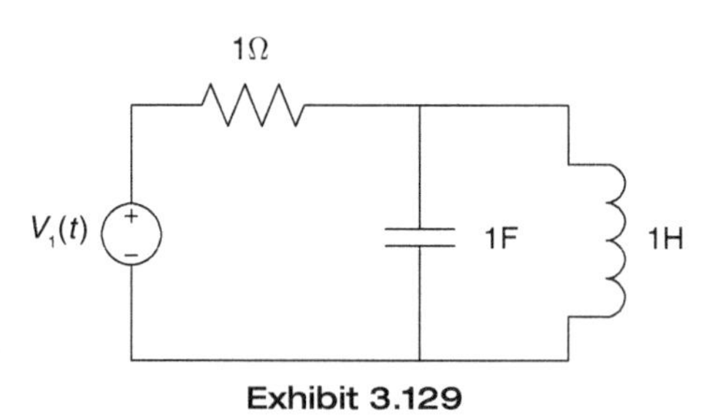

Exhibit 3.129

a. $\dfrac{s}{s^2+1}$

b. $\dfrac{s}{s^2+s+1}$

c. $\dfrac{1}{s^2+s+1}$

d. $\dfrac{1}{s^2+1}$

3.130 Find the transfer function $T(s) = \dfrac{V_0(s)}{V_{in}(s)}$ of the circuit shown in Exhibit 3.130. Assume that the op-amp is ideal.

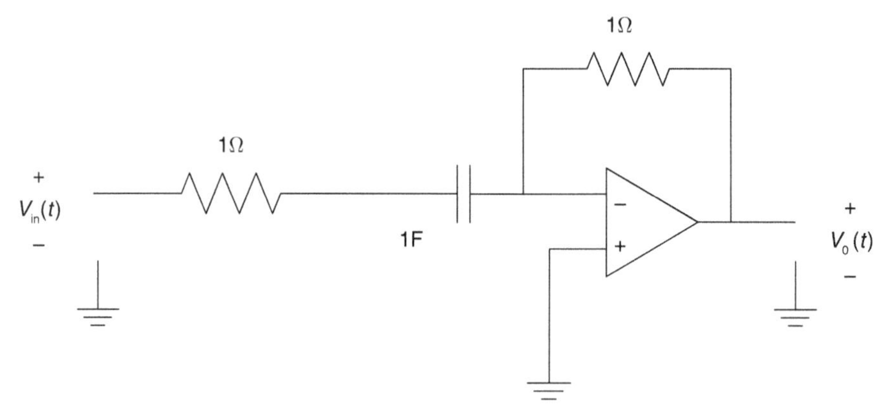

Exhibit 3.130

a. $-\dfrac{s}{s^2+1}$

b. $-\dfrac{1}{s^2+s+1}$

c. $-\dfrac{s}{s+1}$

d. $-\dfrac{1}{s+1}$

3.131 A low frequency ac voltage of 5 volts (rms) is applied to the input of the op-amp circuit in Exhibit 3.131. Determine the magnitude of the output voltage (rms).

a. 5 V
b. 2.5 V
c. 10 V
d. 20 V

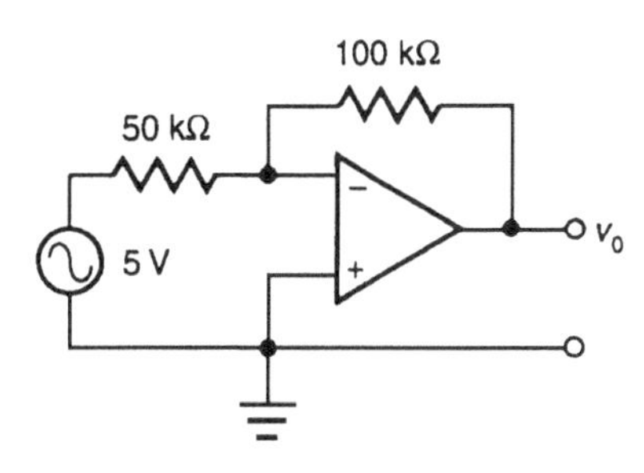

Exhibit 3.131 An op-amp circuit

3.132 A low frequency ac voltage of 5 volts (rms) is applied to the input of the op-amp circuit in Exhibit 3.132. Determine the magnitude of the output voltage (rms).

a. 5 V
b. 2.5 V
c. 10 V
d. 20 V

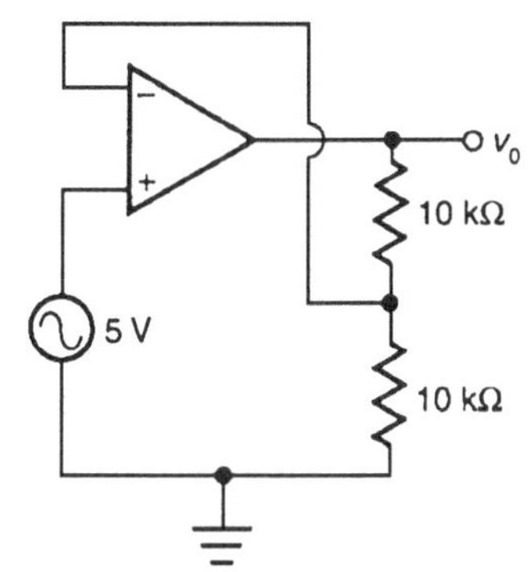

Exhibit 3.132 An op-amp circuit

3.133 The inputs for the op-amp circuits are given in Exhibit 3.133. What is the output voltage from the last op-amp?

a. +1 V
b. −1 V
c. +3 V
d. 3 V

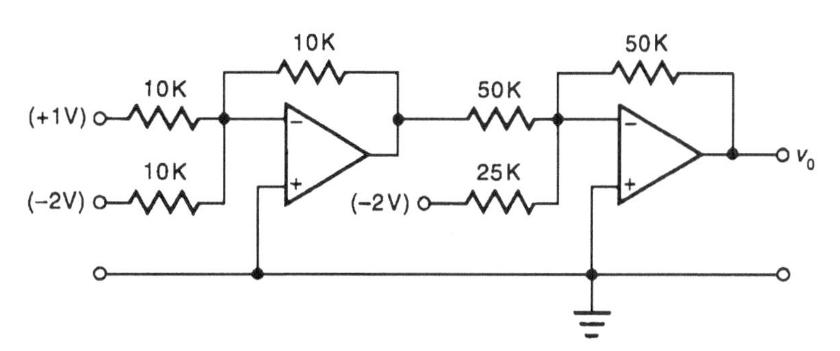

Exhibit 3.133 Summing circuits

3.134 A square wave of 1 V (or 2 V peak-to-peak) is applied to a bridge rectifier circuit where a 1-ohm resistor is connected to the output. What is the average current through the resistor?

a. 0.5 A
b. $\pi/2$ A
c. πA
d. 1.0 A

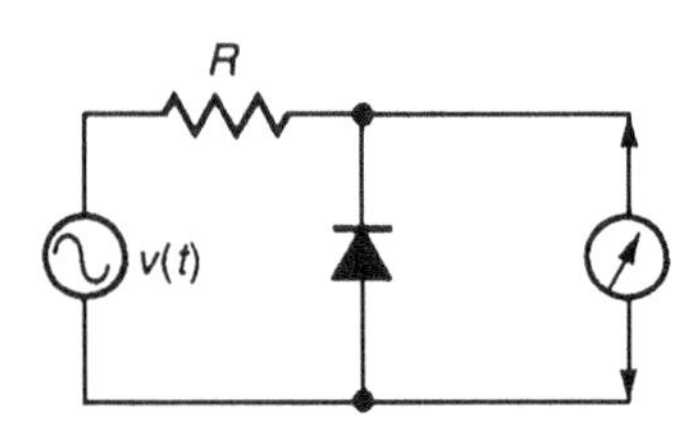

Exhibit 3.135 Clipping circuit

3.135 A sine wave voltage $v(t) = 14 \sin 2\pi ft$ is applied to the clipping circuit of Exhibit 3.135. A dc voltmeter (average reading instrument) is connected to the output. What does the meter read?

a. 2 V
b. 4.5 V
c. 7.5 V
d. 9 V

3.136 For an ideal, two-winding transformer whose nameplate reads 5 kVA, 400:200 V, and 60 Hz, determine the magnitude of a load that could be connected to the low voltage side for rated conditions. Assume the load impedance is $\mathbf{Z}_L = Z\angle 45°$, where Z is:

a. 4 Ω
b. 8 Ω
c. 16 Ω
d. 25 Ω

SOLUTIONS

3.1 c. The density of the plate is:

$$\rho = \frac{\text{mass}}{\text{area}} = \frac{480 \text{ kg}}{\left[(8\times 6)+(4\times 6)\right]\text{m}^2} = \frac{480 \text{ kg}}{72 \text{ m}^2} = 6.67 \text{ kg/m}^2$$

3.2 **a.** Assuming that the density of the plate is 9 kg/m², the mass moment of inertia of the flat plate may be found by:

$$I_x = \frac{1}{3}mh^2 = \frac{1}{3}(\rho A h^2) = \frac{1}{3}\rho(6)(8)^3 + \frac{1}{3}\rho(6)(4)^3$$

$$= \frac{1}{3}(9 \text{ kg/m}^2)(6 \text{ m})(8 \text{ m})^3 + \frac{1}{3}(9 \text{ kg/m}^2)(6 \text{ m})(4 \text{ m})^3$$

$$= 9216 + 1152 = 10{,}368 \text{ kg} - \text{m}^2$$

3.3 **b.** Let the depth of the ditch be h, and set up a vertical s-axis, positive upwards with the origin at A (Exhibit 3.3a). Then, for motion between A and B, $s_0 = 0$, $s = -h$, $v_0 = 4$ m/s, $a = -9.81$ m/s², and $t = 6$ s.
Substituting these values in the relationship $s = s_0 + v_0 t + (at^2)/2$ yields $-h = 0 + 4(6) - 9.81(6)^2/2$ or $h = 152.6$ m.

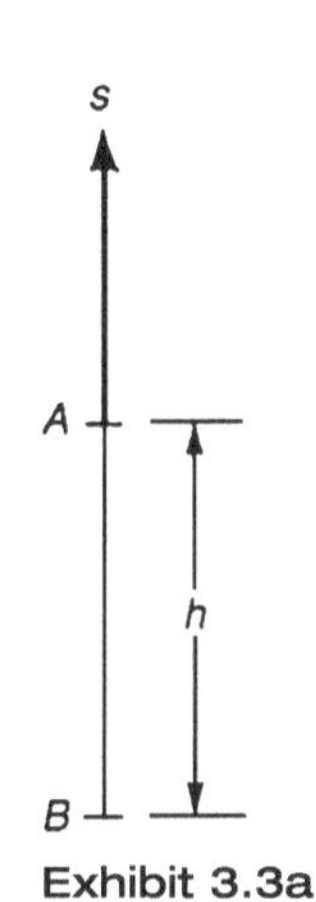

Exhibit 3.3a

3.4 **d**. Set up a horizontal s-axis with origin at O, and positive to the right (Exhibit 3.4a). Then, $sO = 2$ m, and $v = ds/dt$, so

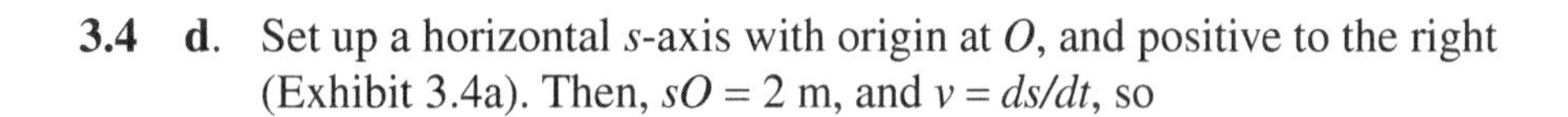

$$\int_{s_0}^{s} ds = \int_0^t v\,dt = \int_0^t (3t^2 - t + 2)\,dt$$

or $s = s_0 + t^3 - t^2/2 + 2t$. Substituting values into the equation yields $s(2s) = 12$ m.

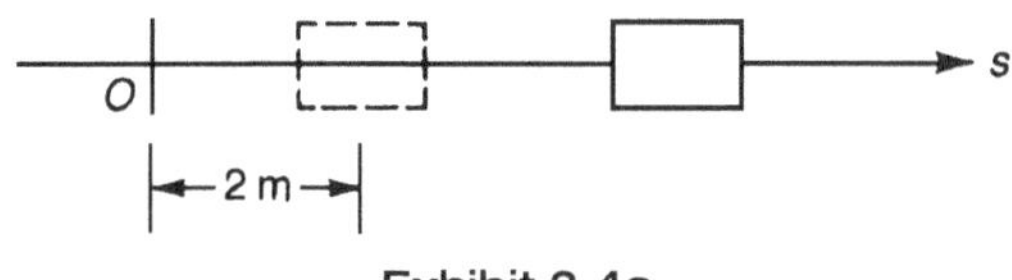

Exhibit 3.4a

3.5 d. The velocity-time curve for this particle is shown below the acceleration curve in Exhibit 3.5a. (Velocity at any given instant t_1 equals the area under the acceleration curve from time 0 to the time t_1, plus the initial velocity). Any area above the $a = 0$ line is counted as positive, and any area below is counted as negative.

The displacement D is the total area under the velocity curve between $t = 0$ and $t = 8$ s. The area of a triangle is $0.5(b)(h)$. Hence, $D = 0.5(6)(8) - 0.5(2)(4) = 20$ m.

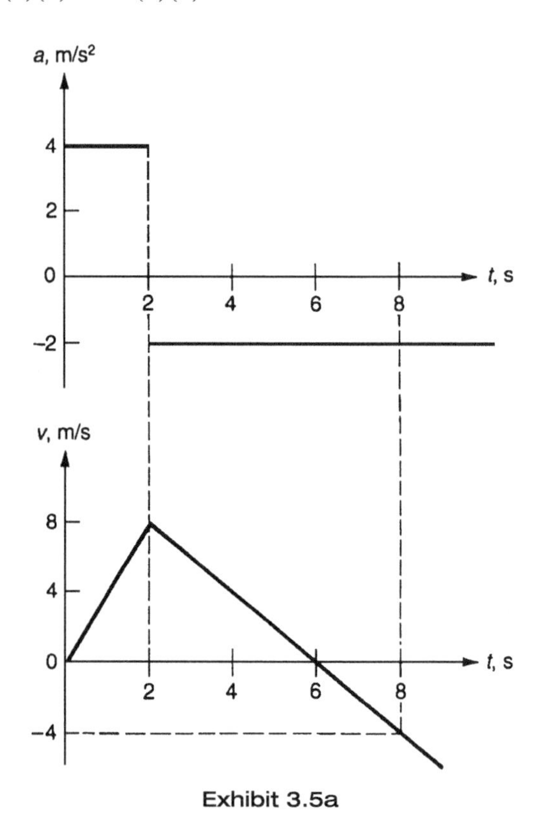

Exhibit 3.5a

3.6 a. When the ball reaches its highest position, B, the vertical component of the velocity of the ball is zero. Applying Equation (3.12) vertically,

$$v_y^2 = v_{yo}^2 + 2a_y(y - y_0)$$
$$0 = (40 \sin 60°)^2 - 2(9.81)(y - y_0)$$

Thus, $y - y_0 = 61.2$ m and $H = 63.2$ m.

3.7 c. Refer to Exhibit 3.7a.

Horizontal Motion:

$$x = x_0 + v_{x0}t + (a_x t^2)/2$$
$$140 = 0 + v_0 t + 0$$
$$t = 1.56 \text{ s}$$

Vertical Motion:

$$y = y_0 + y_{y0}t + (a_y t^2)/2$$
$$-12 = 0 + 0 - 0.5(9.81)t^2$$
$$v_o = 140/\text{t} = 140/1.56 = 89.7 \text{ m/s}$$

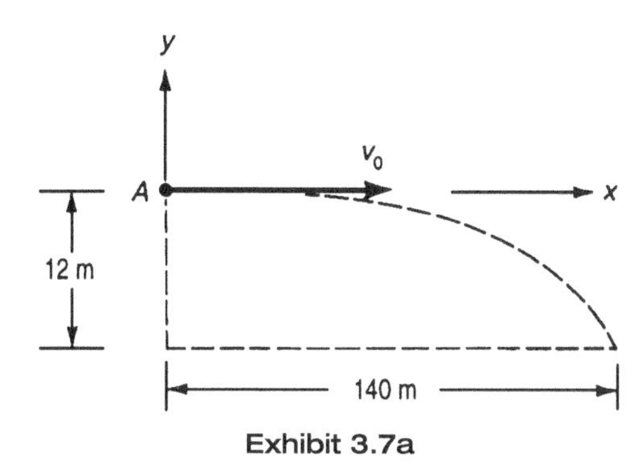

Exhibit 3.7a

3.8 c. In Exhibit 3.8a, we have

$$r = 1 \text{ m}; \qquad \dot{r} = -3 \text{ m/s}; \qquad \ddot{r} = 0$$

Since $\theta = (t^2 + t - 2)$ rad, differentiation gives

$$\dot{\theta} = (2t + 1) \text{ rad/s; at } t = 1 \text{ s, } \dot{\theta} = 2(1) + 1 = 3 \text{ rad/s}$$

Also, $\ddot{\theta} = 2 \text{ rad/s}^2$.

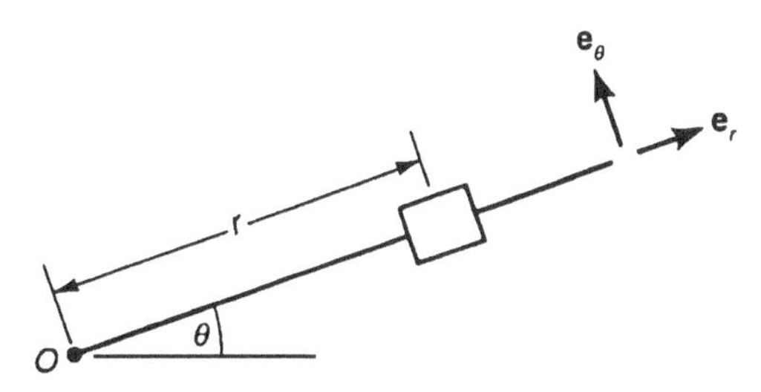

Exhibit 3.8a

Acceleration of the collar is therefore given by

$$\begin{aligned}\mathbf{a} &= (\ddot{r} - r\dot{\theta}^2)\mathbf{e}_r + (r\ddot{\theta} + 2\dot{r}\dot{\theta})\mathbf{e}_\theta \\ &= [0 - 1(3)^2]\mathbf{e}_r + [1(2) + 2(-3)(3)]\mathbf{e}_\theta \ \mathbf{m/s^2} \\ &= [-9\mathbf{e}_r - 16\mathbf{e}_\theta] \ \mathbf{m/s^2}\end{aligned}$$

Hence, $|a| = [(-9)^2 + (-16)^2]^{0.5} = 18.4 \text{ m/s}^2$.

3.9 **d.** Refer to Exhibit 3.9a. The velocity is

$$v = \left(v_r^2 + v_\theta^2\right)^{0.5} \quad \text{where } v_r = \dot{r} \text{ and } v_\theta = r\dot{\theta}$$

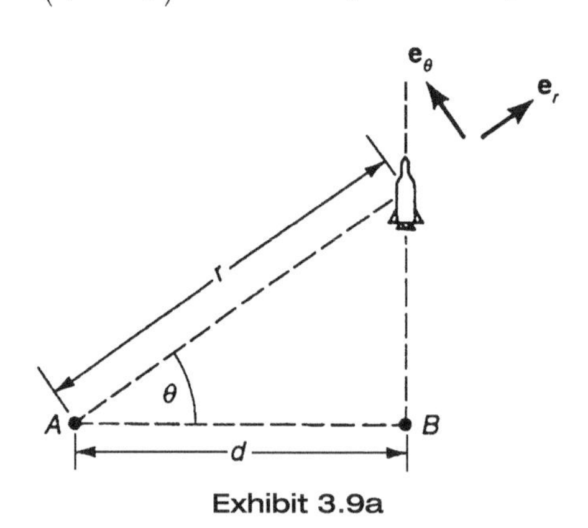

Exhibit 3.9a

Now, $r = d/\cos\theta$ so

$$v_r = \dot{r} = \frac{d\dot{\theta}\sin\theta}{\cos^2\theta} = \frac{d\dot{\theta}\tan\theta}{\cos\theta}$$

and

$$v_\theta r\dot{\theta} = \frac{d\dot{\theta}}{\cos\theta}$$

Thus

$$v^2 = v_r^2 + v_\theta^2 = \frac{d^2\dot{\theta}^2(\tan^2\theta + 1)}{\cos^2\theta} = \frac{d^2\dot{\theta}^2}{\cos^4\theta}; \quad \ddot{r} = 0$$

and

$$v = \frac{d\dot{\theta}}{\cos^2\theta} = \frac{1800(0.1)(2)}{1}\text{ m/s} = 360 \text{ m/s}$$

3.10 **d.** At the highest point, the vertical component of velocity is zero, and the acceleration is normal to the path. Thus, $v = v_x = v_{x0} + a_x t = 50$ cos 30° + 0. Also, the normal component of the acceleration is $a_n = -a_y = 9.81$ m/s². But $a_n = v^2/\rho$. Hence $\rho = v^2/a_n$ = (50 cos 30°)²/9.81 = 191 m.

3.11 **d.** The tangential acceleration is given as $a_t = 3$ m/s²; the normal acceleration is $a_n = v^2/\rho$ = [(20)²/110] m/s² = 3.6 m/s². Hence, the total acceleration is a = [(3)² + (3.6)²]^0.5 m/s² = 4.7 m/s².

3.12 **c.** The normal acceleration is given by $a_n = v^2/\rho$. So, $\rho = v^2/a_n$. Substituting values (converted to consistent units), we obtain

$$\rho = \frac{[800(1000)]^2}{[60(60)]^2(5)(9.81)} = 1007 \text{ m}$$

3.13 d. At any given instant, the angular acceleration of the disk is $\alpha = -k\omega = d\omega/dt$. So,

$$\int_{\omega_0}^{\omega} \frac{d\omega}{\omega} = -k\int_0^t dt$$

$\ln \omega|_{\omega_0}^{\omega} = \ln(\omega/\omega_0) = -kt$ and $t = -(1/k)\ln(\omega/\omega_0) = -(1/1.2)\ln(0.5) = 0.6\,s$

3.14 a. Initially, $\omega_0 = 0$. At $t = 30$ s, $\omega = 300$ rpm $= [300\,(2\pi)/60]$ rad/s $= 10\pi$ rad/s. For uniformly accelerated rotational motion,

$$\begin{aligned}
\omega &= \omega_0 + \alpha t \\
\alpha &= (\omega - \omega_0)/t = (10\pi - 0)/30 \text{ rad/s}^2 = \pi/3 \text{ rad/s}^2 \\
\omega(5) &= \omega_0 + \alpha t = 0 + (\pi/3)(5) \text{ rad/s} = 5\pi/3 \text{ rad/s} \\
v_P(5) &= \omega(5)r = (5\pi/3)\,(2) \text{ m/s} = 10.5 \text{ m/s}
\end{aligned}$$

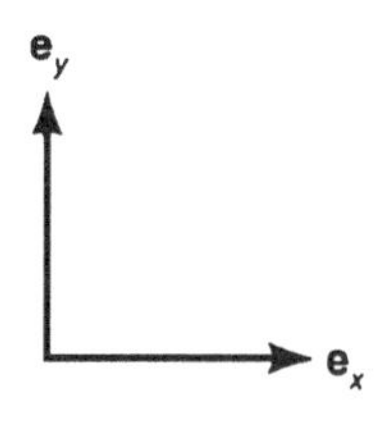

Exhibit 3.15a

3.15 b. Adopt the coordinate system of Exhibit 3.15a. Rolling of the disk without slip on B implies that the velocity of the point A, viewed as a point on D, equals the velocity of A viewed as a point on B. Hence, $\mathbf{v}_A = 2\mathbf{e}_x$ m/s. Since A and O are points of the same rigid body D, $\mathbf{v}_O = \mathbf{v}_A + \omega \times \mathbf{r}_{AO}$, or $-1\mathbf{e}_x = 2\mathbf{e}_x + \omega\mathbf{e}_z \times (1.5)\mathbf{e}_y = (2 - 1.5\omega)\mathbf{e}_x$. Hence, $-1 = 2 - 1.5\omega$ or ω

$= 2$ rad/s. The positive sign indicates a counterclockwise rotation.

3.16 a. In the current configuration, $\mathbf{v}_P = 0$. P and A are points on D, so

$$\begin{aligned}
\mathbf{v}_A &= \mathbf{v}_P + \omega_D \times \mathbf{r}_{PA} = 0 - 3\mathbf{e}_z \times (0.5\mathbf{e}_y - 0.4\mathbf{e}_x) \\
&= (1.5\mathbf{e}_x + 1.2\mathbf{e}_y) \text{ m/s}
\end{aligned}$$

Similarly, because B and A are points on R, $\mathbf{v}_B = \mathbf{v}_A + \omega_R \times \mathbf{r}_{AB}$. Therefore,

$$\begin{aligned}
v_B\mathbf{e}_x &= 1.5\mathbf{e}_x + 1.2\mathbf{e}_y + \omega_R\mathbf{e}_z \times (-2\mathbf{e}_x - 0.5\mathbf{e}_y) \\
&= (1.5 + 0.5\omega_R)\mathbf{e}_x + (1.2 - 2\omega_R)\mathbf{e}_y
\end{aligned}$$

Equating the coefficients of $\mathbf{e}_y$ yields $0 = 1.2 - 2\omega_R$, or $\omega_R = 0.6$ rad/s. The positive sign indicates that ω_R is in the positive $\mathbf{e}_z$ direction, so the rotation is counterclockwise.

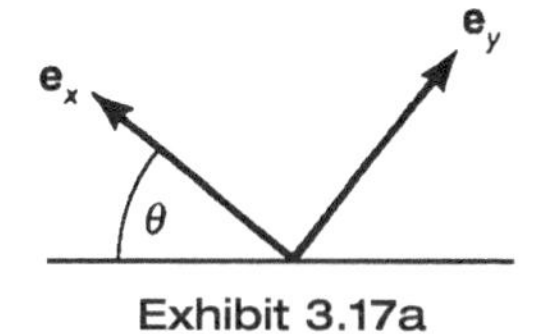

Exhibit 3.17a

3.17 d. Referring to the coordinate system in Exhibit 3.17a, the accelerations are

$$\begin{aligned}
\mathbf{a}_A &= \mathbf{a}_O + \alpha \times \mathbf{r}_{OA} + \omega \times (\omega \times \mathbf{r}_{OA}) \\
\mathbf{a}_O &= 0 \text{ (constant velocity)}
\end{aligned}$$

Since $\theta = (0.5t^2 - t)$ rad, $\dot{\theta} = (t-1)$rad/s and $\ddot{\theta} = 1$ rad/s^2.

So, at $t = 2$ s, $\omega = \dot{\theta}\mathbf{e}_z = 1\mathbf{e}_z$rad/s. Since $a = \ddot{\theta}\mathbf{e}_z = 1\mathbf{e}_z$ rad/s^2,

$$\mathbf{a}_A = 0 + \mathbf{e}_z \times (2\mathbf{e}_x) + \mathbf{e}_z \times [\mathbf{e}_z \times (2\mathbf{e}_x)] = (2\mathbf{e}_y - 2\mathbf{e}_x) \text{ m/s}^2$$

Hence, $|\mathbf{a}_A| = (2^2 + 2^2)^{0.5}$ m/s^2 $= 2.8$ m/s^2.

3.18 b. With the coordinate system in Exhibit 3.18a, we can write

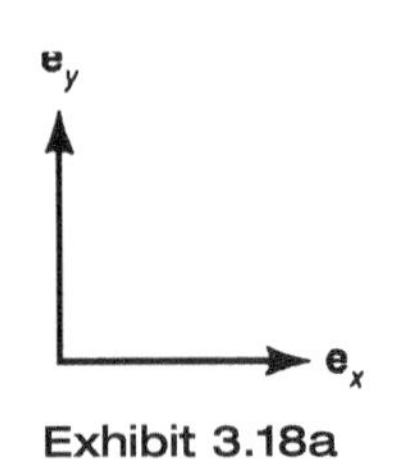

Exhibit 3.18a

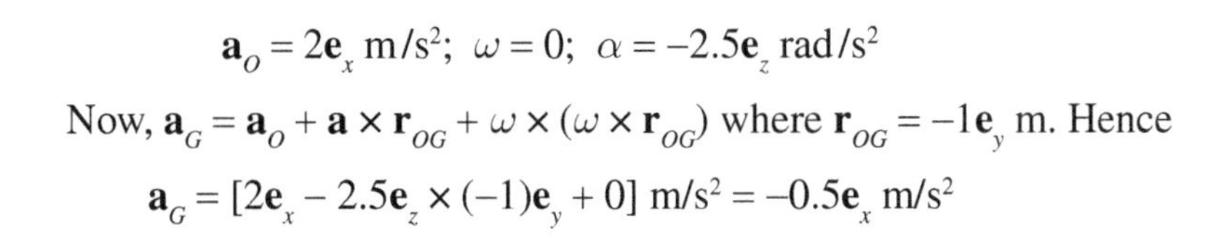

$$\mathbf{a}_O = 2\mathbf{e}_x \text{ m/s}^2;\ \omega = 0;\ \alpha = -2.5\mathbf{e}_z \text{ rad/s}^2$$

Now, $\mathbf{a}_G = \mathbf{a}_O + \mathbf{a} \times \mathbf{r}_{OG} + \omega \times (\omega \times \mathbf{r}_{OG})$ where $\mathbf{r}_{OG} = -1\mathbf{e}_y$ m. Hence

$$\mathbf{a}_G = [2\mathbf{e}_x - 2.5\mathbf{e}_z \times (-1)\mathbf{e}_y + 0] \text{ m/s}^2 = -0.5\mathbf{e}_x \text{ m/s}^2$$

3.19 a. We know that $\mathbf{v}_P = \mathbf{v}_{P/B} + \mathbf{v}_{P'}$ where $\mathbf{v}_{P/B}$ is the velocity of P relative to B and $\mathbf{v}_{P'}$ is the velocity of the point P' of the block that coincides with P at the instant under consideration (coincident point velocity). Here, $|\mathbf{v}_{P/B}| = 6$ m/s and $|\mathbf{v}_{P'}| = 10$ m/s (velocity of block). Because v_P is the vector sum of $\mathbf{v}_{P/B}$ and $\mathbf{v}_{P'}$, as shown in Exhibit 3.19a, we can use the law of cosines,

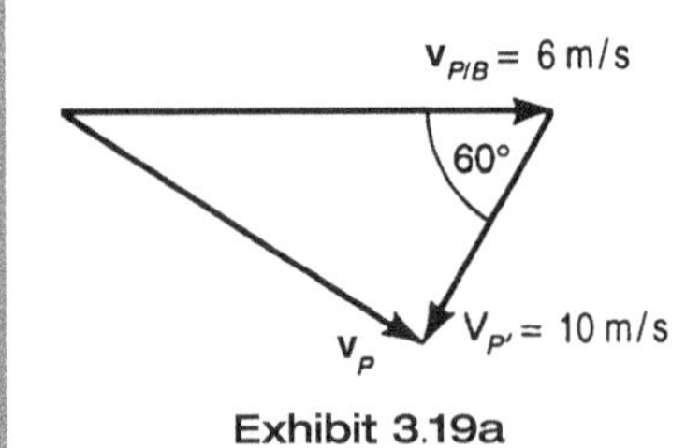

Exhibit 3.19a

$$(\mathbf{v}_P)^2 = 10^2 + 6^2 - 2(10)(6)\cos 60° = 76$$
$$\mathbf{v}_P = 8.7 \text{ m/s}$$

3.20 d.

$$\mathbf{a}_P = \mathbf{a}_{P/D} + \mathbf{a}_{P_e}\rho' + \mathbf{a}_C$$

Here,

$\mathbf{a}_{P/D}$ = relative acceleration = 0

$\mathbf{a}_{P'}$ = acceleration of the point of D that is coincident with P at the instant under consideration:

$$\begin{aligned}\mathbf{a}_{P'} &= \mathbf{a}_O + \alpha \times \mathbf{r}_{OC} + \omega \times (\omega \times \mathbf{r}_{OC})\\ &= 0 + 0 + (-5\mathbf{e}_z) \times [(-5\mathbf{e}_z) \times (0.1\mathbf{e}_y)]\\ &= -2.5\mathbf{e}_y \text{ m/s}^2\end{aligned}$$

$\mathbf{a}_C$ = Coriolis acceleration = $2\omega \times \mathbf{v}_{P/D}$

$\mathbf{a}_C = 2(-5\mathbf{e}_z) \times 2\mathbf{e}_x = -20\mathbf{e}_y \text{ m/s}^2$

Finally,

$$\mathbf{a}_P = [-2.5\mathbf{e}_y - 20\mathbf{e}_y] \text{ m/s}^2 = -22.5\mathbf{e}_y \text{ m/s}^2$$

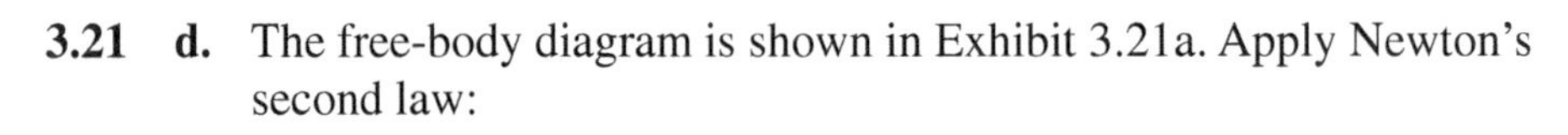

3.21 d. The free-body diagram is shown in Exhibit 3.21a. Apply Newton's second law:

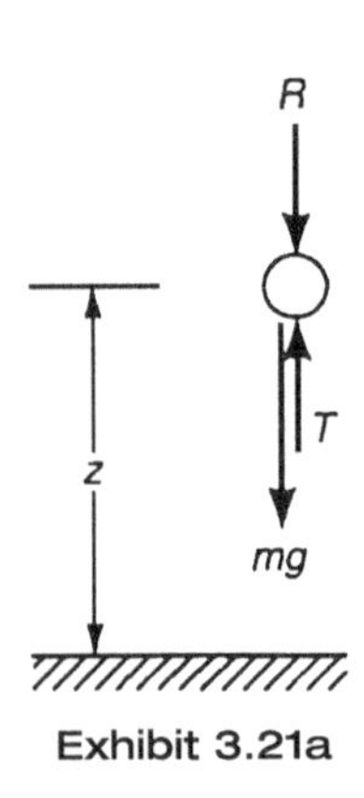

Exhibit 3.21a

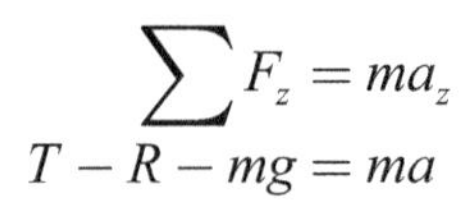

$$\sum F_z = ma_z$$
$$T - R - mg = ma$$

$$a = [(T - R)/m] - g = [(100 - 2z)N/5 \text{ kg}] - 9.81 \text{ m/s}^2$$

so

$$a = 10.2 - 0.4z = \frac{v\,dv}{dz}$$

and

$$\int_0^H (10.2 - 0.4z)\,dz = \int_{10}^0 v\,dv$$

where H is the highest height attained. Note also that $v = 0$ at this height. Integration yields,

$$10.2H - 0.2H^2 = [v^2/2]_{10}^{0} = -50$$

or

$$0.2H^2 - 10.2H - 50 = 0$$

Solving this quadratic (and discarding the negative value) gives the maximum height,

$$H = 55.5 \text{ m}$$

3.22 b. Refer to Exhibit 3.22a. Apply Newton's second law in the x and y directions.

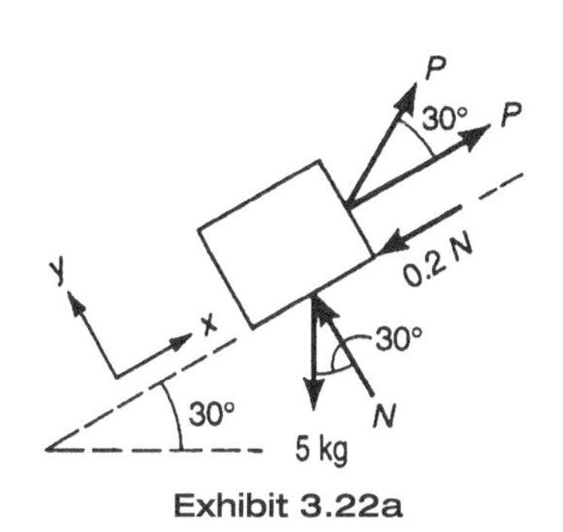

Exhibit 3.22a

$$\sum F = ma_x \qquad \textbf{(i)}$$

$$P + P\cos 30° - 0.2N - 5(9.81)\sin 30° = 5(2)$$

$$\sum F_y = 0$$
$$N + P\sin 30° - 5(9.81)\cos 30° = 0 \qquad \textbf{(ii)}$$

Solving Equations (i) and (ii) simultaneously yields $P = 21.9$ N.

3.23 a. In Exhibit 3.23a, apply Newton's second law in the radial and transverse directions.

$$\sum F_\theta = ma_\theta$$
$$N - mg\cos\theta = m(r\ddot{\theta} + 2\dot{r}\dot{\theta}) \qquad \textbf{(i)}$$
$$\sum F_r = ma_r$$

$$\mu N - P - mg\sin\theta = m(\ddot{r} - r\dot{\theta}^2) \qquad \textbf{(ii)}$$

Substitute values into Equations (i) and (ii):

$$N - 2(9.81)\cos 30° = 2[0 + 2(-1)(5)] \qquad \textbf{(iii)}$$

$$0.1N - P - 2(9.81)\sin 30° = 2[0 - 0.4(5)^2] \qquad \textbf{(iv)}$$

From Equation (iii), $N = -3.0$ newtons. Substituting this value into Equation (iv) gives $P = 9.9$N.

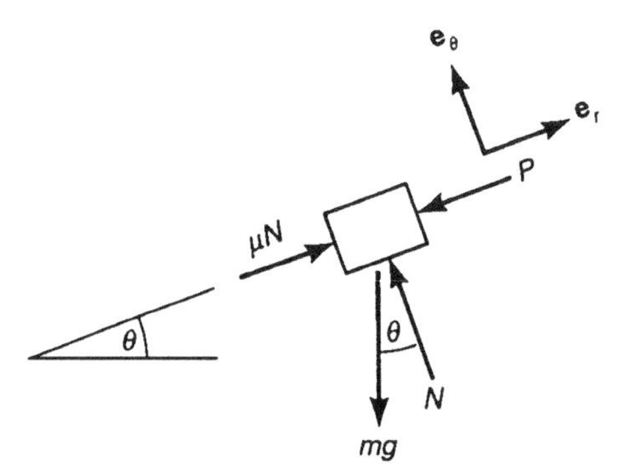

Exhibit 3.23a

3.24 **a.** Apply Newton's second law to the diagrams in Exhibit 3.24:

$$\sum F_n = ma_n$$

$$mg\cos\theta - N = mv^2/\rho$$

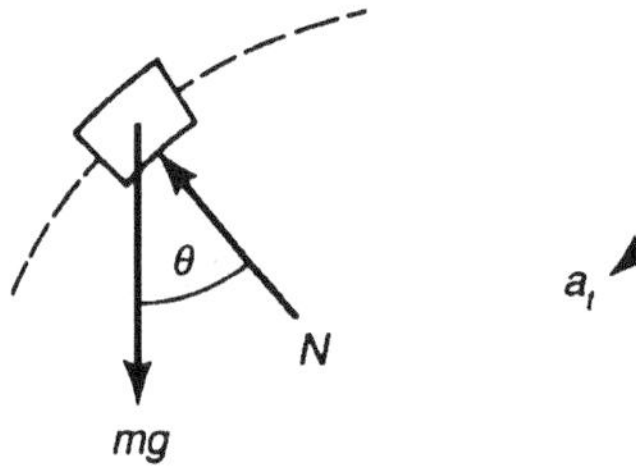

Exhibit 3.24a

or

$$\begin{aligned} N &= m[g\cos\theta - v^2/\rho] \\ &= (3)[9.81\cos 50° - 1.5^2/1] \\ &= 12.2 \text{ N} \end{aligned}$$

The positive sign indicates that N is directed as shown in the free-body diagram, Exhibit 3.24a.

3.25 **a.** At the onset of tipping, the free-body diagram and the inertia force diagram are as shown in Exhibit 3.25a. Take moments about point A:

$$mgb = m(u^2/R)H$$

Hence,

$$u = (Rgb/H)^{0.5}$$

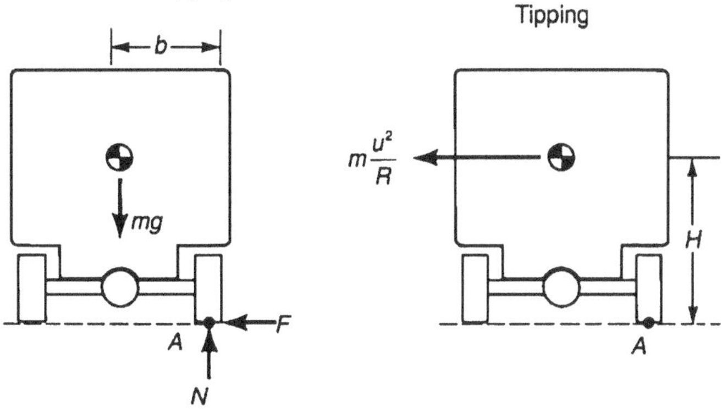

FBD at Onset of Tipping

Inertia Force at Onset of Tipping

Exhibit 3.25a

3.26 c. In Exhibit 3.26a, the sum of the forces in the vertical direction yields

$$N = mg$$

Apply the impulse-momentum principle in the horizontal direction [Equation (3.32)]:

$$\int_{t_1}^{t_2} F_{\text{horiz}}\, dt = mv_2 - mv_1$$

$$F_{\text{horiz}} = (0.25 + 0.5t) - 0.1N$$

Thus

$$\int_0^7 [0.25 + 0.5t - (0.01)(9.81)]dt = (1)v_2 - 0$$

or

$$0.25t + 0.25t^2 - 0.098t\Big|_0^7 = v_2 = 13.3 \text{ m/s}$$

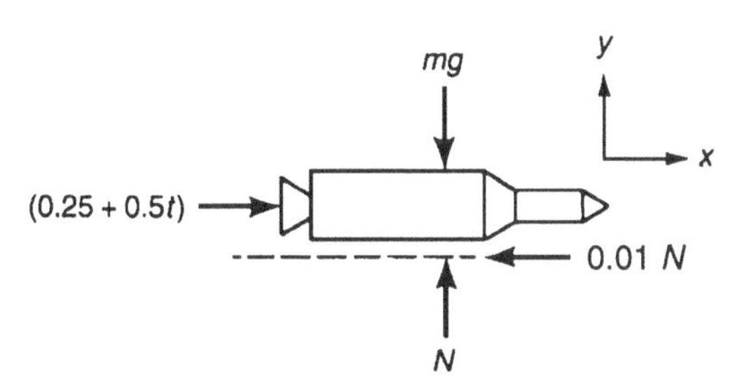

Exhibit 3.26a

3.27 b. Apply the impulse-momentum principle between the instant t_1 when the brakes are applied and the instant t_2 when the truck comes to a stop:

$$\int_{t_1}^{t_2} \mathbf{F}dt = m\mathbf{v}_2 - m\mathbf{v}_1$$

In the direction tangent to the road surface,

$$\int_0^t (mg \sin 10° - F)\,dt = 0 - mv_1$$

or

$$[2000(9.81) \sin 10° - 4000]t = -2000\frac{80(1000)}{60(60)}$$

so that $t = 74.9$ s.

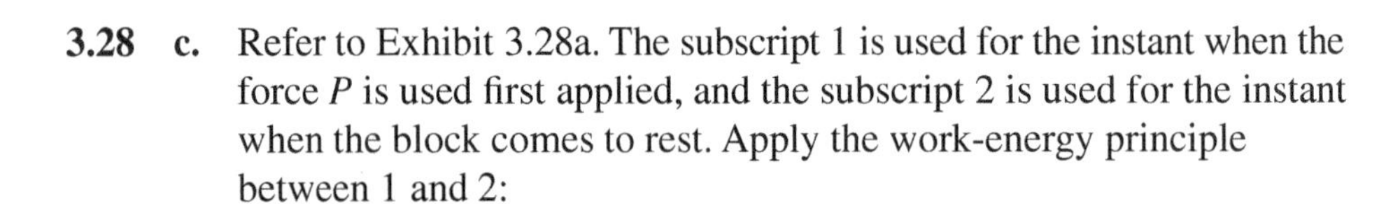

3.28 c. Refer to Exhibit 3.28a. The subscript 1 is used for the instant when the force P is used first applied, and the subscript 2 is used for the instant when the block comes to rest. Apply the work-energy principle between 1 and 2:

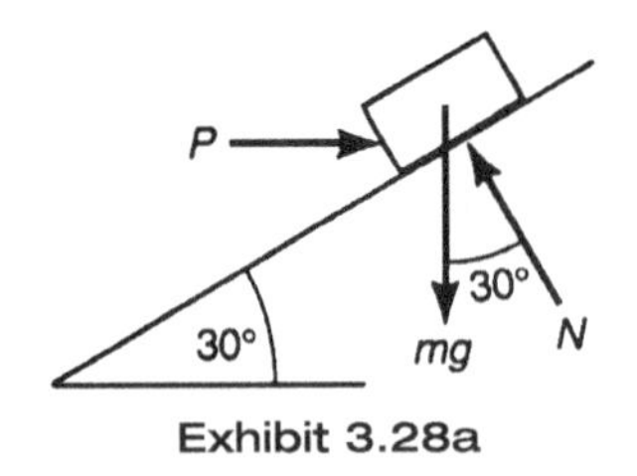

Exhibit 3.28a

$$W_{1-2} = T_2 - T_1$$
$$(mg \sin 30° - P \cos 30°)\Delta x = 0 - \tfrac{1}{2} mv_1^2$$

which yields

$$\Delta x = \frac{\frac{1}{2}mv_1^2}{P\cos 30° - mg \sin 30°} = \frac{0.5(2)3^2}{15\cos 30° - 2(9.81)\sin 30°} = 2.83 \text{ m}$$

3.29 c. Consult Exhibit 3.29a. Apply the work-energy principle between A and B, and then between B and C.

$A \rightarrow B: \quad W_{A-B} = T_B - T_A$

$$mgr = \frac{1}{2}mv_B^2 - 0 \qquad \textbf{(i)}$$

$B \rightarrow C: \quad W_{B-C} = T_C - T_B$

$$N = mg \qquad \textbf{(ii)}$$

$$-\mu NH = 0 - \frac{1}{2}mv_B^2 \qquad \textbf{(iii)}$$

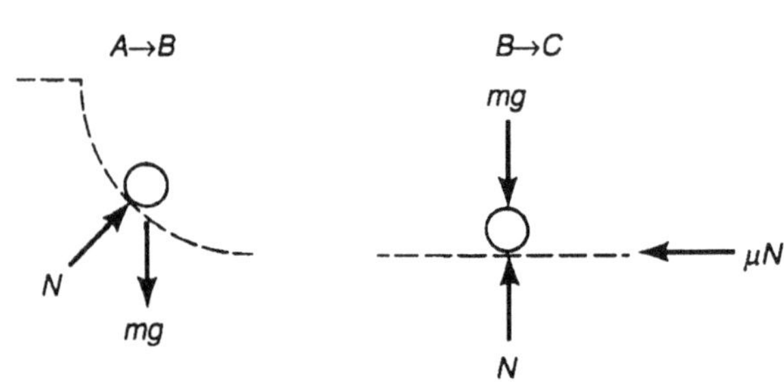

Exhibit 3.29a

Substituting Equations (i) and (ii) into Equation (iii),

$$-\mu mgH = -mgr, \quad \text{or} \quad H = r/\mu$$

3.30 c. Let D be the position at which the spring has its natural (unstretched) length. Apply the work-energy principle (Exhibit 3.30a) from A to D:

$$W_{A-D} = T_D - T_A$$

$$\frac{1}{2}k\Delta^2 - \mu N\Delta = \frac{1}{2}mv_D^2 - 0$$

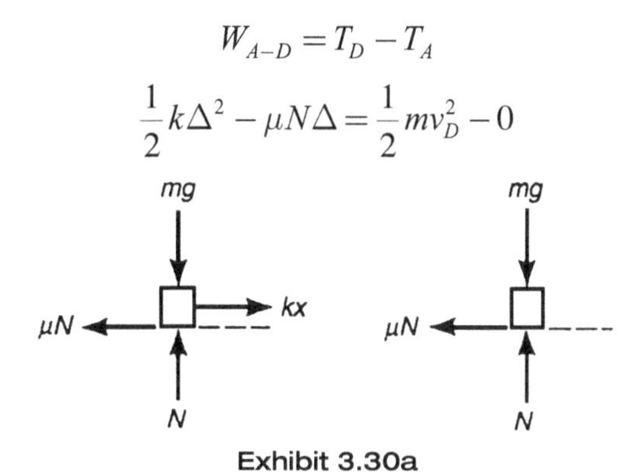

Exhibit 3.30a

Since $N = mg$, we have

$$\frac{1}{2}mv_D^2 = \frac{1}{2}k\Delta^2 - \mu mg\Delta \qquad \text{(i)}$$

Now apply the work-energy principle from D to B:

$$W_{D-B} = T_B - T_D$$

$$-\mu N(1.5 - \Delta) = \frac{1}{2}mv_B^2 - \frac{1}{2}mv_D^2$$

Again, since $N = mg$, and $\frac{1}{2}mv_D^2$ is given by Equation (i), we have

$$-\mu mg(1.5 - \Delta) = \frac{1}{2}mv_B^2 - \frac{1}{2}k\Delta^2 + \mu mg\Delta$$

or

$$\Delta = \sqrt{\frac{2\left(1.5\mu mg + 0.5mv_B^2\right)}{k}} = 0.26 \text{ m}$$

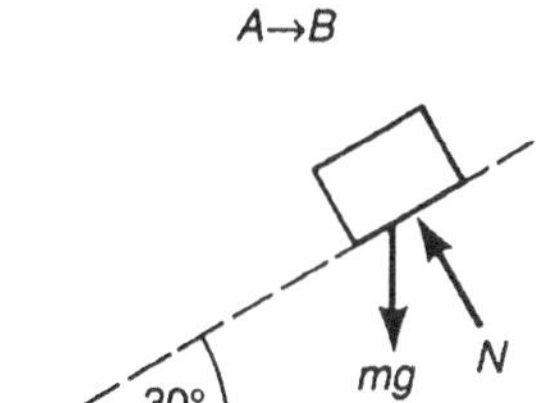

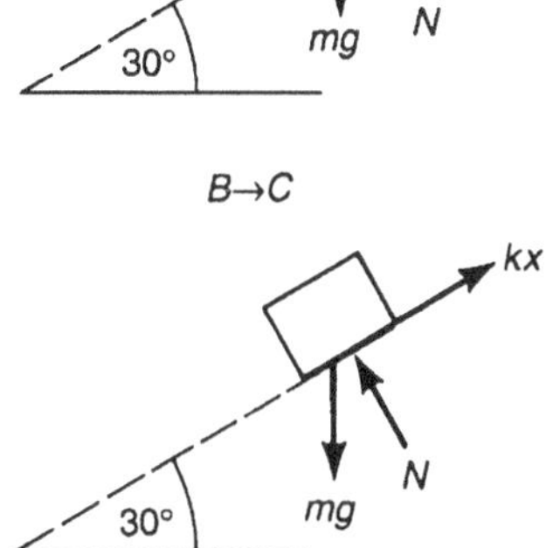

Exhibit 3.31a

3.31 b. Apply the work-energy principle between A and B, and then between B and C.

$$W_{A-B} = T_B - T_A$$

With the forces shown in Exhibit 3.31a,

$$mg\sin 30°(3) = \frac{1}{2}mv_B^2 - 0$$

$$W_{B-C} = T_C - T_B$$

$$-\frac{1}{2}kx^2 + mg\sin 30°(x) = 0 - \frac{1}{2}mv_B^2 = -mg\sin 30°(3)$$

Substituting values, we obtain the quadratic equation $500x^2 - 30x - 90 = 0$, which can be solved to yield $x = 0.45$ m.

3.32 **d.** Using the diagram in Exhibit 3.32a, apply the work-energy principle between B and C:

$$W_{B-C} = T_c - T_B$$

$$-mg \sin 5°(x) - 0.02\, mgx = 0 - \frac{1}{2} m v_B^2$$

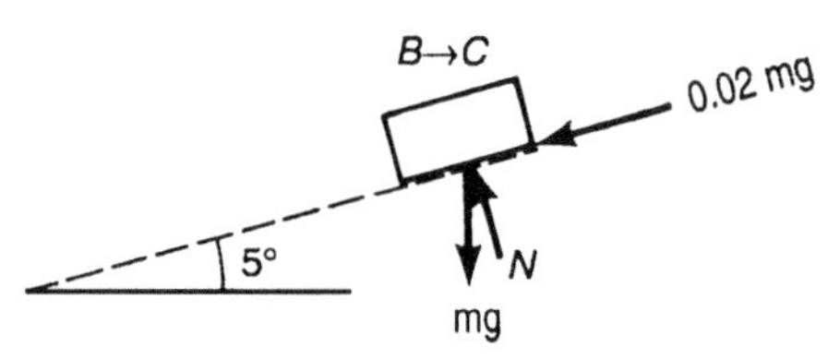

Exhibit 3.32a

or

$$x = \frac{\frac{1}{2} m v_B^2}{mg \sin 5° + 0.02\, mg} = \frac{0.5(500)\left[\frac{(32 \times 1000)}{60 \times 60}\right]^2}{500(9.81)\sin 5° + 0.02(500)9.81} = 37.6 \text{ m}$$

Total distance up the grade is (50 + 37.6) m = 87.6 m.

3.33 **b.** Consult Exhibit 3.33a. The moment of inertia of each rod about its mass center is

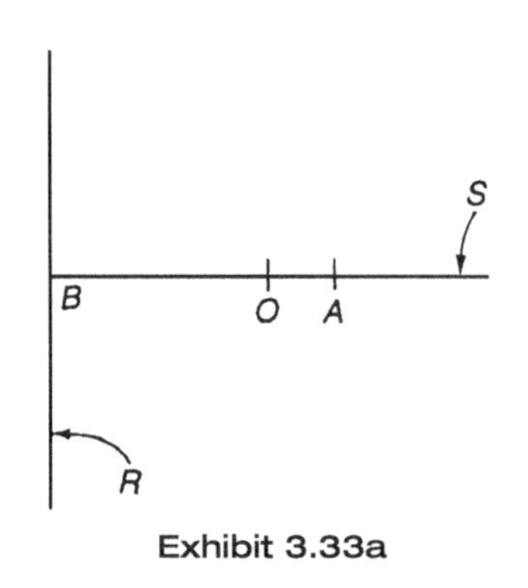

Exhibit 3.33a

$$I_{R/B} = I_{S/O} = \frac{1}{12} ml^2 = \frac{1}{12}(4)(3)^2 = 3 \text{ kg–m}^2$$

Here, O is the mass center of S, and B is the mass center of R. Apply the parallel axes theorem:

$$I_{R/A} = I_{R/B} + m(2)^2 = 3 + 4(2)^2 = 19 \text{ kg–m}^2$$

$$I_{S/A} = I_{S/O} + m(2 - 1.5)^2 = 3 + 4(0.5)^2 = 4 \text{ kg–m}^2$$

And, for the assemblage,

$$I_A = I_{R/A} + I_{S/A} = (19 + 4) \text{kg–m}^2 = 23 \text{ kg–m}^2$$

3.34 **d.** Since $M = I\alpha = (1/2)mr^2\alpha$, $\alpha = M/(0.5mr^2)$. Substituting values, we have $\alpha = 1.67$ rad/s^2. Because this angular acceleration is constant, the final angular velocity is given by

$$\omega = \omega_0 + \alpha t = 1 + 1.67(5) = 9.3 \text{ rad/s}$$

3.35 a. When the desired configuration is achieved, the rod is in translation. The free-body diagram and the inertia force diagram for the rod are shown in Exhibit 3.35a. Taking moments about point O,

$$mg(l/2)\sin\theta = ma(l/2)\cos\theta$$

where l is the length of the rod. Thus,

$$a = g\tan\theta = 9.81\tan 10° = 1.73\ \text{m/s}^2$$

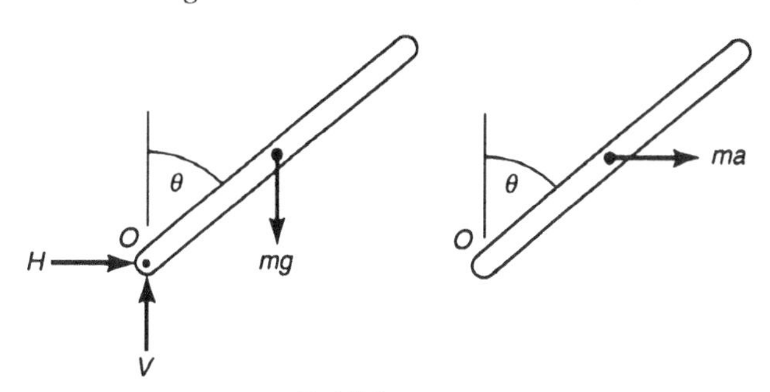

Exhibit 3.35a

3.36 d. From the diagrams in Exhibit 3.36a, and taking moments about O,

$$PH - mgl\ \sin\theta = I_G\alpha + ml^2\alpha$$

so that

$$\alpha = (PH - mgl\ \sin\theta)/(I_G + ml^2)$$

Now,

$$I_C = I_G + m(H-l)^2$$

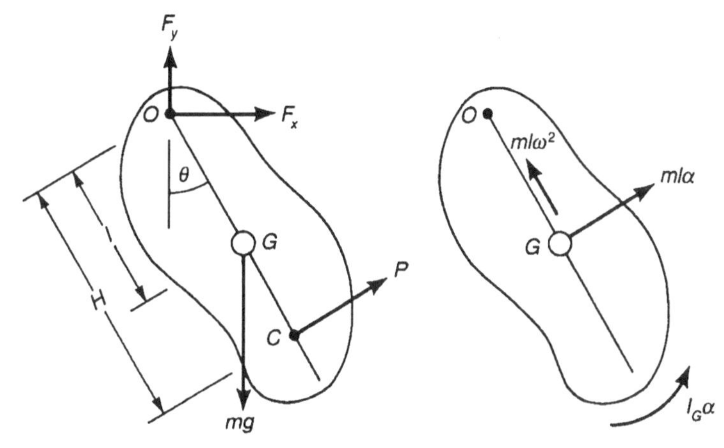

Exhibit 3.36a

from the parallel axis theorem. Thus,

$$I_G = I_C - m(H-l)^2 = mk^2 - m(H-l)^2$$

and

$$\alpha = \frac{PH - mgl\ \sin\theta}{m[k^2 - (H-l)^2] + ml^2}$$

Substitute values to get

$$\alpha = \frac{10(2) - 3(9.81)(1.5)(0.5)}{3[(0.8)^2 - (0.5)^2 + (1.5)^2]} = 0.26\ \text{rad/s}^2$$

3.37 **b.** C is the center of mass of the rod in Exhibit 3.37a. Summing moments about O gives

$$-mg(l/2)\sin\theta = (1/12)ml^2\alpha + m(l/2)^2\alpha + ma(l/2)\cos\theta \qquad \textbf{(i)}$$

Summing forces along the horizontal, gives

$$H = ma + m(l/2)\,\alpha\cos\theta - m\omega^2(l/2)\sin\theta \qquad \textbf{(ii)}$$

From Equation (i),

$$\alpha = -\frac{3}{2}\frac{(g\sin\theta\cos\theta)}{l}$$

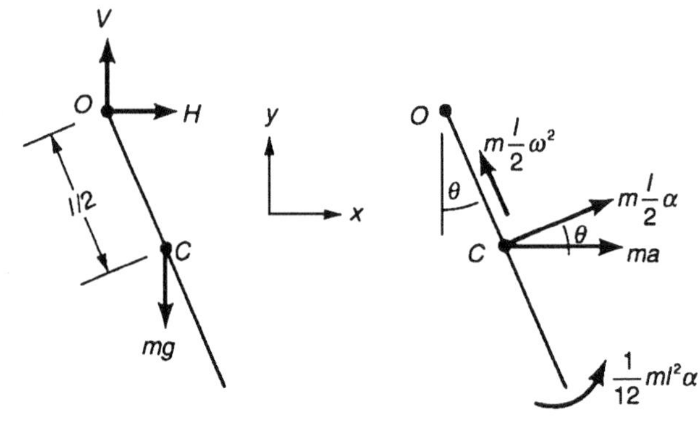

Exhibit 3.37a

Substituting values,

$$\alpha = -4.98\ \text{rad/s}^2 \qquad \textbf{(iii)}$$

Substituting Equation (iii) and the given values into Equation (ii) yields

$$H = (1)(2) + (1)(1)(-4.98)\cos 30° - (1)(2)^2(1)\sin 30° = -4.31\ \text{N}$$

3.38 **a.**

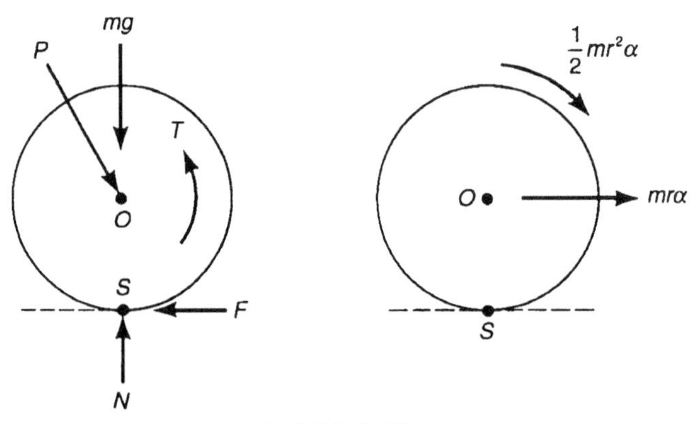

Exhibit 3.38a

Using the free-body diagram shown in Exhibit 3.38a, take moments about the contact point S:

$$rP\cos 60° - T = mr^2\alpha + (1/2)\,mr^2\alpha = (3/2)\,mr^2\alpha$$

or

$$\alpha = (rP\cos 60° - T)/(1.5mr^2) = \text{constant}$$

Substituting values, we find

$$\alpha = 0.83\ \text{rad/s}^2$$

With a constant angular acceleration, the angular velocity is

$$\omega = \omega_0 + \alpha t = 0 + 0.83(10) = 8.33\ \text{rad/s}$$

3.39 **b.** Exhibit 3.39a shows the forces acting on the rod as it swings from position 1 (horizontal) to position 2 (vertical). The work-energy principle gives

$$W_{1\to 2} = T_2 - T_1$$

That is,

$$mg\frac{l}{2} = \frac{1}{2}I_A\omega_2^2 - 0 = \frac{1}{2}\times\frac{1}{3}ml^2\omega_2^2$$

and

$$\omega^2 = \sqrt{\frac{3g}{l}} = \sqrt{\frac{(3)(9.81)}{2}} = 3.84\ \text{rad/s}$$

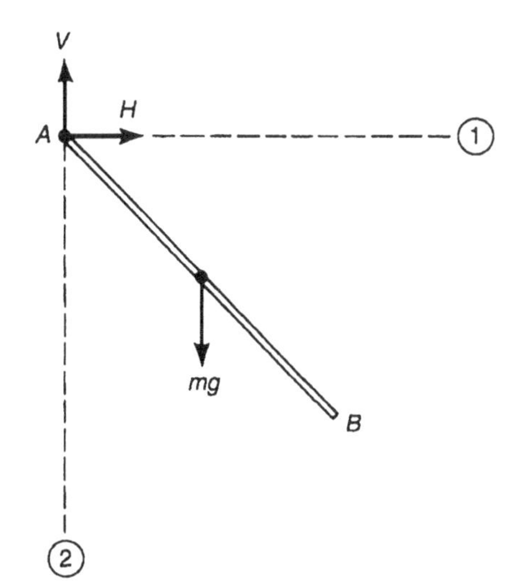

Exhibit 3.39a

3.40 **d.** Exhibit 3.40a shows the forces and torque acting on the rod as it rotates from position 1 (vertical) to position 2 (horizontal). The work-energy relation gives

$$W_{1-2} = T_2 - T_1$$

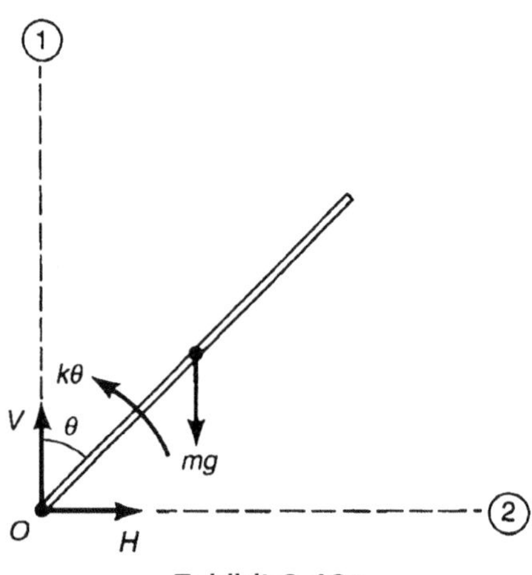

Exhibit 3.40a

That is,

$$mg\frac{l}{2} + \frac{l}{2}k(\theta_1^2 - \theta_2^2) = \frac{1}{2}I_0\omega_2^2 - \frac{1}{2}I_0\omega_1^2$$

Now, $\theta_1 = 0$, $\theta_2 = \pi/2\text{k}$, and $\omega_2 = \omega_2 = 0$. Thus,

$$k = \frac{mgl}{\theta_2^2 - \theta_2^2} = \frac{2(9.8)(3)}{(\pi/2)^2} = 23.8 \text{ N-m/rad}$$

3.41 **a.** The forces acting on the cylinder during this motion are shown in Exhibit 3.41a. Applying the work-energy principle,

$$W_{1-2} = T_2 - T_1$$

F and *R* do no work because their point of application has zero velocity (rolling without slip); *mg* does no work because its point of application moves perpendicular to the force. Work done by the spring force is

$$W_{\text{sp}}\frac{1}{2}k\left(\Delta_1^2 - \Delta_2^2\right) = W_{1-2}$$

where

$$\Delta_1 = [(3^2 + 4^2)^{0.5} - 3]\text{ m} = 2\text{ m}, \quad \text{and} \quad \Delta_2 = 0$$

$$T_1 = 0, \text{ and } T_2 = \frac{1}{2}m(v_P)^2 + \frac{1}{2}I_P\omega^2 = \frac{1}{2}m(\omega r)^2 + \frac{1}{2}\times\frac{1}{2}mr^2\omega^2 = \frac{3}{4}mr^2\omega^2$$

Substituting into the work-energy principle yields

$$W_{1-2} = \frac{1}{2}k\Delta_i^2 = \frac{3}{4}mr^2\omega^2$$

and

$$\omega=\left(\frac{2}{3}\frac{k}{m}\right)^{0.5}\frac{\Delta_1}{r}=\left(\frac{2(2)}{3(12)}\right)^{0.5}\times\frac{2}{0.5}=1.33\text{ rad/s}$$

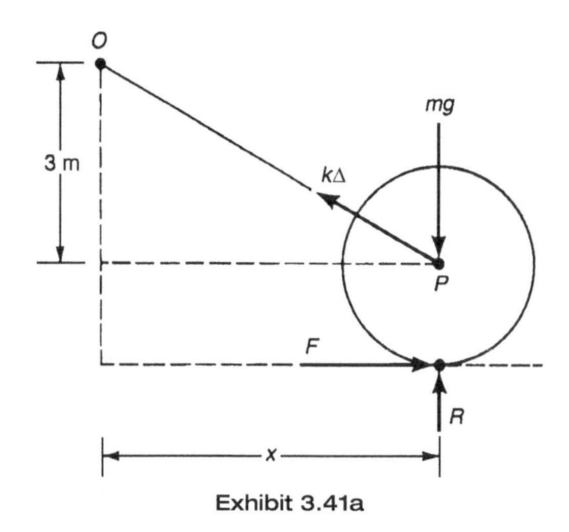

Exhibit 3.41a

3.42 c. Each step through the periodic table introduces an additional proton and electron to a neutral atom.

3.43 a. The number of protons are fixed for an individual element. If the number of protons (and electrons) were varied, the chemical properties would be affected.

3.44 d. To separate H2 into hydrogen atoms, the H-to-H bond would have to be broken, thus requiring energy.

3.45 a. Unlike ions attract. FCC metals possess four atoms per unit cell; BCC metals have two.

3.46 c. The [111] direction passes diagonally through the unit cell. The distance is $a\sqrt{3}$, which equals two repeat distances.

3.47 b. The mass is determined from 27 amu per 6.0×10^{23} atoms. Each cell of four atoms has a volume of $(4r/\sqrt{2})^3$.

3.48 b. Assuming spherical atoms, there are four atoms of radius r per unit cell. The cube edge is $4r/\sqrt{2}$.

3.49 c. There is a double bond between the two carbons. Each hydrogen is held with a single bond.

3.50 c. Density is mass/volume. The mass per FCC unit cell is 4 Au $\times$ (197 g/6.02×10^{23} Au). The volume per FCC unit cell of a metal is (face diagonal/$\sqrt{2}$ $)^3$ or $(4r/\sqrt{2})^3$.

3.51 **b.** Thermoplastic materials are polymerized but soften for molding at elevated temperatures. The polymeric molecules are linear. Thus they include the ethylene-type compounds that are bifunctional (two reaction sites per mer).

Thermosetting materials develop three-dimensional structures that become rigid during processing. For example, phenol is trifunctional and thus forms a network structure. Reheating does not soften them.

3.52 **d.** Benzene is C_6H_6; however, in styrene, one hydrogen is absent at the connection to the C_2H_3–base.

3.53 **d.** Since a cubic crystal has interchangeable *x*-, *y*-, and *z*-axes, the <110> family includes all directions with permutations of 1, 1, and 0 (either + or –). (This is not necessarily true for noncubic crystals.)

3.54 **a.** Since a cubic crystal has interchangeable *x*-, *y*-, and *z*-axes, the {112} family includes all planes with index permutations of 1, 1, and 2 (either + or –). (This is not necessarily true for noncubic crystals.)

3.55 **d.** (b) and (c) involve individual atoms (point imperfections). (a) is a lineal imperfection. Boundaries result from a two-dimensional mismatch of crystal structures.

3.56 **d.** There are two types of dislocations: (1) an edge dislocation may be described as an edge of a missing half-plane of atoms; (2) a screw dislocation is the core of a helix.

3.57 **d.** Unless special efforts are made to grow single crystals, many crystals are nucleated and grow until they encounter neighboring crystals. Each grain is individually oriented.

3.58 **c.** Zinc is sufficiently near copper in size and electrical behavior to proxy for copper in the crystal structure. It is too big for the interstices.

3.59 **b.** The 7.5% copper replaces silver atoms. If it is cooled rapidly, the copper is retained in solid solution. The copper atoms interfere with electron movements within the silver.

3.60 **a.** FCC metals have a higher packing factor than do BCC metals; therefore, with other factors equal, diffusion is reduced.

3.61 **c.**

$$D = D_o e^{-Q/(RT)} = \left(0.022\ \text{cm}^2/\text{s}\right) e^{\frac{-97000\ \text{J/mol}}{8.314\ \text{J/(mol-K)}(1000+273\ \text{K})}} = 2.3 \times 10^{-6}\ \text{cm}^2/\text{s}$$

3.62 **b.** As the grains grow, their volume increases by the third power. Their surface area increases by the square.

3.63 **d.** Metallic crystals are not ionic and do not have discrete ions.

3.64 a. The interstitial sites are smaller than the atoms in metals.

3.65 b. To balance the charge, each missing Fe^{2+} ion must be compensated by two Fe^{3+} ions. Therefore, out of 100 cation sites, two are vacant, four are Fe^{3+}, and thus 94 are Fe^{2+}.

3.66 a. Using a computational basis of four atoms, (1 Mg^{2+} + 1 O^{2-}) + (1 Fe^{2+} + 1 O^{2-}) = (24 + 16) + (56 + 16) = (40 + 72).

3.67 b. The elastic strains between atoms along the boundary follow the same relationships as the strains among atoms within the grains.

3.68 c. Solid solution increases strength (solution hardening). It also decreases conductivity (and increases resistivity). Sterling silver is 92.5Ag–7.5Cu.

3.69 b. When atoms are moved from one site to another, bonds are broken and reconstituted. During transition, an activation energy is required to distort the lattice. Small solute atoms, low-melting-point solvents, and lower atomic packing factors in a lattice all require a lower activation energy. Hence activation energy for diffusion is *directly* proportional to the packing factor.

3.70 a. On cooling, curve CI is encountered at approximately 420°C (790°F). That curve is the solubility limit of Zn in that liquid. Zinc in excess of the solubility limit separates as the intermetallic compound, $MgZn_2$, which is plotted as the vertical line EI.

3.71 d. A ratio of 67 Zn atoms to 33 Mg atoms is 2-to-1; therefore, $MgZn_2$.

3.72 b. The eutectic composition is that of a low-melting liquid saturated with two solids. The 0.8 weight percent composition is a solid, not a liquid, at 727°C.

3.73 d. The (ferrite + carbide) area extends across the lower part of the phase diagram from nil carbon to 6.7 carbon. At the left side there is no Fe_3C; at the right side there is only carbide (Fe_3C contains 6.7% carbon). The amount of carbide between the two extremes may be determined by linear interpolation.

3.74 d. The eutectic temperature is 780°C (1445°F). That is the lowest temperature at which liquid can exist of equilibrium. At 779°C there are approximately equal amounts of α and β, the two solid solutions.

3.75 d. The curves of a phase diagram are solubility limits for the phases within the single-phase regions. Since copper is the solvent for the β structure, β has no upper limit of copper solubility (other than 100%).

3.76 **c.** The diffusion flux is proportional to the concentration gradient. (The other choices cited reduce the activation energy needed for diffusion.)

3.77 **a.** Boundary atoms possess higher energy. Therefore, the boundary is reduced and the grain size is increased at temperatures where the atoms can move. The net movement of the atoms is to the larger grains (with less boundary area).

3.78 **a.** Phase diagrams cannot be used to predict crystalline properties.

3.79 **d.** Above the eutectic temperature, $(\alpha + L)$ can be present concurrently, as can $(L + \beta)$, but not $(\alpha + \beta)$. Below the eutectic temperature, $(\alpha + \beta)$ may be present, but no liquid. As the eutectic temperature is passed during cooling or heating, all three phases may coexist.

3.80 **d.** At 800°C, a 72Ag–28Cu alloy is fully liquid, which therefore has 72–28 composition.

3.81 **d.** During equilibrium cooling, the final liquid for this alloy does not disappear until the eutectic temperature is reached at 780ºC. At that temperature, the liquid composition is 72Ag–28Cu.

3.82 **b.** All α disappears on the right side of the $(\alpha + L)$ field, where the composition is 72Ag–28Cu, or 100g Ag to 39 g Cu.

3.83 **c.** As the temperature is increased, the atoms gain additional energy and can relocate, eliminating the strain energy that accompanies dislocations. *Less time* is required at higher temperature. *Less time* is also required for a highly cold-worked material because there is additional stored energy present.

3.84 **d.** Reduction and polymerization involve chemical reactions. Tempering and recrystallization involve reheating but no shape change.

3.85 **c.** Strength and hardness are increased at the expense of ductility and toughness (opposite of brittleness). The increase is facilitated by microstructures that interfere with dislocation movements. These include a high density of dislocations from plastic deformation, and the presence of many fine, hard particles. Annealing removes dislocations and permits the agglomeration of particles into fewer large particles.

3.86 **d.** The production of tempered martensite is indicated in (c) above. Bainite is formed by isothermally decomposing austenite directly to a microstructure of fine carbide particles within a ferrite matrix. Although the processing differs, the resulting microstructure and properties are nearly identical.

3.87 a. Annealing removes the hardness that was introduced by cold work. Quenched and tempered steels are harder with higher carbon contents, because more hard carbide particles are present. Alloying elements perform several hardening functions: They solution-harden the ferrite matrix; they slow down grain growth; and they delay the formation of pearlite, thus permitting more martensite with slower cooling rates (in turn, more tempered martensite may be realized farther below the quenched surface).

3.88 d. Stresses will relax below the melting temperature. Tensile stresses facilitate the cracking of brittle materials; compression limits cracking.

3.89 d. The reaction occurs only below the eutectoid temperature.

3.90 c. While grain boundaries interfere with slip at low temperatures, they facilitate creep at elevated temperatures.

3.91 d. The processing step of rapid cooling, such as quenching, retains the structures that existed at higher temperatures, even though a solubility limit is exceeded.

3.92 a. Martensite is a transition phase between austenite and ferrite, which retains carbon interstitially. Given an opportunity, the carbon separates as Fe_3C.

3.93 a. Alloying elements can dissolve substitutionally in austenite, which remains FCC.

3.94 b. Hardenability may be described as the ability, or the ease, by which martensitic hardness is obtained at various cooling rates (as located on an end-quenched, or Jominy, test).

3.95 d. The ratio of stress-to-strain is defined as the elastic modulus.

3.96 d. $\sigma_u = F/A_0$

3.97 a. Slip occurs most readily in directions that have the shortest steps, and along planes that are farthest apart. The latter are automatically the planes that are most densely populated.

3.98 c. Cold working—such as rolling, forging, drawing, or extrusion—deforms the material at temperatures below the recrystallization temperature. Strain hardening occurs, increasing both the yield and ultimate strength. Internal strains and minute cracks are introduced as slip or twinning occurs. Ductility, elongation, and the recrystallization temperature are decreased. A preferred grain orientation is introduced in the direction of elongation, and the grains are flattened.

3.99 **d.** Since oxygen is required for rust formation, oxygen-depleted areas become the anode and are corroded. This may lead to pitting, particularly if rust or other corrosion products are accumulated locally to prevent the access of oxygen. Iron and other metals may be protected from corrosion by the presence of a more electropositive metal such as magnesium or zinc. This is the reason for coating steel with zinc to produce galvanized sheet.

3.100 **c.** Although fatigue strength is not sensitive to temperature or loading rates, it is very sensitive to surface imperfections from which cracks originate and propagate.

3.101 **b.** The stress intensity factor is proportional to the applied stress and the square root of the crack depth.

3.102 **b.** Glide occurs parallel to the slip plane.

3.103 **d.** Brittleness exists below T_{DT}.

3.104 **c.** Corrosion commonly occurs in the combined presence of oxygen and water. Protection may be obtained by making a cathode out of the critical part or by avoiding air.

3.105 **c.** With equal strains, $(\sigma_{gl}/\sigma_p) = E_{gl}/E_p = 10$. Likewise with equal areas, $F_{gl} = 10F_p$, and $F_{gl}/(F_{gl} + F_p) = 10/(10 + 1)$.

3.106 **b.** The "mass" of 2 electrons has a charge $Q = 3.2 \times 10^{-19}$ C, thus the electric field is $\mathbf{E} = 500$ V/0.01 m $= 50 \times 10^3$ V/m. The force is then $\mathbf{F} = Q\mathbf{E} = (3.2 \times 10^{-19}) \times (50 \times 10^3) = 1.6 \times 10^{-14}$ N.

3.107 **c.** The magnitude of the length of the resultant vector, R, in the x, y, z plane is

$$R = \sqrt{2^2 + 3^2 + 4^2} = \sqrt{29}$$

The magnitude of the electric field, **E**, is $Q/(4\pi\varepsilon R^2) = (0.3\times10^{-3})/(4\pi \times 8.85 \times 10^{-12} \times 29) = 93{,}000$ V/m.

3.108 **c.** On the b plane (for that plane alone), $E^+ = (-\sigma/2e)_{ax}$, and for the negatively charged plane at the a plane (again for that plane alone, but acting to the right of a), is the same as before, therefore:

$$\mathbf{E} = \mathbf{E}^+ + \mathbf{E}^- = (-\sigma/\varepsilon_0)_{ax} \text{ V/m}$$

3.109 **d.** From a sketch of the vectors in Exhibit 3.109a, the value of $\mathbf{E}_{net}$ is 12.6×10^4 V/m. **F** is then found to be

$$\mathbf{E} = \frac{10\times10^{-6}}{4\pi(8.85\times10^{-12})\,r^2}$$

$$\mathbf{F} = Q_C\mathbf{E}_{net} = 5 \times 10^{-6}\,(12.6 \times 10^4) = 63 \times 10^{-2} \text{ N}$$

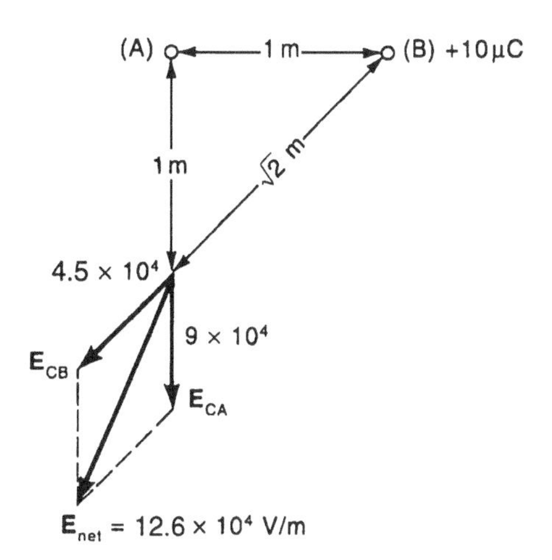

Exhibit 3.109a Vector field

3.110 c. The cross-sectional area of the iron core is $\pi r^2 = \pi \times 10^{-4}$ m^2. The path length is $l = 2\pi R = 2\pi \bullet 5 \times 10^{-2} = 0.1\pi$ m. The permeability is $\mu = \mu_0\mu_r = 4\pi \bullet 10^{-7} \times 10^3 = 4\pi \bullet 10^{-4}$.

Hence, $H = B/\mu = 0.5/(4\pi \bullet 10^{-4}) = 390$ A • t/m, and since H = mmf/length, mmf = $H \times$ length = 390 (0.1π) = 122.5 A • t. Thus, I = mmf/turns = 122.5/100 = 1.225 A.

3.111 c. Assume the two wires are bundled close together (with respect to the 0.2-meter position); then the net current to the right is $I_{net} = 5 - 1 = 4$ A. The flux density is given by $B = (\mu I)/(2\pi r) = (4\pi \times 10^{-7})(4)/(2\pi\ 0.2) = 4 \times 10^{-6}$ tesla.

3.112 c.

$$I_{rms} = \sqrt{(1/T)\int_0^T i^2\,dt} = \sqrt{(1/5)[(1)^2(1) + (3)^2(2) + (1)^2(1) + 0]} = 2 \text{ A}$$

3.113 b. The current at 10 seconds is 10 A. And from the equation for energy storage in an inductor, it is found that W = (1/2) Li^2 = (1/2)(0.2)(10)^2 = 10 J.

3.114 b. From the figure and knowing the effective value for a rectified sine wave, one may compute V_{avg} and convert to I_{avg} from Ohm's law:

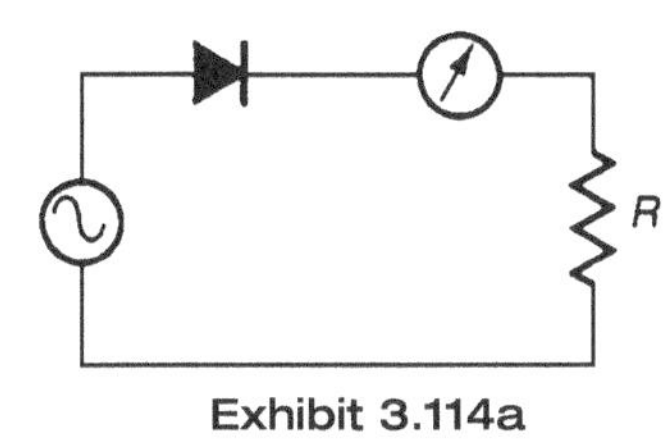

Exhibit 3.114a

$$V_{max} = \sqrt{2}V_{rms} = \sqrt{2}(10 \text{ V}), \quad V_{avg} = V_{max}/\pi = \sqrt{2}(10/\pi) = 4.49$$

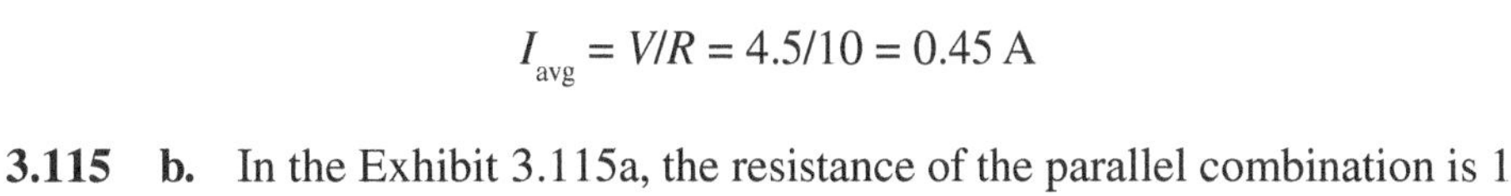

$$I_{avg} = V/R = 4.5/10 = 0.45 \text{ A}$$

3.115 b. In the Exhibit 3.115a, the resistance of the parallel combination is 1 Ω; then the series combination is 2 Ω. The current I_1 is $V/R = 2/2 = 1$ A, and the current for the parallel branch is

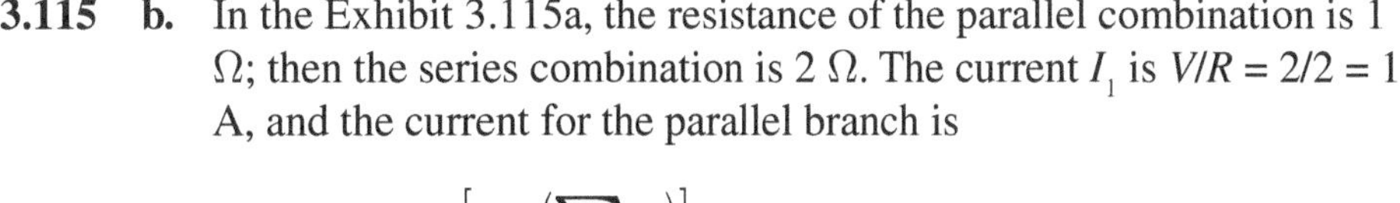

$$I_2 = \left[G_2/\left(\sum G\right)\right]I_1 = (1/2)/[(1/2)(1/2)]1$$

$$= 0.5 \text{ A}, P_2 = I_2^2 R = (0.5)^2\ 2 = 0.5 \text{ watts}$$

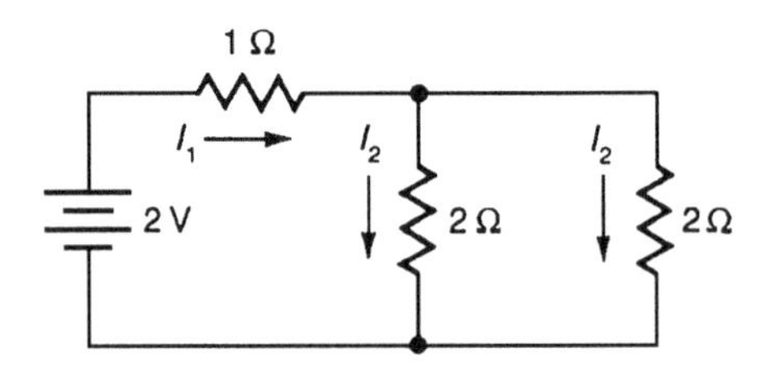

Exhibit 3.115a Circuit configuration

3.116 d. Refer to Exhibit 3.116a. The current is given by $I_2 = [G_2/\Sigma G)]I_{total} = [(1/2)/(1 + 1/2 + 1/3 + 1/4)](2) = 0.48$ A, $P_2 = I^2R = (0.48)^2(2) = 0.46$ watts.

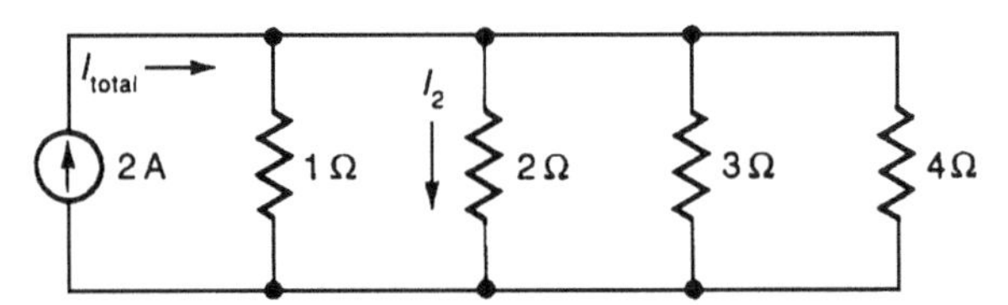

Exhibit 3.116a Circuit configuration

3.117 c. From the Exhibit 3.117a, set up mesh currents, I_1 and I_2, then sum voltages around each loop:

Loop 1: $1 = 2I_1 + 2(I_1 - I_2)$;
Loop 2: $2 = 2(I_2 - I_1) + 2I_2$

By Cramer's Rule,

$$I_1 = \frac{\begin{vmatrix} 1 & -2 \\ 2 & 4 \end{vmatrix}}{\begin{vmatrix} 4 & -2 \\ -2 & 4 \end{vmatrix}} = 2/3\ A \qquad I_2 = \frac{\begin{vmatrix} 4 & 1 \\ -2 & 2 \end{vmatrix}}{\begin{vmatrix} 4 & -2 \\ -2 & 4 \end{vmatrix}} = 5/6\ A$$

$I_{mid\text{-}leg} = I_1 - I_2 = -1/6$ A; $\quad P_{mid\text{-}leg} = I^2R = (-1/6)^2 2 = 0.056$ W

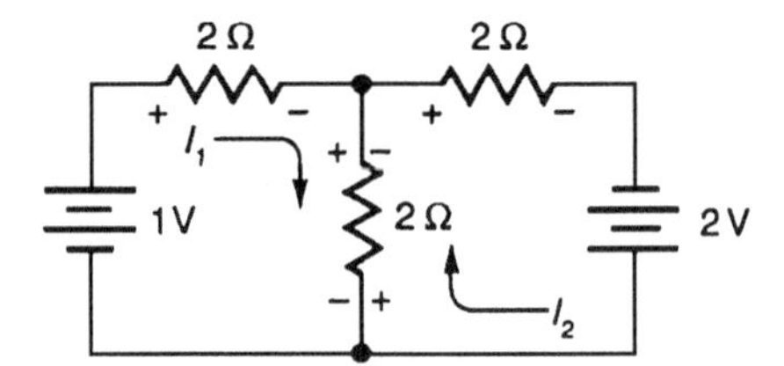

Exhibit 3.117a Circuit configuration

3.118 a. To find the Thevenin voltage, find the voltage across the 4-ohm resistor with R_L removed, which yields 5 volts. Thevenin resistance may be found by either of two methods. One method is to look back into the circuit with all active sources replaced with their own impedances and to calculate this output impedance; the other method is to short out the terminals of interest. For this second method, the impedance is

$$R_{eq} = V_{oc}/I_{sc} = 5/(10/4) = 2\ \Omega$$

3.119 a. First find the node voltage, V_1, at the intersection of the three resistors by summing the currents (start by assuming all currents flow away from the node). Assume that I_1 is to the left, I_3 to the right, and I_2 down.

$$\sum I = 0: (-2) + V_1/3 + (V_1 - 10)/5 = 0,\ 8V_1 = 60,\ V_1 = 7.5\ \text{V};$$

$$V_A = V_1 - I_1R_4 = 7.5 - (-2)4 = 15.5\ \text{V}$$

3.120 a. Refer to Exhibit 3.120a. The time constant, τ, for an R–C circuit is RC. The voltage across the capacitor will discharge at a rate given by

$$v_C = V_C e^{-t/\tau} = 10\ e^{-20/10} = 1.35\ \text{V},\quad i = v/R = 1.35/10^5 = 13.5 \times 10^{-6}\ \text{A}$$

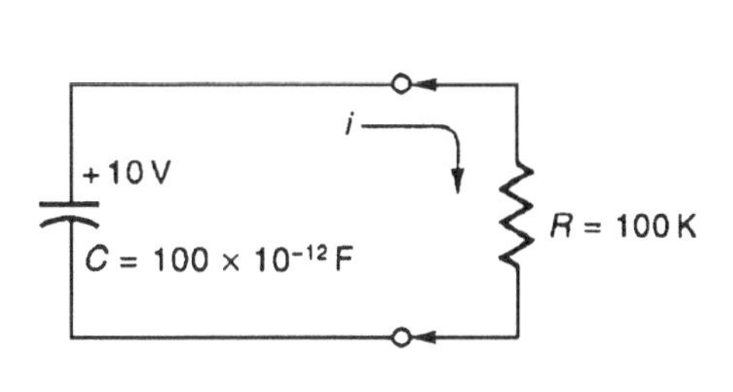

(a) R–C circuit

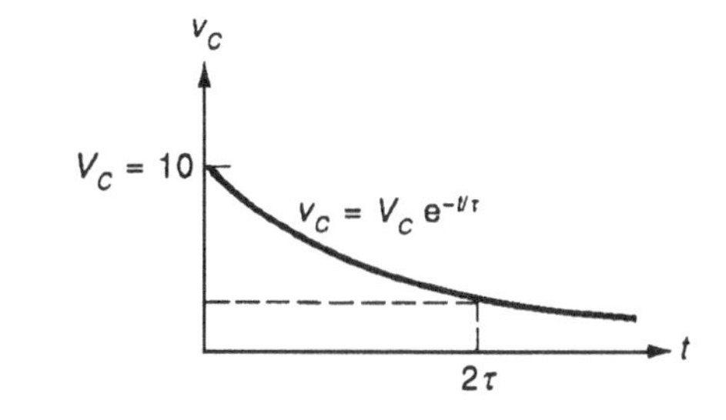

(b) Voltage discharge curve

Exhibit 3.120a R–C circuit

3.121 d. First construct Thevenin's circuit so that all elements will be in series; this makes the R of the RC circuit especially easy to determine. Over one time constant, the voltage will build up to 63% of its final value.

$$V_{oc} = 5\ \text{V},\quad R_{eq} = 5\ \Omega,\quad \tau = R_{ser}\quad C = 3\ (1) = 3\ \text{s}$$

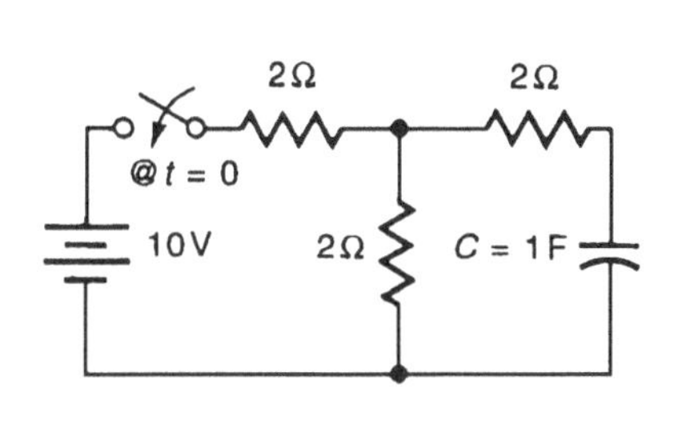

(a) Original circuit

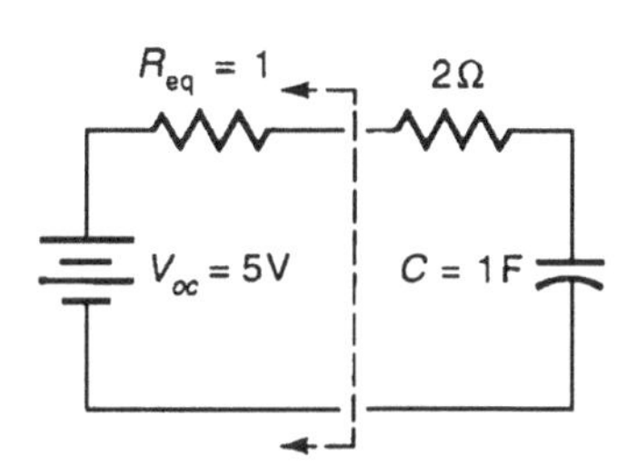

(b) Thevenin's circuit

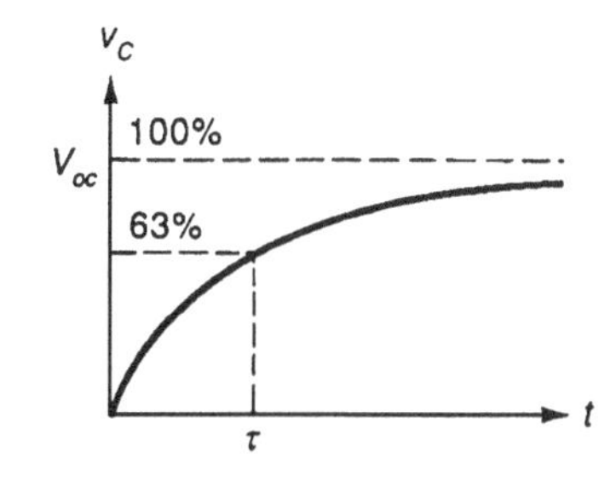

(c) Charging curve

Exhibit 3.121a Simplifying an R–C circuit

3.122 c. Make a Thevenin's equivalent circuit for finding the series resistance. The time constant for an LR circuit is $\tau = L/R = 1/3$ s. For a charging circuit, in one time constant, the current builds up to 63% of its final value.

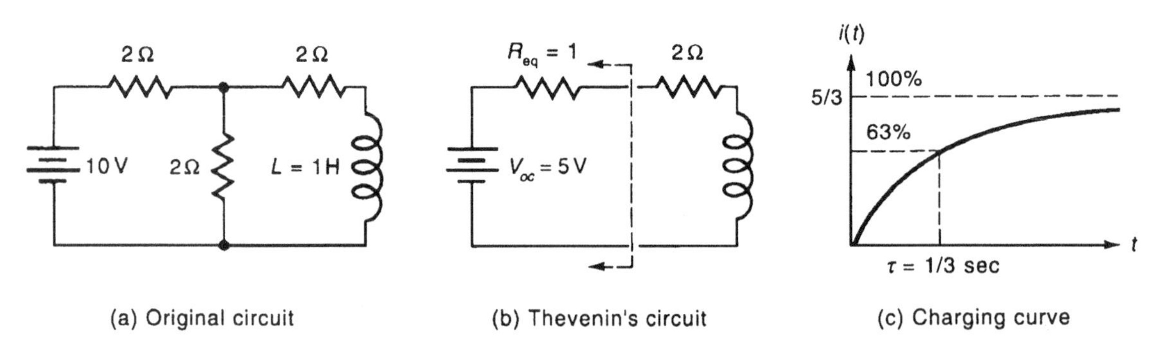

Exhibit 3.122a Simplifying an *L-R* circuit

3.123 b. For a series circuit, $\mathbf{Z}_{\text{total}} = R_2 + \mathbf{Z}_L$, then

$$\mathbf{I} = \mathbf{V}/\mathbf{Z} = 100\angle 0^\circ/(4 + j4) = 100\angle 0^\circ / \left[\sqrt{2}(4\angle 45^\circ)\right] = 17.7 \angle{-45^\circ}$$

$$P_2 = I^2R = (17.7)^2(2) = 625 \text{ watts}$$

3.124 c. First it is necessary to compute the reactances (recall that $2\pi f = 377$ for 60 Hz): $X_L = 2\pi fL = (377)(10/377) = 10\ \Omega$, and $X_C = 1/(2\pi fC) = 10\ \Omega$. Thus

$$\mathbf{Z} = 10 + j10 - j10 = 10\angle 0^\circ$$
$$\mathbf{I} = \mathbf{V}/\mathbf{Z} = 100\angle 0^\circ/10\angle 0^\circ = 10 \text{ A}$$

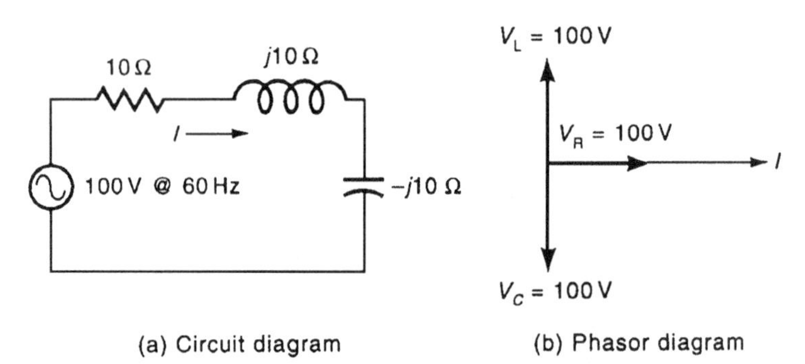

Exhibit 3.124a An *R–L–C* circuit

3.125 d. If the frequency is doubled, the reactances will change to $X_L = 2\ (10) = 20\ \Omega$, $X_C = 0.5\ (10) = 5\ \Omega$; hence, $Z = 10 + j20 - j5 = 18.2\angle 56.3^\circ$. $|I| = |V/Z|$ (answer only requires magnitude) $= 100/18.2 = 5.5$ A.

3.126 d. First find the series equivalent to the parallel portion of the circuit:

$Z_p = (10)(-j10)/(10 - j10) = \left(10/\sqrt{2}\right)\angle -45° = 5\ j5\ \Omega;$

$Z_{total} = j10 + 5 - j5 = 5\sqrt{2}\ \angle 45°\ I = V/Z = 100/\left(5\sqrt{2}\right) = 10\sqrt{2}\ \text{A}$

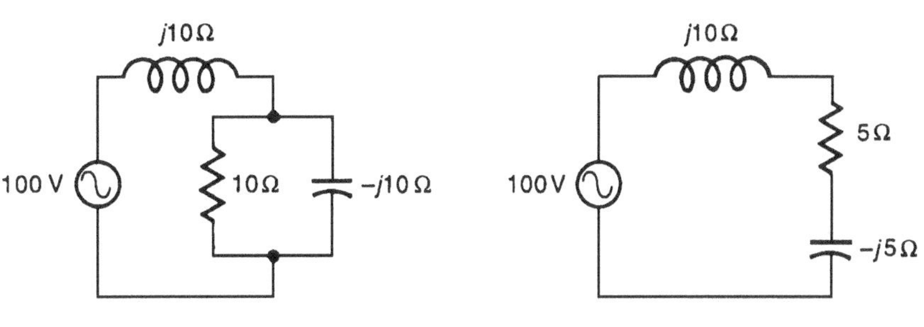

Exhibit 3.126a Reduced circuit

3.127 d. The impedance for the *L–C* portion of the circuit is $\mathbf{Z}_{L\text{-}C} = j5 - j10 = -j5 = 5\angle -90°$; $V_{L\text{-}C} = IZ_{L\text{-}C} = (10)(5) = 50$ V.

3.128 c. Maximum current occurs if X_L and X_C just cancel and Z is left with only resistance: $X_L = 2\pi fL$, $X_C = 1/(2\pi fC)$; $f_{res} = 1/(2\pi\sqrt{LC}) = 1.6$ Hz.

3.129 b. The frequency-domain circuit is shown in Exhibit 3.129a.

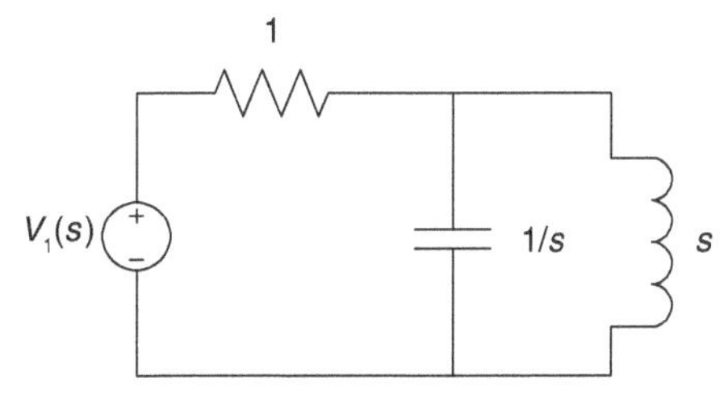

Exhibit 3.129a

Combining the impedances in parallel, $Z_{eq} = \dfrac{s\left(\dfrac{1}{s}\right)}{s + \dfrac{1}{s}} = \dfrac{s}{s^2 + 1}$

Using voltage division rule, $T(s) = \dfrac{Z_{eq}}{1 + Z_{eq}} = \dfrac{s}{s^2 + s + 1}$

3.130 c. As the op-amp is ideal, $T(s) = \dfrac{V_0(s)}{V_{in}(s)} = -\dfrac{Z_2}{Z_1} = -\dfrac{1}{1 + 1/s} = -\dfrac{s}{s+1}$

3.131 c. The op-amp circuit produces a gain of $-(R_f/R_1)$ for an inverting amplifier; and here, only the magnitude is requested so the sign is not important.

$G = (100\text{k}/50\text{k}) = 2;\quad V_0 = 2V_{in},\quad V_0 = 2\,(5) = 10\ \text{V}$

3.132 c. This op-amp circuit produces a gain of $1 + (R_2/R_1)$ for a noninverting amplifier.

$G = 1 + (10\text{k}/10\text{k}) = 2;\quad V_0 = 2\,(5) = 10\ \text{V}$

3.133 **c.** The output from the first summing op-amp is $v_0 = -[(R_f/R_1)v_1 + (R_f/R_2)v_2] = -[(10k/10k)1 + (10k/10k)(-2)] = +1$ V. The output from the second summing op-amp is $v_0 = -[(R_f/R_3)v_3 + (R_f/R_4)v_4] = -[(50k/50k)1 + (50k/25k)(-2)] = +3$ V.

3.134 **d.** Refer to Exhibit 3.134a. The bridge rectifier is a full-wave type, and, for a square wave, the result of taking the average is the maximum of the input itself. $I = V/R = 1/1 = 1$ A (average)

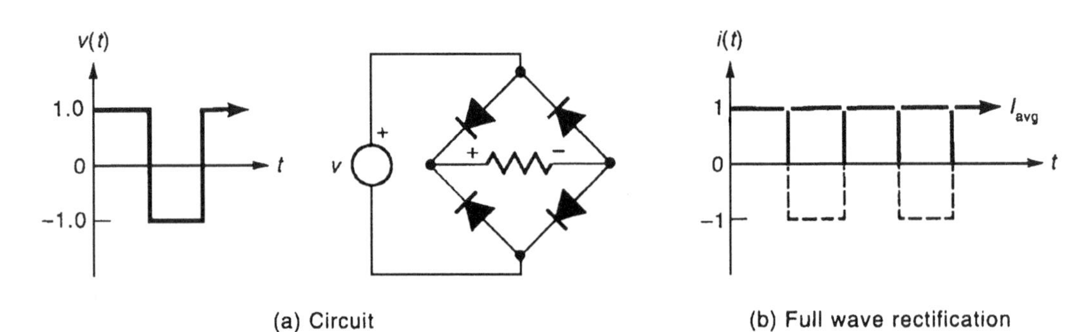

(a) Circuit (b) Full wave rectification

Exhibit 3.134a A full-wave bridge rectifier circuit

3.135 **b.** Since the output is unloaded (that is, the meter resistance is assumed to be >> R), the output is "shorted out" by the ideal diode for the negative half cycle; thus the output voltage appears as an average value. For a half-wave rectifier, the voltage is $V_{avg} = V_{max}/\pi = 14/\pi = 4.5$ V.

3.136 **b.** Assume the load side is V_2, then under full load conditions (rated current at rated voltage), the current is $I_2 = 5000$ VA/200 V = 25 A; $Z_L = 200$ V/25 A = 8 Ω regardless of phase.

CHAPTER 4

Ethics and Business Practices

OUTLINE

The topic of ethics and business practices has 5-8 questions on the FE exam. This is about the same number as the other subjects, so it is clearly an important subject. Questions in the section on ethics and business practices of the exam may cover the following topics:

- Code of ethics and licensure
- Agreements and contracts
- Ethical versus legal behavior
- Professional, ethical, and legal responsibility
- Public protection and regulatory issues

MORALS, PERSONAL ETHICS, AND PROFESSIONAL ETHICS

To put professional ethics for engineers in perspective, it is helpful to distinguish it from morals and personal ethics. *Morals* are beliefs about right and wrong behaviors that are widely held by significant portions of a given culture. Obviously, morals will vary from culture to culture, and though some are common across different cultures, there seems to be no universal moral code.

Personal ethics are the beliefs that individuals hold that often are more restrictive than and sometimes contradictory to the morals of the culture. An example of personal ethics that might be more restrictive than morals might be the belief of an individual that alcohol should not be consumed in a culture that accepts the use of alcohol.

Professional ethics, on the other hand, is the formally adopted code of behavior by a group of professionals held out to society as that profession's pledge about how the profession will interact with society. Such rules or codes represent the agreed-on basis for a successful relationship between the profession and the society it serves.

The engineering profession has adopted several such codes of ethics, and different practitioners may adhere to or be bound by codes that vary by professional discipline but are similar in their basics. Codes adopted by the state boards of registration are typically codified into law and are legally binding for licensed engineers in the respective state. Codes adopted by professional societies are not legally binding but are voluntarily adhered to by members of those societies. The successful understanding of professional ethics for engineers requires an understanding of various codes; for purposes of examining registration applicants, the NCEES has adopted a "model code" that includes many canons common to most codes adopted nationwide.

CODES OF ETHICS

Codes of ethics are published by professional and technical societies and by licensing boards. Why are codes published and why are they important? These fundamental questions are at the heart of the definition of a "profession." Some important aspects of the definition of a profession might include skills and knowledge vital to society; extensive and intellectual education and training important for proper practice in the profession; an importance of autonomous action by practitioners; a recognition by society of these aspects, leading to a governmentally endorsed monopoly on the practice of the profession; and a reliance on published standards of ethical conduct, usually in the form of a code of ethics (Harris, et al., 2005). Such codes are published and followed to maintain a high standard of confidence in the profession by the public served by the practicing professionals, because without high standards of confidence, the ability of a profession to serve the public need may be seriously impaired.

The FE exam questions on ethics and business practices are based on the NCEES code of ethics, a concise body of model rules designed to guide state boards and practitioners as a model of good practice in the regulation of engineering. These rules do not bind any engineer, but the codes of ethics published by individual state boards and of professional societies will be very similar to these in principle. The NCEES handbook has a brief section on ethics. It would be beneficial to read this material before the exam. It is as follows.

NCEES Model Rules of Professional Conduct

A. Licensee's Obligation to Society

1. Licensees, in the performance of their services for clients, employers, and customers, shall be cognizant that their first and foremost responsibility is to the public welfare.
2. Licensees shall approve and seal only those design documents and surveys that conform to accepted engineering and surveying standards and safeguard the life, health, property, and welfare of the public.
3. Licensees shall notify their employer or client and such other authority as may be appropriate when their professional judgment is overruled under circumstances where the life, health, property, or welfare of the public is endangered.
4. Licensees shall be objective and truthful in professional reports, statements, or testimony. They shall include all relevant and pertinent information in such reports, statements, or testimony.
5. Licensees shall express a professional opinion publicly only when it is founded upon an adequate knowledge of the facts and a competent evaluation of the subject matter.
6. Licensees shall issue no statements, criticisms, or arguments on technical matters which are inspired or paid for by interested parties, unless they explicitly identify the interested parties on whose behalf they are speaking and reveal any interest they have in the matters.
7. Licensees shall not permit the use of their name or firm name by, nor associate in the business ventures with, any person or firm which is engaging in fraudulent or dishonest business or professional practices.
8. Licensees having knowledge of possible violations of any of these Rules of Professional Conduct shall provide the board with the information and assistance necessary to make the final determination of such violation. (Section 150, Disciplinary Action, NCEES Model Law)

B. Licensee's Obligation to Employer and Clients

1. Licensees shall undertake assignments only when qualified by education or experience in the specific technical fields of engineering or surveying involved.
2. Licensees shall not affix their signatures or seals to any plans or documents dealing with subject matter in which they lack competence, nor to any such plan or document not prepared under their direct control and personal supervision.
3. Licensees may accept assignments for coordination of an entire project, provided that each design segment is signed and sealed by the licensee responsible for preparation of that design segment.
4. Licensees shall not reveal facts, data, or information obtained in a professional capacity without the prior consent of the client or employer except as authorized or required by law. Licensees shall not solicit or accept gratuities, directly or indirectly, from contractors, their agents, or other parties in connection with work for employers or clients.

5. Licensees shall make full prior disclosures to their employers or clients of potential conflicts of interest or other circumstances which could influence or appear to influence their judgment or the quality of their service.

6. Licensees shall not accept compensation, financial or otherwise, from more than one party for services pertaining to the same project, unless the circumstances are fully disclosed and agreed to by all interested parties.

7. Licensees shall not solicit or accept a professional contract from a governmental body on which a principal or officer of their organization serves as a member. Conversely, licensees serving as members, advisors, or employees of a government body or department, who are the principals or employees of a private concern, shall not participate in decisions with respect to professional services offered or provided by said concern to the governmental body which they serve. (Section 150, Disciplinary Action, NCEES Model Law)

C. Licensee's Obligation to Other Licensees

1. Licensees shall not falsify or permit misrepresentation of their, or their associates', academic or professional qualifications. They shall not misrepresent or exaggerate their degree of responsibility in prior assignments nor the complexity of said assignments. Presentations incident to the solicitation of employment or business shall not misrepresent pertinent facts concerning employers, employees, associates, joint ventures, or past accomplishments.

2. Licensees shall not offer, give, solicit, or receive, either directly or indirectly, any commission, or gift, or other valuable consideration in order to secure work, and shall not make any political contribution with the intent to influence the award of a contract by public authority.

3. Licensees shall not attempt to injure, maliciously or falsely, directly or indirectly, the professional reputation, prospects, practice, or employment of other licensees, nor indiscriminately criticize other licensees' work. (Section 150, Disciplinary Action, NCEES Model Law)

Many ethical questions arise in the formulation of business practices. Professionals should appreciate that expressions like "all is fair in business" and "let the buyer beware" can conflict with fundamental ideas about how a professional engineer should practice. The reputation of the profession, not only the individual professional, is critically important to the ability of all engineers to discharge their duty to protect the public health, safety, and welfare.

The *NCEES Model Rules* addressing a licensee's obligation to other licensees prohibit misrepresentation or exaggeration of academic or professional qualifications, experience, level of responsibility, prior projects, or any other pertinent facts that might be used by a potential client or employer to choose an engineer.

The *Model Rules* also prohibit gifts, commissions, or other valuable consideration to secure work. Political contributions intended to influence public authorities responsible for awarding contracts are also prohibited.

Often, these rules are misunderstood in the arena of foreign practice. Increasingly, engineering is practiced globally, and engineers must deal with foreign clients and foreign governmental officials, many times on foreign soil where laws and especially cultural practices vary greatly. In the United States, the federal Foreign Corrupt Practices Act (FCPA) is a relatively recent recognition and regulation of this problem. Among other purposes, it provides clearer legal boundaries for U.S. engineers involved with international projects.

According to the FCPA, it is not illegal for a U.S. engineer to make petty extortion payments ("grease payments," "expediting payments," and "facilitating payments" are common expressions) to governmental officials when progress of otherwise legitimate projects is delayed by demands for such payments consistent with prevailing practice in that country. It is illegal, however, for U.S. engineers to give valuable gifts or payments to develop contracts for *new business*. In some cultures, reciprocal, expensive gift giving is an important part of business relationships, and the reciprocal nature of this practice can make it acceptable under the FCPA. Most commonly, when the engineer's responsibilities include interactions with foreign clients or partners, the engineer's corporate employer will publish detailed and conservative guidelines intended to guide the engineer in these ethical questions.

The practicing engineer should always be watchful of established and, especially, new business practices to be sure the practices are consistent with the codes of ethics he or she is following.

Example 4.1

The *NCEES Model Rules of Professional Conduct* allow an engineer to do which one of the following?

a). Accept money from contractors in connection with work for an employer or client

b). Compete with other engineers in seeking to provide professional services

c). Accept a professional contract from a governmental body even though a principal or officer of the engineer's firm serves as a member of the governmental body

d). Sign or seal all design segments of the project as the coordinator of an entire project

Solution

Although the other items are not allowed by the *Model Rules*, nowhere does it say that an engineer cannot compete with other engineers in seeking to provide professional services. But, of course, he or she should conduct business in an ethical manner. The correct answer is b).

AGREEMENTS AND CONTRACTS

One aspect of business practice is understanding the concepts and terminology of agreements and contracts.

Elements of a Contract

Contracts may be formed by two or more parties; that is, there must be a party to make an offer and a party to accept.

To be enforceable in a court of law, a contract must contain the following five essential elements:

1. There must be a mutual agreement.
2. The subject matter must be lawful.

3. There must be a valid consideration.
4. The parties must be legally competent.
5. To be a formal contract, the contract must comply with the provisions of the law with regard to form.

A *formal contract* depends on a particular form or mode of expression for legal efficacy. All other contracts are called *informal contracts* since they do not depend on mere formality for their legal existence.

Contract and Related Legal Terminology

Case law—the body of law created by courts interpreting statute law. Judges use precedents, the outcome of similar cases, to construct logically their decision in a given issue.

Changed or concealed conditions—in construction contracting, it is important to specify how changed or concealed conditions will be handled, usually by changes in the contract terms. For example, if an excavation project is slowed by a difficult soil pocket between soil corings, the excavation contractor may be able to support a claim for increased costs due to these unforeseen conditions. When the concealed conditions are such that they should have been foreseen, such claims are more difficult to support.

Common law—the body of rules of action and principles that derive their authority solely from usage and customs.

Damages for delays—in many contracts, completion time is an important concern, and contractual clauses addressing penalties for delays (or rewards for early completion) are often incorporated.

Equal or approved equivalent—terms used in specifications for materials to permit use of alternative but equal material when an original material is not available or an equivalent material can be obtained at lower cost. The engineer is responsible for approving the alternative material.

Equity—system of doctrines supplementing common and statute law, such as the Maxims of Equity.

Errors and omissions—term used to describe the kind of mistakes that can be made by engineers and architects leading to damage to the client. Often, this risk is protected by liability insurance policies.

Force account—a method of work by which the owner elects to do work with his or her own forces instead of employing a construction contractor. Under this method, the owner maintains direct supervision over the work, furnishes all materials and equipment, and employs workers on his or her own payroll.

Hold harmless—clauses are often included requiring one party to agree not to make a claim against the other and sometimes to cooperate in the defense of the other party if a claim is made by a third party.

Incorporate by reference—the act of making a document legally binding by referencing it within a contract, although it is not attached to or reproduced in the contract. This is done to eliminate unnecessary repetition.

Indemnify—to protect another person against loss or damage, as with an insurance policy.

Liquidated damages—a specific sum of money expressly stipulated as the amount of damages to be recovered by either party for a breach of the agreement by the other.

Mechanics' liens—legal mechanism by which unpaid contractors, suppliers, mechanics, or laborers are allowed to claim or repossess construction materials that have been delivered to the worksite in lieu of payment.
Plans—the drawings that show the physical characteristics of the work to be done. The plans and specifications form the guide and standards of performance that will be required.
Punitive damages—a sum of money used to punish the defendant in certain situations involving willful, wanton, malicious, or negligent torts.
Specifications—written instructions that accompany and supplement the plans. The specifications cover the quality of the materials, workmanship, and other technical requirements. The plans and specifications form the guide and standards of performance that will be required.
Statute law—acts or rules established by legislative action.
Statute of limitations—a time limit on claims resulting from design or construction errors, usually beginning with the date the work was performed, but in some cases beginning on the date the deficiency could first have been discovered.
Surety bond—bonds issued by a third party to guarantee the faithful performance of the contractor. Surety bonds are normally used in connection with competitive-bid contracts, namely, bid bonds, performance bonds, and payment bonds.
Workers' compensation—insurance protecting laborers and subcontractors in case of an on-the-job injury; it is often required of contractors.

ETHICAL VERSUS LEGAL BEHAVIOR

Engineers have a clear obligation to adhere to all laws and regulations in their work—what they do must be done legally. But the obligation goes beyond this. Unlike the world of business where cutthroat but legal practices are commonly condoned and frequently rewarded, engineers assume important obligations to the public and to the profession that restrict how they must practice and that often are much more stringent than law or regulation.

When you realize that restricting the practice of engineering to certain licensed professionals by the state is essentially a state-provided monopoly, you may begin to see why there is a difference. Competitive businesses compete in many ways to gain the kind of advantage in their field that engineers and other licensed professionals are given by the state.

Aggressive advertising is one example of a business practice that engineers avoid, even though it is not illegal or prohibited. Before 1978, it was common for professional societies to prohibit or narrowly restrict advertising by their practitioners; however, in 1978 the U.S. Supreme Court ruled such broad restrictions unconstitutional, allowing only reasonable restraints on advertising by professional societies. Since that time, engineering societies have adopted guidelines on advertising. Other professions have been less successful in regulating advertising. For example, the profusion of television advertising by lawyers, and the language of those advertisements, contrasts with the practice of engineering professionals where advertising is more commonly seen in technical journals or trade literature. Many believe the legal profession has suffered a loss of respect as a result, while the profession of engineering still is held in high regard by the public. It is in the interest of the engineering profession to avoid this kind of advertising, even though it is legal, because it can damage the reputation of the profession.

Another example of the importance of self-regulation is the engineer's responsibility to the environment. Although many laws and regulations restrict engineering practices that might damage the environment, there are still many legal ways to accomplish engineering projects that can have adverse environmental effects. Increasingly, codes of ethics are adding requirements for the engineer to consider the environment or the "sustainability" of proposed engineering projects. The engineer's ethical responsibility to work toward sustainable development may go beyond any legal requirements intended to prevent environmental damage.

Conflicts of Interest

A conflict of interest is any situation where the decision of an engineer can have some significant effect on his or her financial situation. It would be a clear conflict of interest for a designing engineer to specify exclusively some component that is only available from a supplier in which that engineer has a significant financial interest, when other components from other suppliers would serve equally well. Engineers must avoid even the *appearance* of a conflict of interest. This is critically important for the reputation of the profession, which the engineer is charged with protecting, in order for engineers to effectively serve the public interest.

An apparent conflict of interest is any situation that might appear to an outside observer to be an actual conflict of interest. For example, if the engineer in the case mentioned above had subsequently divested himself of all interest in the supplier, there is no longer an actual conflict of interest. However, to an outside observer with imperfect information, there might be the appearance of a conflict, resulting in the perception of unethical behavior in the public eye.

The usual remedy for conflicts of interest and apparent conflicts of interest is disclosure and, often, recusal. The engineer's interest must be disclosed in advance, generally to a supervisor, and recusal must at least be discussed. In many cases, recusal may not be necessary, but disclosure is vitally important. In every case, the public perception of the conflict must be considered, with the goal of protecting the reputation of the individuals and the profession.

PROFESSIONAL LIABILITY

Good engineering practice includes numerous checks and conservative principles of design to protect against blunders, but occasional errors and omissions can result in damage or injury. The engineer is responsible for such damage or injury, and it is good practice to carry errors and omissions insurance to provide appropriate compensation to any injured party, whether a client or a member of the public. Such insurance can be a significant cost in some fields of engineering, but it represents a cost of doing business that should be reflected in the fees charged. The most important factor in preventing errors and blunders is to provide adequate time for careful review of all steps in the project by knowledgeable senior licensed engineers. Frantic schedules and unrealistic deadlines can significantly increase the risk.

PUBLIC PROTECTION ISSUES

State boards in all 50 states and the District of Columbia are charged by their states with the responsibility for the licensing of engineers and the regulation of the practice of engineering to protect the health, safety, and welfare of the public. Licensed engineers in each state are legally bound by laws and regulations published by the respective state board. The boards are generally made up of engineers appointed by the state governor; sometimes nonengineering members also are appointed to make sure the public is adequately represented.

State boards commonly issue cease and desist letters to nonengineers who have firms or businesses with names that imply engineering services are being offered to the public or who may actually be offering engineering services without the required state license. These boards also regulate the practice of engineering by their registrants, often sanctioning registrants for inappropriate business practices or engineering design decisions. Many boards require continuing education by registrants for maintenance of proficiency. A weakness of many boards is in the area of discipline for incompetent practices, but this weakness is often offset by tort law whereby incompetent practitioners who cause damage or injury are commonly subject to significant legal damages.

REFERENCES

Harris, Charles E., Jr., Michael S. Pritchard, and Michael J. Rabins. *Engineering Ethics: Concepts and Cases*. Thompson Wadsworth, 2005.

National Council of Examiners for Engineering and Surveying. *Model Rules*, September 2006.

PROBLEMS

4.1 Jim is a PE working for an HVAC designer who often must specify compressors and other equipment for his many clients. He reports to Joan, the VP of engineering. Jim specifies compressors from several different manufacturers and suppliers based on the technical specifications and on his experience with those products in past projects. Joan's long-time friend Charlie, who has been working in technical sales of construction materials, takes a new job with one of the compressor suppliers that Jim deals with from time to time. Charlie calls on Joan, inviting her and any of her HVAC designers to lunch to discuss a new line of high-efficiency compressors; Joan invites Jim to come along. Jim should:

a. decline to attend the lunch, citing concerns about conflict of interest
b. agree to attend the lunch but insist on paying for his own meal
c. agree to attend the lunch and learn about the new line of compressors
d. report Joan to the state board and never specify compressors from that supplier again

4.2 Harry C. is an experienced geotechnical engineer who has many years' experience as a PE designing geotechnical projects and who is very familiar with the rules regarding the requirement for trench shoring and trench boxes to protect construction workers during excavations. During a vacation visit to a neighboring state, he observes a city sewer construction project with several workers in an unprotected deep trench, which, to Harry's experienced eye, is probably not safe without a trench box or shoring. Harry should:

a. remember that he is not licensed in the neighboring state and has no authority to interfere
b. approach the contractor's construction foreman and insist that work be halted until the safety of the trench is investigated
c. advise all the workers in the trench that they are in danger and encourage them to go on strike for safer conditions
d. contact the city engineer to report his concerns

4.3 Engineering student Travis is eagerly anticipating his graduation in three months and has interviewed with several firms for entry-level employment as an electrical engineer. He has received two offers to work for firms A and B in a nearby city, and after comparing the jobs, salaries, and benefits and discussing the choice with his faculty advisor, he telephones firm A whose offer is more appealing and advises them he will accept their job offer. Two weeks later he is contacted by firm C in a different city with a job offer that includes a salary more than 15% higher than the offer he has accepted plus a generous relocation allowance. Travis should:

a. decline the offer from firm C, explaining that he has already accepted a position
b. contact firm A and ask if he can reconsider his decision
c. contact firm A to give them a fair chance to match the offer from firm C
d. advise firm C that he can accept their offer if they will contact firm A to inform them of this change

4.4 EIT Jerry works for a small civil engineering firm that provides general civil engineering design services for several municipalities in the region. He has become concerned that his PE supervisor Eddie is not giving careful reviews to Jerry's work before sealing the drawings and approving them for construction. Jerry asks Eddie to review with him the design assumptions from Jerry's latest design, a steel fire exit staircase to be added to an elementary school building, because he has concerns about the appropriate design loadings. But before the design assumptions are reviewed, Jerry notices the drawings have been approved and released to the fabricator. Jerry should:

a. quit his job and find another employer
b. take a review course in live loadings for steel structures
c. in the future mark each drawing he prepares "Not Approved for Construction"
d. None of the above

4.5 Dr. Willis Hemmings, PE, is an engineering professor whose research in fire protection engineering is nationally recognized. He is retained as an expert witness for the defendant, a structural engineering design firm, in a lawsuit filed by a firefighter who was injured while fighting a fire in a steel structure that collapsed during the fire. The plaintiff's lawyer alleges that the original design of certain components of the fire protection system protecting the steel structural members was inadequate. Hemmings reviews the original design documents, which call for a protective coating that is slightly thinner than is required by the local building code. Hemmings testifies that even though the specified coating is thinner than required, he believes that the design was sound because the product used is applied by a new process that is probably more efficient and the thinner coating probably gave the same level of protection. He bases his testimony on his national reputation as an expert in this field. Such expert testimony is:

a. a commonly accepted method of certifying good engineering design in tort law
b. legal only when given by a licensed professional engineer like Hemmings
c. unethical because it contradicts accepted practice without supporting tests or other data
d. effective only because Hemmings is involved in cutting-edge research

4.6 Jackie is a young PE who works for a garden tool manufacturer that has produced about 100,000 shovels, rakes, and other garden implements annually for more than 20 years. The company recently won a contract to manufacture and supply 5000 folding entrenching tools of an existing design to a Central American military client. The vice president of marketing has been working to develop contracts with other military clients and asks Jackie to prepare a statement of qualifications (SOQ). Jackie is asked to describe the design group (consisting of two engineers, one EIT, one student intern, three CAD technicians, and one IT technician) as a "team of eight tool design engineers," and to describe the company as "experienced in the design, testing, and manufacturing of military equipment, with a recent production history of over two million entrenching tools and related hardware." Jackie should:

a. check the production records to be sure the figures cited are accurate
b. ask the vice president to sign off on the draft of the SOQ
c. object to describing the qualifications and experiences of her group in an exaggerated way
d. be sure to mention that she is a PE and list the states in which she is licensed

4.7 The Ford Motor Company paid millions of dollars to individuals injured and killed in crashes of the Ford Pinto, which had a fuel tank and filler system that sometimes ruptured in rear-end collisions, spilling gasoline and causing fires. While many considered the filler system design deficient because of this tendency, one important factor played a role in the lawsuits. An internal Ford memo was discovered that included the cost-benefit calculations Ford managers used in making the decision not to improve the tank/filler system design. This memo was significant because:

a. it is unethical to use the cost-benefit method for safety-related decisions
b. it is illegal to estimate the value of human life in cost-benefit calculations
c. state law requires estimates of the value of human life be at least $500,000 in such calculations
d. None of the above

4.8 Charles is tasked to write specifications for electric motors and pumps for a new sanitary sewage treatment plant his employer is designing for a municipal client. Charles is concerned that he doesn't have a very good knowledge about current pump design standards but is willing to learn. Charles's fiancée is an accountant employed by a pump distributor and offers to provide Charles with a binder of specifications for all the pumps her firm distributes. Charles should:

a. decline to accept the binder, citing concerns about conflict of interest
b. accept the binder but turn it over to his employer's technical librarian without reviewing it
c. accept the binder and study the materials to gain a better understanding of pump design and specifications
d. ask his fiancée if she knows an applications engineer at her firm who would draft specifications for him

4.9 Professor Martinez is a PE who teaches chemical engineering classes at a small engineering school. His student, Erica, recently graduated and took a job with WECHO, a small firm that provides chemicals and support to oil well drilling operations. WECHO has never employed an engineer and has hired Erica, partly on Prof. Martinez's strong recommendation, in hopes that she will one day become their chief engineer. After she has worked at WECHO for about two years, her supervisor Harry calls Prof. Martinez to explain that WECHO has been required to complete an environmental assessment before deploying a new surfactant, and the assessment must be sealed by a PE. Harry explains that Erica has done all the research to collect data and answer questions on the assessment, and everyone at WECHO agrees that she has done a superb job in completing the assessment, but it still requires the seal of a PE before submission. Harry asks Prof. Martinez if he can review Erica's work and seal the report, reminding him that he had given a glowing recommendation of Erica at the time WECHO hired her. Prof. Martinez should:

a. negotiate a consulting contract to allow him sufficient time and funding to review the report before sealing it
b. require Erica to first sign the report as an EIT and graduate engineer before reviewing it
c. require WECHO to purchase a bond against environmental damage before sealing the report
d. decline to review or seal the report, citing responsible charge issues

4.10 Jack Krompten, PE, is an experienced civil engineer working for a land development firm that has completed several successful residential subdivision developments in WoodAcres, a suburban bedroom community of a large, sprawling, and rapidly growing city. The WoodAcres city engineer, who also served half-time as the mayor, has retired, and the city council realizes that with rapid growth ahead it will be important to hire a new city engineer. They approach Krompton with an offer of half-time city engineer, suggesting that he can keep his current job while discharging the responsibilities of the city engineer—primarily reviewing plans for future residential subdivision developments in WoodAcres. Krompton should:

a. recognize that by holding two jobs he is being paid by two parties for the same work
b. insist that he can only accept the offer if his present employer agrees to reduce his responsibilities to half-time
c. recognize that a 60-hour workweek schedule will take time away from his family
d. recognize that this arrangement will probably create a conflict of interest and refuse the offer

4.11 Willis is an aerospace engineering lab test engineer who works for a space systems contractor certifying components for spacecraft service. He is in charge of a team of technicians testing a new circuit breaker design made of lighter weight materials intended for service in unpressurized compartments in rockets and spacecraft. The new design has passed all tests except for some minor overheating during certain rare electrical load conditions. The lead technician notices that this overheating does not occur when a fan is used to cool the test apparatus and proposes to run the test with the fan to complete the certification process. He points out that the load conditions will only occur during thruster operation in space, which is a much colder environment. Willis should:

a. agree to the lead technician's suggestion, since he has many years of experience in testing and certification
b. agree to run the test as suggested but include a footnote explaining the use of the fan
c. insist on running the test as specified without the use of the fan
d. report the technician to the state board for falsifying test reports

4.12 Shamar is a registered PE mechanical engineer assigned as a project manager on a new transmission line project. He is tasked to build a project team to include several engineers and EITs that will be responsible for design and construction of 7.6 miles of high tension transmission lines consisting of steel towers and aluminum conductors in an existing right of way. He realizes that foundation design and soil mechanics will be an important technical area to his project, and he has never studied these subjects. He wonders if he is qualified to supervise such a project. He should:

a. meet with his supervisor to decline the assignment
b. decline the assignment and contact the state board to report that he is being asked to take responsibility for tasks he is not knowledgeable about
c. accept the assignment and check out an introductory soil mechanics textbook from the firm's technical library
d. accept the assignment and be sure his team includes licensed engineers with expertise in these areas

4.13 Julio is a design engineer working for a sheet metal fabricating firm. He is tasked with the design of a portable steel tank for compressed air to be mass produced and sold to consumers for pressurizing automobile tires. He designs a cylindrical tank to be manufactured by rolling sheet metal into a cylinder, closing with a longitudinal weld along the top, and welding on two elliptical heads. His design drawings are approved by his supervisor, Sonja, a licensed engineer, and by the vice president of manufacturing, but when the client reviews the designs, he asks the VP to change the design so that the longitudinal weld along the top is moved to the bottom where it will not be visible to improve the esthetics and marketability of the product. The VP agrees with this change. Julio learns of this change and objects, citing concerns about corrosion at the weld if it is on the bottom. Sonja forwards Julio's objection with a recommendation against the change to the VP, with a copy to the client, but the VP insists, saying esthetics is very important in this product. Julio should:

a. accept the fact that esthetics governs this aspect of the design
b. write a letter to the client stating his objections
c. put a clear disclaimer on the drawing indicating his objections
d. contact the state board to report that his recommendation has been overruled by the VP

4.14 William is a PE who designs industrial incineration systems. He is working on a system to incinerate toxic wastes, and his employer has developed advanced technology using higher temperatures and chemical-specific catalyst systems that minimize the risk to the environment, workers, and the public. A public hearing is scheduled to address questions of safety and environmental risk posed by the project, and William is briefed by the corporate VP for public affairs about how to handle questions from the public. He is told to buy a new suit, project an air of technical competence, point out that his firm is the industry leader with many successful projects around the world, and describe the proposed system as one with "zero risk" to the public. William should:

a. follow his instructions to the letter
b. insist that his old suit is adequate, because he refuses to appear more successful than he really is, but follow the other instructions
c. follow all instructions, except use the term "minimal risk" rather than "zero risk"
d. resign from his position and look for a different employer who won't ask him to face the public

4.15 Darlene is a metallurgical engineering EIT who works for a firm that manufactures automotive body panels. She has been tasked with improvements to the design of inner fender and trunk floor panels to reduce corrosion damage. After several weeks of study and comparison of alternatives, she submits a new trunk floor panel design utilizing a weldable stainless steel that will significantly reduce corrosion compared to the galvanized carbon steel alternatives she has been considering. The new panels will cost more, however, and after much study and debate, the VP of manufacturing rejects her design and approves an alternative made of a cheaper material. Darlene should:

a. accept the decision and work to finalize a workable design
b. resign from her position, since her employer has lost confidence in her
c. contact the state board to advise that her design decision has been overturned by a nonengineering manager
d. None of the above

4.16 Matt is a young PE who has just started his own consulting practice after six years of work with a small consulting firm providing structural engineering design services to architects. He has worked on steel and timber framed churches, prestressed and reinforced concrete parking structures, and many tilt-up strip center buildings. His expertise has been in the area of design of tilt-up concrete construction, where he has developed some innovative details regarding reinforcement at lifting points. His building designs, when constructed by experienced contractors, have reduced construction times and costs. Because of his expertise, he is approached by lawyer Marlene, who tells Matt that she represents a construction worker who is suing a project owner, contractor, and designer over a construction accident in which a tilt-up wall was dropped during construction, seriously injuring several workers. Marlene asks Matt to serve as an expert witness to assess the design and construction practices in the project and testify as to the causes of the accident. Marlene has taken the case on a contingency fee basis, in which she will earn 40% of any settlement, and she asks Matt if he would rather be paid by the hour for his study and testimony or instead accept 5% of any settlement, which she believes could be as high as $25 million. Matt should:

a. compare the 5% contingency fee with an expected fee based on his hourly rates, realizing that there is some chance he will earn nothing
b. be sure to have Marlene put the contingency fee arrangement in the form of a legal contract
c. decline the contingency fee arrangement and bill on an hourly basis
d. accept the contingency fee arrangement but donate the difference over his hourly rate to charity

4.17 Victor is a consulting engineer who is also in charge of a crew providing land surveying and subdivision design services to developers. He has been contracted to provide a survey of a 14-acre tract where a local developer is contemplating a subdivision, and he realizes that his crew had surveyed this same tract last month for another developer who has abandoned the project. He reprints the survey drawings, changing the title block for the new client. With respect to billing for the drawings, Victor should:

a. bill the new client the same as he billed the original client to be fair to both
b. bill the new client for half of the amount billed to the original client
c. bill the new client only for any work he did to change the drawings and reprint them
d. provide the new client with the drawings without any charge

4.18 Frank is a PE who works for ELEC, an electrical engineering design and construction firm. Frank's job is estimating construction costs, bidding construction projects, and supervision of design of electrical systems for buildings. Frank's bright EIT of five years, Linda, has just received her PE license and has resigned her position to open her own consulting business in a nearby community. Until she is replaced, Frank will have to also do all detail design of electrical systems for their projects. Frank receives a request for proposal (RFP) from a general contractor regarding design of electrical systems for a local independent school district. He realizes that his firm may be in competition with Linda for the engineering design, and knowing that Linda's salary was about half of his, he expects she may have a competitive advantage. Frank should:

a. ask Linda not to bid on this project
b. remind his contact with the general contractor that Linda has just left his firm, that she is inexperienced, and hint that she was sometimes slow to complete her design assignments
c. emphasize his 18 years of experience and subsequent design efficiency in his proposal
d. promote a CAD technician to a designer position so he can show a lower billing rate for engineering design hours

4.19 It is important to avoid the appearance of a conflict of interest because:

a. the engineer's judgment might be adversely affected
b. the engineer's client might suffer financial damages
c. the appearance of a conflict of interest is a misdemeanor
d. the appearance of a conflict of interest damages the reputation of the profession

4.20 The code of ethics published by the American Society of Civil Engineers is:

a. legally binding on all licensed engineers practicing civil engineering
b. adhered to voluntarily by members of the ASCE as a condition of membership
c. legally binding on all engineers with a degree in civil engineering
d. published only as a training guideline for young civil engineers

4.21 Which statement *MOST* accurately describes an engineer's responsibility to the environment?

a. The engineer has a legal obligation to make sure all development is sustainable.
b. The engineer has no obligation to the environment beyond protecting public health and safety.
c. The engineer has a moral obligation to consider the impact of his or her work on the environment.
d. The engineer's environmental responsibility is primarily governed by specific state laws.

SOLUTIONS

4.1 **c.** Jim can accept this invitation. We can assume there is no corporate policy prohibiting or restricting lunch invitations since VP Joan has accepted the invitation; therefore, there is no reason for Jim to decline the invitation. The opportunity to learn more about the new product is useful to him, his employer, and his clients; the cost of the lunch presumably would not be considered a "valuable" gift; and the lunch would not create either a conflict of interest or the appearance of a conflict to a reasonable person. If instead of lunch the offer involved a 10-day elk hunting trip or a vacation in the south of France, the solution would be very different because of the obvious "value" of the gift.

4.2 **d.** Doing nothing (a) is not an option if Harry really believes the trench represents a serious hazard to the workers. His code of ethics requires him to remember that his first and foremost responsibility is the public welfare, which includes the safety of the construction workers. Answers (b) and (c) are not the best way to proceed; his concerns should be reported to an engineer with some authority over the project. Since this is a city-contracted sewer improvement project, the city engineer will have project responsibility and will be the appropriate individual for Harry to take his concerns to.

4.3 **a.** Travis should decline the offer from firm C, explaining that he has already accepted an offer to work for firm A. While the *NCEES Model Rules* don't specifically address issues of personal integrity, it is clear that the engineer's obligation to employer and clients will not be satisfied by any decision that ignores Travis's verbal agreement to employment with firm A. Furthermore, such actions will tarnish his integrity in the eyes of firm A and by implication will harm the reputation and credibility of other students and the profession.

4.4 **d.** None of the first three solutions will address the concerns Jerry has raised about the safety of the particular project in question, so (d) is the correct solution. He should instead meet with Eddie to discuss the details of his design and make a determination if the design is completed safely. If it isn't, he will need to take further action to stop fabrication and construction while the design is reviewed and possibly modified to address any deficiencies. After this is done, he might want to consider all three of the other choices for his future. If he is thwarted in these responsibilities by Eddie, he should contact the state board with his concerns.

4.5 **c.** Hemmings cannot offer expert opinion that is contrary to accepted engineering practice without supporting that opinion with computer modeling, lab test results, or study of the literature. He can offer expert opinion that the design is not in line with accepted engineering practice without any supporting calculations, but he can't maintain that a substandard practice is acceptable without rational supporting evidence. Hemmings's credentials and experience may qualify him as an expert, but they do not relieve him of the requirement to base his professional opinion on facts.

4.6 **c.** Jackie should object to the request to exaggerate the size, qualifications, and experience level of her design group. The *Model Rules* require engineers to "be objective and truthful" in all professional matters, and the suggested exaggerations are clearly in conflict with this requirement.

4.7 **d.** It is not unethical or illegal to use the cost-benefit method, nor are certain values for human life prescribed by law; the answer is none of the above. The assumed values and calculations in the memo may have appeared callous or inflammatory to the juries in the resulting lawsuits, but they were not unethical or illegal. They may have been imprudent—an important lesson is that the public (jury) apparently objected to a design decision that increased the risk of a post-crash fire for such a small net benefit. Even though risk of death had been considered by the designers, the mode of death (burning to death in otherwise survivable crashes) was a factor in the strong reactions by the juries.

4.8 **c.** Studying products from a distributor is perfectly acceptable and can be a good way to gain a better understanding of pump equipment on the market. Accepting the binder does not represent a conflict of interest as implied by choices (a) and (b). Option (d) would be clearly setting himself up for an apparent conflict of interest.

4.9 **d.** Since Prof. Martinez has not been in responsible charge of the development of the assessment, he can't seal it, regardless of the level of review or the capabilities of the EIT who has done the work. Engineering students should be cautious in accepting a position as the sole engineering employee in a small firm; they are first encouraged to gain experience as an EIT under the guidance of an experienced PE and qualify for licensure as a PE before taking a position where licensure might be needed.

4.10 **d.** Krompton should recognize that the proposed arrangement will create a conflict of interest by placing him in charge of reviewing and approving plans from developers and potential developers that are in direct competition with his own employer. Any decision by him that might tend to make development less profitable for other developers could be advantageous to his employer by making their services more cost effective. Even if he were able to make all decisions rationally and without bias, the appearance of a conflict of interest would be very real and would cause a loss of credibility in the city engineer's actions and damage the reputation of the engineering profession.

4.11 **c.** Willis should insist on running the test as specified. The technician's suggestion to use a fan to "fudge" the test is technically indefensible, as well as unethical, in any case—the fan simulates convection cooling, which does not occur in the vacuum of space. Even if the technician had suggested a way to simulate an increased radiative heat flux, changes to a specified test procedure are not made casually. Much more study, documentation, and higher level approvals are involved.

4.12 d. Shamar should accept the assignment and select team members so that all areas of needed expertise are represented by licensed individuals who can seal appropriate portions of the plans. Option (a) is not recommended for an engineer's career advancement—he is expected to accept assignments of increasing responsibility; (b) is detrimental to his career—he will cause unnecessary concern with the board and with his supervisors; and (c) studying an introductory soil mechanics textbook may give him a better understanding of the problem but probably won't qualify him to seal foundation plans for the project described.

4.13 d. Julio is objecting because he knows water accumulates in compressed air storage tanks and causes corrosion, particularly at the bottom where the water will collect. Because of the metallurgy at the weld, corrosion is more aggressive at the weld site, and Julio considers this fundamentally a bad design if the weld is at the bottom. Since his recommendation (based on technical reasons and an increased risk to the public) is overruled under circumstances where the safety of the public is endangered, he is obligated (by the NCEES rules) to "notify his employer or client and other such authority." Having already notified his employer (through Sonja and the VP of manufacturing) and the client, all of whom except Sonja are part of the problem, his next logical step is to contact the state board with his concerns. This is rarely necessary; in most cases, the client and VP will be very interested in Julio's objection as it is based on public safety and will serve to reduce the company's own liability. However, if the employer and VP of manufacturing persist as described here, Julio must take additional action. It is not unreasonable to expect that Julio may face some sort of sanction from a management team that has put him in this position. He may even need a lawyer as this unpleasant situation deteriorates. Whistle-blowing should be considered the solution of last resort, as in this case.

4.14 c. William should avoid the use of the term "zero risk"; he knows that no project has zero risk. He should instead look for ways to quantify the risk that will be informative and meaningful to the public and try to convey the attitude of concern for minimizing the risk consistent with the potential public benefits of the project (increased employment and tax base). He should not do (a), and (b) does not address the problem of misinforming the public about the risk. There is no need to resign; to do so will not help his career or the project.

4.15 **a.** Darlene should accept the decision and work to make the chosen design successful and profitable for her employer. She does not need to resign—the business decision probably does not reflect a lack of confidence. It was made based on costs and profitability, not public safety, so she should not contact authorities to complain that a manager has overruled her. This kind of decision should be considered a "management decision" because it affects business and profits. Decisions that adversely affect public health, safety, or welfare should be considered "engineering decisions," and when these are overturned by nontechnical managers for other reasons, the engineer may be justified or even obligated to report this to authorities.

4.16 **c.** Matt should decline the contingency fee arrangement. While lawyers commonly work on contingency fee arrangements, engineers can't do this. Any contingency fee arrangement would put the engineer in a conflict of interest situation, where his engineering judgment can influence his income. In such a situation, his engineering judgment may not be sound or will at least appear to be conflicted to an outside observer.

4.17 **c.** Victor should be cautious in making sure that no additional surveying or resurveying is needed because of any changes to the tract. If additional fieldwork is not needed, he should bill the new client only for the work required to edit and reprint the drawings. He should not bill the new client the same as the original client, because that would be billing two clients for the same work, which is specifically prohibited by the *Model Rules*. Billing for half of the original amount is the same, just for an arbitrary amount. When two clients require the same survey simultaneously, it may make sense for Victor to facilitate a partnership in the project, but this is not always feasible when one client has already been billed for work done.

4.18 **c.** Option (c) is the only ethical and practical solution listed. Option (a) asking a competitor not to bid is not practical. Option (b) starting rumors about Linda's capabilities is clearly unethical. And option (d) is troublesome—experienced CAD technicians can do some aspects of design if closely supervised by a PE, but it isn't really necessary to make such a promotion just to show a lower billing rate. The billing rate for engineering design could be maintained at the same (competitive) level as when Linda was employed, even if the actual design work is done by Frank at twice the salary until a replacement for Linda can be hired. This in itself is not unethical, but if Frank does not budget sufficient time to do the design work in addition to his other work, it becomes a question of ethics. An engineer must allow sufficient time to do a professional job. The most practical and desirable solution is not listed—Frank should expedite hiring a qualified replacement for Linda, and he might have to consider declining the opportunity to bid on some projects until she is replaced.

4.19 d. Even the appearance of a conflict of interest can damage the reputation of the individuals involved and the profession as a whole, and such situations should be avoided. If there is no actual conflict of interest, the engineer's judgment will not be affected and the client will not suffer damages; nor is it criminal.

4.20 b. Codes of ethics published by professional societies are voluntarily adhered to by membership. They don't carry the weight of law but are much more than training guidelines—members who do not adhere to the society's code can be sanctioned by the society or forfeit their membership.

4.21 c. Many professional societies require the engineer to "consider" the impact on the environment. Some say he or she should consider whether the development is "sustainable." Most legal restrictions only require the engineer to prevent certain kinds of environmental damage and do not require sustainable development. There are, of course, specific state laws that must be followed, but many federal laws and regulations also apply. The engineer's responsibility is broader than laws and regulations in any event, making choice (c) the best answer.

CHAPTER 5

Probability and Statistics

OUTLINE

Expect 10-15 questions on probability and statistics on the Industrial Engineering FE exam. Subtopics include combinatorics, probability distributions, conditional probabilities, sampling distributions, estimation, hypothesis testing, regression, system reliability, and design of experiments.

ENGINEERING STATISTICS

The mathematics chapter of the *FE Supplied-Reference Handbook* contains a section on probability and statistics. This includes formulas for combination and permutation, basic laws of probability, properties of the binomial distribution, properties of the normal distribution, a normal distribution table, a t distribution table, and a table of the F distribution for $\alpha = 0.05$.

The Industrial Engineering chapter of the handbook contains a table of density function formulas with mean and variance for the binomial, hypergeometric, Poisson, geometric, negative binomial, multinomial, uniform, gamma, exponential, Weibull, normal, and triangular distributions. It also contains formulas for one-way and two-way analysis of variance and linear regression.

Do not forget the statistics capabilities of your hand calculator. Make sure you know how to use it to determine sample means, sample standard deviations, linear regression parameters, and correlation coefficients.

COMBINATORICS

For discrete probability calculations it is important to count the number of elements in sets of possible outcomes. The primary method is simply to write down all the elements in a set and count them. For example, count the number of elements in the set S of all possible outcomes of a six-sided die throw. The set definition is $S = \{1, 2, 3, 4, 5, 6\}$. Simple counting gives six elements.

Most useful sets are large, and simple counting is too time-consuming. There are several methods for simplifying this task. One can use the product set concept. If A is a set with n elements and B is a set with m elements, then the product set $A \times B$ has the arithmetic $n \times m$ number of elements. To simplify counting then, first count the sets making up the product set (usually containing a much smaller number of elements) and simply multiply these counts.

Example 5.1

Count the number of possible outcomes for tossing five coins.

Solution

The number of outcomes for a single toss defined by the set $R = \{H, T\}$ is 2 (heads or tails). The result of five coin tosses is the product set $R \times R \times R \times R \times R$. The total number of possible outcomes is then the arithmetic product of the number of outcomes in each of the individual five tosses. This is $2 \times 2 \times 2 \times 2 \times 2 = 2^5 = 32$.

Permutations

If A is a set with n elements, a **permutation** of A is an ordered arrangement of A. Given the set $A = \{a, b, c\}$, the order a, b, c of the elements is one permutation. Any other order—for example, b, c, a—is another permutation.

The set B of all permutations of the set A is defined as the set of all arrangements of the three elements. These are

$$B = \{\{a, b, c\}, \{a, c, b\}, \{c, b, a\}, \{b, a, c\}, \{b, c, a\}, \{c, a, b\}\}$$

There are six permutations. This number also can be derived as follows. The number of ways an element can be chosen for the first space is three. Then there are two elements left. One of these can go in the second space. Then there is one element left. This must go in the third space. This gives the formula $3 \times 2 \times 1 = 6$. In general, the number of ways n distinct elements can be arranged is given by

$$n! = n \times (n-1) \times (n-2) \times \cdots 1$$

and is called the **factorial** of the number n. The factorial of 0 is 1 ($0! = 1$).

For example, count the number of ways a standard playing deck can be arranged. Since there are 52 distinct cards in a deck, there are 52! different arrangements, or permutations.

Now suppose we have the set of letters L in the word *obtuse* so that $L = \{o, b, t, u, s, e\}$. How many two-letter symbols could be made from this set? Notice that the letters are all distinct. We again count the number of ways the letters can be selected. For the first choice it is six; for the second choice it is five. The two selections are now complete. There are therefore $6 \times 5 = 30$ possibilities.

The general formula for the number of permutations, taking r items from a set of n, is given by

$$P(n, r) = n!/(n-r)!$$

Using this equation, one can express the previous example as $P(6, 2) = 6!/(6-2)! = 30$.

Example 5.2

A jeweler has nine different beads and a bracelet design that requires four beads. To find out which looks the best, he decides to try all the permutations. How many different bracelets will he have to try?

Solution

There are $n = 9$ beads. He selects $r = 4$ at a time. The order is important because each arrangement of r beads on the bracelet makes a different bracelet. So the number of different bracelets is

$$P(9, 4) = 9!/(9-4)! = 9 \times 8 \times 7 \times 6 = 3024$$

If the bracelet is a closed circle, there is no discernible difference when it is rotated. Then one observes four identical states for each unique bracelet. This is called ring permutation and is given by the formula

$$P_{\text{ring}}(n, r) = P(n, r)/r$$

There are only $3024/4 = 756$ distinct ring bracelets the jeweler can make.

Example 5.3

(i) In how many ways can four people be asked to form a line of three people?

(ii) In how many ways can the letters of the word BEAUTY be arranged?

(iii) In how many ways can the letters of the word GOOD be arranged?

Solution

(i) $P(4,3) = \dfrac{4!}{(4-3)!} = 24$

(ii) $P(6,6) = \dfrac{6!}{(6-6)!} = 720$

(iii) $P(4;1,1,2) = \dfrac{4!}{1!1!2!} = 12$

Combinations

When the order of the set of r things that are selected from the set of n things does not matter, we talk about combinations.

Again consider the standard playing deck of 52 cards. How many hands of 5 cards can we get from a deck of 52 cards? Count the number of ways the hands can be drawn. The first draw can be any of the 52 cards. The second draw can only be one of the remaining 51 cards. The third draw can only be one of the remaining 50, the next is one of 49, and the last one of 48. So the result is

$$52 \times 51 \times 50 \times 49 \times 48 = 52!/(52-5)!$$

This is the formula for permutations discussed in the last section. But the order in which we receive the cards is not important, so many of the hands are the same. In fact there are 5! similar arrangements of cards that make the same hand. The number of distinct hands is

$$(52 \times 51 \times 50 \times 49 \times 48)/(5 \times 4 \times 3 \times 2 \times 1) = 52!/[5! \times (52-5)!]$$

The general form r items taken from a set of n items when order is not important is written as the binomial coefficient $C(n, r)$, also written $\binom{n}{r}$, and is given by the formula

$$C(n,r) = \frac{n!}{r!(n-r)!}$$

Example 5.4

There are six skiers staying in a cabin with four bunks. How many combinations of people will be able to sleep in beds?

Solution

$C(6, 4) = 6!/[4! \times (6-4)!] = (6 \times 5 \times 4 \times 3 \times 2 \times 1)/[(4 \times 3 \times 2 \times 1) \times (2 \times 1)]$
$= 15$

PROBABILITY

Definitions

An **experiment**, or **trial**, is an action that can lead to a measurement.

Sampling is the act of taking a measurement. The **sample space** S is the set of all possible outcomes of an experiment (trial). An event e is one of the possible outcomes of the trial.

If an experiment can occur in n mutually exclusive and equally likely ways, and if m of these ways correspond to an event e, then the probability of the event is given by

$$P\{e\} = m/n$$

Example 5.5

A die is a cube of six faces designated as 1 through 6. The set of outcomes R of one die roll is defined as $R = \{1, 2, 3, 4, 5, 6\}$. If two dice are rolled, define trial, sample space, n, m, and the probability of rolling a seven when adding both dice together.

Solution

The trial is the rolling of two dice. The sample space is all possible outcomes of a two-dice roll, and the event is the outcome that the sum is 7.

The number of all possible outcomes, n, is the number of elements in the product set of the outcome of two dice when each is rolled independently. The product set is $R \times R$ and contains 36 elements.

The number of all possible ways, m, that the (7) event can occur is (1, 6), (2, 5), (3, 4), (4, 3), (5, 2), and (6, 1) for a total of six ways. The probability of rolling a 7 is $P\{7\} = \frac{6}{36} = \frac{1}{6}$.

Example 5.6

What is the probability of (i) a tail showing up when a fair coin is tossed, (ii) number 3 showing up when a fair die is tossed, and (iii) a red king is drawn from a deck of 52 cards?

Solution

(i) 1/2 (ii) 1/6 (iii) 2/52

General Character of Probability

The probability $P\{E\}$ of an event E is a real number in the range 0 through 1. Two theorems identify the range between which all probabilities are defined:

1. If $\varnothing$ is the null set, $P\{\varnothing\} = 0$.
2. If S is the sample space, $P\{S\} = 1$.

The first states that the probability of an impossible event is zero, and the second states that, if an event is certain to occur, the probability is 1.

Complementary Probabilities

If E and E' are complementary events, $P\{E\} = 1 - P\{E'\}$. Complementary events are defined with respect to the sample space. The probability that an event E will happen is complementary to the probability that any of the other possible outcomes will happen.

Example 5.7

If the probability of throwing a 3 on a die is 1/6, what is the probability of not throwing a 3?

Solution

E is the probability of not throwing a 3, so $P\{E\} = 1 - P\{E'\} = 1 - \frac{1}{6} = \frac{5}{6}$.

Sometimes the complementary property of probabilities can be used to simplify calculations. This will happen when seeking the probability of an event that represents a larger fraction of the sample space than its complement.

Example 5.8

What is the probability $P\{E\}$ of getting at least one head in four coin tosses?

Solution

The complementary event $P\{E'\}$ to getting at least one head is getting no heads (or all tails) in four tosses. So the probability of getting at least one head is

$$P\{E\} = 1 - (0.5)^4 = 1 - 0.0625 = 0.9375$$

Joint Probability

The probability that a combination of events will occur is covered by joint probability rules. If E and F are two events, the joint probability is given by the rule

$$P\{E \cup F\} = P\{E\} + P\{F\} - P\{E \cap F\} \qquad \text{(Rule 1)}$$

A special case of the joint probability rule can be derived by considering two events, E and F, to be mutually exclusive. In this case the last term in Rule 1 is zero since $P\{E \cap F\} = P\{0\} = 0$. Thus, if E and F are mutually exclusive events,

$$P\{E \cup F\} = P\{E\} + P\{F\} \qquad \text{(Rule 2)}$$

Example 5.9

What is the probability of throwing a 7 or a 10 with two dice?

Solution

We will call the event of throwing a 7 A, and of throwing a 10 B. We know from previous examples that $P\{A\} = \frac{1}{6}$, and we can count outcomes to get $P\{B\} = \frac{1}{12}$. Applying the formula,

$$P\{A \cup B\} = P\{A\} + P\{B\} = \frac{1}{6} + \frac{1}{12} = \frac{1}{4}$$

If two events E and F are independent—that is, if they come from different sample spaces—then the probability that both will happen is given by the rule

$$P\{E \cap F\} = P\{E\} \times P\{F\} \qquad \text{(Rule 3)}$$

Example 5.10

What is the probability of throwing two heads in two coin tosses?

Solution

Call the throwing of one head E, the other F. The probability of throwing a single head is $P\{E\} = \frac{1}{2}$, and $P\{F\} = \frac{1}{2}$. The probability of throwing both heads is

$$P\{E \cap F\} = P\{E\} \times P\{F\} = \frac{1}{2} \times \frac{1}{2} = \frac{1}{4}$$

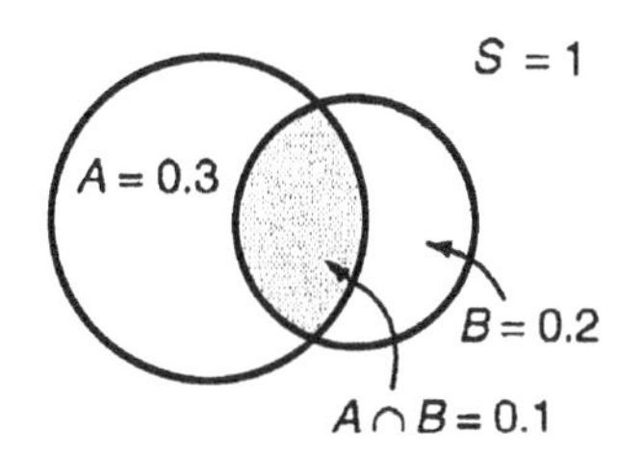

Figure 5.1
Venn diagram of joint probabilities

To visualize joint probabilities, we can use a Venn diagram showing two intersecting events, A and B, as shown in Figure 5.1. Let the normalized areas of each event represent the probability that the event will occur. For example, think of a random dart thrown at the Venn diagram: What are the chances of hitting one of the areas? Assume the areas correspond to probabilities and are given by $P\{S\} = 1$, $P\{A\} = 0.3$, $P\{B\} = 0.2$, and $P\{A \cap B\}$ is 0.6. The probability of hitting either area A or area B is calculated as the sum of the areas A and B minus the overlap area so it is not counted twice:

$$P\{A \cup B\} = 0.3 + 0.2 - .06 = .44$$

The result is also equal to the normalized area covered by A and B. The probability of hitting both A and B on one throw is simply the overlap area $P\{A \cap B\} = 0.1$.

If we throw two darts, the area S is used twice and represents two independent sample spaces. Hence Rule 3 applies.

Example 5.11

A die is tossed. Event A = {an odd number shows up}; event B = {a number > 4 shows up}.

(i) Find the probabilities, P(A) and P(B). (ii) What is the probability that either A or B or both occur?

Solution

(i) P(A) = P(1 or 3 or 5 showing up) = 3 (1/6) = 0.5
P(B) = P(5 or 6 showing up) = 2 (1/6) = 0.333

(ii) Event (A, B) = {5} and P(A, B) = 1/6;
then P(A + B) = P(A) + P(B) – P(AB) = 1/2 + 2/6 – 1/6 = 4/6
Check: Event (A + B) = {1,3,5,6}; then P(A + B) = 4/6

Conditional Probability

The conditional probability of an event E given an event F is denoted by $P\{E \mid F\}$ and is defined as

$$P\{E \mid F\} = P\{E \cap F\}/P\{F\} \quad \text{for } P\{F\} \text{ not zero}$$

Example 5.12

Two six-sided dice, one red and one green, are tossed. What is the probability that the green die shows a 1, given that the sum of numbers on both dice is less than 4?

Solution

Let E be the event "green die shows 1" and let F be the event "sum of numbers shows less than four." Then

$$E = \{(1, 1), (1, 2), (1, 3), (1, 4), (1, 5), (1, 6)\}$$
$$F = \{(1, 1), (1, 2), (2, 1)\}$$
$$E \cap F = \{(1, 1), (1, 2)\}$$
$$P\{E \mid F\} = P\{E \cap F\}/\Pi\{F\} = (2/36)/(3/36) = 2/3$$

The generalized form of conditional probability is known as Bayes' theorem and is stated as follows: If $E_1, E_2, \ldots, E_n$ are n mutually exclusive events whose union is the sample space S, and E is any arbitrary event such that $P\{E\}$ is not zero, then

$$P\{E_k|E\} = \frac{P\{E_k\} \times P\{E \mid E_k\}}{\sum_{j=1}^{n} [P\{E_j\} \times P\{E \mid E_j\}]}$$

RANDOM VARIABLES

The method of random variables is a powerful concept. It casts the set-theory-based probability calculations of previous sections into a functional form and allows the application of standard mathematical tools to probability theory. It is often easy to solve fairly complex probability problems using random variables, although an approach different from the usual one is required.

A random variable, usually denoted by X, is a mapping of the sample space to some set of real numbers. The mapping transforms points of a sample space into points, or more accurately intervals, on the x-axis. The mapping, or random, variable X is called a discrete random variable if it assumes only a denumerable number of values on the x-axis. A random variable is called a continuous random variable if it assumes a continuum of values on the x-axis. The mapping is usually quite easy and intuitive for numerical events but provides no major advantage for nonnumerical discrete sample spaces, where counting remains the major tool.

Example 5.13

Cast the sample space of the outcomes of a roll of a die into random variable form.

Solution

The sample space is the set R defined by $R = \{1, 2, 3, 4, 5, 6\}$. These can easily by written along the x-axis as

$$R = \{1, 2, 3, 4, 5, 6\} \rightarrow 1 \mid 2 \mid 3 \mid 4 \mid 5 \mid 6 \mid x\text{-axis}$$

Probability Density Functions

A probability density function $f(x)$ is a mathematical rule that assigns a probability to the occurrence of the random variable x. Since the random variable is a mapping from trial outcomes, or events, to the numerical intervals on the x-axis, the probability that an event will occur is the area under the probability density function curve over the x interval defining the event.

For a continuous random variable the probability that an event E, mapped into an interval between x_1 and x_2, will occur is defined as

$$\int_{x_1}^{x_2} f(x)\,dx = P\{E\} \qquad \text{for } E \text{ mapped into } (x_1, x_2)$$

For a discrete case the formula is

$$\sum_{i=1}^{n} f(x_i) = P\{E\} \qquad \text{for } E \text{ containing } x_1, x_2, \ldots, x_n$$

It is assumed here that a step interval is associated with each value of x_i; therefore, the equivalent dx in the integral is 1 and is not required in the sum.

Example 5.14

The probability density function of a single six-sided die throw is shown graphically in Exhibit 1. The probability of throwing a 3 is given by the area under the curve over the interval assigned to the numeral 3, which is the step interval from 2.5 to 3.5.

Hence $P\{3\} = f(x) \times 1 = \frac{1}{6} \times 1 = \frac{1}{6}$.

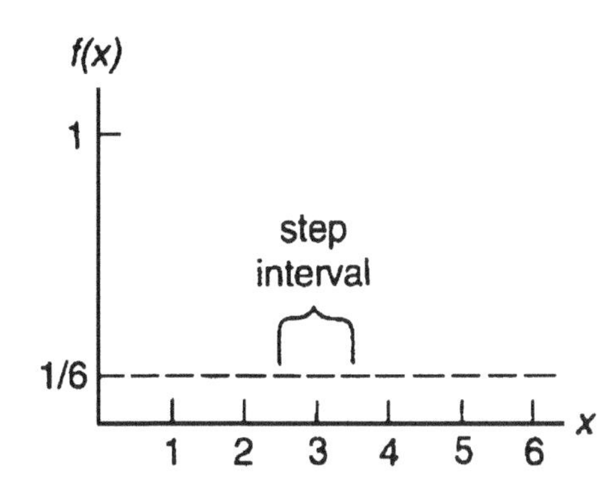

Exhibit 1

Example 5.15

A probability density function is defined as $f(x) = Ax^2$ for $-1 < x < 2$ and zero elsewhere. Find the value of A so that it is a valid density function.

Solution

For a valid density function, $\int_{-\infty}^{\infty} f(x)dx = 1$

Then, $\int_{-1}^{2} Ax^2 dx = A\frac{9}{3} = 3A = 1$; then $A = \frac{1}{3}$

Properties of Probability Density Functions

The expected value $E\{X\}$ of a probability density function is also called the mean, and for a discrete case it is given by

$$E\{X\} = \sum x_i \times f(x_i) = u$$

The expected value of a continuous random variable is

$$E\{X\} = \int_{-\text{inf}}^{+\text{inf}} x \times f(x) \times dx = u$$

The expected value for a discrete random variable of a function $g(X)$ is given by

$$E\{g(X)\} = \sum g(x_i) \times f(x_i)$$

The expected value of a continuous random variable is

$$E\{g(X)\} = \int_{-\inf}^{+\inf} g(x) \times f(x) \times dx$$

Of special interest are the functions of the form

$$g(x) = (x - u)^r$$

These are the powers of the random variables around the mean. The expected values of these power functions are called the rth moments about the mean of the distribution, where r is the power. The second moment about the mean is also known as the variance and is calculated as follows:

$$V\{X\} = E\{(x-u)^2\} = E\{(x^2 - 2xu + u^2)\} = E\{x^2\} - E\{2xu\} + E\{u^2\}$$

Since u is a constant, the second term is $2u^2$ and the third term evaluates to u^2; therefore, the second moment about the mean becomes

$$V\{X\} = E\{x^2\} - u^2 = \sigma^2$$

The square root of the variance is signified by the Greek letter sigma and is called the **standard deviation**.

Example 5.16

Calculate the mean and standard deviation of a single die throw.

Solution

This is a discrete function and can be calculated numerically by the discrete formulas given above. The mean, where $f(x_i) = \frac{1}{6}$ (all outcomes are equally likely), is given by

$$u = E\{X\} = \sum_{i=1}^{i=6} x_i \times f(x_i) = (1+2+3+4+5+6)/6 = \frac{21}{6} = 3.5$$

The standard deviation is given by

$$\sigma = \sqrt{V\{x\}} = \sqrt{E\{x^2\} - u^2} = \sqrt{[(1^2 + 2^2 + 3^2 + 4^2 + 5^2 + 6^2)/6] - 3.5^2} = 1.7$$

STATISTICAL TREATMENT OF DATA

Whether from the outcome of an experiment or trial, or simply the output of a number generator, we are constantly presented with numerical data. A statistical treatment of such data involves ordering, presentation, and analysis. The tools available for such treatment are generally applicable to a set of numbers and can be applied without much knowledge about the source of the data, although such knowledge is often necessary to make sensible use of the statistical results.

In its raw form, numerical data is simply a list of n numbers denoted by x_i, where $i = 1, 2, 3, \ldots, n$. There is no specific significance associated with the order implicit in the i numbers. They are names for the individuals in the list, although they are often associated with the order in which the raw data was recorded. For example, consider a box of 50 resistors. They are to be used in a sensitive circuit, and their resistances must be measured. The results of the 50 measurements are presented in the following table.

Table 5.1 Table of Raw Measurements (Ω)

101	105	110	115	82
86	91	96	117	112
109	103	89	97	98
101	104	99	95	97
85	90	94	112	107
103	94	98	106	98
114	112	108	101	99
93	96	99	104	90
109	106	101	93	92
104	99	109	100	107

Each number is named by the variable x_i, and there are $n = 50$ of them. The numbers range from 82 to 117.

Frequency Distribution

A systematic tool used in ordering data is the frequency distribution. The method requires counting the number of occurrences of raw numbers whose values fall within step intervals. The step intervals (or bins) are usually chosen to (1) be of constant size, (2) cover the range of numbers in the raw data, (3) be small enough in quantity to limit the amount of writing yet not have many empty steps, and (4) be sufficient in quantity so that significant information is not lost.

For example, the aforementioned raw data of measured resistances may be ordered in a frequency distribution table such as Table 5.2. Here the step interval is the event E of a random variable that can be mapped onto the x-axis. The set of eight events is the sample space. If we take a number randomly from the raw measurement set, the probability that it will be in bin 5 is

$$f(E_5) = P\{E_5\} = 10/50 = 0.2$$

Table 5.2 Frequency and cumulative frequency table

Event, E_i	Range, Ω	Frequency	Cumulative Frequency	Probability Density Function, $f(E_i)$
1	80–84	1	1	0.02
2	85–89	3	4	0.06
3	90–94	8	12	0.16
4	95–99	12	24	0.24
5	100–104	10	34	0.20
6	105–109	9	43	0.18
7	110–114	5	48	0.10
8	115–119	2	50	0.04

The last column in Table 5.2 is the probability density function of the distribution. The probability table can be plotted along the x-axis in several ways, as shown in Figures 5.2 through 5.4.

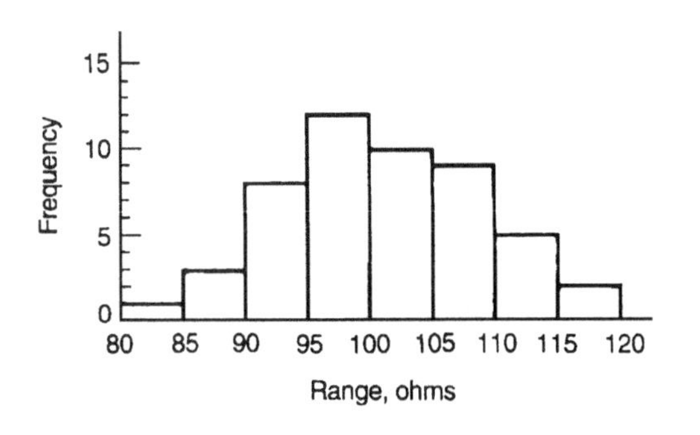

Figure 5.2 Histogram of resistance measurements

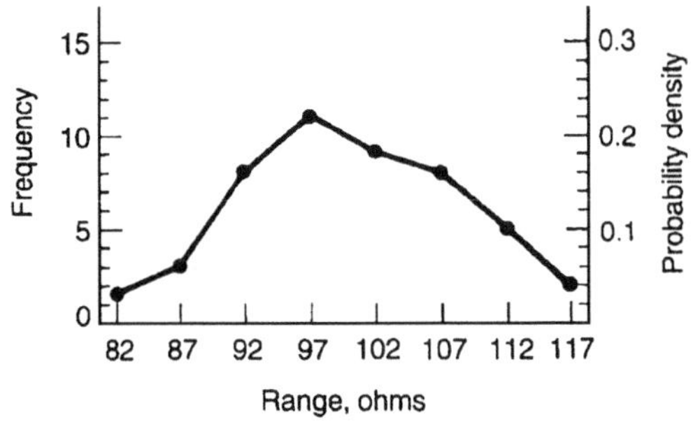

Figure 5.3 Frequency distribution and probability density plot

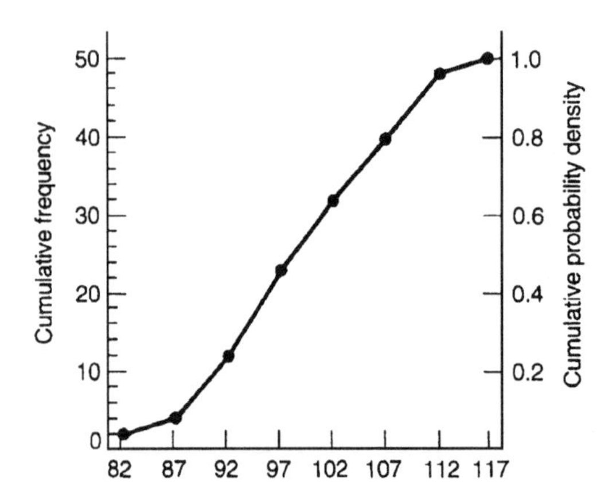

Figure 5.4 Cumulative frequency distribution and cumulative probability density

Standard Statistical Measures

There are several statistical quantities that can be calculated from a set of raw data and its distribution function. Some of the more important ones are listed here, together with the method of their calculation.

Mode The observed value that occurs most frequently; here the mode is bin 4 with a range of 95–99 Ω.

Median The point in the distribution that divides the number of observations such that half of the observations are above and half are below. The median is often the mean of the two middle values; here the median is 4.5 bins, 100 Ω.

Mean The arithmetic mean, or average, is calculated from raw data as

$$\mu = \frac{1}{n}\sum_{i=1}^{n} x_i = 100.6$$

It is calculated from the distribution function as

$$\mu = \sum_{i=1}^{m} b_i \times f(E_i) = 100.4$$

where b_i is the ith event value (for $i = 1$, $b_i = 82$) and m is the number of bins; $f(E_i)$ is the probability density function. (The two averages are not quite the same because of the information lost in assigning the step intervals.)

Standard deviation

(a) Computational form for the raw data:

$$\sigma = \sqrt{\frac{1}{n}\left[\left(\sum_{i=1}^{n} x_i^2\right) - n \times \mu^2\right]} = 8.08$$

(b) Computational form for the distribution function:

$$\sigma = \sqrt{\left\{\left[\sum_{i=1}^{m} b_i^2 \times f(E_i)\right] - \mu^2\right\}} = 8.02$$

Sample standard deviation

If the data set is a sample of a larger population, then the sample standard deviation is the best estimate of the standard deviation of the larger population.

The computational form for the raw data set is

$$\sigma = \sqrt{\frac{1}{n-1}\left[\left(\sum_{i=1}^{n} x_i^2\right) - n \times \mu^2\right]} = 8.166$$

Sample standard deviations and the use of $(n-1)$ in the denominator are discussed in the section on sampling.

Skewness

This is a measure of the frequency distribution asymmetry and is approximately

skewness $\cong$ 3(mean – median)/(standard deviation)

Example 5.17

Two professors give the following scores to their students. What is the mode and arithmetic mean?

Frequency	1	3	6	11	13	10	2
Score	35	45	55	65	75	85	95

Solution

mode = 75; N = 1 + 3 + 6 + 11 + 13 + 10 + 2 = 46

weighted arithmetic mean = $\overline{X_w} = [35(1) + 45(3) + ...\ 95(2)]/46 = 70$

STANDARD DISTRIBUTION FUNCTIONS

In the previous section, we calculated several general properties of probability distribution functions.

To know the appropriate probability density function for an actual situation, two general methods are available:

1. The probability density function is actually calculated, as was done in the last section, by analyzing the physical mechanism by which experimental events and outcomes are generated and counting the number of ways an individual event occurs.
2. Recognition of an overall similarity between the present experiment and another for which the probability density function is already known permits the known behavior of the function to be applied to the new experiment. This work-saving method is by far the more popular one. Of course, to apply this method, it is necessary to have a repertoire of known probability functions and to understand the problem characteristics to which they apply.

This section lists several popular probability density functions and their characteristics.

Binomial Distribution

The binomial distribution applies when there is a set of discrete binary alternative outcomes. Deriving this distribution function helps one understand the class of problems to which it applies. For example, given a set of n events, each with a probability p of occurring, what is the probability that r of the events will occur and $(n - r)$ not occur?

The probability of one event occurring is p.

The probability of r events occurring is p^r.

The probability of $(n - r)$ events not occurring is $(1 - p)^{n-r}$.

The probability of exactly r events occurring and $(n - r)$ not occurring in a trial is given by the joint probability Rule 3:

$$P[r \cap (n-r)] = p^r \times (1-p)^{n-r}$$

However, there are many ways of choosing r occurrences out of n events. In fact, the number of different ways of choosing r items from a set of n items when order is not important is given by the binomial coefficient $C(n, r)$. The total probability of r occurrences from n trials, given an individual probability of occurrence as p, is thus given by

$$C(n,r) \times p^r \times (1-p)^{n-r} = f(r)$$

This is the **binomial probability density function**.

The mean of this density function is the first moment of the density function, or expected value, and is calculated as

$$E\{x\} = \sum_{r=0}^{n} r \times f(r) = \sum_{r=0}^{n} r \times \frac{n!}{(r)!(n-r)!} \times p^r \times (1-p)^{n-r}$$

This can be rewritten as

$$\sum_{r=1}^{n} \frac{n!}{(r-1)!(n-r)!} \times p^r \times (1-p)^{n-r}$$

We can now factor out the quantity $n \times p$ and let $r - 1 = y$. This can be rewritten as

$$n \times p \times \sum_{y=0}^{n-1} \frac{(n-1)!}{(y)!(n-1-y)!} \times p^y \times (1-p)^{n-1-y} = n \times p \times [p + (1-p)]^{n-1}$$

Since the sum is merely the expansion of a binomial raised to a power, and the number 1 raised to any power is 1, the mean is

$$\mu = n \times p$$

A similar calculation shows the variance is

$$\text{var} = n \times p \times (1-p)$$

The standard deviation is

$$\sigma = \sqrt{\text{var}} = \sqrt{n \times p \times (1-p)}$$

Example 5.18

A truck carrying dairy products and eggs damages its suspension and 5% of the eggs break.

(i) What is the probability that a carton of 12 eggs will have exactly one broken egg?

(ii) What is the probability that one or more eggs in a carton will be broken?

Solution

(i) Since an egg is either broken or not broken, the binomial distribution applies. The probability p that an egg is broken is 0.05 and that one is not broken is $(1 - p) = 0.95$. From the equation for the binomial distribution, with $n = 12$ and $r = 1$,

$$p\{1\} = f(1) = C(12,1) \times 0.05^1 \times 0.95^{11} = 12 \times 0.05 \times 0.57 = 0.34$$

(ii) The probability that one or more eggs will be broken can be calculated as the sum of each individual probability:

$$p\{x > 0\} = p\{1\} + p\{2\} + \cdots + p\{12\}$$

However, this requires 12 calculations. The problem can also be solved using the complementary rule:

$$p\{x > 0\} = 1 - p\{0\} = C(12,0) \times 0.05^0 \times 0.95^{12} = 0.95^{12} = 0.54$$

Example 5.19

A biased coin is tossed. Find the probability that a head appears once in three trials.

P(Head) = p = 0.6

Solution

Here q = P(Head not occurring) or P(Tail) = 1 – 0.6 = 0.4.

Then P(1 Head) = C(3,1) 0.6^1 0.4^2 = 0.2880.

Normal Distribution Function

The normal distribution, or Gaussian distribution, is widely used to represent the distribution of outcomes of experiments and measurements. It is popular because it can be derived from a few empirical assumptions about the errors presumed to cause the distribution of results about the mean. One assumption is that the error is the result of a combination of N elementary errors, each of magnitude e and equally likely to be positive or negative. The derivation then assumes $N \to \infty$ and $e \to 0$ in such a way as to leave the standard deviation constant. This error model is universal, since most experiments are analyzed to eliminate systematic errors. What remains is attributable to errors that are too small to explain systematically, so the normal probability distribution is evoked.

The form of the probability density and distribution functions for the **normal distribution** with a mean μ and variance σ^2 is given by

$$f(x) = \frac{e^{-(x-\mu)^2/2\sigma^2}}{\sigma\sqrt{2\pi}} \qquad -\infty < x < \infty$$

$$F(x) = \int_{-\infty}^{x} \frac{e^{-(x-\mu)^2/2\sigma^2}}{\sigma\sqrt{2\pi}}\, dt$$

The normal distribution is the typical bell-shaped curve shown in Figure 5.5. Here we see that the curve is symmetric about the mean μ. Its width and height are determined by the standard deviation σ. As σ increases, the curve becomes wider and lower.

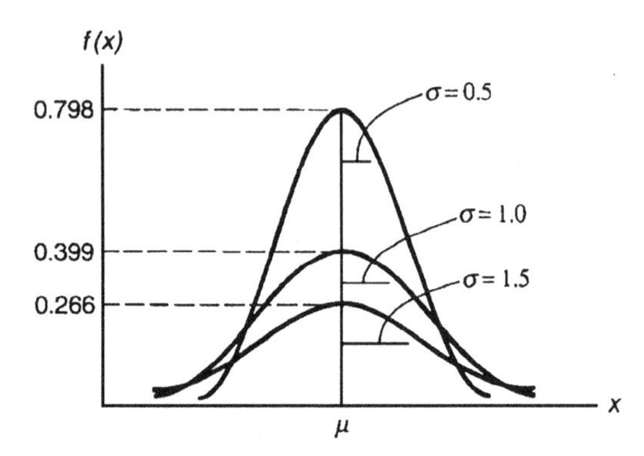

Figure 5.5 Normal distribution curve

Since this function is difficult to integrate, reference tables are used to calculate probabilities in a standard format; then the standard probabilities are converted to the actual variable required by the problem. The relation between the standard variable, z, and a typical problem variable, x, is

$$z = (x - \mu)/\sigma$$

Since μ and σ are constants, the standard probability at a value z is the same as the problem probability for the value at x.

The standard probability density function is

$$f(z) = \frac{1}{\sqrt{2\pi}} \times e^{-z^2/2}$$

Table 5.3 Standard probability table

z	$F(z)$	$f(z)$	z	$F(z)$	$f(z)$
0.0	0.5000	0.3989	2.0	0.9773	0.0540
0.1	0.5398	0.3970	2.1	0.9821	0.0440
0.2	0.5793	0.3910	2.2	0.9861	0.0355
0.3	0.6179	0.3814	2.3	0.9893	0.0283
0.4	0.6554	0.3683	2.4	0.9918	0.0224
0.5	0.6915	0.3521	2.5	0.9938	0.0175
0.6	0.7257	0.3332	2.6	0.9953	0.0136
0.7	0.7580	0.3123	2.7	0.9965	0.0104
0.8	0.7881	0.2897	2.8	0.9974	0.0079
0.9	0.8159	0.2661	2.9	0.9981	0.0060
1.0	0.8413	0.2420	3.0	0.9987	0.0044
1.1	0.8643	0.2179	3.1	0.9990	0.0033
1.2	0.8849	0.1942	3.2	0.9993	0.0024
1.3	0.9032	0.1714	3.3	0.9995	0.0017
1.4	0.9192	0.1497	3.4	0.9997	0.0012
1.5	0.9332	0.1295	3.5	0.9998	0.0009
1.6	0.9452	0.1109	3.6	0.9998	0.0006
1.7	0.9554	0.0940	3.7	0.9999	0.0004
1.8	0.9641	0.0790	3.8	0.9999	0.0003
1.9	0.9713	0.0656	3.9	1.0000	0.0002
			4.0	1.0000	0.0001

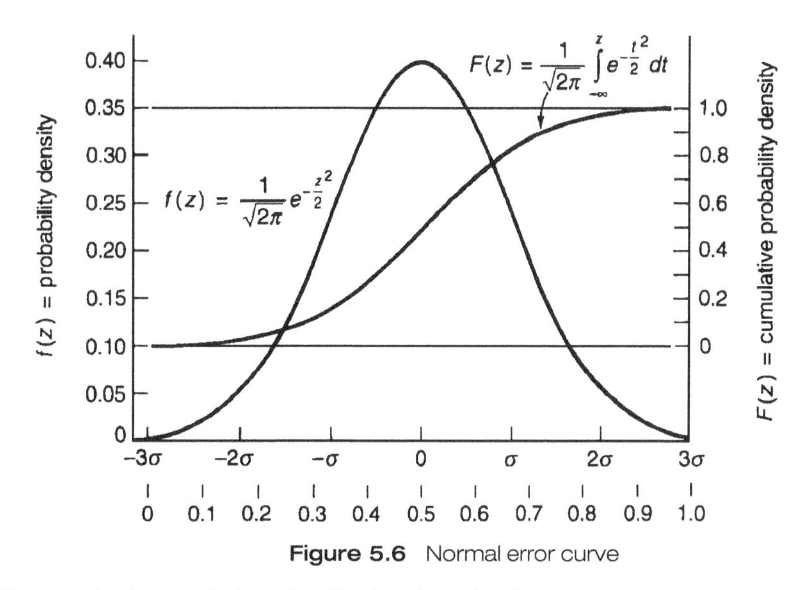

Figure 5.6 Normal error curve

The standard cumulative distribution function is

$$F(z) = \int_{-\infty}^{z} \frac{1}{\sqrt{2\pi}} \times e^{-t^2/2} \times dt$$

The standard probability function is shown graphically in Figure 5.6, and Table 5.3 shows the corresponding numerical values. The standard probability curve is symmetric about the origin and is given in terms of unit *sigma*. To use the table, remember that the function $F(z)$ is the area under the probability curve from minus infinity to the value z. The area under the curve up to $x = 0$ is therefore 0.5. Also, from symmetry,

$$F(-z) = 1 - F(z)$$

Example 5.20

Find the probability that the standard variable z lies within (i) 1σ, (ii) 2σ, and (iii) 3σ of the mean.

Solution

(i) The probability is $P_1 = F(1.0) - F(-1.0)$. From the symmetry of F, $F(-1.0) = 1 - F(1.0)$, so

$$\begin{aligned} P_1 &= 2F(1.0) - 1 = 2(0.8413) - 1 \\ &= 0.6826 \end{aligned}$$

(ii) In this case, the probability is

$$\begin{aligned} P_2 &= F(2.0) - F(-2.0) \\ &= F(2.0) - [1 - F(2.0)] \\ &= 2F(2.0) - 1 = 2(0.9773) - 1 \\ &= 0.9546 \end{aligned}$$

(iii) In the same way,

$$\begin{aligned} P_3 &= 2F(3.0) - 1 \\ &= 2(0.9987) - 1 \\ &= 0.9974 \end{aligned}$$

Example 5.21

A Gaussian random variable has a mean of 1830 and standard deviation of 460. Find the probability that the variable will be more than 2750.

Solution

$P(X > 2750) = 1 - P(X \leq 2750) = 1 - F[(2750 - 1830)/460] = 1 - F(2.0) = 1 - 0.9772 = 0.0228$

t-Distribution

The t-distribution is often used to test an assumption about a population mean when the parent population is known to be normally distributed but its standard deviation is unknown. In this case, the inferences made about the parent mean will depend upon the size of the samples being taken.

It is customary to describe the t-distribution in terms of the standard variable t and the number of degrees of freedom ν. The number of degrees of freedom is a measure of the number of independent observations in a sample that can be used to estimate the standard deviation of the parent population; the number of degrees of freedom ν is one less than the sample size ($\nu = n - 1$).

The density function of the t-distribution is given by

$$f(t) = \frac{\Gamma\left(\frac{\nu+1}{2}\right)}{\sqrt{\nu\pi}\,\Gamma\left(\frac{\nu}{2}\right)\left(1+t^2/\nu\right)^{(\nu+1)/2}}$$

and is provided in Table 5.4. The mean is $m = 0$, and the standard deviation is

$$\sigma = \sqrt{\frac{\nu}{\nu-2}}$$

Table 5.4 t-Distribution; values of $t_{\alpha,\nu}$

Degrees of Freedom, ν	Area of the Tail				
	α = 0.10	α = 0.05	α = 0.025	α = 0.01	α = 0.005
1	3.078	6.314	12.706	31.821	63.657
2	1.886	2.920	4.303	6.965	9.925
3	1.638	2.353	3.182	4.541	5.841
4	1.533	2.132	2.776	3.747	4.604
5	1.476	2.015	2.571	3.365	4.032
6	1.440	1.943	2.447	3.143	3.707
7	1.415	1.895	2.365	2.998	3.499
8	1.397	1.860	2.306	2.896	3.355
9	1.383	1.833	2.262	2.821	3.250
10	1.372	1.812	2.228	2.764	3.169
11	1.363	1.796	2.201	2.718	3.106
12	1.356	1.782	2.179	2.681	3.055
13	1.350	1.771	2.160	2.650	3.012
14	1.345	1.761	2.145	2.624	2.977
15	1.341	1.753	2.131	2.602	2.947
16	1.337	1.746	2.120	2.583	2.921
17	1.333	1.740	2.110	2.567	2.898
18	1.330	1.734	2.101	2.552	2.878
19	1.328	1.729	2.093	2.539	2.861
20	1.325	1.725	2.086	2.528	2.845
21	1.323	1.721	2.080	2.518	2.831
22	1.321	1.717	2.074	2.508	2.819
23	1.319	1.714	2.069	2.500	2.807
24	1.318	1.711	2.064	2.492	2.797
25	1.316	1.708	2.060	2.485	2.787
26	1.315	1.706	2.056	2.479	2.779
27	1.314	1.703	2.052	2.473	2.771
28	1.313	1.701	2.048	2.467	2.763
29	1.311	1.699	2.045	2.462	2.756
inf.	1.282	1.645	1.960	2.326	2.576

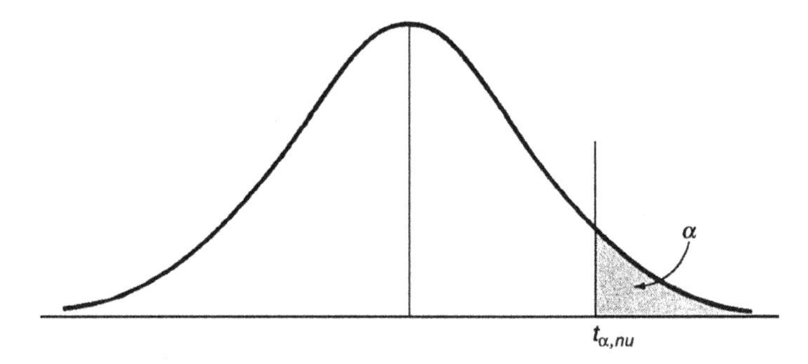

Figure 5.7

Probability questions involving the t-distribution can be answered by using the distribution function $t_{\alpha,\nu}$ shown in Figure 5.7. Table 5.4 gives the value of t as a function of the degrees of freedom v down the column and the area (α) of the tail across the top. The t-distribution is symmetric. As an example, the probability of t falling within ± 3.0 when a sample size of 8 ($\nu = 7$) is selected is one minus twice the tail ($\alpha = 0.01$):

$$P\{-3.0 < t < 3.0\} = 1 - (2 \times 0.01) = 0.98$$

The t-distribution is a family of distributions that approaches the Gaussian distribution for large n.

X² DISTRIBUTION

In probability theory and statistics, the chi-square distribution (also chi-squared or x^2-distribution) is one of the most widely used theoretical probability distributions in inferential statistics (e.g., in statistical significance tests). It is useful because, under reasonable assumptions, easily calculated quantities can be proven to have distributions that approximate to the chi-square distribution if the null hypothesis is true.

The best known situations in which the chi-square distribution is used are the common chi-square tests for goodness of fit of an observed distribution to a theoretical one, and of the independence of two criteria of classification of qualitative data. Many other statistical tests also lead to a use of this distribution.

If $Z_1, Z_2, \ldots, Z_n$ are independent unit normal random variables, then

$$\chi^2 = Z_1^2 + Z_2^2 + \ldots + Z_n^2$$

is said to have a chi-square distribution with n degrees of freedom.

SAMPLE STATISTICS

The *sample mean* and *sample standard deviation* for n independent observations $\{x_1, x_2, \ldots, x_n\}$ are:

$$\bar{x} = \frac{1}{n}\sum_{i-1}^{n} x_i$$

$$s = \sqrt{\frac{\sum_{i=1}^{n}(x_i - \bar{x})^2}{n-1}} = \sqrt{\frac{\sum_{i=1}^{n} x_i^2 - \frac{1}{n}\left(\sum_{i=1}^{n} x_i\right)^2}{n-1}}$$

The sample mean is just the arithmetic average of the observations. As the first form of the standard deviation formula shows, the sample standard deviation is the square root of the *sample variance*. The sample variance is the average square of the difference between an observation and the average of them ($x_i - \bar{x}$). The $n - 1$ denominator is based on the degrees of freedom. Since the nth difference could be calculated from the others, and thus is not independent, our degrees of freedom are $n - 1$.The second form of the formula is more efficient for calculation, but on the exam, you should compute sample means and standard deviations simply by entering the data into the calculator and running the statistics routines.

NORMAL AND *t* SAMPLING DISTRIBUTIONS

The sample mean has sample standard deviation $s/\sqrt{n}$. Let the distribution from which the sample is taken have mean μ and standard deviation σ. Then, unless n is too small, the sample mean has a normal distribution, regardless of the distribution of the sampled variable. We define the *sample mean test statistic*:

$$Z_n = \frac{\bar{x} - \mu}{\sigma/\sqrt{n}}$$

The probability that the sample mean exceeds the true mean by at least $\bar{x} - \mu$ is the one-tail probability associated with the Z_n statistic, and there is an equal probability that the sample mean is less than the true mean by at least $\mu - \bar{x}$; that is, Z_n and $-Z_n$ are associated with equal probabilities in the left and right tails of the symmetrical normal distribution.

For instance, if the sample mean is 3.7, the standard deviation is known or assumed to be 2.1, and the sample mean is based on 30 observations, then $Z_n = (3.7 - \mu)/0.3834$. If we want to know the probability that the true mean is at least 4, we set $\mu = 4$ and solve for Z_n, which is $Z_n = -0.782$. We look up the associated probability on the unit normal table and find a probability of about 22%. Since the normal distribution is symmetrical we know there is also a 22% probability that the true mean is at most 3.4 (the mean, 3.7, minus 0.3). Since the total probability must sum to 100%, we can calculate that there is a 56% probability that the true mean is within the interval from 3.4 to 4.0.

Often the true standard deviation is not known but is approximated by s. In this case, the sample mean follows the t distribution, which is also symmetrical:

$$t_n = \frac{\bar{x} - \mu}{s/\sqrt{n}}$$

CONFIDENCE, HYPOTHESIS TESTING, AND SAMPLE SIZE

The formulas for the Z or t statistics of a sample estimate can be rearranged to suit various purposes. Consider Example 5.22.

Example 5.22

If a robotics system is operating as intended, the average cycle time should be at most 15.0 min. Six independent observations gave cycle times {14.3, 16.0, 14.2, 13.8, 14.1, 13.7} min. From this data, what is the confidence that the system is operating as intended?

Solution

The sample mean is 14.35 min. and the sample standard deviation is 0.840832920. The t_n statistic is $(14.35 - 15.0)/(0.840832920/\sqrt{6}) = -1.89356$. From the *t*-distribution table with $n = 6$ we see that the probability that a *t* variate is at least 1.89356 standard deviations away from its mean (in either direction—the *t*-distribution is symmetric) is a little greater than 0.05, so we are not quite 95% confident that the true average cycle time does not exceed 15.0 min.

In Example 5.22 we asked for a *confidence* (probability), which is obtained by first calculating the value of the *Z* or *t* statistic, then looking up the corresponding one-tail or two-tail probability and subtracting it from 1 to get the confidence. In the handbook, the table for the *t* distribution relates *t*-statistic values to one-tail probabilities, and the unit normal table relates Z-statistic values (called *x* in the table) not only to one-tail probabilities—identified as $R(x)$—but also their complement $F(x)$ and also two-tail probabilities and their complement. The regions of the distribution shaded above the columns in the normal table are helpful in understanding the corresponding probabilities.

In hypothesis testing, the hypothesis to be tested (the *main hypothesis*) is that a suspected effect occurs. The *null hypothesis*, which is the complement of the main hypothesis, is that there is no effect and that the difference between the observed results and those predicted by the null hypothesis is due simply to chance.

We first choose α, the probability of *type-I error* (the acceptable probability of false rejection of the null hypothesis), commonly 1% or 5%, and look up the corresponding *critical value* of the *Z* or *t* statistic, then calculate the actual value of the *Z* or *t* statistic from the observed data under the assumption that the null hypothesis is correct, and conclude that the null hypothesis is rejected if the absolute value of *Z* or *t* exceeds the critical value.

For example, in the situation of Example 5.22, we could have originally suspected that the cycle times were too long—greater than 15 min. The null hypothesis would be that they were not too long; under this null hypothesis the *t* statistic for a 15-minute cycle time would be –1.89356 (see the solution to Problem 5.1). If the probability of error had been set to 5% in advance, then the critical *t* statistic for $\alpha = 0.05$ and $n = 6$ would be 1.943, as obtained from the table for the *t* distribution. Thus, the absolute value of the *t* statistic would have failed to exceed the critical value, and the null hypothesis would not quite have been rejected. Since the probability of falsely rejecting the null hypothesis (type-I error) was set to 5% in advance, a procedure was followed under which there would be at most a 5% probability that the data could mislead us into false conclusions; the data for Problem 5.1 would not quite allow us to claim surely enough that the cycle time was at most 15 minutes; although we would certainly expect that more data would probably support the claim adequately.

The equation for the *t* or *Z* statistic can be solved for the numerator, which is the half-width of the confidence interval. In Example 5.22, for instance, the sample average cycle time is 14.35 min., and the sample standard deviation is 0.840832920. If we chose in advance a type-I error of $\alpha = 0.05$, so that the corresponding *t* statistic for $n = 6$ were 1.943 from the *t*-distribution table, then the confidence interval half-width would be

$$\bar{x} - \mu = t_n s/\sqrt{n} = 1.943 \times 0.840832920/\sqrt{6} = 0.66697$$

We would be 90% confident (5% error on each side) that the true cycle time was within 14.35 ± 0.667 min. The confidence interval would be from 13.683 min. to 15.017 min. A 15-minute cycle time is just inside the confidence interval, which allows 5% error on the high side and 5% on the low. Again, a 15-minute cycle time is almost, but not quite, unusual enough given the data.

If we wanted to estimate how much data would be enough for 95% confidence that the cycle time does not exceed 15 minutes, we could simply look in the *t*-distribution table for *n* that meets the requirements. Recall that the absolute value of the *t* statistic is 1.89356. For $\alpha = 0.05$, the critical value for $n = 6$ is 1.943, so the *t* statistic does not quite exceed the critical value. The critical value for $n = 7$ is 1.895, still slightly too high; for $n = 8$ the critical value is 1.860. Thus, we estimate that if we had two additional observations consistent with the existing sample statistics, the required confidence probably would be obtained.

For the common situation in which the engineer first obtains *m* observations of a variable, determines its *m*-observation mean $\bar{x}$ and its *m*-observation range *r* (or, more accurately, its sample standard deviation *s*), and desires to compute the total number of observations needed, *n*, for a given confidence of estimating the variable to a given relative accuracy *A*, the formula for the *Z* or *t* can be rearranged as follows:

$$n = \left(\frac{zs}{A\bar{x}} \right)^2$$

For estimating the variable within, say, ±5% of its true value, set $A = 0.05$. Assuming that *n* is sufficiently large for the normal distribution to apply, the two-tail confidence interval is given in terms of the standard normal statistic *Z*: for 90% confidence (0.05 probability of error in each direction), $Z = 1.64$; for 95% confidence, $Z = 1.96$; for "two-sigma" (95.44%) confidence, $Z = 2$; for 98% confidence, $Z = 2.33$.

To test whether two variances are equal, the F statistic is used. You should memorize the general definition of the F statistic:

$$F_0 = \frac{s_1^2 / \sigma_1^2}{s_2^2 / \sigma_2^2}$$

This definition is for sample variances and null-hypothesis variances from two populations, the population having the larger ratio of sample variance to null-hypothesis variance being designated as population 1. Also, remember that the null hypothesis (that the two variances are equal) is rejected if

$$F_0 > F_{\alpha/2,\ n_1-1, n_2-1}$$

The *F*-distribution table in the IE chapter of the handbook is for $\alpha = 0.05$ only, so we can test the equality of variances only at the 10% significance level, or the hypothesis that a particular one (chosen before the data are known) is larger only at the 5% significance level.

LEAST SQUARES REGRESSION

Given a sample of *n* independent (*x*, *y*) data pairs $\{x_i, y_i: i = 1, \ldots, n\}$, where *x* is viewed as an independent variable and *y* as a dependent variable, a line $y = a + bx$ can be drawn. For each x_i, the line gives a *y* value $a + bx_i$, and the difference between this and the actual *y* datum y_i is called an error, or residual. The best intercept *a* and slope *b*, in the sense of minimizing the sum of the squares of the errors, are given by formulas listed in the "Least Squares" section of the IE chapter of the handbook.

Example 5.23

Consider a set of data with x values $\{1, 2, 2.5\}$ and corresponding y values $\{1, 2, 3\}$. Verify that your calculator's regression routine yields the intercept a = 1.285714286 and the slope b = –0.357142857. If your calculator allows, determine the associated sums ($\Sigma x = 5.5$, $\Sigma x^2 = 11.25$, $\Sigma y = 6$, $\Sigma y^2 = 14$, $\Sigma xy = 12.5$). Then use these to verify $S_{xy} = 4.5$, $S_{yy} = 6$ and $S_{xx} = 3.5$. (Note: These last three quantities are n times the quantities of the same names usually found in the literature.)

Solution

The sum of the squares of the errors is

$$SS_E = \frac{S_{yy}S_{xx} - S_{xy}^2}{S_{xx}}$$

An unbiased estimator of the variance of these errors is the sum of squared errors divided by $n(n-2)$:

$$S_e^2 = \frac{S_{yy}S_{xx} - S_{xy}^2}{n(n-2)S_{xx}}$$

(Note: The IE chapter of the handbook omits the SS_E formula and had misprints in the S_e^2 formula.)

Given a set of $\{x, y\}$ data, the correlation coefficient is

$$r = \frac{S_{xy}}{\sqrt{S_{xx}S_{yy}}}$$

For example, the x values $\{1, 2, 2.5\}$ and the corresponding y values $\{1, 2, 3\}$ have $r = 0.981980506$. Your hand calculator probably allows r to be obtained directly rather than through the regression parameters.

ANALYSIS OF VARIANCE

The IE chapter of the handbook gives formulas for analysis of variance. The analysis of variance (ANOVA) is the partitioning of variability into component parts. In ANOVA there are *treatments*, which are values of a related factor that we are interested in learning about.

Imagine we want to know which formulation of a fertilizer will produce the best crop yield. We could try k different fertilizers (our treatment) and make n independent observations of an output variable crop yield (x). Let $\bar{x}$ be the mean of x for all observations of all treatments; then associated with the jth observation of the ith treatment is an "error" $x_{ij} - \bar{x}$. The sum of the squares of all these errors is denoted SST in the IE chapter of the handbook and is usually called the total corrected sum of squares. This quantity is partitioned into two parts: $SS(\text{Tr})$ is the *treatment sum of squares*, and SSE is the *error sum of squares*: $SST = SS(\text{Tr}) + SSE$, or $SSE = SST - SS(\text{Tr})$.

Let $T = kn\bar{x}$ denote the sum of all observations, and let T_i denote the sum of observations for the ith treatment. Then, define $C = T^2/kn$. With these definitions we can determine the partitioned quantities as follows:

$$SST = \sum_{i=1}^{k}\sum_{j=1}^{n} x_{ij}^2 - C$$

$$SS(\text{Tr}) = \sum_{i=1}^{k} \frac{T_i^2}{n} - C$$

Now we form a null hypothesis that there is no difference among treatments. An estimate of the variance from $SS(\text{Tr})$ is $SS(\text{Tr})/(k - 1)$. An estimate of the variance from SSE is $SSE/(kn - k)$. According to Cochran's theorem, the ratio of the two estimates of the variance should follow the F distribution with $k - 1$ and $kn - k$ degrees of freedom. We define the test statistic F_0 and reject the null hypothesis at type-1 error level α if the test statistic is greater than that given in the F-distribution table:

$$F_0 = \frac{SS(\text{Tr})/(k-1)}{SSE/(kn-k)} \qquad \text{Reject if} \qquad F_0 > F_{\alpha,k-1,\,kn-k}$$

System reliability is listed as a subtopic in probability and statistics. It is discussed in the "Quality" chapter of this book.

PROBLEMS

5.1 Ten independent observations were made of the time to load a pallet. The observed times, in minutes, were 104, 132, 115, 180, 95, 96, 100, 73, 82, and 113. The number of additional observations needed to be 90% confident that the average will be accurate within ±10% is most nearly:

a. none
b. 5
c. 10
d. 20

5.2 Of five competing formulations of feed, each formulation was fed to six randomly selected piglets for a week. The number of pounds gained during the week was observed for each piglet. The total sum of squared error was 7002, and the sum of squared error between treatments was 3821.5. The implied effect of formulation on weight gain is most nearly:

a. insignificant
b. perhaps significant, but not at the 0.05 level
c. significant at roughly the 0.05 level
d. more significant than at the 0.05 level

5.3 An industrial hygienist has taken periodic readings of the levels of a suspected toxin. Based on generally accepted guidelines, the company has set an acceptable maximum level at 200 units. Below are 10 data readings from a particular part of the facility. Given hypotheses of H_0: $\mu \geq 200$ vs. H_1: $\mu < 200$. What can we say about our chemical levels?

175	215	190	198	184
207	210	193	196	180

a. The null hypothesis is rejected at $\alpha = 0.01$, and the chemical levels appear to be unsafe.
b. The null hypothesis is rejected at $\alpha = 0.01$, and the chemical levels appear to be safe.
c. The null hypothesis is not rejected at $\alpha = 0.01$, and the chemical levels appear to be unsafe.
d. The null hypothesis is not rejected at $\alpha = 0.01$, and the chemical levels appear to be safe.

5.4 What is the probability of drawing a pair of aces in two cards when an ace has been drawn on the first card?

a. 1/13
b. 1/26
c. 3/51
d. 4/51

5.5 An auto manufacturer has three plants (A, B, C). Four out of 500 cars from Plant A must be recalled, 10 out of 800 from Plant B, and 10 out of 1000 from Plant C. Now a customer purchases a car from a dealer who gets 30% of his stock from Plant A, 40% from Plant B, and 30% from Plant C, and the car is recalled. What is the probability it was manufactured in Plant A?

a. 0.0008
b. 0.01
c. 0.0125
d. 0.2308

5.6 There are ten defectives per 1000 times of a product. What is the probability that there is one and only one defective in a random lot of 100?

a. 99×0.01^{99}
b. 0.01
c. 0.05
d. 0.99^{99}

5.7 The probability that both stages of a two-stage missile will function correctly is 0.95. The probability that the first stage will function correctly is 0.98. What is the probability that the second stage will function correctly given that the first one does?

a. 0.99
b. 0.98
c. 0.97
d. 0.95

5.8 A standard deck of 52 playing cards is thoroughly shuffled. The probability that the first four cards dealt from the deck will be the four aces is closest to:

a. 2.0×10^{-1}
b. 8.0×10^{-2}
c. 4.0×10^{-4}
d. 4.0×10^{-6}

5.9 In statistics, the standard deviation measures:

a. a standard distance
b. a normal distance
c. central tendency
d. dispersion

5.10 There are three bins containing integrated circuits (ICs). One bin has two premium ICs, one has two regular ICs, and one has one premium IC and one regular IC. An IC is picked at random. It is found to be a premium IC. What is the probability that the remaining IC in that bin is also a premium IC?

a. $\frac{1}{5}$

b. $\frac{1}{4}$

c. $\frac{1}{3}$

d. $\frac{2}{3}$

5.11 How many teams of four can be formed from 35 people?
a. about 25,000
b. about 2,000,000
c. about 50,000
d. about 200,000

5.12 A bin contains 50 bolts, 10 of which are defective. If a worker grabs 5 bolts from the bin in one grab, what is the probability that no more than 2 of the 5 are bad?
a. about 0.5
b. about 0.75
c. about 0.90
d. about 0.95

5.13 How many three-letter codes may be formed from the English alphabet if no repetitions are allowed?
a. 26^3
b. 26/3
c. $26 \times 25 \times 24$
d. $26^3/3$

5.14 A widget has three parts, A, B, and C, with probabilities of 0.1, 0.2, and 0.25, respectively, of being defective. What is the probability that exactly one of these parts is defective?
a. 0.375
b. 0.55
c. 0.95
d. 0.005

5.15 If three students work on a certain math problem, student A has a probability of success of 0.5; student B, 0.4; and student C, 0.3. If they work independently, what is the probability that no one works the problem successfully?
a. 0.12
b. 0.25
c. 0.32
d. 0.21

5.16 A sample of 50 light bulbs is drawn from a large collection in which each bulb is good with a probability of 0.9. What is the approximate probability of having less than 3 bad bulbs in the 50?
a. 0.1
b. 0.2
c. 0.3
d. 0.4

5.17 The number of different 3-digit numbers that can be formed from the digits 1, 2, 3, 7, 8, 9 without reusing the digits is:
a. 10
b. 20
c. 30
d. 40

5.18 The number of different ways that a party of seven councilmen can be seated in a row is:
a. 1
b. 560
c. 2080
d. 5040

5.19 A student must answer six out of eight questions on an exam. The number of different ways in which he can do the exam is:
a. 8
b. 18
c. 28
d. 48

5.20 Repeat Problem 5.19 if the first two questions are mandatory.
a. 4
b. 8
c. 12
d. 15

5.21 A group of five women wishes to form a subcommittee consisting of two of them. The number of possible ways to do so is:
a. 5
b. 10
c. 15
d. 20

5.22 An integer has to be chosen from numbers between 1 and 100 (both inclusive). The probability of choosing a number divisible by 9 (with a remainder of 0) is:
a. 0
b. 0.01
c. 0.11
d. 0.91

5.23 Four fair coins are tossed. The probability of either one head or two heads showing up is:
a. 1/8
b. 2/8
c. 4/8
d. 5/8

5.24 Two identical bags contain ten apples and five oranges each. The probability of selecting an apple from the first bag and an orange from the second bag is:
a. 1/9
b. 2/9
c. 3/9
d. 4/9

5.25 A bag contains 5 red, 10 orange, 15 green, 20 violet, and 25 black cards. The probability that you will get a black card or a red card if you remove a card from the bag is:
a. 5/85
b. 15/85
c. 25/85
d. 35/85

5.26 Two bags each contain two orange balls, five white balls, and three red balls. The probability of selecting an orange ball from the first bag or a white ball from the other bag is:
a. 0
b. 0.2
c. 0.6
d. 1.0

5.27 A bag contains 100 balls numbered 1 to 100. One ball is drawn from the bag. What is the probability that the number on the ball will be even or greater than 72?
a. 0.64
b. 0.50
c. 0.28
d. 0.14

5.28 A circuit has two switches connected in series. For a signal to pass through, both switches must be closed. The probability that the first switch is closed is 0.95, and the probability that a signal passes through is 0.90. The probability that the second switch is closed is:
a. 0.8545
b. 0.9000
c. 0.9474
d. 0.9871

5.29 Four probability density distributions are shown in Exhibit 5.29. The only valid distribution is:

a. $f_1(x)$
b. $f_2(x)$
c. $f_3(x)$
d. $f_4(x)$

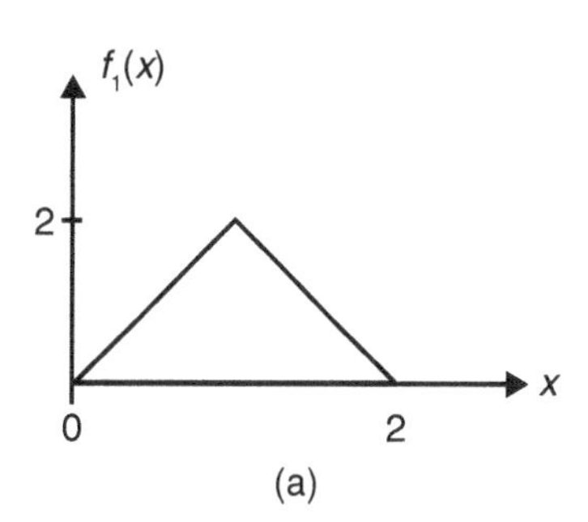

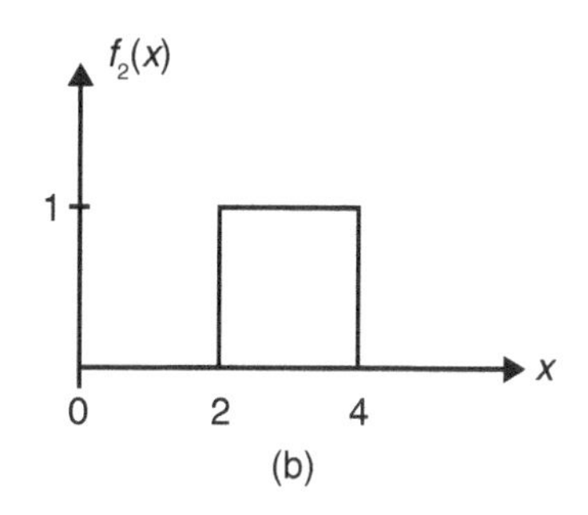

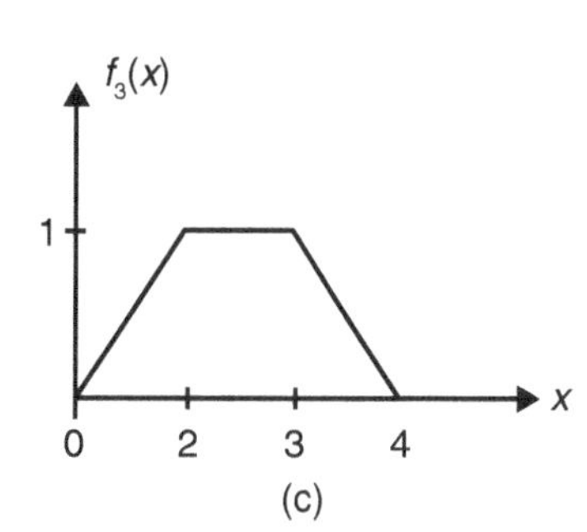

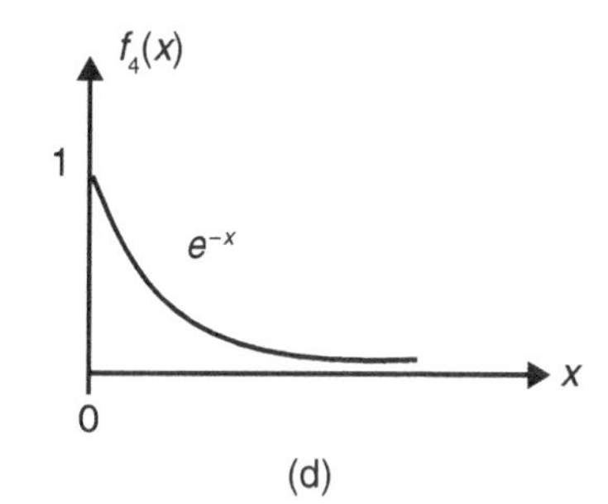

Exhibit 5.29

5.30 Four probability density distributions are shown in Exhibit 5.30. The only valid distribution is:

a. $f_1(x)$
b. $f_2(x)$
c. $f_3(x)$
d. $f_4(x)$

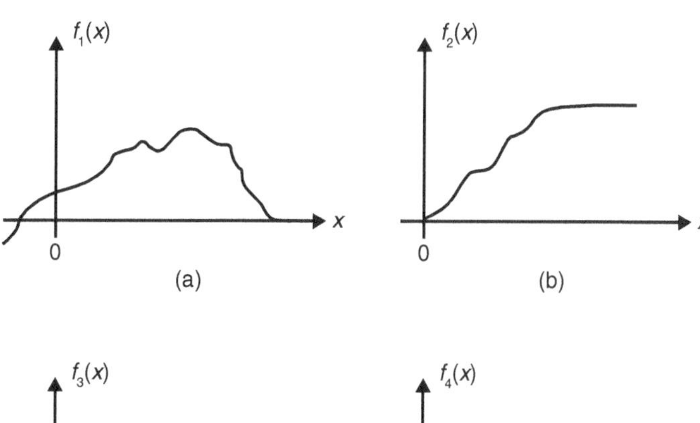

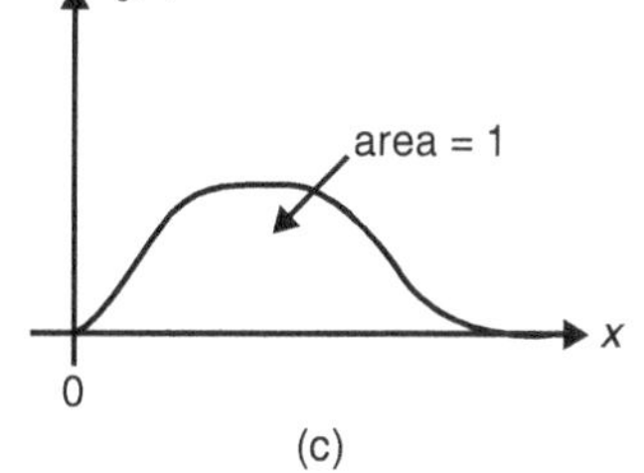

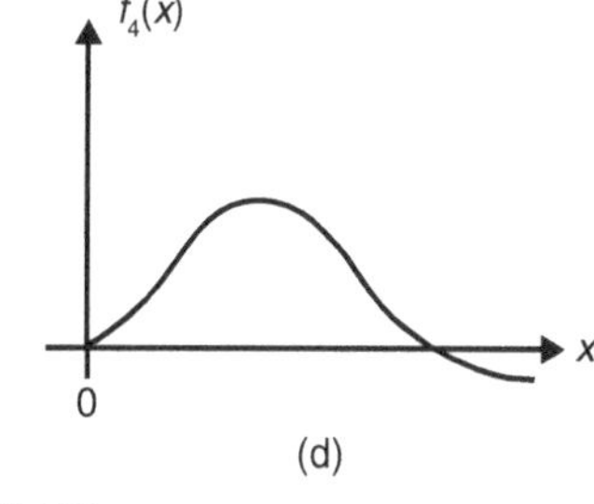

Exhibit 5.30

5.31 Four probability density distributions are shown in Exhibit 5.31. The only valid distribution is:

a. $f_1(x)$
b. $f_2(x)$
c. $f_3(x)$
d. $f_4(x)$

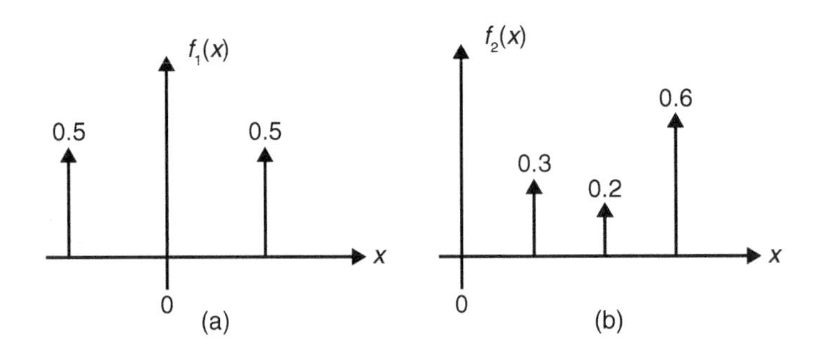

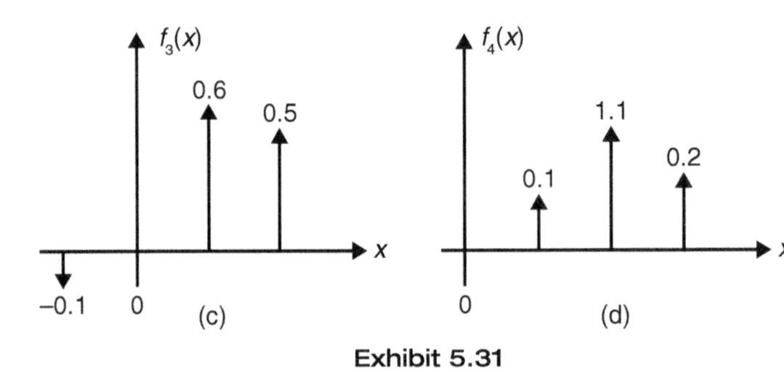

Exhibit 5.31

5.32 Four cumulative probability distribution functions are shown in Exhibit 5.32. The only valid distribution is:

a. $F_1(x)$
b. $F_2(x)$
c. $F_3(x)$
d. $F_4(x)$

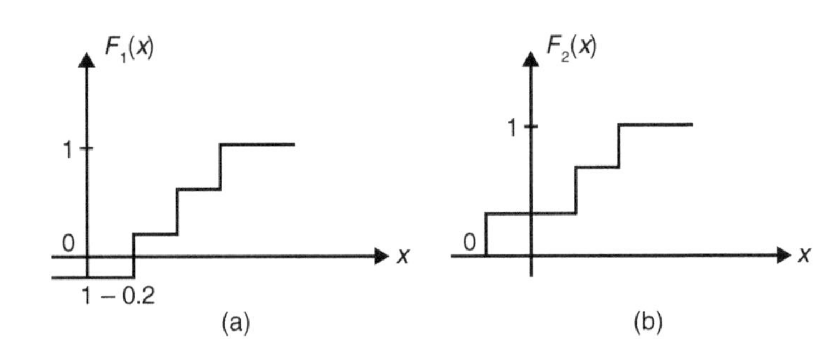

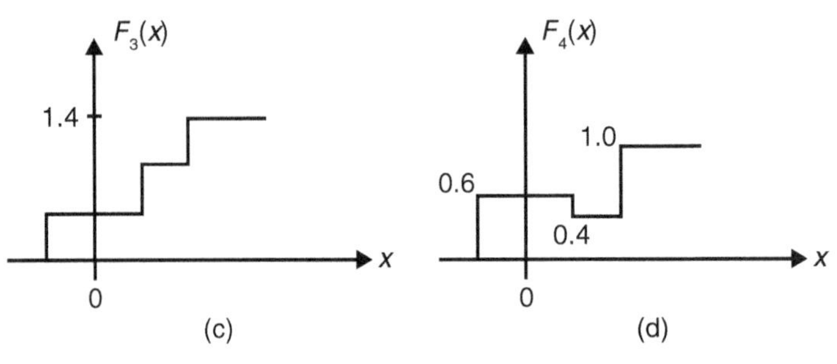

Exhibit 5.32

5.33 A coin is weighted so that heads is twice as likely to appear as tails. The probability that two heads occur in four tosses is:
a. 0.15
b. 0.35
c. 0.45
d. 0.75

5.34 It is given that 20% of all employees leave their jobs after one year. A company hired seven new employees. The probability that nobody will leave the company after one year is:
a. 0.1335
b. 0.2315
c. 0.3815
d. 0.6510

5.35 If four fair coins are tossed simultaneously, the probability that at least one head appears is:
a. 0.1335
b. 0.5635
c. 0.7815
d. 0.9375

5.36 For unit normal distribution, the probability that $(x > 3)$ is:
a. 0.0013
b. 0.0178
c. 0.1807
d. 0.5402

5.37 Scores in a particular game have a normal distribution with a mean of 30 and a standard deviation of 5. Contestants must score more than 26 to qualify for the finals. The probability of being disqualified in the qualifying round is:
a. 0.121
b. 0.212
c. 0.304
d. 0.540

5.38 The radial distance to the impact points for shells fired by a cannon is approximated by a normal Gaussian random variable with a mean of 2000 m and standard deviation of 40 m. When a target is located at 1980 m distance, the probability that shells will fall within ± 68 m of the target is:
a. 0.2341
b. 0.3248
c. 0.5847
d. 0.8710

5.39 The chance of a car being stolen from a residential area is 1 in 120. In one area there are five cars parked in front of the houses. The probability that none will be stolen is:

a. 0.0131
b. 0.3248
c. 0.5847
d. 0.9590

5.40 The standard deviation of the sequence 3, 4, 4, 5, 8, 8, 8, 10, 11, 15, 18, 20 is:

a. 5.36
b. 9.35
c. 15.62
d. 28.75

5.41 Weighted arithmetic mean of the following 50 data points is:

Frequency	3	8	18	12	9
Score	1.5	2.5	3.5	4.5	5.5

a. 1.56
b. 3.82
c. 5.62
d. 8.75

SOLUTIONS

5.1 c. From the Z table the value for a 90% two-sided confidence is Z = 1.64. We calculate the sample mean as $\bar{x}$ = 109.0 and the sample standard deviation as s = 30.0333. The required relative accuracy is stated in the problem as A = 0.10. We then substitute these values into the formula and solve for the result n = 20.41. Thus 21 observations are needed, and the number of additional observations needed is 21 – 10 = 11.

5.2 d. There were n = 6 observations of each of k = 5 treatments. Given SST = 7002 and SS(Tr) = 3821.5, we have SSE = 7002 – 3821.5 = 3180.5. F_0 = (3821.5/4)/ (3180.5/25) = 7.5096. For 4 degrees of freedom in the numerator and 25 in the denominator, the critical F is given as 2.76 in the F-distribution table in the IE chapter of the handbook, which has entries only for α = 0.05. If F_0 were only slightly greater than 2.76, the null hypothesis of no effect on weight gain would be rejected at about the 0.05 level; but since F_0 is much greater than would be necessary for significance at that level, the effect is more significant than at that level.

5.3 c. We find $\bar{x}$ = 194.8 and s = 13.14 for the data points. We calculate a t statistic $t = (194.8 - 200)/(13.14/\sqrt{10}) = -1.25$. Our degrees of freedom are n – 1 or 9. Using the t table in the mathematics section of the handbook, we find value of $t_{.01}$ = 2.821. Our rejection region is $t \leq -t_{.01}$. Since the observed t value (–1.25) is larger than –2.821, the null hypothesis is not rejected and there does not seem to be strong evidence that the mean chemical level is within safety margins.

5.4 c. This is a conditional probability problem. Let B be "draw an ace," and let A be "draw a second ace": $P\{B\}$ = 4/52 (1/13) and $P\{A\}$ = 3/51. Then $P\{A|B\} = P\{A\} \times P\{B\}/P\{B\}$ = 3/51.

5.5 d. This is a Bayes' theorem problem application because partitions are involved. The event E is a recall, with E_1 = Plant A, E_2 = Plant B, and E_3 = Plant C. The conditional probabilities of a recall from Plants E_1, E_2, and E_3 are

$$P(E \mid E_1) = 4/500 = 0.008$$
$$P(E \mid E_2) = 10/800 = 0.0125$$
$$P(E \mid E_3) = 10/1000 = 0.01$$

The probabilities that the dealer had a car from E_1, E_2, or E_3 are $P(E_1)$ = 0.3, $P(E_2)$ = 0.4, and $P(E_3)$ = 0.3. Now applying Bayes' formula gives the probability that the recall was built in Plant A (E_1) as

$$P\{E_1|\text{recall}\} = \frac{P\{E_1\} \times P\{E|E_1\}}{P\{E_1\} \times P\{E \mid E_1\} + P\{E_2\} \times P\{E|E_2\} + P\{E_3\} \times P\{E \mid E_3\}}$$

$$= \frac{0.3 \times 0.008}{0.3 \times 0.008 + 0.4 \times 0.0125 + 0.3 \times 0.01} = 0.2308$$

5.6 d. The problem involves binomial probability. The probability that one item, selected at random, is defective is

$$p_{\text{defective}} = \frac{10}{1000} = 0.01$$

and the probability that one item is good (not defective) is

$$p_{\text{good}} = 1 - p_{\text{defective}} = 0.99$$

The probability that exactly one defective item will be found in a random sample of 100 items is given by the binomial $b(1, 100, 0.01)$, in which

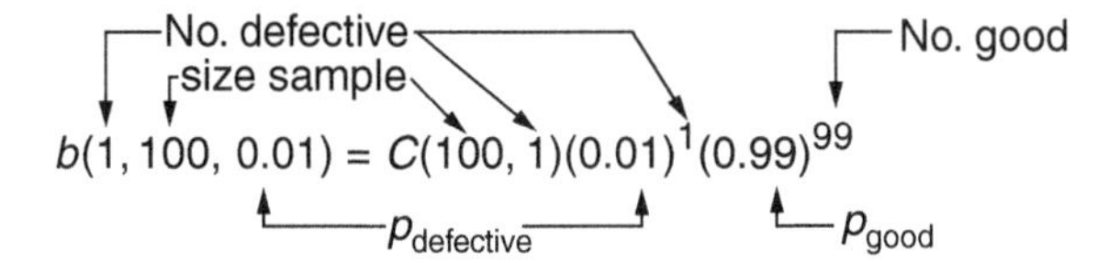

$C(n,r) = \binom{n}{r} = \frac{n!}{(n-r)!r!}$ is the number of combinations of n objects taken r at a time without concern for the order of arrangement.

$C(100, 1) = \frac{100!}{99!1!} = 100$, so $b(1, 100, 0.01) = 100(0.01)(0.99)^{99} = 0.99^{99} = 0.3697$.

5.7 c. Here, $P(S_1) = 0.98$ and $P(S_2 \cap S_1) = 0.95$ are given. Hence the conditional probability $P(S_2 \mid S_1)$ is

$$P(S_2|S_1) = \frac{P(S_2 \cap S_1)}{P(S_1)} = \frac{0.95}{0.98} = 0.97$$

5.8 d. The probability of drawing an ace on the first card is 4/52. The probability that the second card is an ace is 3/51. The probability that the third card is an ace is 2/50, and probability for the fourth ace is 1/49. The probability that the first four cards will all be aces is

$$P = \frac{4}{52} \bullet \frac{3}{51} \bullet \frac{2}{50} \bullet \frac{1}{49} = 0.00\,0037 = 3.7 \times 10^{-6}$$

5.9 d.

5.10 d. Since the first IC that is picked is a premium IC, it was drawn from either bin 1 or bin 3. From the distribution of premium ICs, the probability that the premium IC came from bin 1 is $\frac{2}{3}$, and from bin 3 is $\frac{1}{3}$.

In bin 1, the probability that the remaining IC is a premium IC is 1; in bin 3, the probability is 0. Thus, the probability that the remaining IC is a premium IC is

$$\frac{2}{3}(1) + \frac{1}{3}(0) = \frac{2}{3}$$

An alternative solution using Bayes' theorem for conditional probability is

$$P(\text{bin 1} \mid \text{drew premium}) = \frac{P(\text{bin 1 and premium})}{P(\text{premium})}$$

$$= \frac{P(\text{premium} \mid \text{bin 1}) \bullet P(\text{bin 1})}{\sum_{i=1}^{3} P(\text{premium} \mid \text{bin 1}) P(\text{bin 1})}$$

$$= \frac{1\left(\frac{1}{3}\right)}{1\left(\frac{1}{3}\right) + 0\left(\frac{1}{3}\right) + \frac{1}{2}\left(\frac{1}{3}\right)} = \frac{2}{3}$$

5.11 c. The answer is the binomial coefficient

$$\binom{35}{4} = \frac{35 \bullet 34 \bullet 33 \bullet 32}{4 \bullet 3 \bullet 2 \bullet 1} = 35 \bullet 34 \bullet 11 \bullet 4 = 52{,}360$$

5.12 d. The total number of choices of 5 is $\binom{50}{5}$. Of these, $\binom{40}{5}$ have no bad bolts, $\binom{40}{4} \times \binom{10}{2}$ have one bad bolt, and $\binom{40}{3}\binom{10}{2}$ have two bad bolts. Thus,

$$\frac{\binom{40}{5} + \binom{40}{4}\binom{10}{1} + \binom{40}{3}\binom{10}{2}}{\binom{50}{5}}$$

$$= \frac{\frac{40 \bullet 39 \bullet 38 \bullet 37 \bullet 36}{5 \bullet 4 \bullet 3 \bullet 2} + \frac{40 \bullet 39 \bullet 38 \bullet 37}{4 \bullet 3 \bullet 2} \bullet 10 + \frac{40 \bullet 39 \bullet 38}{3 \bullet 2} \bullet \frac{10 \bullet 9}{2}}{\frac{50 \bullet 49 \bullet 48 \bullet 47 \bullet 46}{5 \bullet 4 \bullet 3 \bullet 2}}$$

$$= \frac{658{,}008 + 913{,}900 + 444{,}600}{2{,}118{,}760} = 0.9517$$

5.13 c. There are 26 choices for the first letter; 25 remain for the second, and 24 for the third.

5.14 a. The probability that only A is defective is

$$0.1 \times (1 - 0.2) \times (1 - 0.25) = 0.06$$

The probability that only B is defective is

$$(1 - 0.1) \times (0.2) \times (1 - 0.25) = 0.135$$

The probability that only C is defective is

$$(1 - 0.1) \times (1 - 0.2) \times (0.25) = 0.18$$

Now add to find the final probability, which is

$$0.06 + 0.135 + 0.18 = 0.375$$

5.15 d. Simply multiply the complementary probabilities $(1 - 0.5) \times (1 - 0.4) \times (1 - 0.3) = 0.21$.

5.16 a. Apply the binomial distribution. The probability of 0 bad is $(0.9)^{50}$; of 1 bad, $\binom{50}{1}(0.1)(0.9)^{49}$; and of 2 bad, $\binom{50}{1}(0.1)^2(0.9)^{48}$. Adding these,

$$(0.9)^{48}[(0.9)^2 + 5.0(0.9) + 1225(0.1)^2] = 0.112.$$

5.17 b. This is the permutation of arranging 3 objects out of 6

$$P(6,3) = \frac{6!}{(6-3)!} = 20$$

5.18 d. This is the permutation of arranging 7 persons out of 7:

$$P(7,7) = \frac{7!}{(7-7)!} = 5040$$

5.19 c. This is the selection (or combination) of 6 out of 8:

$$C(8,6) = \frac{8!}{(8-6)6!} = 28$$

5.20 d. Since two questions are mandatory, only four questions have to be selected out of six.

Then, $C(6,4) = \dfrac{6!}{(6-4)!4!} = 15\cdot$

5.21 b. This is the selection (or combination) of 2 out of 5:

$$C(5,2) = \frac{5!}{3!2!} = 10$$

5.22 c. Since there are 11 integers that are exactly divisible by 9, probability = 11/100 = 0.11.

5.23 d. $P(1\text{ head}) = \dfrac{C(4,1)}{2^4} = \dfrac{4}{16}$; $\quad P(2\text{ heads}) = \dfrac{C(4,2)}{2^4} = \dfrac{6}{16}$

Since P(1 head AND 2 heads) is 0, P(1 head or 2 heads) = (4/16) + (6/16) = 5/8.

5.24 b. Let event A = (Apple from a bag) and event B = (Orange from the other bag).

Then, P(A) = 10/15 and P(B) = 5/15.

Since the events are independent, P(A AND B) = P(A) P(B) = (10/15) (5/15) = 2/9.

5.25 d. P(Black OR Red) = P(Black) + P(Red) – P(Black AND Red) = (5/85) + (30/85) – 0 = 35/85.

5.26 c. Let event A = (Orange from a bag) and event B = (White from the other bag).

P(A) = 2/10 = 0.2 and P(B) = 5/10 = 0.5

P(A OR B) = P(A) + P(B) – P(A AND B) = 0.2 + 0.5 – (0.2)(0.5) = 0.6

Note: P(A AND B) = P(A) P(B) as events A and B are independent.

5.27 a. Let event A = (number is even) and event B = (number > 72).

P(A) = 50/100 = 0.5 and P(B) = 28/100 = 0.28

Event (A and B) = (number is odd and > 72); P(A AND B) = 14/100 = 0.14

P(A OR B) = P(A) + P(B) – P(A AND B) = 0.50 + 0.28 – 0.14 = 0.64

5.28 c. P(both closed) = P(1 is closed) P(2 is closed)

0.90 = (0.95) P(2 is closed); then, P(2 is closed) = 0.90/0.95 = 0.9474

5.29 d. For a valid probability density function, $\int_{-\infty}^{\infty} f(x)\,dx = 1$, or the total area under the curve should be 1. For $f_1(x)$, the area is (1/2)(2)(2) = 2; for $f_2(x)$, it is 1(4 – 2) = 2; for $f_3(x)$, it is (1/2)(4 + 1) = 2.5. For $f_4(x)$, $\int_0^{\infty} e^{-x}\,dx = 1$.

So, $f_4(x)$ is the only valid distribution.

5.30 c. For a valid probability distribution function, both $f(\infty)$ and $f(-\infty)$ should be zero and $f(x) \geq 0$ for all x. Also, $\int_{-\infty}^{\infty} f(x)\,dx = 1$. Then, $f_1(x)$ and $f_4(x)$ are not valid since $f(x) < 0$ for certain values of x. $f_2(x)$ is not valid because $f(\infty)$ is not 0. $f_3(x)$ is the valid function as it satisfies all the conditions.

5.31 a. The distribution is discrete, but the rules are similar to those of Problem 5.29. $f_1(x)$ is the only valid distribution. $f_2(x)$ is not valid since the sum of the densities (equivalent to integrating) is more than 1. $f_3(x)$ has a negative value. $f_4(x)$ has a value more than 1.

5.32 b. For a valid cumulative probability distribution function $F(x)$, the following rules apply: $F(-\infty) = 0$, $F(\infty) = 1$, $0 \leq F(x) \leq 1$, and $F(x_1) \leq F(x_2)$ if $x_1 < x_2$. For $F_1(x)$, $F(x)$ has a negative value; for $F_3(x)$, $F(\infty)$ is more than 1; for $F_4(x)$, the rule $F(x_1) \leq F(x_2)$ if $x_1 < x_2$ fails. Only $F_2(x)$ obeys all the rules.

5.33 a. Let p = P(head on the first toss); then, P(tail on the first toss) = $1 - p$

But, $p = 2(1 - p)$; solving, $p = 0.667$.

P(two heads in four tosses) = C(4,2)$(0.667)^2(1 - 0.667)^2$ = 0.1481

5.34 **a.** P(leaving the job) = 0.25; P(none will leave the job) = $C(7,0)(0.25)^0 (1 - 0.25)^7 = 0.1335$

5.35 **d.** P(head) = $p = 0.5$; P(at least one head) = 1 – P(no head) = $1 - C(4,0)(0.5)^0(1 - 0.5)^4 = 0.9375$

5.36 **a.** Using the normal distribution table, P(X > 3) = 1 – F(3) = 0.0013.

5.37 **b.** $P\{X \le 26\} = F(26) = F\left(\frac{26-30}{5}\right) F(-0.8) = 1 - F(0.8) = 1 - 0.7881 = 0.2119$

5.38 **d.** $P\{1980 - 68 < x \le 1980 + 68\} = F(2048) - F(1912)$

$$= F\left(\frac{2048-2000}{40}\right) - F\left(\frac{1912-2000}{40}\right) = F(1.20) - F(-2.2)$$

$$= 0.8849 - \{1 - 0.9861) = 0.8710$$

5.39 **d.** $C(5,0)(1/120)^0(119/120)^5 = 0.9590$

5.40 **a.** $\text{mean} = \overline{X} = \frac{\sum x}{n} = \frac{114}{12} = 9.5$

$$\text{variance, } \sigma^2 = (1/12)\left[(3-9.5)^2 + (4-9.5)^2 + \ldots\right] = 28.75$$

standard deviation $\sigma = 5.36$

5.41 **b.** $\frac{3(1.5) + 8(2.5) + \ldots + 9(5.5)}{3 + 8 + 18 + \ldots + 9} = 3.82$

CHAPTER 6

Modeling and Computation

OUTLINE

Expect 8-12 questions on algorithm and logic development, databases, decision theory, optimization modeling, linear programming, mathematical programming, stochastic models, and simulation on the Industrial Engineering FE exam.

LOGIC FLOW DIAGRAMS AND PSEUDOCODE

When FE exams mention flow charts, they usually mean logic flow diagrams, not the process flow diagrams familiar to industrial engineers through various simulation languages such as SIMAN. A process flow diagram depicts the flow of transactions (customers or jobs) through a set of facilities. Instead of programming a computer to simulate a system, the process flow diagram describes the system being simulated. It is non-procedural programming. Logic flow diagrams and pseudocode, by contrast, are ways of documenting step-by-step procedures. The entity that "flows" in a logic flow diagram is the "attention" or focus of the logic engine. A logic flow diagram and a set of lines of pseudocode are two ways of specifying what the computer should do and where it should go next, where it will again be told what to do.

To read and understand a modern logic flow diagram, you need only understand six standardized symbols and the connections among them. Examine Figure 6.1.

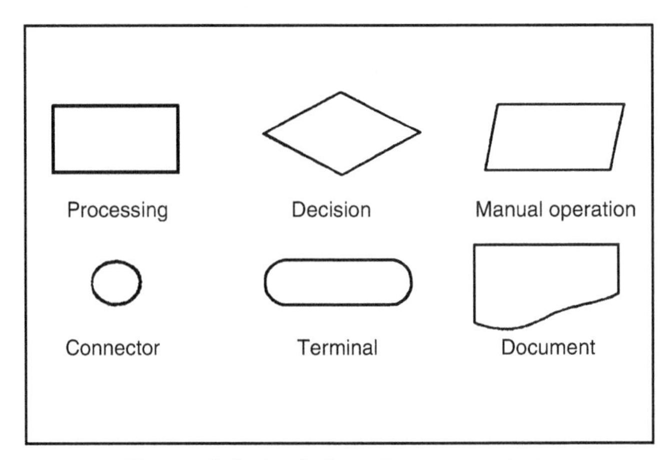

Figure 6.1 Logic flow diagram symbols

The processing block, a rectangle, contains an instruction—simple, or complex enough to be expanded into a lower-level logic flow diagram elsewhere, but in any case, logic flow that enters the block cannot temporarily exit and return. The manual operation block is a special processing block for procedures in which the computer system does not take part (e.g., "Verify that the picture on the badge matches the person"). The decision diamond contains a question, usually binary, and there is a different exit for each of the possible answers to the question. The connector allows logic flow to skip clutter; if the logic goes into the connector labeled *A* and elsewhere comes out of a connector also labeled *A*, it is the same as if a logic flow line were drawn explicitly between the two places. The terminal symbol does not represent a computer terminal; rather, it is a symbol labeled (e.g., START or STOP) that indicates the beginning or end of logic flow. Finally, the document symbol indicates production of a report.

Logic flow is generally downward. Figure 6.2 shows the basic components of logic flow.

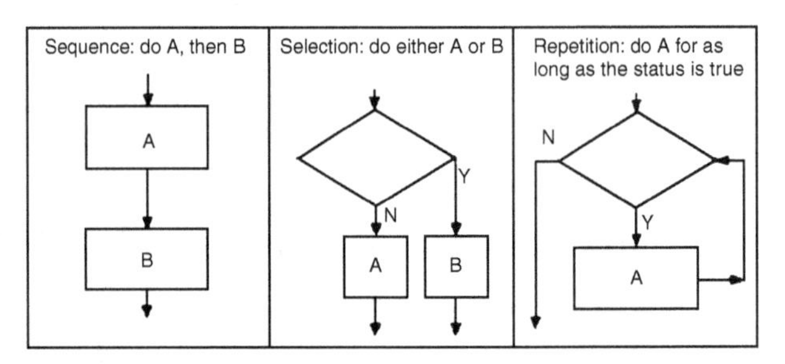

Figure 6.2 Basic components of logic flow

In the left-most section of Figure 6.2, we see an example of processing blocks. In this case, the activity in block B will follow the activity in block A every time. In the center we see the application of the decision diamond. If the answer to decision is no, then the activity in block A will be performed. If it is yes, then the activity in block B will be performed. The configuration to the right represents a continuous loop where A is repeatedly performed until the answer to the decision question is no. On the FE exam, you may be asked to interpret a logic flow diagram, or a problem statement may use one to tell you the logic of a procedure. Clean logic flow is the aim for both logic flow diagrams and pseudocode. Every procedure can be depicted both ways, but a diagram sometimes makes it easier to detect logic flows.

Figure 6.3 illustrates pseudocode and logic flow diagramming, and also, along with Figure 6.4, detection of logic flaws. Consider the automated preparation of a standard solution of a hard-to-dissolve salt, starting after the salt has been added. The procedure is first to alternately add water and stir until there is the amount called for; then add heat, stir, add water if necessary (if too much water has evaporated); and repeat until there is complete dissolution.

Now, what is wrong with the documentation of the procedure? One clue is the "go-to-ish" appearance of the diagram, with logic entry into the STIR box in the middle of a loop of adding water and checking if more water is needed. You can't isolate parts of the diagram such that the logic goes in only once to a part and comes out only once from the part.

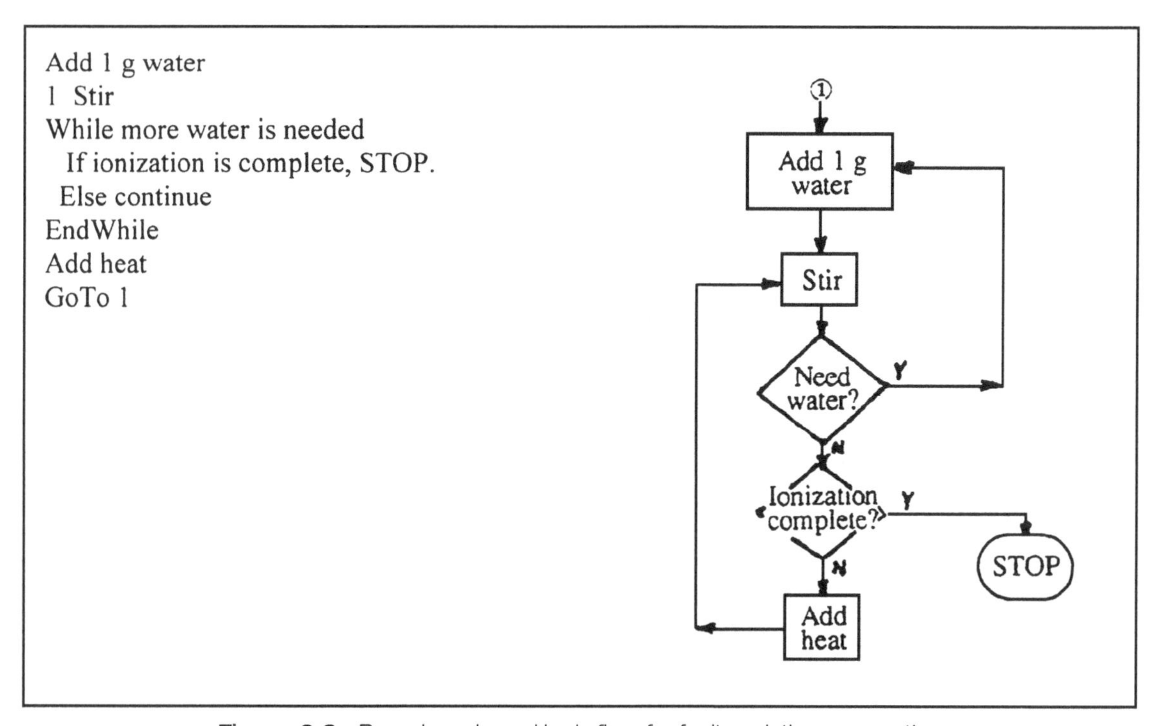

Figure 6.3 Pseudocode and logic flow for faulty solution preparation

The root cause of the difficulty is that there are really two phases of the procedure—adding the initial charge of water and heating to dissolve. The heating-to-dissolve phase has a subprocedure whereby if too much water has evaporated, some more is added, but it is a mistake to make a single "Add 1 g water" box and a single "More water needed?" diamond serve what are really two different purposes. Figure 6.4 shows the fix. Note that the two phases are separated; logic goes into the first phase, out of it, and into the second phase, cleanly.

Industrial engineers should be able to detect when a procedure has poorly organized logic and when a procedure does not include all contingencies. For example, if a clerk takes orders by telephone only from customers whose data are on file, then the order-taking procedure must include a subprocedure to collect customer data, and that subprocedure must include a return to order taking. Further, the procedure should be organized into a customer verification phase and an order-taking phase rather than beginning with an order, then finding out that the customer's data is not on file, etc.

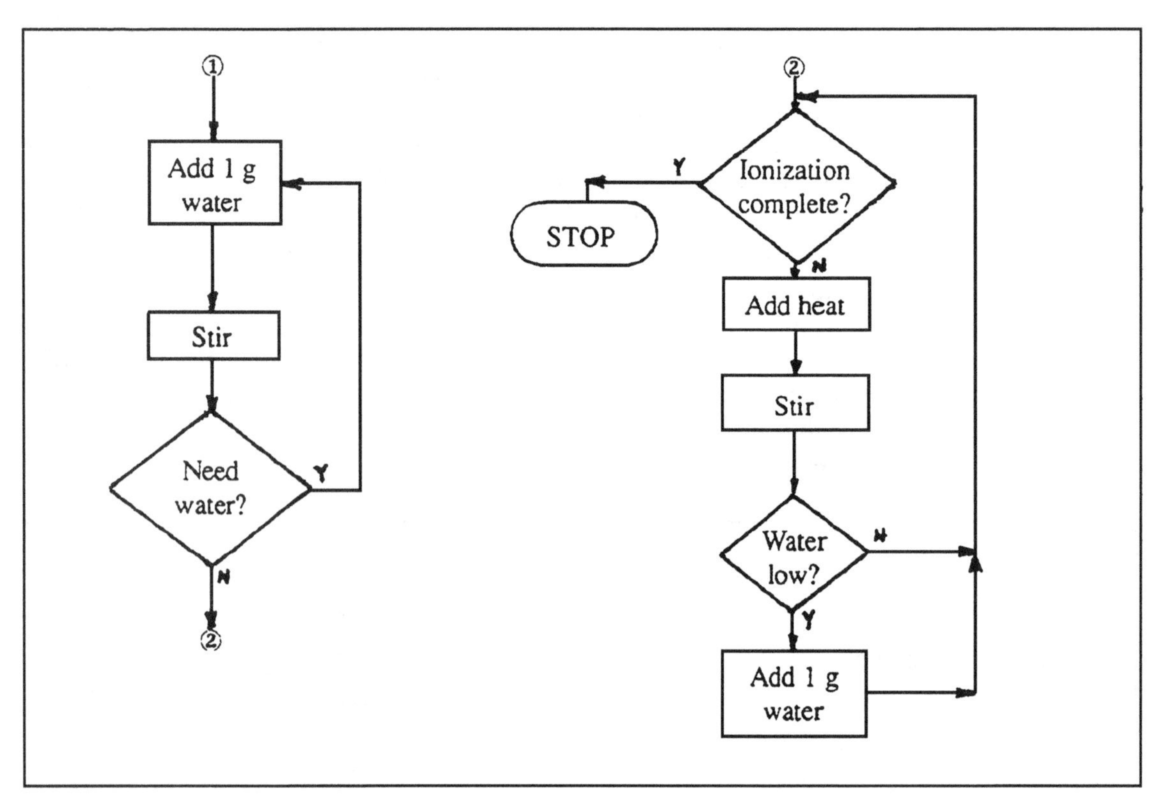

Figure 6.4 Improved logic flow for automated solution preparation

DATABASES

A relation is a table, structured like a flat file. It has columns (like fields), each with a fixed column heading, data type (character, number, date) and size. It has a variable number of rows (like records). A relational database is a collection of relations (tables). It follows rules that maximize data integrity, the first one being that no fact is kept redundantly. For example, in a genealogical database, it would be out of the question to store data explicitly saying that one member was another's first cousin because that fact follows directly from data relating parents and children. Among a set of facts some of which imply others, an attempt is made to store the more basic ones. For example, in an organization chart, we store which jobs report to which jobs, not which people report to which people.

Every table has a key, which is one or more attributes such that there can be only one row having a given value of the key. In the following restaurant database consisting of three tables, the keys of the tables are the underlined attributes:

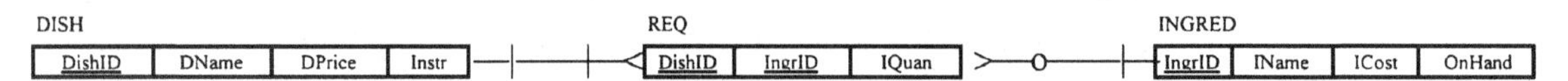

Figure 6.5 Tables and keys for restaurant database

The DISH table has a row for each dish, identified by the value of its DishID, for example, "HE" for a hamburger with everything. Every dish has a name (the value of DName in the dish's row) for human convenience since the dish identifier may not be very descriptive; DName is a candidate key. Dprice, the price charged for the dish, is in this table because it is functionally dependent on this table's key: We can write the functional dependency as DishID → DPrice; the meaning is that if a value of DishID is known, the database must contain a unique value of DPrice. All attributes kept in a table must be functionally dependent on its key.

The database has a table for ingredients, and a table for the many-to-many relationships between dishes and ingredients that can be called requirements. For example, if dish "HE" requires two units of the ingredient "TO" (two slices of tomato), then there is this row in the REQ table: {HE, TO, 2}. Since the IQuan value depends on both dish and ingredient, the REQ table was designed to provide a place to store it. Note the double key {DishID, IngrID}.

The links between tables in the relational diagram show the minimum and maximum cardinalities of the relationships between tables. Minimum cardinalities are indicated by an oval for *zero* or a hash mark for *one*, and maximum cardinalities are indicated by lack of a crow's foot for *one* or a crow's foot for *many*. Going from left to right, we read the following facts from the links: There is exactly one row of DISH for each row of REQ; there are from one to many rows of REQ for each row of DISH; there are from zero to many rows of REQ for each row of INGRED; and there is exactly one row of INGRED for each row of REQ. (Interpreted, these facts are that each requirement refers to one dish, each dish has at least one requirement, an ingredient can be in many requirements but might not be in any, and each requirement refers to one ingredient.)

Null values are not outlawed (e.g., a table of purchases can have a delivery-date column, where a null value would mean that the purchase had not yet been delivered). Existence or non-existence of rows is meaningful (e.g., if eggs ("EG") are not required in hamburgers with everything, there would exist no row {HE, EG, 0} in *REQ*).

Normalization rules help ensure good relational design. A table is said to be in first normal form (1NF) if it is indeed a relation having a key and doesn't have more than one fact in a column. For example, if dishes could have more than one name, we couldn't simply put all of them in the DName column. Also, we wouldn't provide enough different columns to accommodate the maximum number of names. We would do this:

DISH

DishID	DPrice	Instr	
HE	2.40	On a hot griddle, ...	
WO	3.75	Chop the onions...	

DISH_NAME

DishID	Dname
HE	Hamb w/Every
HE	Mayo Wonder
WO	Western Omelet

If an address field in a table includes ZIP code, then the table fails 1NF unless the applications will never query ZIP codes or use them in any way. A catalog number that has facts crammed into it violates 1NF unless the facts are elsewhere and the database never gets them from the catalog number:

PRODUCT

CatNo	Descrip	Type	Mfr	...
17Ref4197-31	Refrigerator, Maytag, 31 cu ft, white, LH	Refrigerator	Maytag	

We can't eliminate the Type or Mfr columns and try to get the information from parsing the catalog numbers or descriptions without violating 1NF. The apparent redundancy, (e.g., that there are three ways to tell that the row concerns a refrigerator), is not really a redundancy since the database management system (DBMS) will not parse CatNo or Descrip. There's nothing wrong with descriptive, human-interpretable catalog numbers, so long as the DBMS doesn't have to deal with them.

Second normal form (2NF) exists if a table is in 1NF and its attributes depend on all of, not just part of, the key. For example, in the restaurant database, we would never put the ingredient name along with the ingredient identifier in the requirement table: REQ(DishID, IngrID, ~~IName~~, IQuan).

Third normal form (3NF) is freedom from transitive dependencies. If every customer chooses a haircut style and every haircut style has a price, we don't keep a table (CustID, Style, Price); instead, we keep a table (CustID, Style) and another table (Style, Price). Boyce-Codd normal form (BCNF), whereby every determinant in a table is a candidate key, is a more general version of 3NF. (If there is a functional dependency $A \rightarrow B$, then A is a determinant.).

The highest normal form that an industrial engineer needs to deal with is fourth normal form (4NF), which requires that a table be free of multivalued dependencies (MVDs). MVDs exist when, given the value of attribute X, there is a set of values of Y that the database must yield, the set of values being independent of the values of any other attributes. For example, if delivery route 1 must be run on Tuesdays and Saturdays and can use either vehicle 1 or vehicle 2, and other delivery routes have similar data, we can't store the data as on the left below, where the table is not even a relational one, but if we fix it by putting in all the combinations, you can see the redundancy that constitutes 4NF failure:

DELIV1 (1NF failure)

Route	Weekday	Vehicle
1	Tuesday	1
1	Saturday	2

Not relational; interpreted relationally, this would say that route 1 uses vehicle 1 on Tuesday and vehicle 2 on Saturday.

DELIV2 (4NF failure)

Route	Weekday	Vehicle
1	Tuesday	1
1	Tuesday	2
1	Saturday	1
1	Saturday	2

The symptom that identifies 4NF failure is the listing of combinations whose components are redundant. The fix is to list the independent sets separately:

ROUTE_WEEKDAY

Route	Weekday
1	Tuesday
1	Saturday

ROUTE_VEHICLE

Route	Vehicle
1	1
1	2

Some relational design tips: Don't represent data as structure (e.g., there shouldn't be columns for breakfast, lunch, and dinner if there can be a meal-type column so that the data, not the column structure, tells which meal type it is).

Represent many-to-one relationship by putting the identifier of the "parent" in the table for the "child" (e.g., list the dorm in the student table, not the students in the dorm table). Unary relationships are a special case of this; for instance, in the member table for a genealogy, let there be a mother column and a father column (not columns for children).

DECISION THEORY

We are often required to make decisions under uncertainty—which job offer to take, where to invest our money, or which research project to fund. Decision makers select between alternatives, or options, when making a decision. The outcome of the decision is affected by factors outside the control of the decision maker. Each possible situation is referred to as a possible state of nature. An example would be that the demand for our new product could be high, medium, or low. The decision maker will generally have some information about the likelihood of the possible states of nature, this is called prior probabilities. The quantitative measure of the outcome is called the payoff. If our decision is whether or not to drill for oil, the states of nature are hitting oil or coming up dry. The respective payoffs might be making $1 million or losing $100,000. A decision maker can use various criteria to make a decision, potentially resulting in different decisions. Maximax is the decision criterion of the eternal optimist; it selects the alternative with the maximum possible payoff. Maximin is the decision criterion of the total pessimist; it identifies the minimum payoff from the various states of nature for each alternative. It then selects the maximum of these minimum payoffs. Laplace is a decision criterion that considers the average. It averages the payoffs for all of the states of nature for each alternative and then selects the one with the highest average (assuming we are seeking to maximize the payoff such as profits gain instead of avoid losses). Bayes decision rule is similar except instead of a simple average, it weights the states of nature with the associated probabilities. Our final decision criterion is maximum likelihood, which identifies the state of nature with the largest prior probability and chooses the decision alternative that has the largest payoff for that state of nature.

Example 6.1

Consider the following four alternatives and the payoffs under three states of nature. Using the different decision criteria, select the associated decision.

Alternatives	Great Economy (30%)	Good Economy (50%)	Bad Economy (20%)
Get MSIE	$110,000	$90,000	$50,000
Get MBA	$90,000	$60,000	0
Take job offer	$80,000	$80,000	$80,000
Start business	$150,000	$75,000	–$20,000

Solution

Maximax—The maximum payoffs for the four alternatives are 110K, 90K, 80K, **150K**. Select the maximum of these. Start a business.

Maximin—The minimum payoffs for the four alternatives are 50K, 0K, **80K**, –20K. Select the maximum of these. Take the job offer.

Laplace—The average for the four alternatives are **83.3K**, 50K, 80K, 68.3K. Select the maximum of these. Get your MS in industrial engineering.

Bayes—The weighted average for the first alternative is 0.3 × 110K + 0.5 × 90K + 0.2 × 50K = 93K. The values for all four alternatives are **93K**, 57K, 88K, 76.5K. Select the maximum of these. Get your MS in industrial engineering.

Maximum likelihood—The state of nature with the highest probability is "good economy" at 50%; the four alternatives' payoffs are **90K**, 60K, 80K, 75K. Select the maximum of these. Get your MS in industrial engineering.

As you can see different decision criteria can result in different decisions.

DECISION TREES

A decision tree is a graphical representation of the decision process. They consist of *decision nodes*, represented by a square, indicating a decision is to be made. The *branches* leaving the node represent the possible decisions. *Event nodes* are random events shown by circles. The branches leaving the nodes represent possible outcomes of the random event. Imagine you are considering entering your prize chili recipe in the local chili cook-off. If you win you get $1,000. It costs $25 to enter and the chance of winning is 10%. The decision tree for the decision is shown in Figure 6.6.

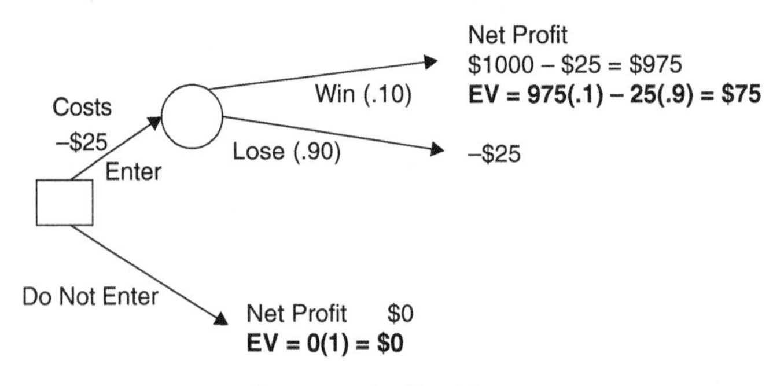

Figure 6.6 Decision tree

The expected values for the cook-off decision alternatives are shown as EV. For entering the contest, it is the profit from winning ($1000–$25) times the probability (10%) plus the cost of entering and losing (–$25) times the probability (90%) of this happening. So if you enter the contest, the expected value is $75, which is better than the EV of not entering, $0.

OPTIMIZATION MODELING

There is a general procedure for any optimization problem.

1. Isolate the decision variables (values that can be varied to generate potential solutions) and adopt symbols for them.
2. Express the objective function (the value you try to maximize or minimize) in terms of the decision variables.
3. Express constraints (conditions that must be satisfied) as functions of the decision variables.

4. Examine the formulation to determine whether it fits a common model, such as linear programming or one of the network models.
5. Simplify the problem (for example, by using equality constraints to eliminate variables, or eliminating redundant constraints).
6. Solve the problem to find values of the decision variables that optimize the object function and meet the constraints.
7. Interpret the result. If the method yields the values of decision variables but not the value of the objective, it is nearly always beneficial to determine the objective value. Verify the results back to the context of the problem statement (e.g., $x_6 = -1.5$ may look good at first glance, until you realize that it says to build a negative fractional number of airplanes at plant 6). This would lead you to realize that you forgot the non-negativity constraint and require you to rework the problem.

For a multiple-choice exam, before simplifying and solving, you should check to see whether you can avoid calculating a solution and simply test the listed responses for optimality and feasibility. Perhaps you can easily determine whether a given response obeys the constraints and gives a larger value of the objective (or smaller, if you are minimizing) than other responses that also obey the constraints.

LINEAR PROGRAMMING

Linear programming questions on the FE Exam can include the following (and probably many more):

- Formulation—Given a problem statement (a word problem), recognize a correct or incorrect objective function or constraint, or identify the incorrect element in a given formulation
- Solution literacy—Given output from computer software such as EXCEL Solver recognize slack variables, basic variables, objective-function coefficients, solution values, feasibility, and so forth
- Graphical solution—Given a linear program with only two decision variables, answer questions that require you to perform a graphical solution to determine the answer or perform a sensitivity analysis.

The *FE Supplied-Reference Handbook* gives a general linear programming formulation and a few remarks about slack and surplus variables (variables added to an inequality to transform it into an equality).

To discuss linear programming, let us consider the following example. *Scenario LP-1*: A nutritionist at a grade school cafeteria has decided to make candied broccoli to fill out the last 50-g portion of a meal that needs at least 30 but not more than 100 additional calories (kcal) and at least 20 mg of additional vitamin C. Per gram, the candy has 3.6 kcal and 0 mg of vitamin C, and the broccoli has 0.44 kcal and 1.2 mg of vitamin C. Customer satisfaction is proportional to the ratio of candy to broccoli. Determine the optimal number of grams of each ingredient.

Formulation

Let x_1 and x_2 represent the number of grams of candy and broccoli, respectively. The objective as stated is to maximize x_1/x_2. The stated need for a 50-g portion gives the constraint $x_1 + x_2 = 50$, $x_2 \geq 50$, or $x_2 \leq 50$, depending on whether the statement is interpreted as calling for a portion that weighs exactly 50 g, at least

50 g, or not more than 50 g; we will explore all three interpretations. The number of calories (kcal) in the portion is $3.6x_1 + 0.44x_2$, and the lower and upper limits for calories give the constraints $3.6x_1 + 0.44x_2 \geq 30$ and $3.6x_1 + 0.44x_2 \leq 100$. The number of milligrams of vitamin C in the portion is $1.2x_2$, and the vitamin C requirement gives the constraint $1.2x_2 \geq 20$. Finally, the variables are nonnegative, $x_1 \geq 0$ and $x_2 \geq 0$.

Recall the general principle of optimization that if two objective functions are strictly increasing with each other, then to maximize one is to maximize the other. Thus we will not let the objective function x_1/x_2 throw us out of the realm of linear programming. Since x_2 and x_1 strictly decrease with each other through the constraint $x_1 + x_2 = 50$, it follows that x_1 strictly increase with x_1/x_2, so we can define $z = x_1$ as the objective function.

Interpreting the statement as calling for a portion that weighs no more than 50 g, the formulation is

$$\text{Maximize } z = x_1 \quad [0]$$

$$\text{subject to } x_1 + x_2 \leq 50 \quad [1]$$

$$3.6\,x_1 + 0.44\,x_2 \geq 30 \quad [2]$$

$$3.6\,x_1 + 0.44\,x_2 \leq 100 \quad [3]$$

$$1.2x_2 \geq 20 \quad [4]$$

$$x_j \geq 0\ \forall j$$

Dual

Every linear programming model has two forms, the primal and the dual. The primal is the original form of the problem, for example, maximizing profit. The dual is an alternative form of the model derived from the primal. In our example, the dual would provide information such as shadow prices for the resources and profit coefficients. The handbook provides notation on transforming a primal form into a dual form.

Graphical Solution

Recall scenario LP-1 and its formulation. *Scenario LP-1*: A nutritionist at a grade school cafeteria has decided to make candied broccoli to fill out the last 50-g portion of a meal that needs at least 30 but not more than 100 additional calories (kcal) and at least 20 mg of additional vitamin C. Per gram, the candy has 3.6 kcal and 0 mg of vitamin C, and the broccoli has 0.44 kcal and 1.2 mg of vitamin C. Customer satisfaction is proportional to the ratio of candy to broccoli. Determine the optimal number of grams of each ingredient. ("50-g portion" interpreted as up to 50 g.)

$$\text{Maximize } z = x_1 \quad [0]$$

$$\text{subject to } x_1 + x_2 \leq 50 \quad [1]$$

$$3.6\,x_1 + 0.44\,x_2 \geq 30 \quad [2]$$

$$3.6\,x_1 + 0.44\,x_2 \leq 100 \quad [3]$$

$$1.2x_2 \geq 20 \quad [4]$$

$$x_j \geq 0\ \forall j$$

Since there are only two decision variables, the problem can be solved graphically. A solution sketch such as you might make in the white space of your exam booklet is shown in Figure 7.9. The decision variables x_1 and x_2 are plotted as abscissa and ordinate, respectively. The boundary for each constraint is plotted by plotting points that exactly meet it. For example, constraint [2] is exactly met, or tight, at any point on the line $3.6x_1 + 0.44x_2 = 30$, and points can be plotted by assuming a value for one variable (often zero is convenient) and solving for the other. This constraint is shown graphically by the line that passes through points *A* and *B*. Since the constraint is an inequality, its infeasible side is identified, in this case the area to the left of the line labeled "too few calories." Hash marks are typically put on the feasible side, in our case to the right of the line. If the constraint is an equality equation, the feasible region is restricted to points on its line. Once all of the constraints are plotted, we have the feasible region with boundary points labeled *A*, *B*, *C*, and *D*. If the region is not closed, it is called unbounded. If all of the constraints cannot be met at the same time it is an infeasible problem. This becomes readily apparent when you plot the constraints graphically.

An iso-objective line can be plotted by assuming a value *C* for the objective function and plotting the line $z = C$. All iso-objective lines are parallel. In Figure 6.7, the iso-objective line is shown as a dashed line.

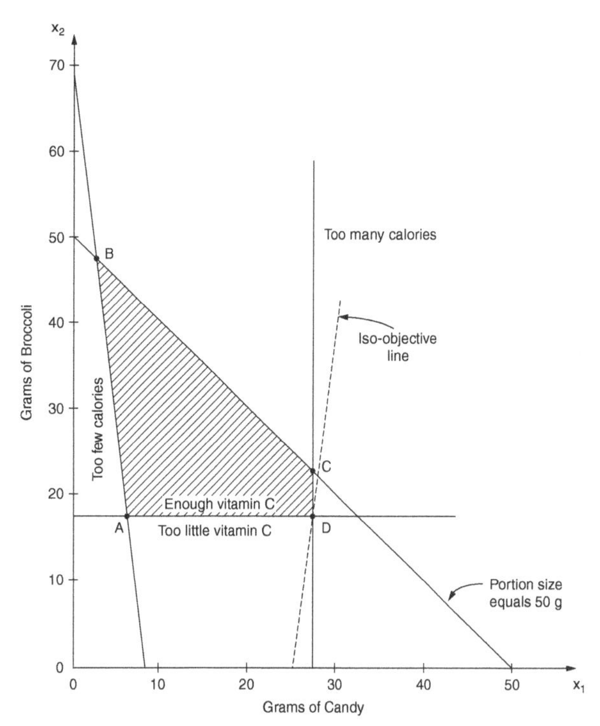

Figure 6.7 Graphical solution

The figure is drawn with the "at most 50 g" interpretation of the first constraint. For the "exactly 50 g" interpretation, the constraint is the line passing through points *B* and *C* and the optimal solution must be on the line. For the "at least 50 g" interpretation, the constraint would be above the line.

In this problem, the iso-objective lines ($x_1 = C$) are vertical. In general, to identify the optimal solution, you can draw an iso-objective line, identify on which side of it the objective function improves, and then slide it in the normal (perpendicular) improving direction until with further movement it would no longer touch the feasible region. Here the improving direction is to the right and up, since we are maximizing. For a minimization problem, the improving direction would be down and to the left. The optimal solution can be accurately determined by noting which lines intersect at the optimal solution and solving algebraically for the values of x_1 and x_2 that satisfy their equations. For the "at most 50 g" interpretation shown in Figure 6.8, the optimal solution is 25.74 g of candy and 16.67 g of broccoli (point *D*), and the objective function value is 25.74. The weight constraint is slack at that point; the weight of the portion is 25.74 + 16.67 = 42.41 g. Under the other interpretations (that the weight is 50 g or ≥ 50 g), the optimal solution is 24.68 g of candy and 25.32 g of broccoli, point *C*, with an objective function value of 24.68.

NETWORK MODELS

Many network optimization models related to linear programming should be in an industrial engineer's repertoire. Some, such as the transportation model, the transshipment model, and the general network flow model, have manual solution procedures that are too tedious to test on the exam. Others are covered in the following sections.

Assignment Algorithm

An assignment problem is the problem of forming pairs that match entities of one class with entities of another at minimum total cost, where the individual pair costs $\{c_{ij}\}$ for pairing entity *i* from the first class with entity *j* from the second class are given. The assignment problem is useful for personnel assignments (e.g., forming project teams in a matrix-structured organization at a construction firm), for logistics problems (e.g., assigning trucks to convoys), and for location problems, such as the following

Scenario AS-1: Suppose that five machines (*M*1–*M*5) are to be located at five stations (*A*–*E*) with one machine per location. The machines process different materials. The estimated weekly material handling costs when machine *i* is located at station *j* are as follows:

i \ *j*	*A*	*B*	*C*	*D*	*E*
*M*1	9	7	8	0	9
*M*2	6	1	3	5	5
*M*3	8	6	7	1	5
*M*4	9	5	7	6	8
*M*5	7	5	2	4	9

Figure 6.8 Material handling cost matrix

To minimize the sum of weekly material handling costs for all the machines, which location should be assigned to each machine?

The Hungarian assignment algorithm can be performed by hand to assign items to various locations at minimal costs. To apply the Hungarian assignment algorithm, we first find relative costs. To do this, we first diminish each row by its smallest element and then diminish each column by its smallest element. Then, if a full set of assignments at zero relative costs can be made, it is optimal. Otherwise, we proceed further with the same steps. We could convert the matrix in Figure 7.10 to the matrix in Figure 6.9. Since the *M*1 row already has a zero, nothing is done to this row. The smallest value in the *M*2 row is 1, so we subtract 1 from every value in the row and get the resulting row in the Figure 6.9 matrix. We continue in the same fashion for the other three rows and get the matrix on the left. We then repeat the process with the columns and subtract 4 from the first column and 3 from the last column. Since the other columns already have zero values, we don't do any transformation.

Diminishing each row by its smallest element:

9	7	8	0	9
5	0	2	4	4
7	5	6	0	4
4	0	2	1	3
5	3	0	2	7

→

Diminishing each column:

5	7	8	0	6
1	0	2	4	1
3	5	6	0	1
0	0	2	1	0
1	3	0	2	4

Figure 6.9 Relative cost matrix

Based on the zero values in the matrix, we attempt to assign each machine to a unique location. This results in *M*1 at *D*; *M*2 at *B*; *M*3 at *D*; *M*4 at *A*, *B*, or *E*; and *M*5 at *C*. We cannot have two machines at *D*, so we are unable to make a complete set of assignments all at zero relative cost. So we cover all the zeros with the least number of lines (illustrated in Figure 6.9 to the right). We then determine the minimum uncovered value, in our case, 1, and diminish the uncovered elements by this minimum. At the same time we increase the doubly covered elements by the same amount, so the 4 value at *M*2 assigned to *D* is increased to 5. This results in the values shown in Figure 6.10. We again attempt to make a complete set of assignments at zero relative cost. Now a complete set of assignments at zero relative cost is possible, as shown by the boxed zeros in Figure 6.10.

4	6	7	0	5
1	0	2	5	1
2	4	5	0	0
0	0	2	2	0
1	3	0	3	4

Figure 6.10 Optimal machine assignment

The optimal assignments are shown on the original assignment cost matrix in Figure 6.11.

i \ *j*	*A*	*B*	*C*	*D*	*E*
M1	9	7	8	0	9
M2	6	1	3	5	5
M3	8	6	7	1	5
M4	9	5	7	6	8
M5	7	5	2	4	9

Figure 6.11 Optimal machine assignment and associated costs

The total of estimated weekly material handling costs is the sum of the shaded costs, which is 17 for this set of assignments. Note that machines *M*3 and *M*4 are not assigned to their respective minimum-cost stations as you might have initially expected. If we had sequentially assigned the machine without considering the relative costs, our nonoptimal solution would have been *M*1 at *D*, *M*2 at *B*, *M*3 at *E*, *M*4 at *C*, and *M*5 at 7 with a cost of 20.

The assignment model can be used when the numbers of entities of the two classes are not equal. For example, if there were a sixth machine, we would have six rows, all containing material handling costs. We would add a dummy column containing zero costs and solve as usual. The machine assigned to the dummy location would be the machine that it is best not to assign. Conversely, if there were more locations than machines, dummy rows would be added with zero elements, and the solution would identify the locations where it is best not to put a machine.

Shortest and Longest Route Algorithms

Given a set of locations and the travel times or distances between locations that are directly linked, the shortest-route problem is finding a sequence of locations (a route) such that the total travel time or distance is minimized between a specified source location S and a specified sink (also called a final destination) location T. A shortest route is the quickest, shortest, or least costly way to get from S to T.

Shortest-route problems can be displayed on a directed graph whose nodes represent locations and whose arcs represent direct links or transitions. Associated with each arc, say the arc from node i to node j, is an arc cost c_{ij}.

There is a primal linear program that can be defined for a shortest-route problem. Its variables—the primal variables—represent the flow, from 0 to 1, of a test element that goes from S to T. Actually the flow is either 0 or 1 because it turns out that the problem structure automatically rules out noninteger flows. There is a primal variable for each arc, and an arc is said to be basic if its flow is 1; the flow in nonbasic arcs is zero. The primal constraints provide that the sum of the flows on all arcs leaving a node equals the sum of the flows on all arcs entering the node; the flow entering node S and the flow leaving node T is set to 1.

The dual of this program, which is particularly easy to solve using the dual algorithm, has a variable y_i for each node i, and the variable represents the total travel time (or distance or transition cost) from node i to the sink node T. The dual algorithm happens to solve for the shortest route from every node to T, regardless of which node is designated as the source node S.

The dual algorithm for the shortest-route problems is the following procedure:

1. Set $y_T = 0$. Set $y_i = \infty$ for all other nodes $i \neq T$.
2. In some order (most efficiently, from sink generally back to source), replace each existing value y_i with a new value $y_i = \min\ \{c_{ij} + y_j\}$. That is, for each arc that leaves node i, compute the sum of its arc cost and the existing dual variable value for the destination node of the arc; set y_i equal to the smallest such sum.
3. Repeat step 2 until a pass occurs in which no y_i values are changed.
4. Recover the primal solution as follows: Mark a basic arc from (leaving) each node except T. The basic arc is the one (or one of the ones, in case of ties) for which $y_i = c_{jj} + y_j$. (For nonbasic arcs that are not tied with the basic arc, the sum $c_{ij} + y_j$ will exceed y_i). The shortest route from S (or from any node) is the unique path starting at S (or other node) and proceeding entirely through basic arcs to T.

A longest-path problem is similar to a shortest-route problem and can be solved with the analogous dual algorithm that maximizes rather than minimizes.

The forward-pass, backward-pass procedure used in project scheduling is a variation of the longest-path dual algorithm. A longest-path network must be free of cycles. Shortest-route and longest-path problems can occur not only when nodes represent locations but when they represent events or states of a system.

Max Flow Algorithm

The max flow algorithm, also known as the max-flow, min-cut algorithm, solves the problem of determining the maximum flow from node S to node T in a capacitated network. The data of a max flow problem is the network structure (a set of nodes and arcs, with S and T designated) and a capacity of each arc. The max-flow, min-cut algorithm has two parts.

In the max-flow part of the algorithm, you seek to show that a flow of at least H units can pass from S to T. Try the most capacious path you can find. Assert a trial flow equal to the minimum of the capacities along the path. Then subtract the trial flow from each capacity to determine the unused capacity of each arc. Find a path that has no zero unused capacities and repeat; that is, assert a second trial flow increment and again subtract to determine unused capacities. During the procedure, you are free to use any zero-unused-capacity arc in the opposite direction from its use for prior steps, and you increase its unused capacity by the amount of the flow increment that you put into it. Continue until it appears that no more flow can be put through the network. The sum of flow increments is a candidate for the maximum flow. Call it H.

In the min-cut part of the algorithm, you seek to show that a flow of at most J units can pass from S to T. Draw a cut through what appears to be the bottleneck part of the network; the cut can wander anywhere, except that it must separate S from T, cannot cross any arc twice, and must enter the network from one side of the S–T axis and leave only once, on the other side. Call the sum of the flows across the cut J.

If $J = H$, then this is the maximum flow—the capacity of the network. Otherwise $J > H$, and you have missed either a further flow increment or a more restrictive cut. It is guaranteed that the two can be made equal.

Minimal Spanning Trees

Given a collection of nodes and potential arc distances or costs from every node to every other node, a spanning tree is a set of arcs such that there exists a path (sequence of arcs), no matter how indirect, that connects every pair of nodes. A minimal spanning tree is a spanning tree that among all spanning trees has the least sum of arc distances or costs.

The minimal spanning tree problem is a connectivity problem, having applications in situations where capacity is irrelevant. Your home telephone may have a capacity of only one call, but it is nice to know that from that telephone you can reach every other telephone in the network, even if your call has to be routed very indirectly. This is achieved because your telephone, like every other, is connected to the nearest part of the network.

The minimal spanning tree algorithm is extremely simple: Begin by connecting the two nodes that are closest to each other; this forms a partial tree. Among nodes in the partial tree, identify the one that is closest to an unconnected node. Connect it to the unconnected node. Continue until there are no unconnected nodes. The result is a minimal spanning tree, and automatically each node will have only one connection—one arc that links it to the world.

Example 6.2

You have been contracted to deliver natural gas via a pipeline to the ten customer locations shown in the map in Exhibit 1. You want to minimize cost by having the least amount of pipes as possible. Determine the connections from each node that will satisfy this goal, using the minimal spanning tree method.

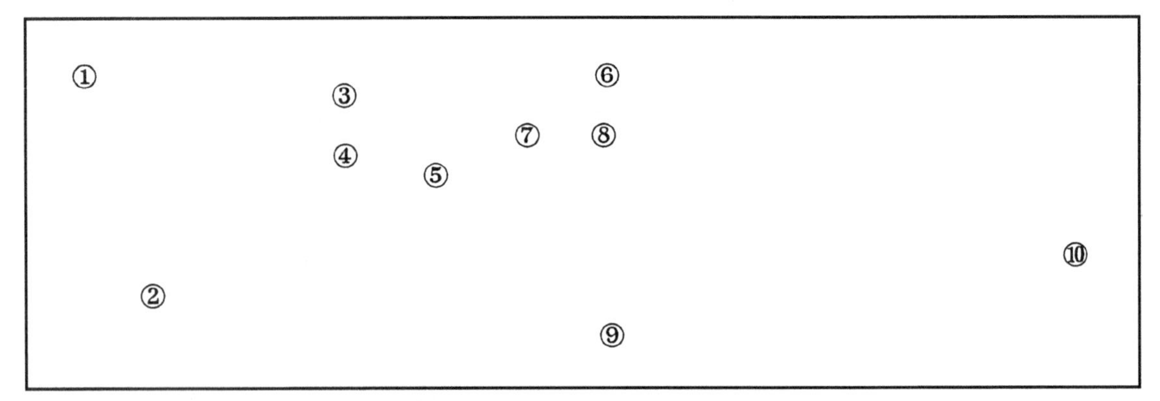

Exhibit 1 Map of customer locations

Solution

The two customers closest together are 3 and 4, so we connect them. Then we examine the remaining 8 customers to see which one is closest to our partial network (3-4), which is 5 and we bring it into the network. In the next iterations we add 7, 8 and 6. At that point we have the partial network 3-4-5-7-8-6, as shown in Exhibit 2.

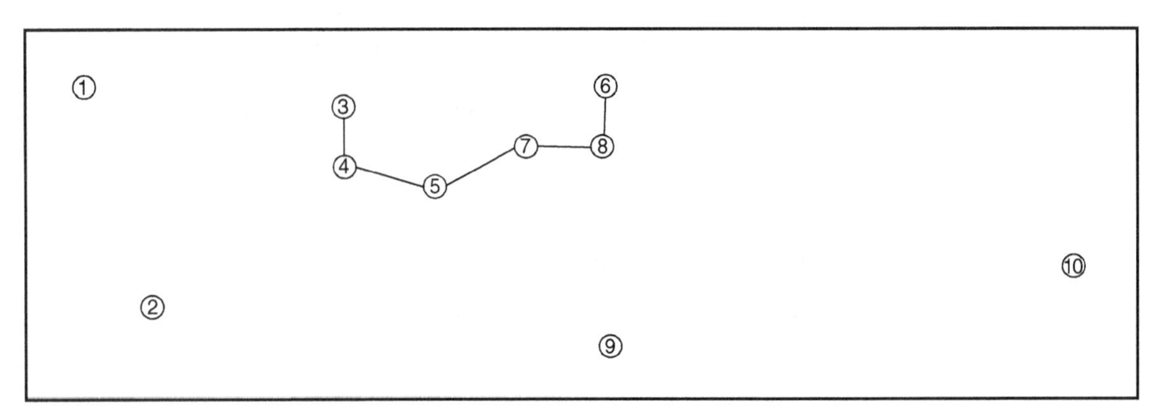

Exhibit 2 The partially connected pipeline network using the minimal spanning tree algorithm

At this point, customers 1, 2, 9, and 10 are not included in the network. The shortest distance to a nonconnected customer is from 8 to 9, so that connection is added to the network. We continue adding the final three customers and have the network shown in Exhibit 3.

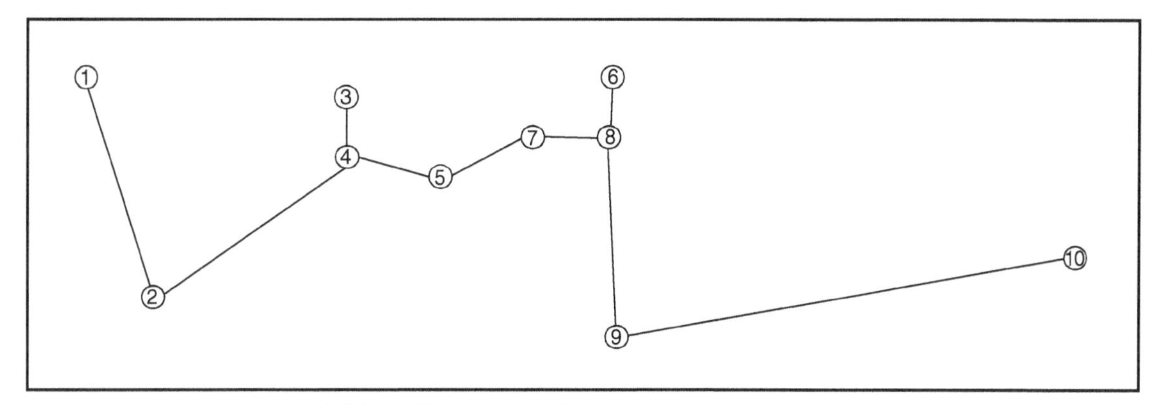

Exhibit 3 The completely connected pipeline network

In summary, your solution should have the following connections:

Node	Connected to	Node	Connected to
1	2	6	8
2	1, 4	7	5, 8
3	4	8	6, 7, 9
4	2, 3, 5	9	8, 10
5	4, 7	10	9

MATHEMATICAL PROGRAMMING

Stochastic Models

Markov Chains

Let p_{ij} be the conditional probability of state j being the next epoch after state i. A *transition matrix* of these conditional probabilities can be written as a stochastic matrix with all rows summing to 1. Left-multiplying a row vector of state probabilities onto P solves the total probability formula.

Example 6.3

If a two-state system has the transition matrix P shown below and a 40% probability of starting in the first state, what is its probability of being in that same state the next epoch?

$$P = \begin{pmatrix} 0.50 & 0.50 \\ 0.90 & 0.10 \end{pmatrix}$$

Solution

The probability of starting in state 1 (p_1) is given as 0.40. We know p_2 is 0.60 since probabilities of all possible states must sum to 1. We write this as a probability vector (0.40, 0.60). Multiplying this onto P gives us the probabilities for the next epoch.

$$(0.40 \quad 0.60)\begin{pmatrix} 0.50 & 0.50 \\ 0.90 & 0.10 \end{pmatrix} = (0.74 \quad 0.26)$$

Thus we have a 74% probability of being in state 1 during the next epoch.

We can find the long-run probabilities (p_i) of being in each state by solving the matrix-multiplication equation, as shown in Example 6.4.

Example 6.4

Given the P matrix shown, what is the long-range probability of the system being in state 1?

$$P = \begin{pmatrix} 0.40 & 0.60 \\ 0.90 & 0.10 \end{pmatrix}$$

Solution

We can set $p_2 = 1 - p_1$. Then solving for p_1.

$$(p_1 \quad 1 - p_1)\begin{pmatrix} 0.40 & 0.60 \\ 0.90 & 0.10 \end{pmatrix} = (p_1 \quad 1 - p_1)$$

$$0.40p_1 + 0.90(1 - p_1) = p_1 \Rightarrow p_1 = 0.60$$

In the long run, our system is in the first state 60% of the time.

Queuing Notation

The handbook contains queuing definitions, fundamental relationships, and single-server and multiple-server queuing models. The Kendall Notation is a commonly used method of denoting the parameters of a queuing system. In full form there are six symbols:

A is for arrival time distribution

S is for service time distribution

m is the number of servers

B is the number of buffers (also called system capacity)

K is the population size

SD is the service discipline (FCFS, first come, first serve—as an example)

Each of these is separated by a slash (/) in the notation. If a parameter is infinite, we omit it from the notation. The handbook provides a simplified Kendall notation for describing a queuing system with four symbols. The first two symbols identify the type of arrival and service distributions. In specifying these two parameters, "M" means Markovian. Remember, in a Markovian process the time between successive events has the exponential distribution, and the number of events in a given time interval has the Poisson distribution. "E_k" is used for Erlang-k and "D" is for a deterministic distribution. In the notation, "*G*" means general, that is, any distribution. The third symbol is the number of servers, which is denoted with the parameter *s*. The fourth symbol in the simplified handbook notation is the capacity—the value of the parameter *M*.

Thus an M/M/1 queuing system means we have an exponentially distributed interarrival time, exponentially distributed service time, and one server. An infinite population can be assumed with this notation.

The system in problem 6.1 at the end of this chapter would be an M/M//3/3. Nothing is shown for the renumber of servers, because *s* is defined as the number of servers in parallel, with equal performance rates. An M/M/3/3/3 system would be similar except that the service rates would be μ, 2μ, and 3μ when 1, 2, and 3 servers were busy, respectively (not arbitrary values μ_1, μ_2, and μ_3).

Queuing Models

The fundamental relationships $L = \lambda W$ and $L_q = \lambda W_q$ are known as Little's Law and are truly fundamental, but here λ is the accepted arrival rate (often called λ_{eff} when the offered rate is λ). This distinction is important for finite queues (M small), since in the finite queue formulas in the handbook, the server utilization factor $\rho = \lambda/s\mu$ is based on the offered arrival rate. Remember that the accepted arrival rate is the rate at which customers both enter and leave ($x\mu$ is the rate they leave while x servers are busy, but with probability P_0 no server is busy).

For example, barbers in a six-station barber shop can each cut hair at an average rate of three haircuts per hour. If customers spend an average of 40 minutes in the shop, the average waiting time before cutting is $W_q = W - 1/\mu = 40/60 - 1/3 = 1/3$ hr.

The single-server model M/M/1/∞/∞ is the most widely used queuing model. Since P_n (the long-run probability that n customers are in the system) decreases rapidly with n, the M/M/1/∞/∞ model can be used for finite-capacity M/M/1/M/∞ queues that rarely get full (P_n near zero). The server utilization factor, since $s = 1$ for this model, is $\rho = \lambda/\mu$. The long-run probability distribution is $P_0 = 1 - \rho$ and $P_n = P_0\rho^n$. The formula for L is $L = \rho/(1 - \rho)$. The formula for $L_q = \lambda^2/[\mu(\mu - \lambda)]$ is given in the handbook. You may wish to remember the more convenient forms $L_q = \rho^2/(1 - \rho)$, $L_q = \rho L$, or the formula that is valid for any single-server model: $L_q = L - \rho$.

Take care with the finite-queue model because ρ here is the offered arrival rate. The assumption behind the model is that a customer that arrives while the system is full (probability P_M) is turned away, so the rate at which service is actually performed is $\lambda_{eff} = \lambda(1 - P_M)$. Another point not mentioned in the handbook for this model is how to handle the special case $\lambda = \mu$ or $\rho = 1$. The formula for P_0 is simply the reciprocal of the finite geometric series formula:

$$P_0 = \frac{1}{1+\rho+\rho^2+\ldots\rho^M}, \qquad \text{which when } \rho \neq 1 \text{ is } \frac{1-\rho}{1-\rho^{M+1}}.$$

Also, note that for this model $P_n = P_0\rho^n$. The special case $\lambda = \mu$ or $\rho = 1$ gives $P_0 = P_1 = \cdots = P_M = 1/(1 + M)$.

The handbook lists the long-run formulas for the important single-server M/G/1/∞/∞ queue, which has Markovian arrivals (called "Poisson input") and arbitrary service times. Recall that σ^2 is the variance of the service time $1/\mu$ (not of the service rate μ). As before, $\rho = \lambda/\mu$ for a single-server system. Define the coefficient of variation of the service time as $c = \sigma\mu$ (not sigma over μ, because the mean is $1/\mu$). Then the formula given in the handbook can be expressed in dimensionless form:

$$L_q = \frac{\lambda^2\sigma^2+\rho^2}{2(1-\rho)} = \frac{\rho^2(c^2+1)}{2(1-\rho)}$$

Since $c = 1$ for the M/M/1/∞/∞ queue, the formula reduces to $L_q = \rho^2/(1 - \rho)$ for $c = 1$. This is sometimes viewed as an upper limit on L_q, although with group arrivals we can have $c > 1$, giving greater L_q. The lower limit, for constant service time, is $c = 0$, leading to $L_q = \rho^2/[2(1 - \rho)]$, which is half of the average queue length for exponential (Markovian) service time.

The handbook gives formulas for multiple-server M/M/s/∞/∞ queuing systems. The formulas for $s = 2$ and $s = 3$ are explicitly listed. The finite-capacity ($M < \infty$) versions of the formulas are not given.

SIMULATION

When the probability distribution of a result cannot conveniently be determined from probability distributions and other data for quantities that determine the result, Monte Carlo simulation can be used. If the system to be simulated is a queuing system or other system involving discrete state changes, discrete-event simulation can be used. Industrial engineers are expected to demonstrate knowledge of some of the basic concepts and techniques of simulation.

Random Generation

To generate a random variate x from any population, the first step is to generate one or more (0, 1) uniform variates $\{u_i\}$. These are variates from the continuous uniform distribution with range 0 to 1, which has the convenient property that $F(u) = u$. This means, e.g., that 90% of the variates are 0.90 or less, 20% are 0.20 or less, etc.

Instead of generating truly random variates, we generate pseudorandom variates by a method that gives a population of values that has the correct mean and variance and seems (according to statistical tests) to yield a series of uncorrelated values. To generate a variate u, the method that industrial engineers are expected to recall is the multiplicative congruential method:

$$Z_i = aZ_{i-1} \bmod m$$

Starting with a positive integer seed or starting value Z_0, we multiply it by the multiplier a and let Z_1 be the remainder when the result is divided by the modulus m. The first variate is $u_1 = Z_1/m$. Then Z_2 is computed from Z_1 in the same manner, and the second variate is $u_2 = Z_2/m$, and so forth. If the computer has words of length b bits, and $m = 2^b - 1$ is the largest integer representable in the computer, then overflow can be used instead of division. If a and m are chosen carefully, this method will give all integers 1, 2, ..., m–1 exactly once before repeating.

As an illustration, if a mythical multiplicative generator had seed 3217, multiplier 293, and modulus 9941, then $Z_1 = 293 \times 3217 \bmod 9941 = 8127$ and $u_1 = 0.8175234$. Continuing, $Z_2 = 5312$ and $u_2 = 0.5343527$.

If a cumulative distribution function $F(x)$ has an inverse F^{-1} (that is, if it can be solved for x), then a random variate from the population can be generated, using only one (0, 1) variate u, by substituting u for F in the inverse formula. This is the inverse transform method of generating variates. For example, the exponential distribution (not given in the handbook, but expected to be remembered) has $F(x) = 1 - e^{-x/\theta}$, where θ is the mean, and this can be solved for x. With u substituted for F, the result is $x = -\theta \ln (1 - u)$. As an illustration, if we have $u = 0.8175234$, then for a negative exponential variate with mean 6.00 we have $x = -6.00 \ln (0.1824766) = 10.207$. In practice, since if u is a (0, 1) uniform variate then $1 - u$ is also a (0, 1) uniform variate from a mirroring population, u is usually substituted for $1 - u$, giving $x = -\theta \ln u$; in this case the negative exponential variate from 0.8175234 would be 1.209.

For an arbitrary distribution, if we plot $F(x)$, we can get a variate x from a (0, 1) uniform variate u by reading the abscissa that corresponds to $F = u$. The algebraic equivalent of this is obvious: If $F(b)$ is the smallest F that equals or exceeds u, then let $x = b$.

Monte Carlo Simulation

Recall that the main idea of simulation is that a large group, or *run*, of simulation experiments is performed, where in each experiment all the parameters take on definite values so that the result has a definite value. If the definite values of parameters are random variates from their respective distributions, the sample population of results will approach the sought distribution of results.

Discrete Event Simulation

One testable skill for discrete-event simulation is the ability to handle the operating logic of a queuing system—processing events and advancing the simulated time. When an event occurs, three things are done immediately:

- The system state is updated. (1) If the event is an arrival, a customer is added to a queue, or (2) if the event is a service completion, a customer leaves a queue and starts a service or enters another queue or departs.
- A record of the event is preserved. (Formerly this was done by updating statistical counters; now it is more common to read an image of the event to a file.)
- It is determined whether the simulation run is over (e.g., whether the number of customers, or the simulated time, has reached a preset limit).

If the simulation run is not over, the next event is determined, (it is the arrival or departure that has the minimum occurrence time that equals or exceeds the current simulation time), the simulation time is advanced to the time of the next event, and the next logical cycle begins (updating the system state).

Another testable skill is that of computing statistics from a simulation history. For example, given a stepwise graph of the number of customers in a subsystem as a function of time, an estimate of the long-run contents of the subsystem is computed as a time-weighted average: If D_0, D_1, … are the total durations that 0, 1, … customers were present, and the sum of the durations is T, then the weighted sum $(0 \times D_0 + 1 \times D_{1+} \dots)/T$ gives an estimate of the long-run contents.

PROBLEMS

6.1 Three machines each have a Markovian failure rate $\lambda = 0.20$ failures per hour, and the operators can fix them at Markovian rates $\mu_1 = 0.50$, $\mu_2 = 0.80$, and $\mu_3 = 1.00$ fixes per hour, respectively, when 1, 2, or 3 machines are down. For an infinitesimal time interval δ, the Markov transition matrix for this situation is as follows (where the states are 0, 1, 2, or 3 machines down):

$$P = \begin{pmatrix} 1-\lambda\delta & \lambda\delta & 0 & 0 \\ \mu_1\delta & 1-\lambda\delta-\mu_1\delta & \lambda\delta & 0 \\ 0 & \mu_2\delta & 1-\lambda\delta-\mu_2\delta & \lambda\delta \\ 0 & 0 & \mu_3\delta & 1-\mu_3\delta \end{pmatrix}$$

The long-run probability that no machines are down is most nearly

a. 0.26
b. 0.38
c. 0.55
d. 0.66

6.2 A university lab is for students enrolled in lab-validated classes only. When a student arrives and is identified, the assistant asks the student for the class name; queries the database, if necessary, to verify that the class is a lab-validated one; and queries the database to verify that the student is enrolled in the class. The most wasteful thing about this procedure is that:

a. The assistant should remember the authorized classes without querying the database.
b. The lab should be open to students in any classes, not just certain ones.
c. Its focus shifts from student to class and back to student.
d. Asking the student for the class name is unnecessary.

6.3 Consider the following linear programming model:

$$\text{Maximize } Z = 4X_1 + 2X_2$$

$$\text{Subject to } X_1 + X_2 \le 3 \qquad \textbf{[Resource A]}$$

$$2X_1 + 5X_2 \le 10 \qquad \textbf{[Resource B]}$$

$$X_1, X_2 \ge 0$$

The increase in Resource A that will make its constraint redundant is most nearly:

a. 0
b. 2
c. 3
d. 6

6.4 The figure shows a small example of an automated baggage handling system. Identifiers of loading/unloading stations are shown inside the node symbols. Identifiers of guideways are not shown; they are of the form exemplified by "H-1" for the guideway that goes from station H to station 1. Arrows show the directions of guideways. Travel times are shown next to the guideway symbols; for example, the travel time for guideway H-1 is 16.

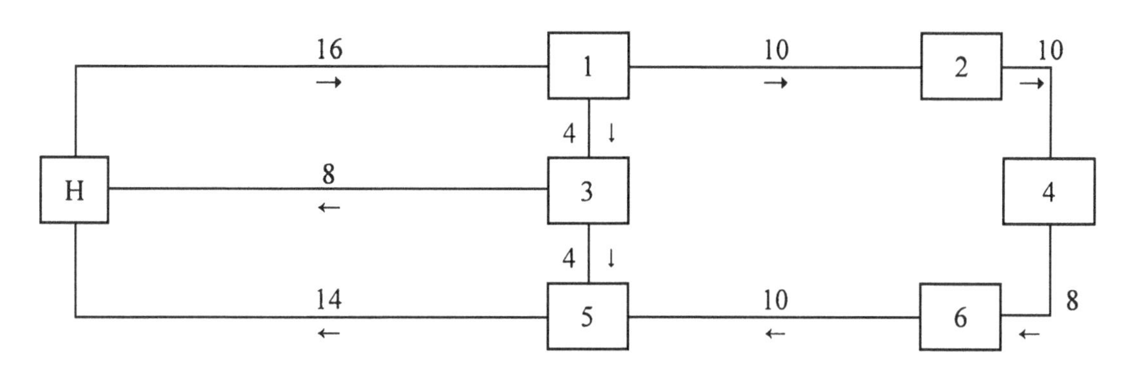

In a relational database for the system, which data could be omitted without violating relational database principles?

a. Explicit listing of guideway identifiers
b. Explicit listing of station identifiers
c. Explicit listing of arrow directions
d. All of the above

6.5 Consider the following problem.

Max $3X_1 + X_2$

Subject to: $X_1 + X_2 \leq 10$

$$2X_2 < 16$$

$$X_1 - 0.5X_2 < 7$$

$$X_1, X_2 > 0$$

The optimal solution for the problem is:

a. 0, 0
b. 7, 0
c. 8, 2
d. 9, 1

6.6 Shipping-time data in hours from mid-lake to mid-lake for a 13-lake inland waterway system has been compiled and written on a network sketch as shown in Exhibit 6.6.

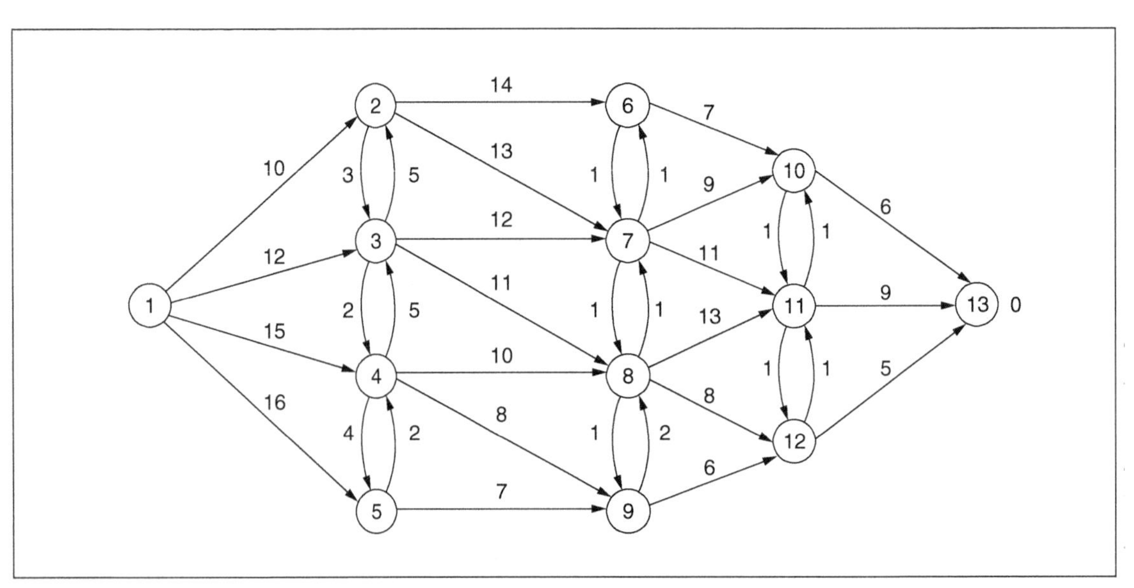

Exhibit 6.6 Shipping times network

The total shipping time from lake 1 to lake 13 is:

a. 30 hours
b. 31 hours
c. 32 hours
d. 33 hours

6.7 In the street network shown in Exhibit 6.7, the line thicknesses denote capacities of 50, 100, and 200 vehicles per minute, and the double-thick line denotes an infinite capacity, which represents not a real street but the equivalence of more than one entry point and exit point:

Exhibit 6.7 Street network diagram

If this area is fenced off from the outside world except along the double-thick lines, the number of vehicles per minute that can pass through it from the S end to the T end is most nearly:

a. 400
b. 450
c. 800
d. 900

6.8 Which is the shortest route between node A and H in the following network?

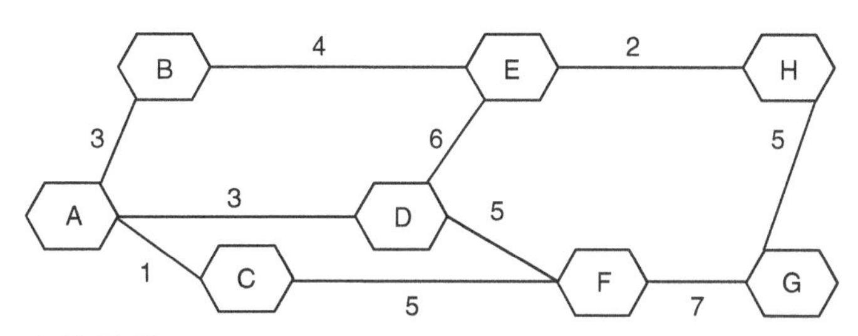

a. A-B-E-H
b. A-D-E-H
c. A-C-G-H
d. A-D-F-G-H

6.9 Julie can decide to build a small, medium, or large restaurant. The probability of a high customer demand is 30% and of low customer demand is 70%. Which alternative would be expected to have the highest profit?

Restaurant	High Demand	Low Demand
Large	$200,000	–$20,000
Medium	$150,000	$20,000
Small	$100,000	$60,000

a. Large restaurant
b. Medium restaurant
c. Small restaurant
d. Not enough information is provided to determine

6.10 The state has a single drive-through toll booth. They have determined that 75 customers/hour can go through it. The mean arrive rate is 45 customers/hour. Which of the following statements are true?
a. Probability the system is empty = 0, mean number of cars in queue = 1
b. Probability the system is empty = 0, mean number of cars in queue = 0.9
c. Probability the system is empty = 0.4, mean number of cars in queue = 0.9
d. Probability the system is empty = 0.4, mean number of cars in queue = 0.75

6.11 Suppose that there are two doctors in a country town. Records show that from year to year:

- of Dr. Adams's patients, there is a probability of 70% that a patient stays with him, and thus 30% of the patients will move to Dr. Bond; and
- of Dr. Bond's patients, there is a probability of 80% that a patient stays with her, and thus 20% of the patients will move to Dr. Adam.

At the beginning of the year, 55 patients go to Dr. Adam and 45 go to Dr. Bond. How many patients are estimated to be at each doctor at the end of year 1?
a. Adam = 38.5, Bond = 36
b. Adam = 47.5, Bond = 52.5
c. Adam = 78.5, Bond = 82.5
d. Adam = 55.5, Bond = 44.5

SOLUTIONS

6.1 d. Let the row vector of long-run state probabilities be (p_0, p_1, p_2, p_3). This vector times P should give itself. Any three of the four equations from this operation can be used, the fourth being linearly dependent because the row sum is one. Let us arbitrarily omit the equation that gives p_2 and multiply the matrix given in the problem by probability vector P. That results in the values for p_0, p_1, and p_3 shown in the table below. We now determine the probabilities of moving from p_0 to p_1, from p_1 to p_2, and remaining in p_3. To do this we multiple by the ratio of the failure rate divided by the fix rate for the number of machines involved. This results in the right-hand side of the table.

$$(1 - \lambda\delta)p_0 + \mu_1\delta p_1 = p_0 \Rightarrow p_1 = \frac{\lambda}{\mu_1} p_0$$

$$\lambda\delta p_0 + (1 - \lambda\delta - \mu_1\delta)p_1 + \mu_2\delta p_2 = p_1 \Rightarrow p_2 = \frac{\lambda^2}{\mu_1\mu_2} p_0$$

$$\lambda\delta p_2 + (1 - \mu_3\delta)p_3 = p_3 \Rightarrow p_3 = \frac{\lambda^3}{\mu_1\mu_2\mu_3} p_0$$

Now, since we know the probabilities must sum to one, we have

$$p_0 + p_1 + p_2 + p_3 = 1 \Rightarrow p_0 + \frac{\lambda}{\mu_1} p_0 + \frac{\lambda^2}{\mu_1\mu_2} p_0 + \frac{\lambda^3}{\mu_1\mu_2\mu_3} p_0 = 1$$

$$p_0 = \frac{1}{1 + \frac{\lambda}{\mu_1} + \frac{\lambda^2}{\mu_1\mu_2} + \frac{\lambda^3}{\mu_1\mu_2\mu_3}} = \frac{1}{1 + 0.4 + 0.1 + 0.02} = \frac{1}{1.52} = 0.6578947$$

6.2 d. If a database query can determine whether a given class is on the authorized list, and another database query can determine whether a given student is enrolled in a given class, then it is certain that a single database query could determine whether a given student is enrolled in any class that is on the authorized list. Asking the student for the class name is therefore unnecessary, and it is wasteful since human time is spent doing this. (Response A is incorrect because the assistant queries the database to verify a class as lab-validated only when necessary. Since response B begs the question (fails to accept the question as asked), it is incorrect. Since reversing the last two queries in the procedure would eliminate the alternation of themes yet save no time, response C is incorrect even though unnecessary shifting of focus is generally undesirable.)

6.3 b. If you draw the two constraints, resource A is a line from (0,3) to (3,0). The constraint for resource B is a line from (5,0) to (0,2). When the line for resource A passes through (5,0), it is redundant. This is accomplished by increasing the resource by 2.

6.4 d. A relation design for the system would be as follows:

GUIDEWAY

GuidewayID	From Station	To Station	TravelTime
H–1	H	1	16
1–2	1	2	10
2–4	2	4	10
4–6	4	6	8
6–5	6	5	10
5–H	5	H	14
1–3	1	3	4
3–5	3	5	4
3–H	3	H	8

Explicit listing of guideway identifiers could optionally be omitted by letting the key of the "GUIDEWAY" table be the compound (FromStation, ToStation). Explicit listing of station identifiers could optionally be omitted by omitting the "STATION" table entirely, since there is no data for that table other than the identifiers, and they are already listed elsewhere. (The question "Which stations exist?" could be answered by the SQL query SELECT FromStation FROM GUIDEWAY UNION SELECT ToStation FROM GUIDEWAY). Explicit listing of arrow directions must be omitted; the information conveyed by them in the figure is equivalent to the information conveyed in the GUIDEWAY table by the placement of each guideway's to/from station identifiers.

6.5 c. This is a simple two-dimensional problem that can be solved graphically. However, a faster solution would be to examine each of the potential solutions and determine if it maximizes the objective function and if it meets all of the constraints. The objective function is to maximize $3X_1 + X_2$. The four answers have values of 0, 21, 26, and 28, respectively. Since D has the great value we check to see if it meets all of the constraints. When we substitute (9,1) into $X_1 - 0.5X_2 < 7$ we get 8, which violates the constraint. Thus answer D is not feasible. Answer C meets all of the constraints and has the maximum objective function value (26).

6.6 d. The dual variable at node 13 is marked as 0, and the other dual variables are understood as infinite. Passing right to left and up to down (in the order 10, 11, 12, 6, 7, 8, 9, 2, 3, 4, 5, 1), the dual variable marked at each node is the minimum of the possible sums of an outgoing arc cost and the dual variable it leads to. The result of the first two passes is shown in Exhibit 6.6a.

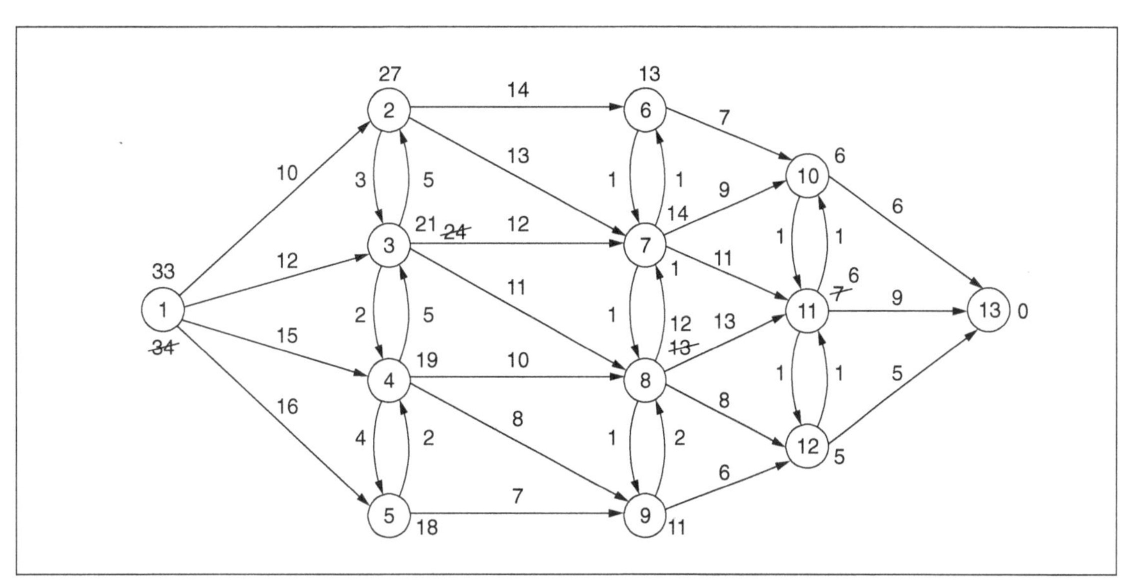

Exhibit 6.6a Shipping times network

The third pass leads to the following values, and no change occurs on the fourth pass. The basic arcs are marked in Exhibit 6.6b. (Those for which the outgoing arc cost plus the value it leads to are equal to the dual variable, with arbitrary breaking of ties):

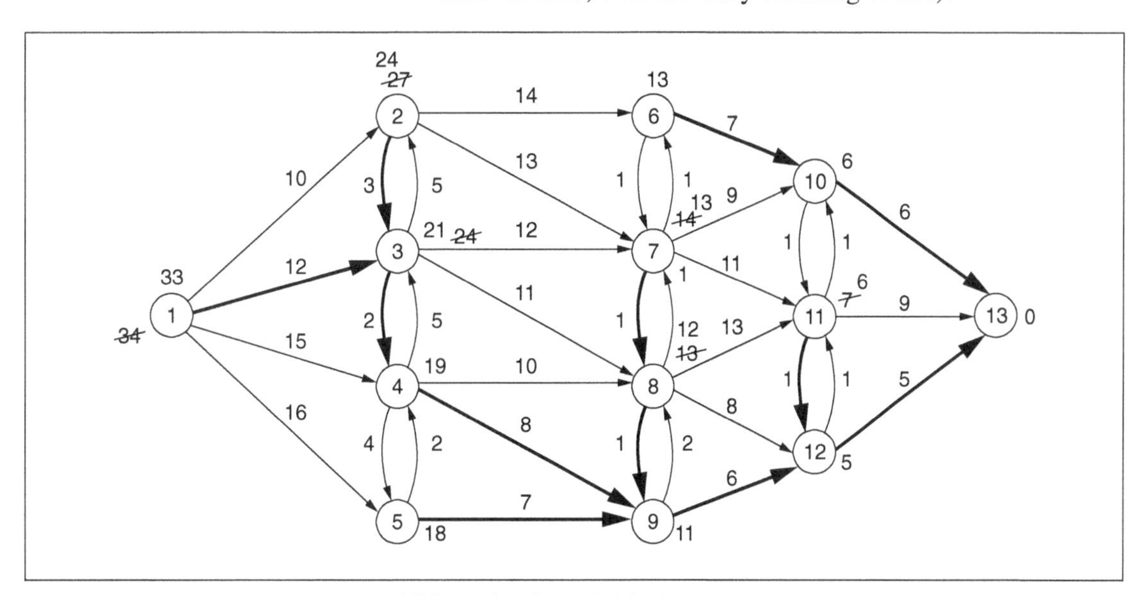

Exhibit 6.6b Solved shipping network

The optimal path from lake 1 to lake 13 passes through lakes 3, 4, 9, and 12. The total shipping time is 33 hours.

6.7 **a.** At each end, as shown in Exhibit 6.7a, it is possible to make a cut that has a capacity of only 400 vehicles per minute. (The capacities of the cut streets are shown for the cut at the S end.) There are several ways to reach the maximum flow, except that they all use the cut streets at each end to full capacity.

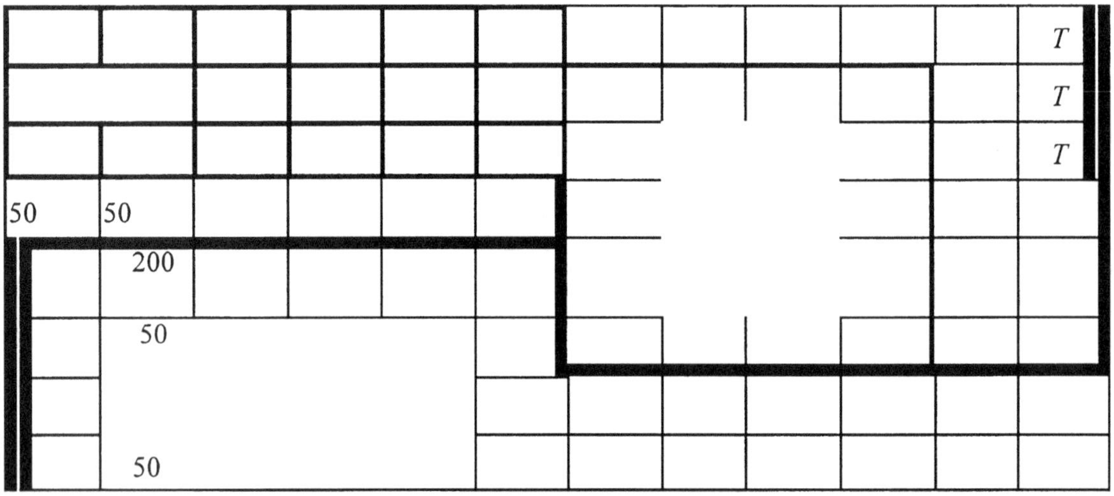

Exhibit 6.7a Solved network diagram

6.8 **a.** You can solve this problem by identifying the shortest path in the network or you can evaluate the four answers and select the shortest. Answer C is not a feasible solution. The values for the other three answers are (9, 11, & 20) respectively.

6.9 **c.** The expected value for a large restaurant is (0.3 × \$200,000) + (0.7 × –\$20,000) = \$46,000. A medium is (0.3 × 150,000) + (0.7 × \$20,000) = \$59,000. A small is (0.3 × \$100,000) + (0.7 × \$60,000) = \$72,000, which is the largest profit.

6.10 **c.** μ = (75 customers/hour)/(60 minutes/hour) = 1.25 customers/minute can be determined from the problem. $\lambda = 45/60 = 0.75$ arrivals / minute. $P_0 = 1 - (\lambda/\mu) = 0.40$. L_q = expected number of cars in the queue $= \lambda^2 / [\mu(\mu - \lambda)] = (0.5625/0.625) = 0.90$.

6.11 **b.** Dr. Adam starts with 55 patients; after a year he retains 70% (38.5 patients). He lost 30% (16.5 patients). Dr. Bond starts with 45 patients; after a year she retains 80% (36 patients) and loses 20% (9 patients). Dr. Adams has 38.5 retained patients and 9 gained patients for a total of 47.5 patients. Dr. Bond has 36 retained patients and 16.5 gained patients for a total of 52.5 patients. Since no patients are allowed to die in the problem, we can double check the answer to make sure we have the same number of patients in the system (55 + 45 = 100 at year 0 and 47.5 + 52.5 = 100 at year 1).

CHAPTER 7

Industrial Management

OUTLINE

In the Industrial Engineering FE Exam, expect 8-12 questions on various aspects of industrial management. Management is a broad topic; given the multiple choice nature of the exam, there tends to be numerous questions related to the definition of terms. Management consists of planning, organizing, staffing, leading, and controlling an organization (a group of one or more people or entities) to accomplish a goal.

MANAGEMENT PRINCIPLES

Management operates through various functions, often classified as planning, organizing, staffing, leading, and controlling.

- Planning: Determining goals, strategies, and activities
- Organizing: Implementing plans (what, who, when)
- Staffing: Job analysis, recruitment and hiring for appropriate jobs
- Leading: Motivating and directing others to get work done
- Controlling: Monitoring performance against plans

Engineering managers oversee finances, technology, and other resources.

To empower people is to give them the opportunity to be self-directing. A trend in management is moving from direct supervision to enabling the group actually doing the work to adopt the details of its work procedure. Employees are empowered when they can change their own work for the better. It has been found that

- workers want to use several skills, not one—job variety;
- they want to know how their own task adds value to the goods or services being produced—recognition of the significance of their work;
- they value autonomy—freedom to take responsibility for the outcome of their own work; and
- they want feedback on the actual results of their work.

Jobs can be restructured to meet these aims in the following ways:

- Job rotation—several people learn several jobs. This gives more staffing flexibility to management but demands more skill of workers, often without more pay. Variety is not really enhanced very much, because a worker may do one boring job in the morning and a different boring job in the afternoon. Job rotation does not contribute to autonomy. It can be used to reduce an employee's exposure to a hazardous or difficult job. In some situations it is used to reduce ergonomic risk in highly repetitive tasks.
- Job enlargement—increasing a job's scope by adding to the types of tasks performed. For example, instead of having three workers screen tax forms, seven workers do data entry from the tax forms, and one worker carry forms to and from stations, you could let 11 data-entry workers do their own screening and transporting forms. In this way, the workers' jobs have been enlarged since they process a tax form completely; however, they have not increased their autonomy with this change.
- Job enrichment—giving a worker more responsibility. For example, letting a machine operator, rather than a maintenance worker, perform simple, routine preventive maintenance tasks that the machine requires. Job enrichment does contribute to autonomy.
- Group work—combining jobs, allowing workers to allocate tasks among themselves. Group work contributes to autonomy, responsibility, recognition, and feedback, but it can be inefficient if the tasks are highly structured. This is also called self-directed work teams.

STRUCTURE OF ORGANIZATIONS

An organization or enterprise is a structured social entity formed to achieve definite purposes. To call it an organization is to emphasize its structure; to call it an enterprise is to emphasize its purposes. A family may lack structure and purpose and therefore not be an organization or enterprise; on the other hand, a church, corporation, or governmental unit would have the necessary structure and purpose. An organization or enterprise is managed, that is, controlled, by people (managers) who hold authority as defined by the structure of the organization.

In the usual hierarchical authority structure, also called a bureaucratic structure or function-oriented structure, people are placed in the organizational structure according to the function they perform. For example, accountants are in the accounting department. Industrial engineers are expected to recognize the usual job titles and the hierarchy they imply, and to be able to display a hierarchy in the form of an organization chart. The span of control of a manager is the number of employees that the manager directly supervises. The primary advantage of this structure is the consolidation of specialties; all of the designers are in the same department and are supervised by a manager who is also a designer. The drawback is that people are in technical silos, and communications between people working on related tasks can suffer.

The opposite of a function-oriented structure is a pure project-oriented structure, such as a pickup band in music or an all-star team in sports. In this structure, people with various expertise work together on a common goal (often called a project) and report to a manager. The primary advantage of this structure is the ability of people with different abilities to work closely together with others on the same project. Designers, manufacturing engineers, marketing experts, and others would all work together under the same manager. Drawbacks include potential duplication of efforts and isolation of workers from others in the same specialty area.

A hybrid alternative to a function-oriented structure and the pure project-oriented structure is a matrix structure, in which project teams are formed and dissolved. For example, an electrical engineer would be permanently assigned to the electrical engineering department (as in a function-oriented structure), but would usually be assigned to a project team, being supervised by the project manager for the project. Her supervisor in the electrical engineering department, using input from her temporary project supervisor, would decide on her promotions, manage her professional growth, and control or influence her project assignments.

The main advantage of matrix organization is that it provides better control over the schedule, performance, and cost of projects. It simplifies communication, both among people working on a project and between the project and its client. It renews the Hawthorne effect (see the following section) by shifting the reporting responsibilities of workers. It helps to prevent power structures from becoming entrenched and poisoning the professional atmosphere. It provides opportunities for junior managers to develop professionally.

The main disadvantage of matrix organization is that it introduces uncertainty and discontinuity in employees' evaluations. It can cause professional conflicts between doing the job well (as might be emphasized by a functional manager) and doing it quickly (as might be emphasized by a project manager). It can interfere with professional development by isolating a specialist from others in the same field. It can cause discomfort by requiring employees frequently to prove themselves anew.

Motivation

Industrial engineers are expected to be familiar with the basic principles of motivation, since one of their main roles is that of productivity facilitator. Leadership—stimulating others to perform—requires knowledge of motivation—the psychological factors that make a person self-directed, so that the person will begin work, work well, and keep working. (These three volitions are called arousal, direction, and persistence.)

Scientific management was articulated by Frederick W. Taylor in the 1920s in his book of the same name. It emphasized positive reinforcement (the carrot is better than the stick), daily pay incentives, and fair and consistent work standards. Taylor's views were opposite of Deming's TQM view (see chapter 12 for more on TQM). Taylor held the workers in low opinion and felt the managers needed to clearly define the work method for the workers.

Pay incentives actually did not succeed well as motivators. By the 1930s, scientific management was considered dehumanizing and was supplanted by the human relations movement. Much of the impetus for this was the discovery of the Hawthorne Effect. Harvard researchers, trying to determine the effect of lighting level on productivity at the Hawthorne plant where Western Electric Company built relays, found that productivity was increased by either increasing or decreasing the illumination of workstations, and an eventual consensus emerged that paying attention to workers, in almost any manner, had a motivating effect.

In 1960, Douglas McGregor's Theory X and Theory Y sparked the realization that people tend to behave according to others' expectations. If you manage a worker under the assumption that he or she lacks integrity, is lazy, avoids responsibility, is uninterested in achievement, is incapable of self-direction, is indifferent to organizational needs, is passive, avoids making decisions, and is stupid, you are managing under Theory X. Theory Y is the opposite.

The first popular content theory of motivation—a theory that purports to explain why motivation strategies work or do not work—was Maslow's Hierarchy of Needs (1943). In Maslow's hierarchy, the top level is self-actualization. If all lower-level (more basic) needs are satisfied, a worker will be motivated only by self-actualization—the joy of the work itself, work as play, work as a source of personal pride.

The next level down from self-actualization is ego—personal status, recognition, achievement. The next level down from ego is status (Maslow called this level "social"). The ego and status levels are closely related, the distinction being that ego is one's own opinion of oneself, and status is others' opinions. The next lower level is security (Maslow called this level "safety"). Finally, the lowest level is the physiological level—adequate pay. Thus, going from the bottom, a worker might pay primary attention to earning money until there is enough money; then pay primary attention to job security until there is enough job security; then pay primary attention to job status and recognition until there is enough; then pay primary attention to pride of achievement; and finally, having already secured enough pay, security, recognition, and pride, then pay primary attention to whatever is the most fun or rewarding.

Maslow's Hierarchy of Needs

Self-actualization—personal fulfillment, self-direction
Ego—personal pride in achievement, self-satisfaction
Social—status, honors, recognition by public, peers, superiors
Safety—job security
Physiological—adequate pay

A direct consequence of Maslow's hierarchy of needs is that money is not a strong motivator unless a worker drastically lacks money. According to the two-factor theory based in part on Maslow's hierarchy of needs and also known as the motivation-hygiene theory, the factors in a worker's environment can be classified into satisficers (or motivators) and dissatisficers (or hygiene factors):

Table 7.1 The Two-Factor (Motivation-Hygiene) Theory

Satisficers—Improvement in a Satisficer Enhances Motivation	Dissatisficers—Improvement in a Dissatisficer Enhances Motivation Only Until a Satisfactory Level is Reached
Advancement	Salary
Responsibility	Benefits
The work itself	Personnel and other policies
Recognition	Supervision
Achievement	Working conditions

The two-factor theory offers a diminishing-returns approach to investing in employee motiviation: Spend available resources first to bring dissatisficers up to a satisfactory level, and then spend on satisficers. Satisficers are not considered to have any satisfactory or unsatisfactory level.

Another was to classify motivation is as extrinsic and intrinsic motivators. Extrinsic motivation is anything outside yourself that you can obtain that will increase your motivation; examples would be money, a nice company car, a high GPA in school, or an award. Intrinsic motivation is the opposite; it is internal. Examples would include your joy in work, a desire to learn, or the satisfaction of a job well done.

Incentives

An incentive is something of value offered to workers specifically to motivate them to work more effectively. The most commonly used incentives are merit-based promotion and merit-based pay raises. Another is annual bonuses. These are long-term incentives, and in accordance with behavioral theory, short-term incentives, such as wage incentives and recognition (awards), are more powerful.

Most short-term incentive plans are based on relative improvement and fit the following definitional structure:

$$[\textit{Relative improvement}] = \frac{[\textit{Actual level}] - [\textit{Acceptable level}]}{[\textit{Incentive level}] - [\textit{Acceptable level}]}$$

It has been found, however, that it is best to set the incentive level to the standard production rate and to set the acceptable level to zero. If the incentive level is different from the standard production rate, it in effect becomes the standard production rate. If the acceptable level is other than zero, then the relative improvement is not proportional to production rate, and employees deem the resulting incentive unfair. Thus modern incentives have a simplified definitional structure:

$$[\textit{Relative improvement}] = \frac{[\textit{Actual level}]}{[\textit{Standard production rate}]}$$

The participation ratio, R, is a multiple of the relative improvement. The base rate is an hourly rate; base pay is the pay for a number of hours (commonly 8 or 40) at the base rate. Total pay for that number of hours can be defined for modern incentive plans as follows:

$$[\textit{Total pay}] = \min\{[\textit{Base pay}], [\textit{Base pay}] + R \times [\textit{Relative improvement}] \times [\textit{Base pay}]\}$$

Finally, modern incentive plans let $R = 1$. This value of the participation ratio means that, on a piecework basis, the worker receives the same amount of money per piece when producing faster than the standard rate as when producing at the standard rate. (Note the minimum operator: The total pay cannot be less than the base pay.)

$$[\textit{Total pay}] = \min\{[\textit{Base pay}], [\textit{Base pay}] \times (1 + [\textit{Relative improvement}])\}$$

This pay can be for an individual or a group.

Records for individual piecework incentives are commonly kept in the following manner, which is equivalent to the preceding equation:

$$[\textit{Earned time}] = [\textit{Number of pieces completed}] \times [\textit{Standard time per piece}]$$

$$[\textit{Pay}] = [\textit{Base rate}] \times \max\{[\textit{Earned time}], [\textit{Number of hours worked}]\}$$

If the worker works overtime (more than 40 hours per week), the earned regular time is computed on the basis of the number of pieces completed in the first 40 hours and the regular pay is computed as shown; then the earned overtime is separately counted on the basis of the number of pieces completed during overtime, and the overtime pay is computed separately, with the overtime rate (1.5 times the base rate) used instead of the base rate.

PROJECT MANAGEMENT

Project management differs from traditional management due to the nature of projects.

A project is **temporary** group activity to produce a unique product, service, event or result. Projects have defined beginnings and ends; they are defined in scope and resources. Traditional industrial managers supervise routine operations, and project managers face challenges managing a specific set of operations designed to accomplish a singular goal rather than ongoing routine operations. A project team often includes people who do not usually work together; this can be done using the matrix organization structure discussed above or virtual teams (where project team members are at various locations and interact using computer technology such as e-mail and Web-based meetings). Examples of projects range from development of a new software package, the installation of a new packaging line, the relief effort after a natural disaster, to hosting a fund-raiser dinner for the relief effort. Project managers must make trade-offs to balance the goal of completing a project on time, on budget, and with the required performance. An example of a trade-off decision would be whether to authorize overtime to complete a project on schedule with the consequences of exceeding the project's budget.

Project managers (PM) have several tools at their disposal. The work breakdown structure (WBS) is a hierarchical list of tasks that must be completed during the course of the project. At the lowest level, the subtasks are work packages that can be completed in about one shift. Once the tasks and subtasks have been determined, the PM can estimate costs and assign resources to develop a budget and schedule. A project laid out graphically with a time scale on the x-axis is called a Gantt Chart. As time passes, the PM is able to track the completion of tasks using this chart and determine if the project is on track with respect to the schedule. Another scheduling tool used by PMs is the PERT/CPM chart (Project Evaluation and Review Technique and Critical Path Method).

PERT/CPM

The following are the steps in constructing these charts:

1. Identify the tasks to be completed (often using the WBS).
2. Determine precedents (such as tasks A and B both must be completed before task C can be initiated).
3. Draw these graphically from left-to-right, from the start of the project to the end.
4. Working from the left, determine the earliest start (ES) and earliest finish (EF) time for each task. This is dependent on the time at which all preceding tasks are completed. This is called a "forward pass" in PERT/CPM.

5. Once this has been completed, you have the earliest completion time for the project.
6. Using this completion time, work from the end of the project (backwards pass) to determine the latest possible start (LS) and end time (LE) for each task that will not delay the overall project.
7. For each task calculate the difference between the latest start and earliest start times; this is called "slack."
8. Any task with a zero slack time value is on the critical path.

As the name suggests, tasks on the critical path are the most important tasks when managing a project. Any delay in these tasks will delay the overall project. It is possible for a project to have multiple critical paths. If a task that is not on a critical path is delayed by more than the available slack time, that task and the tasks that follow it will also become critical.

Example **7.1**

Kim is the project manager opening a new faculty in Seoul, Korea. She has determined a set of major task, the time they will require, and what needs to be completed in what order. This information is shown in the table. She must determine the critical path.

Task	Time Required	Precedence	ES	EF	LS	LF
A	10	—				
B	8	A				
C	2	A				
D	3	C				
E	5	C				
F	7	D, E, G				
G	4	B, D				
H	1	F				

First she draws the precedence diagram. Since A has no precedence, it is the first task in the network. Once it is done, B and C can start and so on.

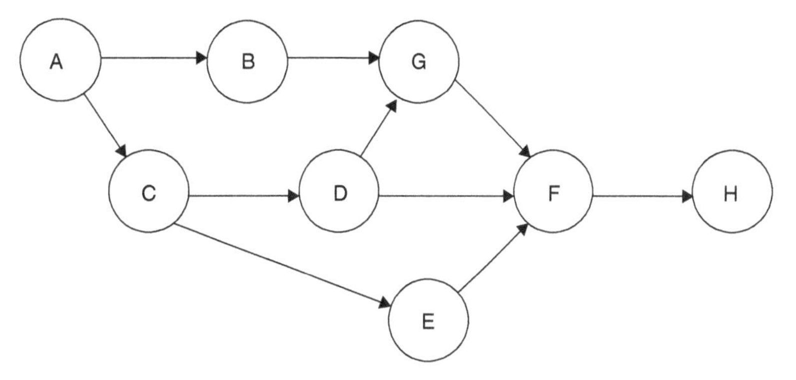

Exhibit 1 Precedence diagram

Then doing the forward pass starting at time zero, determine which jobs can be started. In this case, only task A can start. Since it would take ten weeks, the earliest it could finish is ten. Once A is completed, one can do B and/or C. No other jobs are to be started since they have precedence that are not yet completed. The earliest start for both B and C is ten, the competition of task A. We add the duration of each job (8 and 2) to the get the earliest completion time (18 and 12) respectively. At time 12 (the earliest completion of C), we can start both D and E. At time 18 (the earliest completion of B), we can start G, if and only if, D is also completed at that time. Since B has the latest early finish time (week 18), G must wait until then to be started. Task F must also wait on more than one task to complete before it can start. The latest EF time for its three precedence (D, E, G) is 18. We continue to the end of the network, and the earliest completion time for the project is week 30. That completes the forward pass.

Task	Time Required	Precedence	ES	EF	LS	LF
A	10	—	0	10		
B	8	A	10	18		
C	2	A	10	12		
D	3	C	12	15		
E	5	C	12	17		
F	7	D, E, G	22	29		
G	4	B, D	18	22		
H	1	F	29	30		

Now we start the backwards pass. It is the same method for the prior pass, only backwards. Instead of starting at time 0, we start at time 30, which is the week when the entire project is finished if we starting every task as early as possible. Working from the right-hand side, the only task we can finish at week 30 and not delay the project is task H. So its latest finish time is 30 weeks. We know it takes only one week to complete, so the latest we can start it and not extend our over project is week 29. The precedence for H is task F. It must finish at time 29 weeks; the latest it can start is 22 weeks (finish time – time required = 29 – 7 = 22). Three tasks must finish before F; they are D, E, and G. The latest any of them can finish is 22. We subtract their time required (3, 5, and 4) from their latest finish time (22) to get their latest start times (19, 17, and 18) respectively. We continue this process for the rest of the table. The latest we can start A is week 0; this is the start of our network and a good double check we've done the math correctly. This finishes the backwards pass. We now determine the slack for each task. Since the time required is constant for each task, we can use LS – ES or LF – EF and get the same number. So for task A, it is 10 – 10 = 0 – 0 = 0. With a slack of 0, we know it is on the critical path. For task B, the slack is also 0. For task C, it is 5 (15 – 10 or 17 – 12). We continue this for the remaining task and complete the table. Any task with 0 slack is on the critical path. In our case, that is A-B-G-F-H. The remaining tasks have slack associated with them.

Task	Time Required	Precedence	ES	EF	LS	LF
A	10	—	0	10	10	0
B	8	A	10	18	18	0
C	2	A	10	12	17	5
D	3	C	12	15	22	7
E	5	C	12	17	22	5
F	7	D, E, G	22	29	29	0
G	4	B, D	18	22	22	0
H	1	F	29	30	30	0

Example 7.2

What if Kim learns from a contractor that Task C cannot start until week 15? What if it can't start until even later? We recalculate the time values for C and the tasks that follow. Task C now has ES = 15 and EF = 17. The times for task D are adjusted to ES = 17 and EF =20. The times for task E are adjusted to ES = 17 and EF = 22. Task F is dependent on the finish of D, E, and G. All three still finish at time 22. The remainder of the network is unchanged.

This means if Kim reschedules task C to week 15, the project can still finish on time, but task E is also on the critical path. She will have parallel critical paths. If the task is delayed even more than the whole project will be delayed, it will not finish by week 30.

Suppose we want to draw a Gantt chart for the earliest schedule. It would be:

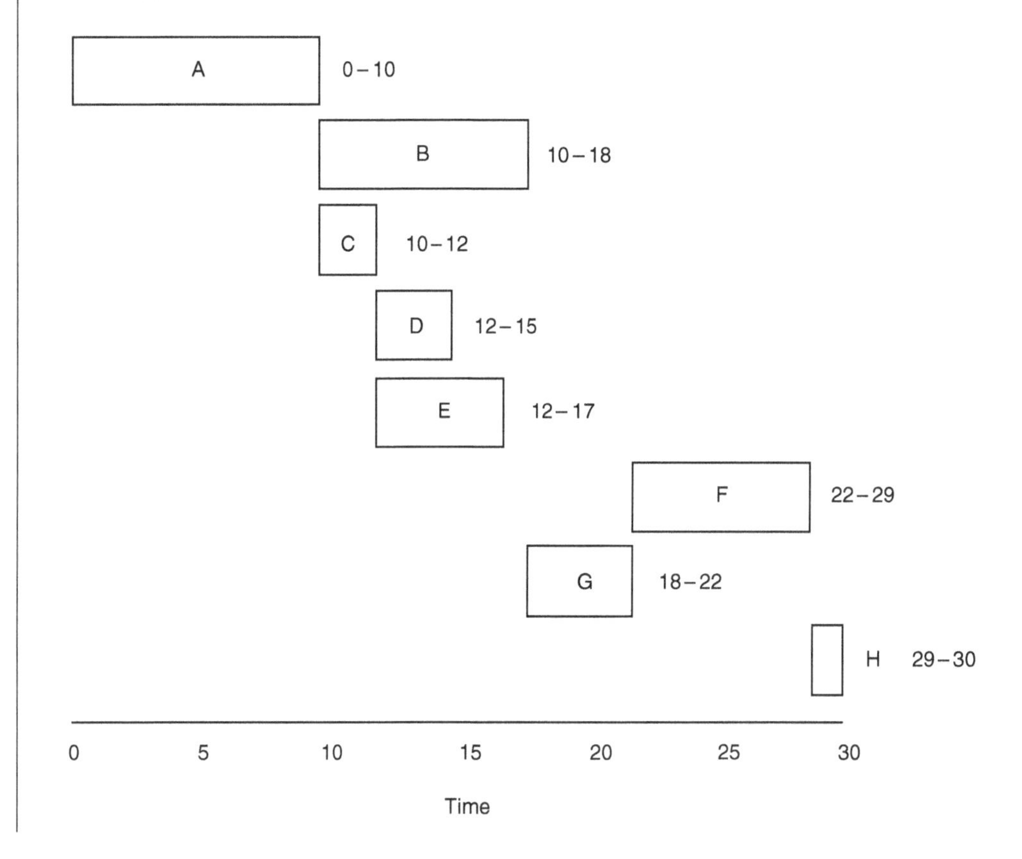

JOB EVALUATION

In order to determine base pay fairly, an enterprise uses either seniority, job evaluation techniques, or a combination. Job classification, as in civil service pay grades, is too arcane to be of interest to industrial engineers. Pairwise comparison, the basis of several decision software packages, is too inefficient and arbitrary to be of interest to industrial engineers.

The two job evaluation methods that industrial engineers should know, both of which allow a committee to determine pay scales on an objective basis, are job factor analysis and point systems.

Job factor analysis requires the committee to select a set of key jobs and a set of job factors. There should be no more than 250 key jobs for a large organization; usually there are no more than about 40. There should be no more than a dozen job factors; usually there are six or eight.

A key job is an actual job that is considered typical of many actual jobs. Its performance expectations must be well documented and stable. Collectively the jobs must cover the ranges of factor values fairly evenly; for example, if physical effort is a factor, then among the jobs there must be some key jobs that have high, moderate, and low values of that factor.

Most job factors are aspects of what the employee has to put into the job (education or experience required, responsibility, working conditions, etc.); although one or two can be aspects of the value of the job to the organization. Job factors must be attributes of the job, not of the people who fill them. For example, neither contribution to racial balance nor physical beauty is an appropriate job factor. Job factors should be independent; for example, working conditions and noise level cannot be two job factors.

Job factors and key jobs are arrayed in a job-factor money apportionment table:

Key Job Title	Education and Experience	Mental Effort	Physical Effort	Responsibility	Working Conditions
Bricklayer	$10.20	$3.00	$2.50	$4.00	$3.00
Hod carrier	$3.00	$1.22	$7.50	$0.80	$3.00

The entries are nominal base pay for each factor. The row sums are nominal base pay for the job; for example, a bricklayer is paid $22.70/hour and a hod carrier is paid $15.52.

Consistency checking is what makes the process fair. For example, if a hod carrier's mental effort was questioned as too low, the committee would examine other key jobs having similar values for the mental-effort factor and adjust the $1.22 upward if there were jobs in the table that paid more for mental effort, yet were perceived by the committee as having less mental effort than is required for a hod carrier.

Job point systems have predetermined job factors and predefined levels for each factor. For example, the Midwest Industrial Management Association (MIMA) publishes a point-system plan for office jobs and one for shop jobs. The plan for shop jobs has 11 factors, each with five degrees or levels. To evaluate a job, the user (again, usually a committee, not an individual) chooses a level for each factor for the job, using published guidance. For example, if an office job requires a high school diploma or equivalent, it is second-degree for education, but if it requires a Ph.D., it is fifth-degree for education. The plan lists a predetermined point quantity for each level for each factor, and the sum of these overall factors is the point score for the job. As an illustration, here is a hypothetical point table:

	Education and Experience	Physical and Mental Effort	Responsibility for People and Materials	Job Conditions and Hazards
High	10	14	14	12
Medium	5	7	7	6
Low	1	2	1	1

For example, a job judged to be medium-degree in every factor would earn a point score of 25; a job judged to be high-degree in every factor would earn the maximum point score of 50. Pay scales could be tied directly to point scores, or indirectly through pay grades, each one corresponding to a range of point scores. In the MIMA system, which has a maximum point score of 500, there are 12 pay grades.

As for job-factor analysis, consistency checking is what makes a point system fair. If a job seems to rate too high or low, its degree for each factor can be checked against that for other jobs.

STRATEGIC PLANNING

Strengths, Weaknesses, Opportunities, and Threats (SWOT Analysis) are often used in strategic planning to determine strategic niche that an organization can exploit for success. In this analysis, organizations determine their core competency, their unique skills or resources that provide a competitive edge. Business process re-engineering (BPR) is a business management strategy, originally pioneered in the early 1990s. It analyzes and improves workflows and processes within an organization. The BPR aimed to help organizations rethink how they do things in order to dramatically improve customer service, cut operational costs, and become world-class organizations.

Many organizations set performance objectives in an effort to achieve organizational goals. Using management by objectives (MBO), subordinates and their supervisors determine specific performance objectives together. Progress towards these objectives is periodically reviewed, and performance is evaluated against them. The goal is to motivate workers using MBO rather than controlling them. Balanced scorecard is a strategic planning and management system used to align business activities to the vision and strategy of the organization, improve internal and external communications, and monitor organization performance against strategic goals. The MBO method is a performance assessment tool at the individual level, and balanced scorecard is at the organizational level.

PRODUCTIVITY MEASURES

Performance measures are key to industrial engineering. Applications of performance measures have been presented in other chapters in the book. Questions on the exam could be written with a variety of applications including work measurement and plant layout. Productivity measures are often ratios of what actually happened over what could have or should have happened. The result gives you a percentage to evaluate how well something is doing, much like a grade on an exam.

Efficiency is how well something is using the inputs (time, effort, or resources) for an intended purpose. It is the ratio of resources produced over resources used. Utilization is similar in that it is the actual output over the design capacity. There can be a difference between "effective capacity" and "design capacity." For example, say the city public transportation department has 40 buses but 2 of them are broken due to mechanical problems. When only 36 bus drivers show up to work we have:

$$\text{Efficiency} = 36 \text{ actual bus runs} / 38 \text{ bus run effective capacity} = 94.7\%$$

$$\text{Utilization} = 36 \text{ actual bus runs} / 40 \text{ bus run design capacity} = 90.0\%$$

If we are evaluating workers' output, the ratio of standard hours produced over the actual hours required is called "realization."

PROBLEMS

7.1 Which step in Maslow's Hierarchy of Needs Theory deals with achieving one's potential?
a. Safety
b. Self-actualization
c. Social
d. Esteem

7.2 In management by objectives it is critical participative goals are set that are:
a. approved by management
b. easy to attain
c. tangible, verifiable, and measurable
d. general in nature

7.3 An organization has among its job titles those of board chair, president, vice president–manufacturing, vice president–marketing, vice president–finance, manager–material control, manager–quality control, manager–plant 1, manager–plant 2, manager–plant 3, manager–advertising, manager–purchasing, and manager–sales. Ignoring the effects of incompleteness of the list of job titles, the spans of control of employees in the order mentioned are:
a. 0, 1, 1, 1, 1, . . .
b. 1, 1, 3, 8
c. 1, 3, 6, 2, 0, . . .
d. 1, 3, 5, 2, 1

7.4 A tenured, well-paid high school chemistry teacher asks her principal for three things with seemingly equal emphasis: more money, a reduced teaching load (letting her assistant cover the makeup lab period), and permission to make one of her senior classes an honors class. Which of the following reactions would best maintain her motivation?
a. A pay increase and a reduced teaching load
b. A pay increase and permission to offer the honors class
c. A reduced teaching load and permission to offer the honors class
d. A double pay increase

7.5 The fastest pipefitter welder in the Goldberg Spool Shop is paid by the weld. This week he produced 87 good welds in the first 40 hours and 17 good welds in 10 hours of overtime. The standard time per weld is 0.500/hour. His base pay is $18.00/hour. His pay for the week is most nearly:
a. $1030
b. $1050
c. $1070
d. $1090

7.6 Project management often has the following characteristics *EXCEPT*:
a. matrix organization structure
b. reoccurring tasks
c. budget constraints
d. deadlines and milestones

7.7 Which of the following is *NOT* a useful management option to address to varying demand in a service environment?
a. Using overtime
b. Hiring workers
c. Laying off workers
d. Adjusting inventory levels

7.8 Jack inherits a struggling pizza business from his aunt. To improve the company's financial performance, he studies the competition, determines what sets his pizzas apart from the others in town, and installed a computer system to take and track orders over the phone. He has done all of the following *EXCEPT*:
a. SWOT analysis
b. determine core competencies
c. develop a mission statement for his business
d. business process re-engineering

7.9 A critical path in PERT/CPM does *NOT*:
a. help determine the amount of slack time
b. identify important activities
c. calculate project performance with respect to budget
d. calculate the duration of the whole project

7.10 Which of the following is *NOT* an example of a management function?
a. Working with and through people
b. Balancing efficiency against effectiveness
c. Obtaining the most from limited resources
d. Setting goals

7.11 Which answer corresponds to a person's internal desire to do something, due to such things as interest, challenge and personal satisfaction?
a. Theory X
b. Theory Y
c. Intrinsic motivators
d. Extrinsic motivators

7.12 In project management, the critical path is the sequence of activities which has the:
a. most activities
b. shortest time
c. longest time
d. greatest variance

SOLUTIONS

7.1 b. See definitions in Maslow's Hierarchy of Needs.

7.2 c. All goals should be tangible, verifiable, and measurable rather than general (answer D) since a worker's performance will be compared against them. Goals should challenge and motivate a worker (rather than answer B). Answer A (approved by management) is true for MBO, but it is not a critical aspect compared to answer C.

7.3 c. The board chair supervises the president (span of control 1), who supervises three vice presidents (span of control 3); the vice president for manufacturing supervises the managers for material control, quality control, the three plants, and purchasing (span of control 6); the vice president for marketing supervises the managers for advertising and sales (span of control 2); the vice president for finance supervises no listed managers (span of control 0, ignoring effects of incompleteness of the list); and the spans of control of the managers are not asserted in the responses.

7.4 c. By either Maslow's hierarchy of needs or the two-factor theory, the answer is clear if the request for a reduced teaching load is properly interpreted. Such a request might seem to be low-level; one might reason that with time off the pay per unit time is greater and place the request at the physiological level (Maslow's lowest level), or reason that workload is part of working conditions and classify a reduction in workload as removal of a dissatisficer in the two-factor theory. But such a request more likely stems from a desire to do a better job; the teacher would undoubtedly use the free period of good professional advantage rather than as an excuse to loaf. The free period would allow more opportunity for achievement and satisfaction and therefore belongs at the top Maslow level or as a satisficer in the two-factor theory. Given this, the choice is among two-part combinations of which only in response C are both parts at a high Maslow level or are both parts satisficers.

7.5 b. In the first 40 hours, the earned time is $87 \times 0.5 = 43.5$ hour, and the pay is $\$18.00 \times 43.5 = \$783.00$. In the 10 hours of overtime, the earned time is $17 \times 0.5 = 8.5$ hour, and the pay is $(1.5 \times 18.00) \times \max\{8.5, 10\} = \$270.00$. The pay for the week is $\$783.00 + \$270.00 = \$1053.00$. [If separate accounting of production in regular time and overtime had not been done, then the earned hours would have been $(87 + 17) \times 0.5 = 52$ hour, the average pay during the 50 hours would have been \$19.80/hour, and the pay would have been $19.80 \times 52 = \$1029.60$.]

7.6 b. Project management is used for unique tasks, not reoccurring tasks. The matrix organization structure often provides the flexibility that is needed since projects have a set duration. Budget and schedule constraints are key to project management, along with technical performance.

7.7 **d.** In the service sector inventory (i.e., haircuts, oil changes, massages) cannot be stored. Inventory management is distinctly different between manufacturing and service organizations. The other responses can be done to adjust the number of workers to match variations in demand.

7.8 **c.** Studying others and what sets his pizzas apart is part of SWOT and determining core competencies. Improving the operations by adding computer technology could be considered business process re-engineering. A mission statement is the purpose of an organization. He didn't do this for his pizza business.

7.9 **c.** Important activities are those on the critical path. Having zero slack put an activity on the critical path by definition. Calculating the project's duration is a step in determining the critical path. Budget performance is not included in PERT/CPM. Once a project is started, it can be used to monitor schedule performance.

7.10 **b.** Managers plan (answer D), organize, staff (answer A), lead, and control (answer C).

7.11 **c.** Theory X is the term for the assumption that the workers dislike work and are lazy, where Theory Y is the opposite, assuming that workers are creative, seek responsibility, and can be trusted to exercise self-direction. Answer D is external desires rather than internal (answer C).

7.12 **c.** The critical path in a PERT/CPM network is the path that requires the longest time regardless of the number of tasks on the path.

CHAPTER 8

Manufacturing, Production, and Service Systems

OUTLINE

Expect 8-12 questions in this area on the Industrial Engineering FE Exam. Subtopics include manufacturing processes, manufacturing systems, process design, inventory analysis, forecasting, scheduling, aggregate planning, production planning, lean enterprises, automation concepts, sustainable manufacturing, and value engineering.

MANUFACTURING PROCESSES

Machining Formulas

Material removal rate (MRR) formulas for material drilling, slab milling, and face milling, are given in the handbook.

Manufacturing Tolerances

Tolerances for dimensions are typically cited with a plus and minus value, such as 24.0 ± 0.34 cm, which means that work will be rejected if its dimension turns out to be less than 23.66 cm or more than 24.34. If a process is working to an established tolerance, it is assumed that there is a 3-sigma risk, 0.0013 (from the normal table in the handbook) of rejection on either side, or a 99.74% probability of meeting the tolerance. A tolerance of ± 0.34 cm implies an as-built standard deviation of 0.34/3 = 0.1133 cm.

The tolerance for a sum of n independent dimensions having respective tolerances $\{t_i: i = 1, \ldots, n\}$ is the square root of the sum of the squares of the respective tolerances:

$$\sqrt{\sum_{i=1}^{n} t_i^2}$$

Tool Economy

The Taylor tool life formula is provided in the handbook as $VT^n = C$. Machining labor, excluding setups, is inversely proportional to machining speed V, but tool life T typically decreases faster than linearly with speed (a typical value of n is 0.8), requiring more tool replacements and setups.

Another economy issue often dealt with in touch-labor manufacturing is that of investing in jigs or other devices to speed up the work or make it more accurate. Long manufacturing runs can economically justify elaborate expenditures. Economic calculations should be made on the basis of profit per year if demand is not limited and the aim is to optimize the overall production rate; otherwise, and more commonly, economic calculations should be made on the basis of cost per piece, under the assumption that machines and labor can be utilized elsewhere if time is saved on the operation under question.

MANUFACTURING SYSTEMS

Work-in-Progress (WIP) Inventory

Work-in-progress (WIP), also known as work-in-process, is the total of parts and assemblies in the production flow. Prior to processing, the inventory is considered raw material, and upon completion, it is called finished goods inventory. When two stations are in series and are producing at the same average production rate (λ) and there is no WIP inventory, then an interruption in one causes an interruption in the other. If the upstream station is interrupted, the downstream station is starved for input and has to wait for the first station to return to work. If the downstream station is interrupted, the upstream station is blocked. The first station must stop working and wait for the second station to open before the work can move and the first station can return to work. Such shutdowns can be very expensive, but having WIP also has associated expenses.

If there is too little WIP, the overall production rate diminishes due to starvation. On the other hand, WIP ties up capital, requires floor space, risks pilferage or deterioration, causes extra material handling, and increases the makespan (the time a unit spends in the system).

Little's Law states that the average number of items in the system is the product of the average rate at which items leave the system and the average time each item spends in the system. This law can be used for calculations related to WIP and throughput. For example, if the average content in a WIP buffer is 50 units and the throughput is 300 units/hour, then 50/300 hour or 10 minutes extra time in the system is spent by each unit.

There is a strong connection between manufacturing system design and facility layout. Types of systems including cellular, group technology, and flexible manufacturing systems are covered in Chapter 9.

PROCESS DESIGN

Line Balancing

The handbook gives a brief set of definitions and formulas related to line balancing. Let an operation be a collection of n tasks, numbered $i = 1, 2, \ldots, n$, each having a processing time t_i. These tasks are performed in a production line that is a collection of N stations, each station performing a subset of the tasks. The line is considered perfectly balanced if the total processing times are the same at every station. In practice, it is uncommon for a line to be perfectly balanced because tasks typically can only be divided at certain points.

Assume OR is the required production rate (output rate per shift) and the total of all the processing times is Σt minutes. Then, the minimum number of operating minutes required per shift is $OR \times \Sigma t$. If the shift length is OT minutes, then at least $N_{min} = (OR \times \Sigma t)/OT$ stations are required in the case where the number of lines is one. If there are to be L lines operating in parallel, then $N_{min} = (OR \times \Sigma t)/(OT \times L)$. We can hope to find a balance so that the actual number of stations N is not much greater than N_{min}. While the calculation of N_{min} often results in a fractional number, in practice, N has to be a whole number.

Example 8.1

Consider the following set of nine tasks having processing times {5, 3, 6, 8, 10, 7, 1, 5, 3}, in minutes, and precedence relationships as shown in Exhibit 1. Let the required production rate be $OR = 28$ operations per shift, and let a shift have $OT =$ 480 operating minutes. What is the minimum number of stations required? What arrangement of work stations minimizes idle time?

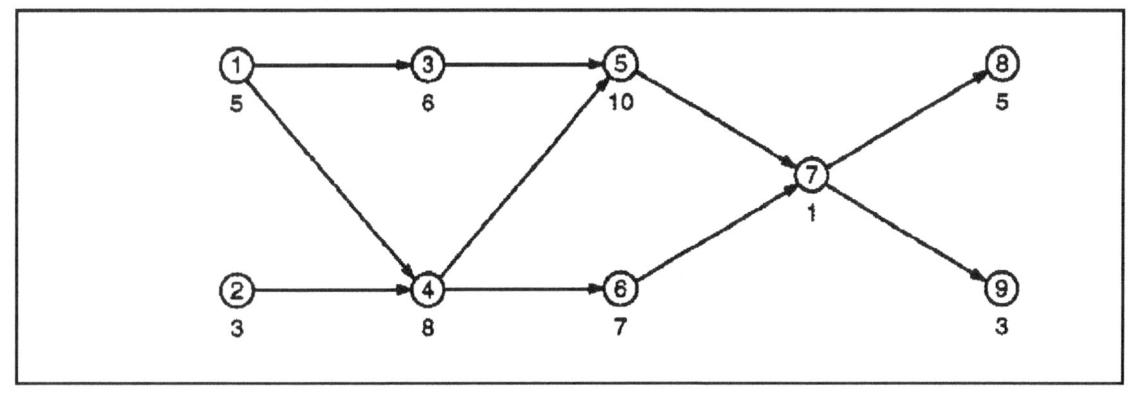

Exhibit 1 Work task relationship diagram

Solution

Since the sum of processing times is $\Sigma t = 48$ min., then if there is to be only one line we have $N_{min} = 28 \times 48/480 = 2.8$. With a very good balance, we can hope to have only three stations.

A line balance must collect tasks into stations obeying the precedences so that it will not be necessary for a partially completed piece to leave a station and later return. There are heuristics for performing line balancing. They are beyond the scope of this review, but you should remember that, while obeying precedences, you want to make the sum of processing times within each station to be as close to $\Sigma t/N$ as possible.

Here $\Sigma t/N = 48/3 = 16$. Given some trial and error, it is possible to get a perfect balance as shown in Exhibit 2.

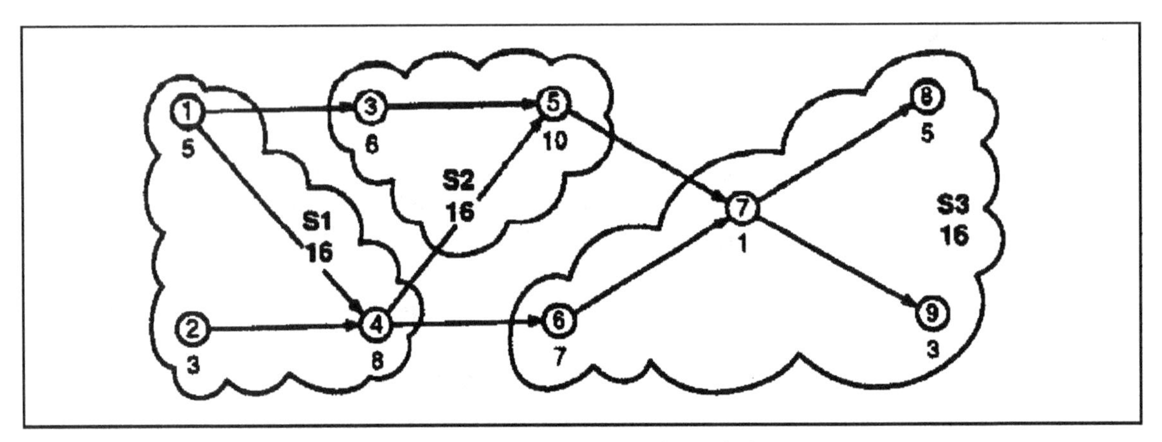

Exhibit 2 Three workstation solution

The cycle time, denoted *CT* in the handbook, is the maximum of the sums of processing times within each station. It is found by summing the processing time for the tasks at each station. In this perfectly balanced line, *CT* is 16 min.

To illustrate cycle times and idle times, let us consider what would happen with a greater production rate, say $OR = 42$ operations per shift, so that (again, for only one line) we have $N_{min} = 42 \times 48/480 = 4.2$. Exhibit 3 illustrates a balance for five stations:

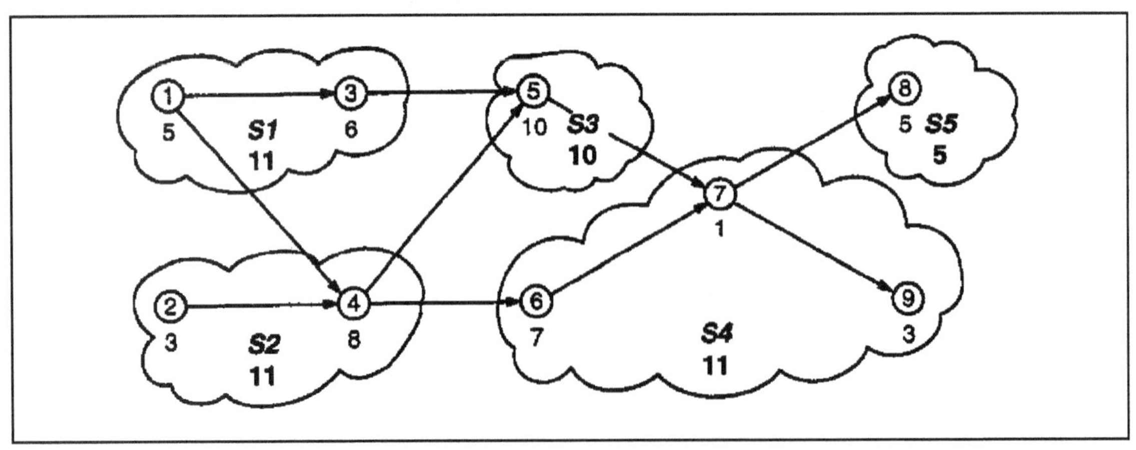

Exhibit 3 Five workstation solution

Here the cycle time is $CT = 11$ min (the maximum processing time value for the five stations). The balance is poor, so let us check if we can make the required production. The actual production rate is $OT/CT = 480/11 = 43.636$ which is adequate given $OR = 42$. (You should remember the requirement $OT/CT \geq OR$, which is not listed in the handbook. This implies that it is acceptable to exceed the required output rate, but not to fall below it.)

The idle time at a station is $CT - ST$, where ST is the sum of processing times at the station. For the above stations, $S1$ through $S5$, the idle times are $\{0, 0, 1, 0, 6\}$ min. The idle time per cycle is the sum of these, or 7 min. Finally, the percentage idle time for the line is the idle time per cycle divided by the available time. This is given in the handbook as $[\Sigma(CT\text{-}ST)/(N_{\text{actual}} \times CT)] \times 100$. In our case it would be $7/(5 \times 11) = 12.727\%$ idle time.

INVENTORY ANALYSIS

The handbook gives the formulas for economic order quantity (EOQ) and economic manufacturing quantity (EMQ) also known as EPQ for economic production quantity. The EOQ is the order quantity that minimizes total inventory holding costs and ordering costs. As an illustration of the EMQ model, suppose that it costs $A = \$50$ to place and receive an order, the expected demand is $D = 3{,}000$ units, the holding cost (including capital opportunity cost, space rental, utilities, pilferage, and deterioration) is $h = \$40$ per unit per year, and the replenishment rate is $R = 4{,}000$ units per year, then the economic lot size is:

$$EMQ = \sqrt{\frac{2AD}{h(1 - D/R)}} = \sqrt{\frac{2 \times 50 \times 3000}{40 \times (1 - 3000/4000)}} \approx 173 \text{ units per lot}$$

Some vendors offer quantity discounts, lower prices for larger orders. For this type of problem, calculate the EOQ for each price range and then compare costs for each solution. Select the alternative with the lowest total holding and ordering cost.

Safety stock is a buffer; it is the extra level of stock that is maintained to avoid stockouts (shortages in raw materials or other inventory) due to uncertainties or lead times. The amount of safety stock an organization has on hand can dramatically affect cost. Too much safety stock can result in high inventory holding costs and waste due to spoilage or expiration. Too little safety stock can result in lost sales and higher rates of customer dissatisfaction. Reorder point is the inventory level when an order for additional material should be placed. It is the product of demand and lead time.

Example **8.2**

ABC tire company sells tires at the rate of 5,000 tires/year, and it takes nine working days after the placement of an order for the tire to arrive. Tires are delivered in lots of 1,000, and there are 250 working days/year. If no planned shortages are allowed, what is the reorder point?

$$\text{Reorder point} = \text{Demand} \times \text{Lead time}$$

It is important to watch units in problems on the exam. We need to get demand per day; 5,000 tires/year / 250 working days/year = 20 tires/working day. With a lead time of 9 working days the reorder point is 20 tires/day × 9 days = 180 tires. Thus when the tire inventory reaches 180, an order will be placed for another 1,000 tires.

FORECASTING

Forecasting is making estimates about future events. The IEs often forecast future demand based on past demand to plan production. Qualitative forecasting techniques are subjective, based on the opinion and judgment of experts. When past data are available, several quantitative forecasting techniques can be used. Assume we had the following demand data and need to forecast demand for July.

January	February	March	April	May	June
52	47	63	66	72	85

Our first forecasting technique is the Naïve Approach, also called the Last-Value Forecasting Method. It is simply using the most recent time period's value; in our example it would be 85 units. This approach does not concern itself with variation or trends over time. Our next method is taking a simple average. In our example, this approach results in (52 + 47 + 63 + 66 + 72 + 85) / 6 = 68.5. You may want to emphasize more recent periods rather than treating them all equally. In that case, you can use a moving average where you only consider the last n time periods. In our example, a three-month moving average would be calculated with the last three months (66 + 72 + 85) / 3 = 75.5. The exponential smoothing forecasting method provides a more sophisticated version of the moving-average method. It gives the greatest weight to the last month and then progressively smaller weights to the older months. It uses a smoothing constant (a) between 0 and 1. This method places a weight of a on the last value and $(1 - a)$ on the next-to-last value.

$$\text{Forecast} = a\,(\text{Last value}) + (1 - a)\,(\text{Next to last value})$$

For example, when $a = 0.7$, the method places a weight of 0.7 on the last value and 0.3 on the next-to-last value. In our example that would be 0.7(85) + 0.3(72) = 81.1. A larger value of a places more emphasis on the more recent values, and a smaller value places more emphasis on the older values. Formulas for moving average, exponentially weighted moving average, and least squares linear regression are provided in the handbook.

There are two measures of forecasting error. The first is mean absolute deviation (MAD) which measures the average forecasting error.

$$\text{MAD} = (\text{Sum of forecasting errors}) / (\text{Number of forecasts})$$

The mean square error (MSE) measures the average of the square of the forecasting error.

$$\text{MSE} = (\text{Sum of square of forecasting errors}) / (\text{Number of forecasts}).$$

The MSE increases the weight of large errors relative to the weight of small errors.

SCHEDULING, AGGREGATE PLANNING, AND PRODUCTION PLANNING

In scheduling, we deal with three levels of schedules: long range planning (considers strategic focus), intermediate planning (creates aggregate plans), and short range planning (work schedules for specific products). Planning helps synchronize flow throughout the supply chain; it affects costs, equipment utilization, employment levels, and customer satisfaction. The inputs in the planning process are resources (workers, equipment, etc.), demand forecasts, policies (overtime, inventory levels, back ordering, subcontracting, etc.), and costs (inventory carrying, back orders, overtime, etc.). The outputs of production plans are the total cost, projected inventory levels, production output, subcontracting, backorders, and related items. Organizations can use a level capacity strategy which maintains a steady rate of regular-time output while meeting variations in demand with inventory levels, overtime, part-time workers, subcontracting, and/or back orders (orders taken in one period and deliveries promised for a later period). It has the advantage of stable output rates and workforce levels but disadvantages of greater inventory costs, increased overtime, increased idle time, and varying resource utilization. Another strategy is chase demand strategy which attempts to match capacity to demand. The advantages of this strategy are lower inventory costs and high labor utilization. The disadvantage is the cost of adjusting output rates and/or workforce levels. This strategy is used in the service section since inventory (services) typically cannot be stored or back ordered.

Priority Rules

There are several dispatching rules that prioritize the order jobs are performed. The differing rules improve performance with respect to different performance measures. The most basic rule, often used by default, is first-come-first-served (FCFS). Under this rule, jobs are processed in the order they arrive, like customers waiting in a line at a drive-through. It has average performance on most scheduling criteria and the sense of "fairness" appeals to customers. Next is shortest processing time first (SPT); when a machine becomes available, the jobs waiting in the queue are reviewed and the one with the shortest processing time on that machine is selected. This rule has the benefit of quickly moving jobs through that workstation which can be useful if several other steps remain in the production process. It excels at minimizing job flow and number of jobs in the system. The next rule is earliest due date (EDD); for this rule, when the machine becomes available, the waiting jobs are reviewed and the one whose due date is the soonest is selected. This rule strives to reduce tardiness; however it performs poorly on many scheduling criteria. Critical ratio is a dispatching rule where the ratio of the time remaining to required work time remaining is calculated and jobs are scheduled in order of increasing ratio. This rule performs well on average job lateness criteria.

Performance Measures

How "good" a schedule is can be evaluated by a variety of performance measures:

- Average completion time = Flow times / number of jobs
- Utilization = Process times / Flow times
- Average number of jobs in the system = Flow times / Process times
- Average job lateness = Late times / number of jobs

Johnson's Rule

This scheduling rule optimizes the schedule when there are only two machines and *n* jobs. All jobs are listed with the time on each machine. Select the job with the shortest time; if the shortest time is for processing on the first machine, the job is scheduled first, and if it is on the second machine, it is scheduled last. Once a job is scheduled, it is eliminated. Continue this process for the remaining jobs.

Example **8.3**

Job	Time on Machine 1	Time on Machine 2
A	2	5
B	6	3
C	7	8
D	5	9
E	6	4

The shortest time is 2 (Job A on Machine 1). Since it is on the first machine, it is placed first in the schedule of jobs. The next shortest is 3 (Job B on Machine 2). Since it is on the second machine, it is placed last in the schedule of jobs. Our schedule so far is A,…, B. Of the remaining jobs, Job E has the shortest time (4). Since it is on machine 2, it is placed towards the end of the schedule, which is now A,…, E, B. Two jobs remain (C & D). The shortest time is 5 for D on machine 1. It is placed next at the beginning of the schedule and C is placed in the remaining spot, resulting in a schedule of A, D, C, E, B. If there had been a tie, it would be broken arbitrarily.

Just-in-Time (JIT) Operations

Just-in-time (JIT) is a management philosophy with the goal of reducing waste in general and reducing WIP as much as possible. Within a plant, a JIT strategy saves space and reduces tied-up capital; more importantly, perhaps, it focuses attention on causes and cures for delays. The key characteristics of JIT are as follows:

- Uses a pull system for material flow
- Has small lot sizes
- Emphasizes high quality processes
- Minimizes setup requirements
- Balances workstation loads
- Requires close ties with suppliers

A manual method of achieving JIT operation for a stable staged process is the use of kanbans (cards or other signals used to control the flow of production through a factory). This is an example of a pull system, in which material handling actions are taken only on demand, as contrasted to the usual push system, where production is in advance of customer demand. Material is handled and stored in standard containers, each of which can be moved only when a kanban is attached to it. When material is needed, a worker attaches a kanban to an empty container and takes it to the preceding workstation, where the arrival constitutes an order that in turn may trigger the preceding workstation to order materials for a station farther upstream. The full container is taken to the ordering station, and the kanban is removed. The basic balance relation is:

$$[\text{Demand in cycle}] \approx [\text{Number of kanbans}] \times [\text{Size of standard container}]$$

Example 8.4

A work station has a cycle demand of 330 units and the container size is 24 units. Given that only full containers are moved and the station has 330 units on hand or requested, if the station receives an order for 170 units from downstream, how many units in excess will it have on hand after the request is filled?

Solution

First we find the number of kanbans that will be needed. Number of kanbans = 330 / 24 = 13.75, which we round up to 14. Given the 330-unit demand, a request for 336 units (14 containers) will be made. After covering the 170 units requested downstream, the station has 166 units remaining (336 – 170).

Material Requirements Planning (MRP)

If a switchboard requires two days of assembly time and is assembled from six units of a component that has a lead time of four days and one unit of a component that has a lead time of eight days, it is relatively simple to determine what quantities need to be ordered and when so that a delivery date can be met. The MRP computer programs, originally developed to produce material procurement schedules given a bill of materials and cost data, now include forecasting, inventory control, and other functions. Their main function is to produce a master production schedule that coordinates all operations. On the FE exam, MRP questions can have the structure of project scheduling or production planning problems.

It is widely believed that adoption of JIT philosophies will simplify MRP requirements and allow MRP programs to be run on personal computers; most current MRP programs run on workstation-level computers. Enterprise resource planning (ERP) is business management software that allows an organization to integrate various facets of an operation, including product planning, development, manufacturing processes, sales and marketing. A key strength of ERP is that it facilitates information flow between business functions inside the organization.

LEAN ENTERPRISES

A lean enterprise is a flexible operation that uses considerably less resources than a traditional system. Lean systems tend to achieve greater productivity, lower costs, shorter cycle times, and higher quality. The goal of lean is a balanced system where production is smooth and material flows rapidly through the system. This is done by eliminating disruptions, making the system flexible, and eliminating waste (called "Muda"). Waste represents unproductive resources. There are seven sources of waste identified in lean systems:

1. Inventory
2. Overproduction
3. Waiting time
4. Unnecessary transporting
5. Processing waste
6. Inefficient work methods
7. Product defects

An organization can become more lean by improving product design (i.e. using interchangeable parts), process design (i.e. smaller lot sizes, shorter setup time), and empowering workers (i.e. cross-training, emphasizing continuous improvement). Value stream mapping is a lean technique used to analyze and design the flow of materials and information required to bring a product or service to a consumer.

Automation Concepts

Computer-integrated manufacturing (CIM) is an approach that uses computers to control the entire production process. Sensors are used to monitor the status of production and material handling equipment. When this technology is applied in inventory control, it is called an automated storage and retrieval system (AS/RS). Computers monitor the demand and request materials from a specific storage location as needed. Automated guided vehicles (AGV, mobile robots that follow set paths) can be used for material handling in these systems. Examples of this technology are discussed in Chapter 9.

Sustainable Manufacturing

Sustainable is literally the capacity to endure. In engineering, it has come to mean reducing the negative impact on the environment. Sustainable manufacturing considers all aspects of the life cycle from the raw materials that are used, to the energy required for production, to how the product will be used, to whether it can be remanufactured or recycled.

Value Engineering

Value engineering is a systematic approach to improve the "value" of products or services by examining function. Value is defined as the ratio of function to cost. Value can therefore be increased by either improving the function or reducing the cost. The key to value engineering is identifying and removing unnecessary expenditures.

PROBLEMS

8.1 A 3/8-inch bit is to drill four six-inch deep holes per minute in a hardwood slab. If it rotates at 1750 rpm, the required feed rate in inches per revolution is most nearly:
a. 0.001
b. 0.010
c. 0.100
d. 1.000

8.2 Two stations produce 300 units/hour of furniture legs that are trimmed and smoothed at the first station and coated at the second. The second station requires 12-min. reloading operations two or three times a day. The amount of WIP required to protect the first station from these interruptions is most nearly:
a. 66 units
b. 120 units
c. 126 units
d. 132 units

8.3 A factory produces the following mixture of small consumer products: A (900 units/day), B (200 units/day), and C (25 units/day) at four work stations (W, X, Y, and Z). The parts have the following routing: A goes to W, X, and Z; B goes to W, X, Y, returns to X, and then to Z; and C goes to W, X, Y, and Z. What is the best combination of material handling equipment?
a. Use conveyors throughout the facility
b. Use a conveyor between W and X and forklifts for all other transfers
c. Use AGVs throughout the facility
d. Use AS/RS throughout the facility

8.4 In order to submit a bid for manufacturing a control panel for a traffic control center, a proposal team will prepare a preliminary production plan. The control panel's final assembly will take three weeks. There are four subassemblies that go into the final assembly; each of the subassemblies takes two weeks. Two of the subassemblies are from immediately available parts; the other two must each be preceded by a raw material procurement requiring an eight-week lead time. There is room enough in the shop to allow no more than one subassembly job to proceed at a time. The lead time for manufacturing the control panel will be most nearly:
a. 13 weeks
b. 15 weeks
c. 17 weeks
d. 19 weeks

8.5 Which of the following is *NOT* a benefit of small lot sizes in lean systems?
a. In-process inventory is considerably less.
b. Each product is produced less frequently.
c. Carrying costs are reduced.
d. There is less clutter in the workplace.

8.6 Which of the following is *NOT* a measure for judging the effectiveness of a schedule sequence?

a. average number of jobs at the work center
b. total number of jobs at the work center
c. average flow time
d. average job tardiness

8.7 Scheduling in service systems may involve scheduling:

a. equipment
b. workers
c. customers
d. all of the above

8.8 Given the processing times and due dates listed, the average completion time using the earliest due date schedule is most nearly:

Job	Time on Machine 1	Time on Machine 2
A	17	15
B	10	25
C	5	20
D	7	32
E	11	38

a. 10
b. 24
c. 30
d. 32

8.9 A barber has tracked the number of customers he has had over the last six weeks.

Week 1	Week 2	Week 3	Week 4	Week 5	Week 6
83	110	95	80	65	50

Which statement is *TRUE* about the forecasted number of customers for week 7?

a. The naïve approach forecasts a greater number than a three-week moving average forecast.
b. The three-week moving average forecasts a greater number than the grand average.
c. The forecast with a smoothing constant of 0.8 is greater than the naïve approach forecast.
d. The naïve approach forecasts a great number of customers than the grand average.

8.10 An EOQ value of 85 for a product in an import/export business was calculated. Which of the following is a *TRUE* statement?

a. If the ordering cost increases, the EOQ will remain unchanged.
b. If the ordering cost increases, the EOQ will increase.
c. If the ordering cost increases, the EOQ will decrease.
d. A change in the ordering cost will have no effect on the EOQ.

SOLUTIONS

8.1 b. The required material removal rate can be calculated from the volume of material that must be removed, the hole area × number of holes × depth. This gives a $MRR = (\pi/4)(3/8)^2 \times 4 \times 6$ in.^{3}/min. From the handbook formula, $MRR = (\pi/4)(3/8)^2 f \times 1750$, where f is the feed rate in inches per revolution. Equating and solving for f, we have $f = 24/1750 = 0.0137$ inches per revolution.

8.2 a. We assume that the production rate is a standard one and that the reloading times are included. At 300 units/hour or 5 units/min., 12 min. is the standard production time for 60 units. In practice, we would provide for slightly more than 60 units of WIP.

8.3 b. There is heavy volume between stations W and X (1150 units/day) and light volume between Y and Z (25 units/day). The flow between stations X and Y is in both directions. Conveyors are expensive and require heavy volume in fix paths to be cost justified. The low volume of product C and the two-way flow between X and Y makes option A impractical. Automated Guided Vehicles (AGV), option C, are expensive and would be more common in other industrial settings. Option D, AS/RS stands for automated storage and retrieval systems. These are typically used for storage of finished goods inventory or other situations when you have extended storage periods and high levels of WIP. The best solution is a combination that uses the efficiency of conveyors for the high volume moves and the flexibility of forklifts for the other moves.

8.4 b. The sequencing problem here has the structure of a resource-constrained project scheduling problem. Each raw material procurement must be completed before its respective subassembly can begin; all four subassemblies must be complete before the final assembly. None of the subassemblies can be simultaneous. A Gantt chart can be used to make the scheduling requirements clear:

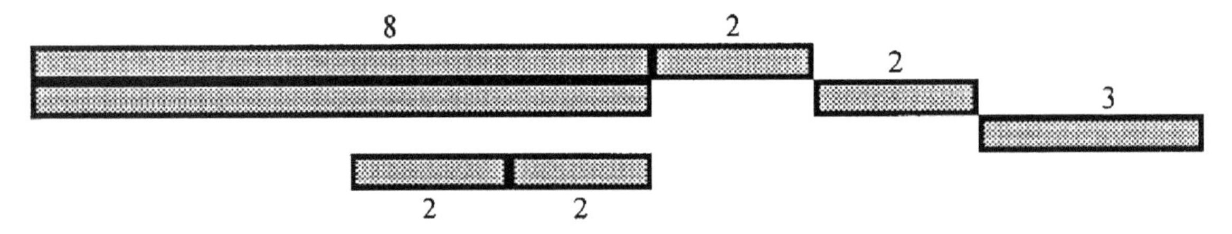

The schedule finishes in 15 weeks.

8.5 b. In lean we make products more frequently in small lots resulting in less WIP. Thus in-process inventory is considerably less, and the carrying costs are reduced. Smaller lot size also often results in less clutter in the workplace since WIP is not stored throughout the facility. In lean, products are made more often in smaller quantities.

8.6 b. "Total number of jobs at the work center" is not a common performance measurement.

8.7 d. All of the answers are true. Scheduling in the service section tends to have higher variability and can be done by an appointment or reservation system.

8.8 d. The EDD schedule is A-C-B-D-E. Following that schedule, the flow time and hours late is shown in the table below. Flow time is the sum of processing to the completion of that job. "Hours Late" is the difference between "Flow Time" and "Hour Due." We then add the flow times (17 + 32 + 22 + 39 + 50) and get 160. Divide by 5 (number of jobs) to get the average 32 hours.

Job	Procession Time (Hours)	Flow Time	Hour Due	Hours Late
A	17	17	15	17 – 15 = 2
B	10	22 + 10 = 32	25	32 – 25 = 7
C	5	17 + 5 = 22	20	22 – 20 = 2
D	7	32 + 7 = 39	32	39 – 32 = 7
E	11	39 + 11 = 50	38	50 – 38 = 12

8.9 c. The naïve forecast is the forecast for the last week, 50 customers. The grand average for the six weeks is 80.5 customers. The three-week moving average considers the last three weeks (80, 65, 50) and results in a forecast of 65 customers. The forecast with a smoothing constant is [(0.8 × 50) + (0.2 × 65)] and results in a forecast of 53 customers.

8.10 b. $EOQ = \sqrt{\frac{2AD}{h}}$; as the cost to place an order (A) goes up, the EOQ goes up, which could also be determined with an understanding of the logic: if it costs more to place an order, you will want to do it less often.

CHAPTER 9

Facilities and Logistics

OUTLINE

Expect 8-12 questions in this area on the Industrial Engineering FE Exam. Subtopics include flow measurements and analysis, layouts, location analysis, process capacity analysis, material handling capacity analysis, and supply chain management and design. In addition, questions on mathematical optimization can have a facility location context (see Chapter 6, Modeling and Computation) and there is also overlap with topics in Chapter 8, Manufacturing, Production, and Service Systems.

FACILITY DESIGN

The Industrial Engineering chapter of the *FE Supplied-Reference Handbook* includes sections on Equipment Requirements, People Requirements, Plant Location, and Material Handling.

Equipment Requirements

The basic formula given in the handbook for M_j, the required number of machines of type j, applies to the case where, for each machine type j and product i, there is a required production rate P_{ij} (e.g., widgets per year on machine type 1), a given operation time (duration) T_{ij} (e.g., machine-1 hours per widget), and a given product-type-specific machine availability C_{ij} (e.g., hours per year that widgets can be processed on machines of type 1):

$$M_j = \sum_{i=1}^{n} \frac{P_{ij}T_{ij}}{C_{ij}}$$

More commonly, there is availability $C_j = \sum_i C_{ij}$ for each machine type (e.g., total machine-1 hours available for processing widgets and other products) so that the formula becomes:

$$M_j = \frac{1}{C_j} \sum_{i=1}^{n} P_{ij} T_{ij}$$

In most situations the actual number of machines required is M_j rounded up to an integer. Time units must be compatible: T_{ij} can be expressed in standard hours or normal hours, but C_{ij} must correspondingly be the number of standard hours or normal hours available per year. If a machine can operate unattended, standard hours are the appropriate basis. C_{ij} must take any machine downtime into account; for example, if there are 2000 standard hours per year and the machine has a 92% availability, C_{ij} is 1840 hr.

When an operation on product i by machine j has a scrap rate q_{ij}, then the required production rate P_{ij} refers to the production started, not finished. For example, if the production requirement is for 7000 good pieces per year and the scrap rate is 3%, then the appropriate value of P_{ij} is 7000/(1 – 0.03) = 7216.5.

People Requirements

The basic formula given in the handbook for A_j, the required number of operators for an operation of type j, applies to the case where for each product i and operation type j there is a required production rate P_{ij}, a given operation duration (usually the standard time) T_{ij}, and a given product-type-specific operator availability C_{ij}:

$$A_j = \sum_{i=1}^{n} \frac{P_{ij} T_{ij}}{C_{ij}}$$

More commonly there is availability $C_j = \sum_i C_{ij}$ for each operator type (e.g., total operator-1 hours available for processing widgets and other products), so that the formula becomes:

$$A_j = \frac{1}{C_j} \sum_{i=1}^{n} P_{ij} T_{ij}$$

In most situations, the actual number of operators required is A_j rounded up to an integer. Time units must be compatible: T_{ij} can be expressed in standard hours or normal hours, but C_{ij} must correspondingly be the number of standard hours or normal hours available per year.

If operators are qualified for more than one operation, the total number of operators A for a set of operations is the sum of the A_j values for the set; we do not round the A_j values, but usually round A to next integer.

If the machine and people requirements M_j and A_j for every operation or machine j in a department are less than the corresponding numbers of machines and people actually provided—call these M_{j*} and A_{j*}—then the department has overcapacity. Let R^* be the minimum ratio among all the individual resource overcapacity ratios M_{j*}/M_j and A_{j*}/A_j in the department. Define P as the finished production rate from which all the required production rates $\{P_{ij}\}$ were derived. Then the finished production rate that can be achieved given the overcapacity is R^*P.

FACILITY LOCATION

Industrial engineers are expected to be able to solve the median location problem—the single-facility minimum location problem with a rectilinear distance metric. For example, let us locate an air compressor that will supply the air demands in the amounts listed to each of the five numbered locations shown in Figure 9.1. The compressed air supply lines are to be laid out parallel to the x and y axes. To minimize costs (assumed to be proportional to demand-weighted distances), the optimal x coordinate for the compressor is at the x median, which is the x coordinate such that both the sum of the demands to its left and the sum of the demands to its right are less than half the total demand. Similarly, the optimal y coordinate is at the y median. For this example, since the x median is the x coordinate of demand ③ and the y median is the y coordinate of demand ②, the compressor should be placed as shown in Figure 9.2.

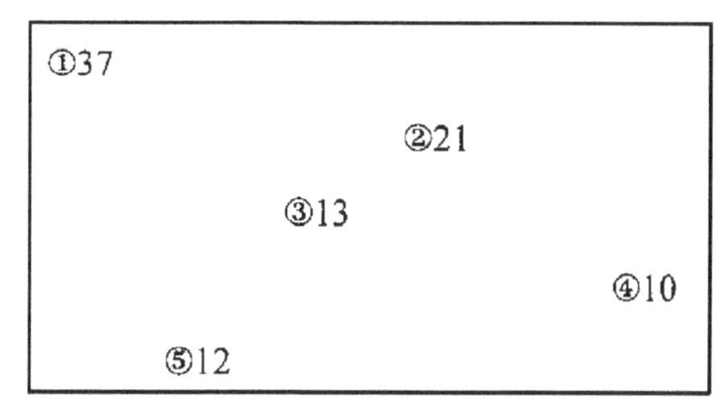

Figure 9.1 Locations of Demands

When the metric is Euclidean (beeline) distance, the same problem (called the Steiner-Weber or Fermat problem) can be solved simply. You will not be asked to use formulas, but you should remember that this is the problem that can be solved by tying a bunch of strings together at one end, threading the other ends through holes in a horizontal scale model, hanging weights proportional to the demands from the strings, and noting the equilibrium position of the knot.

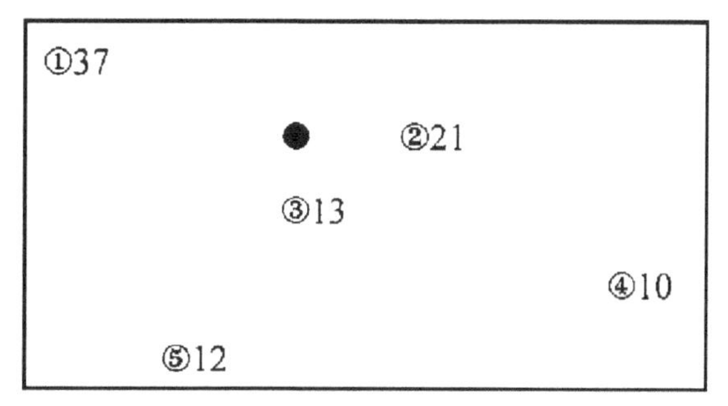

Figure 9.2 Location of Equilibrium Position for Air Compressor

One version of the Steiner-Weber problem could appear on the FE exam: If there are exactly four demanding locations and all four demands are equal, then there are two cases—if one location is inside a triangle formed by the others, it is the optimal location; otherwise the demanding locations form a convex tetrahedron, and the optimal location is where its diagonals cross.

If the objective of locating a facility were to minimize the demand-weighted sums of squares of Euclidean distances, the optimal location would be at the center of gravity of the demands. If there are m demands $\{w_i: i = 1, \ldots, m\}$ at locations whose x and y coordinates are $\{a_i\}$ and $\{b_i\}$, then the optimal x and y coordinates of the serving facility are:

$$x = \frac{\sum_{i=1}^{m} w_i a_i}{\sum_{i=1}^{m} w_i} \qquad y = \frac{\sum_{i=1}^{m} w_i b_i}{\sum_{i=1}^{m} w_i}$$

FLOW MEASUREMENT AND ANALYSIS

To/from charts show the relationship between a set of points; they are a tool used in location departments within a facility. They are similar to the mileage between cities charts shown on most road maps. In industrial applications, they show the flow of material between functional areas of a plant (i.e. receiving, paint department, milling). Rows and columns have identical titles in a corresponding sequence. The entries in the chart may be the distance between departments, number of material handling trips made between departments in each day, or the total material move by weight. Often the bottom of the chart is blank since the distance between department x and y is the same as from y to x. If there is a difference due to one-way aisles, they can be different. This chart is often used with the activity relationship chart which shows if it is desirable to have departments close together (for example if WIP flows between them) or undesirable (for example the paint and welding departments near each other could be a fire hazard). The notation for this chart is the vowels in order of importance alphabetically (A = absolute, E = especially, I = important, O = occasional, U = unlikely, and X = do not locate nearby).

LAYOUTS

There are different types of layouts that are common in different applications. A product layout is organized around the flow of work in a production and/or assembly process. A classic example is the automobile assembly plant. It is used for high volume, high repetition operations. It is costly to set up or change and often has limited flexibility. The process layout, also called a "functional layout," is organized by grouping similar machines together (i.e. lathes, presses, grinders). It is often used for machine shops. It provides a high degree of flexibility, but material handling costs may be higher with this layout. It would be used for high variety, low volume production. A cellular layout is used with group technology and cellular manufacturing where products with similar designs and/or fabrication steps are produced in the cell. It is similar to the product layout but offers more flexibility. Flexible Manufacturing Systems (FMS) is a special case of group technology that integrates computer aided manufacturing systems and automated guide vehicles (AGV) in the production process. It is very expensive and has limited applications. Aircraft manufactures have used this technology in their fabrication process when they can process similar parts and the need for tight tolerances justifies the cost of automation. A rare layout design is the fixed location when the product is stationary and the workers and equipment come to its location. Examples would include building a large ship or construction project.

Supply chain management (SCM) is the management of interconnected entities that span the movement and storage of materials that spans from raw materials, through work-in-process, to finished goods from point of origin to the point of consumption. Supply chain managers are often faced with location and distribution challenges.

PROBLEMS

9.1 Among many other operations, we must produce 4,000 widgets and 6,000 wadgets per year. A casting well accommodates two widgets or four wadgets; the normal time for casting a batch of two widgets is 35 min., and the normal time for casting a batch of four wadgets is 20 min. Four holes will be drilled in each widget and two holes in each wadget. The normal time for drilling a hole is 13 sec. The plant operates 2,000 hr per year, and there is a 15% shift allowance. The required number of wells and drills is most nearly:

a. one casting well and a small share of a drill
b. one casting well and one drill
c. two casting wells and a small share of a drill
d. two casting wells and a drill

9.2 A switchboard is assembled in four operator-attended operations whose normal times are 0.55, 0.84, 1.10, and 0.75 hr, respectively. Each operator works, including overtime, no more than 2,200 hours/year. If the shift allowance is 15%, how many people qualified for switchboard assembly must be on the payroll in order to produce 4,000 switchboards per year?

a. 5
b. 6
c. 7
d. 8

9.3 An inspection department has four analyzers, two counters, and three inspectors. It also has an old-style analyzer, still in good condition, though not currently being used, that is equivalent to 0.74 analyzers. The department's minimal requirements for the desired production rate of 700 inspections per shift are for 3.64 analyzers, 1.733 counters, and 2.4 inspectors. The achievable production rate of inspections per shift is most nearly:

a. 770
b. 800
c. 900
d. 910

9.4 On the flat desert floor, four villages are at respective coordinates (2, 4), (1, 9), (3, 5), and (4, 4). All are the same size, and all of their wells have gone dry. The best place for the government to provide a water supply tank is most nearly at coordinates:

a. 3, 4
b. 3, 5
c. 2, 6
d. 2, 4

9.5 Which of the following are advantages of the process layout?

a. Better utilization of machines
b. High degree of flexibility of machine assignment
c. Lower investment in machinery required compared to other layout designs
d. All of the above

9.6 Which type of layout would likely have the greatest WIP?

a. Product layout
b. Process layout
c. Cellular layout
d. There is no relationship between layout and WIP.

9.7 Murray Manufacturers has three products (X, Y, and Z). Each product flows through the four manufacturing units (A, B, C, and D) in the plant. Each product visits the units in different orders and has a different demand (see table). The factory has horizontal and vertical aisles. The travel distance is 10 meters between 1-2, 1-3, 2-4 and 3-4 and 20 meters between the routes 1-4 and 2-3. Which layout minimizes the overall flow distance?

Product	Route	Demand per Week
X	A-B-C-D	100
Y	A-B-C-B-D	50
Z	A-C-D	25

1	2
3	4

a. 1 is A, 2 is B, 3 is C, 4 is D
b. 1 is A, 2 is C, 3 is B, 4 is D
c. 1 is A, 2 is D, 3 is B, 4 is C
d. 1 is A, 2 is D, 3 is C, 4 is B

9.8 Which of the following material handling equipment has both continuous delivery and a fixed path?

a. Jib crane
b. Bridge crane
c. Roller conveyor
d. Fork lift truck

9.9 Mike, the material manager, is responsible for filling 80 orders per day. The work day is 480 minutes. Each order requires an average of 30 minutes of forklift time. What is the theoretical minimum number of forklifts that Mike needs to meet demand?

a. 2.67
b. 5
c. 16
d. 24

9.10 An assembly line is to be designed for a product that requires five tasks. The task times are 2.3 minutes, 1.4 minutes, 0.9 minutes, 2.1 minutes, and 1.7 minutes. The maximum cycle time is _______ and the minimum cycle time is _______ minutes.

a. 2.30; 0.90
b. 2.30; 1.68
c. 8.40; 1.70
d. 8.40; 2.30

SOLUTIONS

9.1 a. Let the product types be 1 and 2 for widgets and wadgets, respectively, and let the machine types be 1 and 2 for casting wells and drills, respectively. $P_{1,1} = 4000$. $P_{1,2} = 4000 \times 4 = 16{,}000$. $P_{2,1} = 6000$. $P_{2,2} = 6000 \times 2 = 12{,}000$. In normal hours, $T_{1,1} = 35/2/60 = 0.2916$, $T_{1,2} = T_{2,2} = 13/3600 = 0.0036$, and $T_{2,1} = 20/4/60 = 0.0833$. In normal hours, $C_1 = C_2 = 2000(1 - 0.15) = 1700$.

$$M_1 = \frac{1}{C_1}\sum_{i=1}^{2} P_{i1}T_{i1} = \frac{1}{1700}(4000 \times 0.2916 + 6000 \times 0.0833) \approx 0.98 \text{ well}$$

$$M_2 = \frac{1}{C_2}\sum_{i=1}^{2} P_{i2}T_{i2} = \frac{1}{1700}(16{,}000 \times 0.0036 + 12{,}000 \times 0.0036) \approx 0.059 \text{ drills}$$

(Unless there is some special condition—drills extremely specialized, drills very cheap, no excess drill capacity available among the "many other operations," or material handling very expensive—it seems wasteful to acquire a separate drill for a set of operations that requires only 6% of the capacity of one drill.)

9.2 c. There is only one product. The production rate is 4000/year for every operation. The total operator availability for all operations is 2200 per year, and operators are evidently qualified for all operations. The total standard time per switchboard is

(0.55 + 0.84 + 1.10 + 0.75)/(1 – 0.15) = 3.81 hr. $A = 4000 \times 3.81/2200 = 6.93 \rightarrow 7$ operators.

9.3 b. The overcapacity ratios are (4 + 0.74)/3.64 = 1.302 for analyzers, 2/1.733 = 1.154 for counters, and 3/2.4 = 1.25 for inspectors. As limited by counters, the achievable production rate is $1.154 \times 700 = 807.8$.

9.4 b. The flat desert floor implies a Euclidean metric, and all four demands are equal. This is an instance of the four-location, equal-weight Steiner-Weber problem. When the coordinates are plotted, we see that the village at coordinates (3, 5) is inside a triangle formed by the other villages and is therefore the location that minimizes the sum of beeline distances of pick up or delivery of water.

9.5 d. Grouping similar machines together provides the flexibility in the assignment of work to a specific machine (i.e., milling can be done on any of several machines). This can result in higher utilization and the need for fewer machines.

9.6 b. For both product and cellular layouts, the machines are arranged in the order of operations which minimize both travel time and WIP. This is not true for the process layout.

9.7 c. First you should determine the flow between stations using a to/from chart. The distance traveled between A and B is 150 (100 from product X and 50 from product Y). The value for the distance traveled between B and C is 200 (100 for B to C for product X and 100 for C to B for product Y) and so on. Next calculate the flow-distance score for each layout by summing the product of distance covered and number of trips for each combination.

From/To	A	B	C	D
A	—	150	25	0
B		—	200	50
C			—	125
D				—

For option A = (10 × 150) + (10 × 25) + (20 × 200) + (10 × 50) + (10 × 125) = 7,500
For option B = (10 × 150) + (10 × 25) + (20 × 200) + (10 × 50) + (10 × 125) =7,500
For option C = (10 × 150) + (20 × 25) + (10 × 200) + (20 × 50) + (10 × 125) = 6,250
For option D = (20 × 150) + (10 × 25) + (10 × 200) + (10 × 50) + (20 × 125) =8,250

9.8 c. A jib crane is mounted to the floor and pivots loads in a circular path. A bridge crane is mounted to overhead beams and follows a fixed path with a rectangular work area. All conveyors (roller, belt, etc.) provide continuous material handling, unlike the other answers that make trips to deliver material. Conveyors follow a fixed path, unlike a forklift truck that can drive to various locations via a variety of paths.

9.9 b. First calculate how much at best each forklift can move per day (480 minutes per day / 30 minutes per load = 16 loads per day). The number needed is the ratio of demand over capacity (80 loads per day / 16 loads per forklift = 5 forklifts per day). Thus 5 forklifts is the minimum number required.

9.10 d. The maximum cycle time is when there is only one workstation. It is the sum of all of the times, 8.40 minutes. The minimum cycle time is driven by the bottleneck; if you have five workstations the longest task is 2.30 minutes.

CHAPTER 10

Human Factors, Ergonomics, and Safety

OUTLINE

Expect 8-12 questions on hazard identification and risk assessment, environment stress assessment, industrial hygiene, design for usability, anthropometry, biomechanics, cumulative trauma disorders, systems safety, and cognitive engineering.

The Industrial Engineering chapter of the *FE Supplied-Reference Handbook* contains an "Ergonomics" section that includes the NIOSH lifting formula, a review of basic biomechanical equations, and a chart of OSHA's permissible noise exposures. The handbook also contains a table of human body dimensions, charts on hearing and hearing loss, and a chart on heat stress.

HAZARD IDENTIFICATION AND RISK ASSESSMENT

Hazard identification is the process of identifying hazards (a potential to cause harm and/or damage to an individual or property), engineering methods to avoid them, and mitigating their impact. Categories of hazards include biological, chemical, electrical, environmental, ergonomic, and physical. Risk assessment is the determination of risk related to a hazardous situation or threat. Risk is assessed based on the magnitude of the potential loss (L) and the probability (p) that the loss will occur. The figure below is a risk matrix (also called a fever chart) that shows the interaction between impact and probability. The risk of being hit by a meteorite would be catastrophic, and the likelihood is very, very low, so it would not be a hazard that would get many resources for contingency planning. A paper cut would be at the other extreme of the matrix (high likelihood but low impact). Risks in the upper right hand corner would be the focus of risk mitigation.

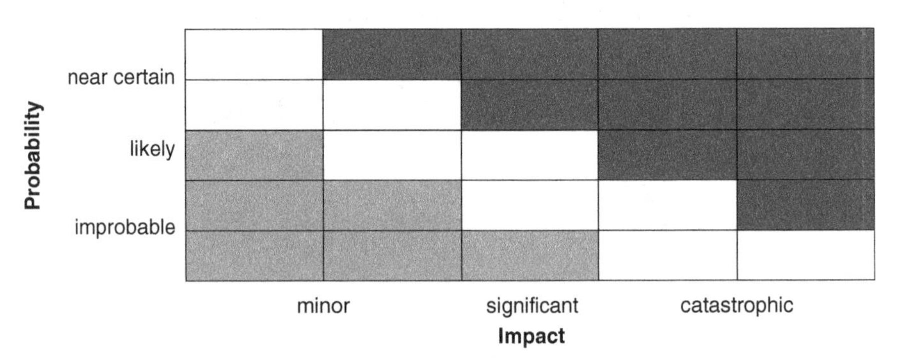

Figure 10.1 Risk Matrix

ENVIRONMENT STRESS ASSESSMENT

An environmental stress assessment would consider hazards in the work environment such as noise, heat, vibration, or chemicals. The handbook has a table listing permissible noise exposures based on the associated OSHA standard. Like other environmental hazards, it considers exposure time and exposure level. Notice that as the noise level increases, the amount of time allowed decreases. This is a permissible exposure limit, a value which is allowed for day after day work exposure based on regulations. A higher value without a margin of safety is the threshold limit value, a level which is believed acceptable for daily exposure without adverse health effects. A peak exposure or ceiling limit is the absolute exposure limit that should not be exceeded at any time. For noise, the permissible exposure limit is 85 dBA; below this, no noise protection plan is required. The threshold limit value is 90 dBA for an eight-hour day, while the peak exposure is 140 dBA which should never be exceeded even for a brief sound such as a weapon firing. These limits vary with the hazard. Since no data other than noise and temperature is provided in the handbook, do not expect specific quantitative questions on other hazards.

INDUSTRIAL HYGIENE

Industrial hygiene is synonymous with occupational hygiene. It is the science of preventing illness from hazards in the workplace (i.e., asbestos exposure). By comparison, safety is concerned with preventing accidents (i.e., electrocution), and ergonomics is concerned with cumulative trauma disorders (i.e., carpal tunnel syndrome). Industrial hygienists often use sampling equipment to measure the exposure level to various hazards (noise, dust, chemicals, etc.)

For any environmental hazard, the following protective steps can be taken:

1. Replace the hazard material with something less hazardous.
2. Protect the worker from exposure by personal protective equipment (PPE).
3. Manage the exposure administratively, such as total time of exposure.
4. Require training on the hazard and appropriate precautions.

MSDS, material safety data sheets, are an important part of industrial hygiene. They are intended to provide workers and emergency personnel with procedures for working with the substance safely, physical data (i.e., melting point), toxicity, health effects, first aid, reactivity, storage/disposal, protective equipment to be used, and other information to meet right to know requirements. There is a duty to label dangerous substances and to educate workers and community members on the hazards.

DESIGN FOR USABILITY

Usability is a measure of how easy a product or system is to use; "user-friendly" is another term for this concept. It can be measured by various factors, including learnability (how easy it is for users to accomplish their goals the first time they encounter the product or system), efficiency, memorability (how easy it is for return users to re-establish proficiency), accuracy, and user satisfaction. A key concept for designing for usability is user-centered design, which is about defining who the users are, defining their tasks and goals, their experience levels, what functions they want and need from a system, what information they want and need, and understanding how the users think the system should work. Steps involved include user analysis, design/redesigning, prototyping, and testing. Using this approach, the designer focuses on the users' needs. Users are involved at all stages of the process.

Usability considerations include the following:

- Who are the users, what do they know, and what can they learn?
- What do users want or need to do?
- What context will it be used in?
- What must be left to the machine?
- Can users, including the disabled, easily accomplish intended tasks?
- How much training do users need? What documentation or other supports are needed?
- What type of errors will users make? Can the user recover from these errors?
- What expectations do users bring to the product or system?

ANTHROPOMETRY

The table of human body dimensions in the handbook is really two separate tables, one for males and one for females. Dimensions may be assumed to be normally distributed, and the standard deviations are given (the 50th-percentile value of each dimension is also the mean). Values for other percentiles can be found by using the given standard deviation and the corresponding Z value from the normal table. Do not assume that dimensions of a mixed male-female population are normally distributed.

The popliteal height sitting (height from foot sole to the back of the knee) determines the seat height from the floor beyond which a footrest is required. The seat height plus thigh clearance height determines the minimum height of an obstruction above the knee; the seat height can be less than popliteal height sitting if the legs are allowed to extend forward.

Ideal height for hand work is 5 cm below the elbow rest height sitting or the elbow height standing; often hand work is above the worktable surface (e.g., a computer keyboard is 2 or 3 cm high). The worktable surface should not be below knuckle height for standing work, which would require the worker to bend. The ideal working range for standing work is between knuckle height and shoulder height which lessens the stress on the shoulders and back. The last line of the table gives body weights.

BIOMECHANICS

As the basic equations for "Biomechanics of the Human Body" in the handbook shows, the up, down, and sideways components of force exerted by the hands (and body weight) are balanced by equal and opposite force components at the feet; similarly, the torques (moments) exerted at the hands are balanced by equal and opposite torques at the feet. Torques and forces are also balanced at each point between the hands and feet.

Recall that the moment about point A due to a force acting at point B is computed by multiplying the length of the moment arm (distance between the points) by the force component perpendicular to the moment arm (line between the points). For example, with the elbow held at a right angle, if the elbow flexor acts at a point two inches in front of the elbow fulcrum and exerts a 400-lb forward force, the moment is 800 inch pounds, and will cause the palm of the hand, say 12 inches from the elbow fulcrum, to exert a downward force F on an external object such that $12 \times F = 800$, or $F \approx 67$ lb. At the same time, the force of the feet on the floor is 67 lb less than if the downward force were not being exerted.

Recall that the component of a force at an angle α from the line of the force F has the value $F \cos\alpha$. For example, if a worker is exerting a 20-kg force in pushing a cart up a 5% slope ($\tan \alpha = 0.05$, so that $\alpha = \tan^{-1} 0.05 = 2.8624°$), then the horizontal component is $20\cos\alpha$, or 19.975 kg (and the vertical component is $20\sin\alpha = 0.99875$ kg).

If the coefficient of friction between an object and a plane is μ, and if the component of force perpendicular to the plane is F, then the frictional force that resists movement of the object parallel to the plane is μF. Commonly μ is 0.3 to 0.4 for cartons sliding on floors, 0.05 to 0.10 for things on wheels, and 0.7 to 0.8 for shoes on floors. Formulas for these calculations are given in the handbook.

CUMULATIVE TRAUMA DISORDERS

Ergonomics is the science of conforming the workplace and all of its elements to the worker. Its goal is to eliminate physical injuries and cumulative trauma injuries. Cumulative trauma disorders (CTD) are injuries, temporary or permanent, to the soft tissues in the body (nerves, tendons, muscles, etc.) that result from accumulation of repetitive motions stress. These are also called repetitive strain injuries (RSI). Examples include back injuries, carpal tunnel syndrome, tendonitis, and trigger finger. The following are risk factors that increase the chances of developing a CTD:

- Repetitive motion
- Excessive force applications
- Unnatural posture
- Prolonged static exertion

- Fast movement
- Vibration
- Cold environment
- Pressure or sharp edges on soft tissues

Carpal Tunnel Syndrome

Carpal tunnel syndrome (CTS) is the most common CTD. The carpal tunnel is comprised of tendons and bones in the wrist. The median nerve passes through this tunnel and branches into the fingers. When there is inflammation in this area and pressure compresses the median nerve, it is called CTS. Symptoms include pain, numbness, and loss of sensitivity. Treatment for CTS includes anti-inflammatory drugs, splints, and surgery.

The most expensive CTD are injuries to the lower back. These are often associated with unnatural work postures and/or manual material handling work tasks. Other disorders include bursitis (inflammation of the bursa, small sacs of that help create smooth movements in various joints in the body) and tendinitis (inflammation of a tendon).

Workstation design is a key tool used by ergonomists to prevent work-related injuries. The goal is to have workers in a neutral posture (a natural, comfortable position that minimizes stress on the body). The posture of the worker should be considered from head to toe. The head should be facing forward to reduce twisting of the neck. The nature line of sight is 10 degrees below the horizon; this is key when positioning a computer monitor in front of the worker, at an appropriate viewing distance. Next the shoulders should be at a comfortable position, and having the computer keyboard or work object at the appropriate height will help maintain this posture. You want to avoid extreme angles for the elbow, hips, knees and ankles. Wrist rest, properly adjusted chairs, and footrests are some examples of items that can be added or adjusted in workstations to allow people of varying shapes and sizes to work comfortably. Back posture is another very important consideration. You want to avoid twisting and provide good support in the lumbar region.

Lifting

The National Institute for Occupational Safety and Health (NIOSH) has developed a lifting formula for manual, two-handed lifts. The equation is

$$RWL = LC \times HM \times VM \times DM \times AM \times FM \times CM \text{ or}$$

$$RWL = 51(10/H)(1 - 0.0075|V{-}30|)(0.82 + 1.8/D)(1{-}0.0032A) \times FM \times CM$$

where values for the frequency multiplier (FM) and the coupling multiplier (CM) are found in tables. The handbook gives the NIOSH lifting formula as:

$$RWL = 51(10/H)(1 - 0.0075|V - 30|)(0.82 + 1.8/D)(1 - 0.0032A)$$

This is a modification of the full NIOSH lifting formula. It assumes that the frequency is no greater than one lift every five minutes and that the coupling is rated as "good," which results in an FM and a CM of one. The formula is in U.S. customary units. The maximum recommended weight in pounds for a two-hand lift is the RWL. The load constant (LC) is 51 pounds. H is the distance that the load is in front of the body, measured from the mid-point of the line joining the inner ankle bones to the mid-point of the hand grasps. The minimum value of *H* (when an object is next to the worker's body) is 10 inches. The vertical distance (*V*) is calculated at both the origin and at the destination of the limit. The difference between these two is the vertical distance traveled (*D*).

The calculated *RWL* will always be equal to or less than 51 pounds because the multipliers (horizontal, vertical, distance, asymmetry, frequency, and coupling) are always between zero and one. The lifting index (*LI*) provides an estimate of the physical stress associated with a manual lifting task.

$$LI = \text{Load Weight (lbs)}/RWL\ \text{(lbs)}$$

If the *LI* is greater than one, the task exceeds the recommended weight limit, and the task should be redesigned. The *LI* and *RWL* should be calculated both at the origin and at the destination of the lift.

SYSTEMS SAFETY

A system is group of interrelated elements or parts that interact to achieve a common goal. Systems safety is a risk management strategy that identifies and analyzes hazards and addresses mitigation of them in complex systems. It is different from traditional safety in that it uses a systems-based approach. Traditionally tools such as failure mode and effect analysis (FMEA) look at the impact of component failures as the cause of accidents and the resulting effects. As systems have become more complex, new approaches to risk management were needed. The concept of systems safety is to improve the design safety-critical subsystems in complex systems (such as spacecrafts, weapons systems, and transportation systems) by conducting hazard analyses to identify hazards, verify designs, assess safety features, and procedures risk mitigation strategies before the system is built.

COGNITIVE ENGINEERING

Cognitive engineering is concerned with how mental processes, such as perception, memory, reasoning, and motor response, affect humans' interactions with products or systems. The relevant topics include information processing, situation awareness, mental workload, decision-making, human error, and mental models. It is a subset of the larger field of human factors. Miller's Magic Number was introduced in a 1956 paper by George Miller. The number is actually a range, seven plus or minus two. It is the limit of our capacity for processing information. It is influenced by the capacity of an average human's working memory. If a person is asked to make an absolute judgment on a single dimension, he or she can typically only distinguish around seven different levels. An application of this human limitation is the Likert scale; most people can easily rate something on a 1-5 or 1-7 scale but a 1-10 or even 1-100 is too great. What would it mean if you rated a song as a 93 on a "dance-ability scale"? Could you really distinguish it from a 92 or 94? Another application of this is presenting information in chunks (data groups rather than in a string) improves memory performance. A phone number presented as 1-800-555-1234 is easier to remember than 18005551234. This also applies to part numbers, credit card numbers, and license plates.

Situation awareness is the perception of position in the environment with respect to time and/or space and the ability to project this status into the future. It is critical in complex environments loaded with stimulus, such as an aircraft pilot flying in combat. An accurate mental model is a prerequisite for situation awareness. A mental model is a set of well-defined, highly-organized, yet dynamic knowledge structures developed over time from experience. It is our representation of how something works. For example, an engineer's mental model of a microwave oven may be rather complex, but for someone else, it may simply be "a magic box we put food into to heat quickly." Both models would work well for heating a bowl of soup, but the engineer's mental model would likely be more useful when trying to troubleshoot a problem with the microwave. Our mental models are influenced by schemas (a mental structure of preconceived ideas or a cognitive framework). A schema influences how we process information. We do better with information that fits our schemas than contradictory information; we process it faster and more accurately. "Spud" Webb is a former NBA basketball player who won a slam dunk contest. You likely have an image of what a great slam-dunking basketball player looks like (based on your schema). "Spud" is five feet, seven inches tall: Are you surprised?

A related concept is bias, an inclination towards a particular perspective. We all suffer from bias. When identical resumes were reviewed with the only difference being the person's name (male vs. female; common name vs. unusual name), they were evaluated differently by potential employers. This has led to the use of blind auditions and blind reviews.

We all have learned expectations about how things work; these are called "population stereotypes." For example, in America we push the light switch up to turn on the light, turn the volume knob clockwise to make it louder, and red means "stop" or "hot." We expect the hot water at a sink to be controlled by the knob on the left and if we move into a house plumbed backwards, we will frequently turn the wrong knob. That is a population stereotype, users' learned expectations. How well a design matches our expectations is called "compatibility." Poor compatibility in the design of controls, displays, and other devices will increase response time and errors, and reduce user satisfaction.

PROBLEMS

10.1 Which of the following statements is *FALSE* about human hearing?
a. The hearing threshold in decibels tends to shift higher with age.
b. The ability to detect sounds tends to differ between men and women.
c. Our primary aspect of a sound that determines whether a human can hear it is its frequency (measured in hertz).
d. OSHA requires hearing protection for workers in loud environments.

10.2 Once or twice an hour, a 40-lb carton of finished goods is lifted from a table that is 33 in. high to a conveyor that is 27 in. high. The worker twists 30 degrees to one side to pick up the carton and 30 degrees to the other side to lower the carton in one smooth continuous motion. The horizontal distance is 18 in. The carton has optimally designed handles. The *RWL* is most nearly:
a. 21.8 lbs
b. 24.5 lbs
c. 40 lbs
d. 51 lbs

10.3 A 5th-percentile woman exerts a 15-kg horizontal force to close a door. The minimum coefficient of friction required for her shoes against the floor is most nearly:
a. 0.1
b. 0.3
c. 0.5
d. 0.7

10.4 A worker sits at a 24-in. desk on an adjustable chair without a footrest. For approximately what percentage of a 50%-female workforce will the worker's elbows clear the desktop?
a. 25%
b. 45%
c. 65%
d. 85%

10.5 Steve is a baggage handler for an airline. He spends his day outside loading airplanes in Alaska. Which of the following is *NOT* a risk factor for a cumulative trauma disorder?
a. Cold temperature
b. Repetition of lifting bag after bag
c. Twisting required to lift bags from a cart onto the plane
d. All of the above are cumulative trauma risk factors.

10.6 Which of the following statements is an example of a mental model?
a. "I don't like classical music!"
b. "The gas pedal is always on the right and the brake is on left in a car."
c. "It is slippery outside after a storm."
d. "I can't remember my account number; it is too long!"

10.7 When designing a usability test, we should consider:

a. the novice user
b. the disabled user
c. what the user is trying to do
d. all of the above

10.8 Risk is:

a. a function of probability and consequence
b. a random, unpredictable occurrence
c. a concern for design engineers but not manufacturing engineers
d. an unmanageable cost of doing business

10.9 Abdul is analyzing what would happen if a chemical storage tank ruptures and spills hazardous content before the operator can contain it. Which analysis technique is he using?

a. User analysis
b. Industrial hygiene assessment
c. Failure Mode and Effect Analyses (FMEA)
d. Human error analysis

10.10 Mohammed wants to determine the correct height to install peepholes in the doors of his apartment complex to allow renters to see who is at the door before opening it. Which anthropometric data should he use?

a. 95th percentile male eye height
b. 50th percentile male eye height
c. 5th percentile male and 95th percentile female eye height
d. 5th percentile female eye height

SOLUTIONS

10.1 c. Statements A and B are true. Graphs providing specific values related to both appear in the handbook. Both frequency (measured in hertz) and pressure (measured in decibels) affect our ability to hear. This is also shown graphically in the handbook. The OSHA requires hearing protection for workers in loud environments as listed in a table in the handbook.

10.2 b. HM = 10/18 = 0.55. Since V = 27 or 33 in., the VM (1-0.0075|V-30|) is 0.97. Since D = 6 in., the DM (0.82 + 1.8/D) is 1 (not 1.12; the factor cannot exceed 1, and is 1 for any lift distances 10 inches or less). Since the A is the same for both the origin and destination (30 degrees), the AM (1 – 0.0032A) is 0.90 for both. Thus the RWL = 51 × 0.55 × 0.97 × 1.0 × 0.90 = 24.5 lbs.

10.3 c. From the table of human body dimensions given in the IE chapter of the handbook, the woman's weight is 46.2 kg. This force acts perpendicularly to the floor and will induce a frictional force 46.2μ kg to resist horizontal movement of her shoe. For her shoe not to slide, the frictional force must resist 15 kg (the equal and opposite reaction to the force on the door), and μ must be at least 15/46.2 ≈ 0.325.

10.4 d. Let x_p represent the popliteal height, sitting, and let x_e represent the elbow rest height, sitting. The elbows clear the desktop while the feet can touch the floor (barefoot) if (in cm) $x_p + x_e \geq$ 24 in. × 2.54 cm/in. = 60.96 cm. From the body dimensions table, for females at the percentile corresponding to normal variate z, the sum of popliteal and elbow height is $39.8 + 23.3 + \sigma z$ cm, where $\sigma = (2.6^2 + 2.9^2)^{1/2} = 3.895$ is the standard deviation of the sum. Thus for females, we have $63.1 + 3.895z \geq 60.96 \text{ cm} \Rightarrow z \geq -0.55$, and from the normal distribution table in the handbook, we see that approximately 29th-percentile women or bigger (71% of women) would be big enough for their elbows to clear the desktop. For males, we have $44.2 + 24.3 + \sigma z \geq 60.96$ cm, where $\sigma = (2.8^2 + 3.0^2)^{1/2} = 4.104 \Rightarrow z \geq -1.84$, so that 3rd percentile men or bigger (97% of men) are big enough. The 71% of women and 97% of men is 84% of the mixed workforce.

10.5 d. Cold temperature is a risk factor for CTD. By definition, highly repetitive tasks increase the risk of a cumulative trauma. As shown in the NIOSH lifting guideline, twisting is a hazardous component of manual material handling.

10.6 b. Statement A is an opinion. Statement C is an observation. Statement D is the effects of 7 ± 2 on short term memory capacity. Statement B is a learned expectation of how one thinks a car should be designed.

10.7 d. In usability testing, you should consider the variety of users, the environments they will be in, and the tasks they are attempting.

10.8 a. The definition of risk is a function of probability and consequence. It can be predicted with some probability level. It should be a concern for both the designer and manufacturing engineer because hazards (defects) can be introduced in both the design and the manufacturing. Managers develop contingency plans to mitigate risks.

10.9 c. User analysis is a human factors technique to identify the characteristics, abilities, needs, etc., of the user. Industrial hygiene would explore the health risks from exposure to the chemical. Failure Mode and Effect Analyses (FMEA) evaluate component failures (storage tank) and the effects. Human error analysis looks at what would happen if the operator made an error.

10.10 d. The 50th percentile male would be designing for the average, one size rarely fits all, and this should be avoided. Designing for the range is preferred, but 5th percentile female to 95th percentile male is not the option given. In this unusual application, the concern is if short users (5th percentile female) can see out of the hole. Tall users would bend down to see out.

CHAPTER 11

Work Design

OUTLINE

Expect 8-12 questions on methods analysis, time study, predetermined time standard systems, work sampling, and learning curves on the Industrial Engineering FE exam.

METHODS ANALYSIS

Industrial engineering has its roots in methods analysis, the study of work methods to improve productivity, quality, and worker well-being. It consists of decomposing work tasks and identifying non-productivity aspects to eliminate and productivity aspects to be improved. Productivity can be improved by changing the product design, the process design, and/or the workstation layout. Frank and Lillian Gilbreth were early pioneers. They developed 17 fundamental motions they named "Therbligs," their last name spelled backwards—almost.

Therbligs

Therbligs include the following:

- Transport empty (reach): reaching for an object with empty hand (effective motion)
- Grasp: grasping an object to gain control over it (effective motion)
- Transport loaded (move): moving an object using a hand motion (effective motion)
- Hold: holding an object (ineffective motion)
- Release load: release control of an object (effective motion)

- Preposition: positioning and/or orienting an object for the next operation and relative to an approximation location (effective motion)
- Position: positioning and/or orienting an object in the defined location (ineffective motion)
- Use: manipulate a tool in the intended way during the course working (effective motion)
- Assemble: joining two parts together (effective motion)
- Disassemble: separating multiple components that were joined (effective motion)
- Search: attempting to find an object using the eyes and hands (ineffective motion)
- Select: choosing among several objects in a group (ineffective motion)
- Plan: deciding on a course of action (ineffective motion)
- Inspect: determining the quality or the characteristics of an object using the eyes and/or other senses (ineffective motion)
- Unavoidable delay: waiting due to factors beyond the worker's control and included in the work cycle (ineffective motion)
- Avoidable delay: waiting within the worker's control causes idleness that is not included in the regular work cycle (ineffective motion)
- Rest to overcome a fatigue: a pause in the motions of the hands and/or body (ineffective motion)

The Gilbreths' classification as effective or ineffective motions may seem confusing at first. They were studying high repetition, short duration tasks. They felt the work should be well defined, eliminating the need for ineffective therbligs such as plan or rest. Therbligs that are effective can be used in an ineffective manner. An operator disassembling a subassembly from packaging material is maybe a value-add operation, while an operator disassembling a part for rework would not be value-added; yet, both are effective therbligs. For example, for how therbligs can be used to improve a work method, consider a worker who reaches for a tool, grasps it, moves it, and uses it in the production of a part. If the worker is reaching and moving a tool a great distance, it could be moved closer to the worker reducing these two motions (improving workstation layout). If the part design is changed, all of these therbligs might be eliminated entirely (improving product design). Another improvement could be replacing a hand tool with a power tool (improving work method).

Motion Economy

The principles of motion economy are guidelines that can be used in work analysis to improve efficiency and reduce worker fatigue by working with the way the body moves rather than against it. There is no one single list of principles. The following are some commonly listed principles to work with the human body:

- Utilize both hands; do not have one hand idle or holding an object for an extended time.
- Two hand movements should be synchronized and symmetrical (like rowing a boat).

- Work should be designed for both left- and right-handed people.
- Work with the object's momentum and gravity whenever possible.
- Manual operations should be automated or mechanized when appropriate.
- Use natural motions (i.e., continuous curves) instead of unnatural motions (i.e., moving in a straight line with abrupt changes in direction).
- Minimize the need to focus on an object with the eyes to find it or move it.
- Utilize the worker's feet and legs when appropriate (i.e., foot pedals).
- Tools and materials should be located in fixed positions close to the worker and where they will be used.
- Provide a good working environment (i.e. lighting, ventilation, chair).
- When possible, combine operations/motions.

Additional Methods Analysis Techniques

A variety of analysis techniques exist that are graphical depictions of the steps required to produce a good or provide a service. Given the multiple choice nature of the exam, it would be difficult to ask questions about creating such charts, but a student would be well served to understand their purpose. At the lowest level of detail is the two-hand process chart (also called left-hand/right-hand chart or simultaneous motion chart). This chart graphically compares the actions of both hands, at the therblig level of motion, on a vertical time scale. It is a useful analysis tool to identify method problems related to the principles of motion economy. The operation process chart shows the chronological sequence of all operations, inspections, and materials used in a manufacturing process. The flow process chart is in greater detail. It lists each event sequentially and classifies each of them as operation, transportation, delay, storage, or inspection. This chart is useful in identifying hidden costs such as delays, temporary storage, or excessive travel distances. The flow diagram is a floor plan layout with lines drawn on it to depict the movement of material and WIP (work in progress) through the work station. If the diagram looks like someone dumped a pot of spaghetti on a plate, it is a strong indication that you have a material flow problem. The gang process chart (or man-machine chart) is a depiction of how balanced a work cell is; it shows the relationship between operating time and idle time for workers and machines. Other analysis techniques are included in the quality chapter of this book (Chapter 12).

TIME STUDY

Industrial engineers are expected to know the basic definitions and concepts of work measurement without needing to consult reference materials.

A task is a sequence of elements. Each task or element must have a definite end event that is also the start event for the next task or element. A time standard is the amount of time that a thoroughly trained and experienced operator working at a normal pace (discussed in more detail later in this chapter) will require to complete the task using a set work method, and under specific working conditions. Changes in a task, such as adding work elements or changing equipment, will change the corresponding time standard. Time allowances are commonly included in the time standard. This is additional time that is not part of a task or element—time can be allowed for personal activity (i.e., breaks, phone calls), fatigue, and unavoidable delays.

Time studies yield observed times. Stopwatches or video recording are used to time from the start to the end of an event to establish observed times (after subtracting any interruptions from the time duration). These times must be adjusted to a normal pace and be increased with time allowances before they can be considered time standards.

As an example, let the ith replication of the observed time (duration) of task or element j be denoted O_{ij}. Each observed time O_{ij} is converted to a normal-time datum N_{ij} by multiplying by the pace rating, R_{ij}, which is a subjective estimate of the pace of the work. Normal pace (R_{ij} = 1.00) is defined as the speed at which a typical experienced worker would be expected to work. An R_{ij} = 0.85 would be given when a worker is working at a slower pace such that the number of completions per unit time would be 85% of those at normal pace. For a worker working at a faster than normal pace, an effort rating of greater than 1.00 would be given. R_{ij} = 1.20 is the fastest effort rating usually seen in practice. A normal-time estimate is the average:

$$N_j = \frac{1}{n}\sum_{i=1}^{n} N_{ij} = \frac{1}{n}\sum_{i=1}^{n} O_{ij}R_{ij}$$

If a task consists of J elements, its normal time estimate is the sum of those for the elements:

$$N = \sum_{j=1}^{J} N_j$$

If a task or element is paced by machinery rather than the operator, a pace rating of 100% is assumed.

Let S be the time standard for a task or operation for which N has already been estimated. The allowance time is usually expressed either based on the job time or on the workday; the formulas for each are given in the handbook. The task allowance—additional time provided in the time standard for personal needs, recovering from fatigue, and unavoidable delays—is included with the time estimate to establish a time standard. For any element or task, the time standard is the normal time multiplied by the allowance factor ($ST = NT \times AF$ is given in the handbook). The estimated production rate per shift, F, is the ratio of the number of time units available for work per shift over the time standard expressed in the same time units.

Example 11.1

A time study of an assembly operation produced an average observed time value of 6.70 minutes and a pace rating of 110%. The company uses a 15% allowance. Calculate the time standard for the operation and the expected number of parts assembled per 450-minute shift.

Solution

The normal time for that task would be 6.70 × 1.10 = 7.37 minutes. The time standard would be 7.37 + (0.15 × 7.37) ≈ 8.48 minutes. If the shift duration is T = 450 min., the number of parts assembled during a shift would be $F = T/S$ or 450 minuntes per shift / 8.48 minutes per part = 53 parts per shift.

It is important to pay attention to the units in a problem. You may need to convert between seconds, minutes, and shifts, for instance. Including the units in your calculations can help prevent conversation errors. Time standards can be established using various techniques. Stopwatch time is a common approach, but it is labor intensive and the effort ratings involved in converting time estimates into time standards can be controversial due to their subjective nature.

Predetermined Time Standard Systems

Starting in the 1930s, several companies realized a more cost-effective way of establishing time standards. Industrial engineers noticed that similar jobs often had key elements that were used repeatedly. By building a database of these common work elements and their associated time values, IEs only had to study the unique aspects of a work task to develop a time standard rather than studying the entire work task. This work measurement approach is called standard data. Over time these databases were generalized into predetermined time standard systems, a system of standard time values that could be applied to a wide variety of tasks. One of these earliest systems was called Methods-Time Measurement (MTM). Films of workers operating presses were studied and rated by a group of IEs. Tasks were broken into very detailed elements based on Therbligs, distance travel, and degree of care required.

To establish a time standard, the IE would observe a work task and break it into its elemental components (reach nine inches, grasp object, transport object six inches, etc.). The engineer would then look up the associated time value for each of the component motions in the MTM database. To simplify the MTM system, a time unit was created, time measurement unit, or TMU. One TMU = 36 milliseconds (0.036 seconds) and one hour = 100,000 TMU. The component times are summed to create the normal time for the task. Then the appropriate time allowance is added. Since the original time values were rated and adjusted to 100%, the IE does not have to use effort rating when using predetermined time standard systems. Over time the early predetermined time standard systems have been simplified. Ranges of distances have been combined. Common elements have been grouped into larger work elements. Today other systems, including MOST and MODAPTS, are used in various industries.

Other methods of establishing time standards include the use of historical data. This approach is similar to standard data in that a database of time values is used to create time standards. Production records are used to create the database of times. One drawback to this approach is that it is dependent on the method and equipment used in the past. If equipment is updated or methods revised, changes will not be directly captured in the time standards. Also, historical data is based on how long it did take to do a task, not on how long it should take to do it. Allowances factors are not included with this work measurement technique since delays and worker breaks are already included in the data.

Time standards can also be established by estimation, but this approach is very subjective. It tends to be applied at the aggregate level and is often used as a preliminary estimate for planning and cost estimation.

Work Sampling

Work sampling is a statistical technique to determine the proportion of time spent in different activities based on observations. It is also called "activity sampling" or "occurrence sampling." In a work sampling study, a large number of observations are made of the workers over an extended period of time. The observations are taken at random times during the study. This technique can be used in service and manufacturing operations. The handbook provides formulas for calculating the level of error for a set of observations.

Example 11.2

Karli observed a work crew 100 times. She classified their efforts as productive 70% and idle 30% of the time. What is the absolute error in her observations at a 95% confidence level (which equals a Z value of 1.96)?

Solution

For this problem $p = 0.7$ (the percent observed as productive) and $1 - p = 0.3$ (the percent idle). The other values in the formula are $n = 100$ and $Z = 1.96$. $D = Z\sqrt{[p(1 - p)/n]} \approx 0.09$

LEARNING CURVE

As an individual or organization performs many repetitions of a process or operation, learning occurs and productivity grows. The most widely used model for this is summarized in the handbook. The direct labor hours T_N required to perform repetition N are considered to decline according to a learning rate ϕ (less than 1; a typical value is 0.85) such that $T_{2N} = \phi T_N$. Let $T_1 = K$, the direct labor hours for the first repetition; then, according to this learning-curve model, the second repetition requires $T_2 = \phi K$ hours, the fourth requires $T_4 = \phi^2 K$ hours, the eighth requires $T_8 = \phi^3 K$ hours, and so forth: for each doubling of experience, the effort is multiplied by ϕ. We define the exponent $s = \log_2\phi = \ln\phi/\ln 2$ (for example, $\phi = 0.85$ gives $s = -0.234465254$), and express the learning curve as $T_N = KN^s$.

For example, if the first unit takes 304 hr and the learning rate is 85 percent, then the second unit takes 258.4 hr (304×0.85) and the fourth unit takes 219.64 hr (258.4×0.85). The formula is used to calculate that the 15th unit will take 161.1 hr ($T_{15} = 304 \times 15^{-0.234465254}$).

When two points on a learning curve are given, (e.g., the fourth and seventh units took 20.2 and 17.1 hr, respectively), the parameters K and s can be obtained by solving the two resulting $T_N = KN^s$ equations together (e.g., $20.2 = K \times 4^s$ and $17.1 = K \times 7^s$). When a learning curve is to be fitted to more than two points, linear regression may be used on the logarithmic form of the model: $\ln T_N = \ln K + s\ln N$. In the regression, the (x, y) data points are pairs $(\ln N, \ln T_N)$ and the regression gives intercept $\ln K$ (whence you can get $K = e^{\ln K}$) and slope s. Another possibility is to plot $\ln N$ as an abscissa and $\ln T_N$ as an ordinate, estimating the intercept and slope visually.

Although the learning curve never levels off, its slope becomes very small. Customarily it is assumed that after a given $N = N^*$ no further learning takes place, and T_N remains constant thereafter: $T_N = T_{N^*}$ for $N > N^*$.

A frequent use of the learning curve is to estimate the effort to produce a certain number of additional units, say W units, after Y units have already been produced. Since it is inconvenient during the FE Exam to sum a long series, the handbook gives an approximation for T_{avg}, the average time for the first N units:

$$T_{avg} = \frac{K}{N(1+s)}[(N+0.5)^{(1+s)} - 0.5^{(1+s)}]$$

The total effort (total time) for the first N units is obtained by suppressing N in the denominator. To estimate the total effort to produce a certain number of additional units, say W units, after Y units have already been produced, you can use the formula to calculate the total effort for both numbers of units and take the difference; in doing this, you can save time by noting that the 0.5^{1+s} terms add out, and the initial factor can be multiplied after $(N + 0.5)^{1+s}$ for the smaller N (that is, Y) is subtracted from that for the larger N (that is, $Y + W$)

A linear approximation to the average effort to produce W units after Y units have already been produced, computed by averaging T_Y and T_{Y+W}, gives an overestimate, since it averages two points on a convex decreasing curve. For large N, as the curve flattens, the error becomes smaller. The error for the approximation given in the T_{avg} formula is always small.

PROBLEMS

11.1 Workers were observed packaging software kits. The observed time per kit averaged 26.2 seconds, and the observed workers were judged to be working at a 105% pace. If a 14% allowance is used for the task, the number of kits expected to be produced per worker per 450-minute shift is most nearly:

a. 844
b. 861
c. 886
d. 901

11.2 In methods analysis, a "therblig" is a:

a. charting technique
b. job enrichment technique
c. basic elemental motion
d. fraction (.0006) of a minute

11.3 In a work sampling study, the absolute accuracy is a function of:

a. the variability of observed times
b. the desired confidence for the estimated job time
c. when the observations are taken
d. none of the above

11.4 In a stopwatch time study, the average time it takes a given worker to perform a task a certain number of times is the:

a. observed time
b. normal time
c. standard time
d. performance rating time

11.5 A job had an observed cycle time of four minutes, a performance rating of 80 percent, and an allowance factor that was 20 percent of job time. Normal time for the job in minutes is:

a. 3.20
b. 3.84
c. 4.00
d. 4.80

11.6 A job took 123 hours for the second repetition. Assume a 90% learning curve. Which of the following is *NOT* true?

a. The first repetition took approximately 137 hours.
b. The fourth repetition will take approximately 111 hours.
c. The time required will level off at approximately 90 hours.
d. We can calculate the total time required for the first 90 repetitions of this job.

11.7 A job had an observed time of 10 minutes, a performance rating of .90, and an allowance factor of 20 percent of job time. Twenty-five cycles were timed. Standard time for the job in minutes is:

a. 10.0
b. 10.8
c. 12.5
d. dependent on the number of cycles observed

11.8 A fast-food restaurant manager wants to determine the number of workers to staff the drive-thru window, soft drink dispenser, fryer, and cash registers. Which chart would be most useful for this analysis?

a. Micromotion analysis chart
b. Flow diagram
c. Operation process chart
d. Gang process chart

11.9 Which chart is not designed to show delays?

a. Micromotion analysis chart
b. Flow diagram
c. Flow process chart
d. Gang process chart

11.10 An allowance factor would normally *NOT* include time for which of the following factors?

a. Noise levels
b. Resting to overcome fatigue
c. Personal phone calls
d. Restroom allowances

SOLUTIONS

11.1 b. N = (26.2 sec./kit) × 1.05 = 27.51 sec./kit S = 27.51 sec./kit × (1.14) = 31.3614 sec./kit
$F = T/S$ = (450 min/shift × 60 sec./min.) /31.3614 sec. ≈ 861 kits/shift.

11.2 c. Response D is the definition of a TMU.

11.3 b. Variation is not in the formula for calculating absolute error. The "time of day" may influence the data but care should have been taken to randomize the observations and this factor cannot be represented in the formula for absolute error. Z in the formula is an estimate of our confidence.

11.4 a. The normal time includes an adjustment for effort. Standard time would include the allowance for the job. Performance rating is a percentage value.

11.5 a. The observed time is 4 minutes. The normal time is 4 minutes × the rating (80%) = 3.2 minutes. If this is multiplied by the allowance (1.2), then the time standard is 3.84 minutes.

11.6 c. Every time the number of repetitions is doubled the time changes by 90%. Thus going from the 1st to 2nd rep is X × 0.90 = 123 hours. Solving for X (the time required for the first rep), you get 137 hours. The 4th is 123 × 0.90 or 111 hours. Response D is true; the formula is given in the Handbook. Response C is correct because it is a false statement.

11.7 b. Since you are given observed time, the number of cycles is extra information you don't need to calculate standard time.

$ST = OT \times PR \times AF$, thus 10 minutes × 0.90 × 1.20 = 10.80

11.8 d. The gang process chart shows the activities of the workers and equipment (i.e., fryer). Using the chart the manager would see if the workers/equipment are working or idle over time.

11.9 b. The micromotion analysis chart shows delays at the therblig level. The flow process chart and gang process chart show delays at a higher level. The flow diagram is concerned solely with movement of materials.

11.10 c. Allowances are given for personal (restroom allowance), fatigue (noise levels and resting to overcome fatigue), and unavoidable delay. A personal phone call is an avoidable delay and time allowances are not commonly given for these.

CHAPTER 12

Quality

OUTLINE

Expect 8-12 questions on Six Sigma, quality management and planning tools, control charts, process capability and specifications, sampling plans, design of experiments for quality improvement, and reliability engineering on the Industrial Engineering FE exam.

Quality is the ability of a product or service to consistently meet or exceed customers' expectations. There are both qualitative and quantitative techniques that are used in quality. This topic has overlap with both statistics (Chapter 5) and industrial management (Chapter 9). The mathematics chapter of the *FE Supplied-Reference Handbook* contains a section on probability and statistics. This includes formulas for combination and permutation, basic laws of probability, properties of the binomial distribution, properties of the normal distribution, a normal distribution table, a t distribution table, and a table of the F distribution for $\alpha = 0.05$. The Industrial Engineering chapter of the handbook contains a table of density function formulas with mean and variance for the binomial, hypergeometric, Poisson, geometric, negative binomial, multinomial, uniform, gamma, exponential, Weibull, normal, and triangular distributions. It also contains formulas for one-way and two-way analysis of variance and linear regression. When you use the handbook during the exam, remember useful material for quality questions can be found in both sections. Don't forget the statistics capabilities of your hand calculator. Make sure you know how to use it to determine sample means, sample standard deviations, linear regression parameters, and correlation coefficients. Also be sure it meets the NCEES requirements for a calculator before the day of the exam.

SIX SIGMA

Six Sigma is a business process for improving quality, reducing costs, and increasing customer satisfaction. The name comes from statistics, where this high level of quality represents no more than 3.4 defects per million. The goal of Six Sigma is to identify and remove causes of defects. Plan-do-check-act (PDCA) or plan-do-study-act (PDSA) is an iterative method of continuous improvement. It is related to the improvement cycle DMAIC (Define, Measure, Analyze, Improve, and Control). The DMAIC cycle is used for existing products, related is DMADV (Define, Measure, Analyze, Design, Verify) which is used during product development.

The term *5S* stands for five Japanese terms that translate into sort, straighten, shine, standardize and sustain (or similar terms). It is used to organize and reduce waste in the workplace. Affinity diagram is a three-step process to sort a large number of items (such as potential causes of defects) into manageable subgroups. The first step is to record each idea on a card or small piece of paper, look for ideas that are related, and sort cards into groups until all cards have been assigned to a group. Many quality techniques also have applications in lean enterprise, which is discussed in Chapter 8.

TOTAL QUALITY MANAGEMENT

Common themes shared by most modern theories of management are consensus—that everyone should share the same goals—and empowerment—that everyone is enabled to work effectively toward those goals. Total quality management (TQM) is an approach to implement consensus and empowerment through fundamental organizational change. In an organization, TQM occurs not as a philosophy but as a program for organizational change with a definite commitment made in advance by top management. A TQM program can significantly change the way things are done and can be expensive. It emphasizes commitment to change and continuous improvement.

Recognition of the many involved constituencies or stakeholders is essential:

Owners	Creditors	Suppliers
Managers	Unions	Regulators
Employees	Customers	Society

The newer stakeholder capitalism movement involves making decisions on behalf of other stakeholders as opposed to only the owners. The TQM movement is not so concerned with decision making—in fact, the main function of managers becomes enabling, making sure that workers have the right resources and environment, rather than decision making.

Barriers to consensus include adversarial relationships (for example, union versus management), some traditional management ideas, fear (especially fear of superiors), and lack of communication. A TQM applies consensus-building techniques to create and maintain a shared vision of where the organization should go and how everyone should contribute.

In addition to consensus-building and vision-building techniques, TQM programs use benchmarking—observing in detail how successful firms solve problems. This is usually done by sending employees to study at different plants (not necessarily in the same industry) that have succeeded in a way that the first plant wants to emulate. For example, employees from U.S. Repeating Arms (the manufacturer of Winchester firearms) visited the Saturn division of General Motors to benchmark Saturn's successful implementation of JIT manufacturing.

To empower people is to give them the opportunity to be self-directing. Direct supervision must be replaced by enabling the group actually doing the work to adopt the details of its work procedure. Employees are empowered when they can change their own work for the better. A TQM program generally results in some added degree of job enrichment and group work. These and other tools that can be used in TQM are discussed in Chapter 9.

W. Edwards Deming summarized his management philosophy in his 14 Points, which are often quoted as guiding principles for TQM. Deming's 14 Points are listed as the following, with commentary in parentheses.

1. Create constancy of purpose for improvement of product and service.
2. Adopt the new philosophy. (Mistakes and negativism are unacceptable.)
3. Cease dependence on mass inspection. (Inspection does not improve process quality.)
4. End the practice of awarding business on price tag alone. (Develop long-term quality relationships.)
5. Improve constantly and forever the system of production and service. (Strive for continuous improvement.)
6. Institute training.
7. Institute leadership.
8. Drive out fear.
9. Break down barriers between staff areas. (Departments should work together, not compete.)
10. Eliminate slogans, exhortations, and targets for the workforce.
11. Eliminate numerical quotas.
12. Remove barriers to pride of workmanship.
13. Institute a vigorous program of education and retraining.
14. Take action to accomplish the transformation.

Point number 11—elimination of numerical work quotas; substitution of aids and helpful supervision for quotas—has been widely debated because it seems to say that work standards should not be set or that work measurement should not be done. Actually, Dr. Deming himself clarified this point many times, saying that he meant elimination of arbitrary quotas such as sales quotas (i.e., "We sold 20,000 last year, and we can achieve 25,000 this year if we all . . .").

The TQM philosophy emphasizes continuous improvement and the importance of quality. Fundamental is the belief that workers want to do a good job and will if management provides an environment to encourage them. The TQM philosophy believes that ideas can be generated at all levels of the organization.

QUALITY FUNCTION DEPLOYMENT (QFD)

Quality function deployment is a method of translating user demands into design quality. It is used during the design of products and services. It is used to capture the voice of the customer (VOC), what the customer wants or needs and transforming those into engineering characteristics. For example, a new car buyer wants a quiet ride. The QFD method would be used to identify manufacturing requirements (such as tolerances for the fit of the car doors or the size of gasket around the windows) to meet this want. A key tool in the QFD process is the House of Quality, which is a diagram shaped like a house that shows the relationship between customers' desires and organization or product capabilities. It shows the "whats" and "hows" of meeting customers' wants. A Kano model can be used in product development to classify features into delighters/exciters, satisfiers, and dissatisfiers.

Basic Tools of Quality

These are basic tools used to identify problems and improve quality. Check sheets, also called tally sheets, are among the simplest. A check sheet is a form used to collect quality data as it happens. It is a quick tool to assess a process. The results of this tool are often used in more detailed analysis later. Histograms and flow charts can be used to document processes as well.

Pareto Principle

The Pareto Principle is known as the 80-20 rule or "the vital few and the insignificant many." It was suggested by Joseph Juran and named for Vilfredo Pareto who observed that 80% of land in Italy was owned by 20% of the population. In business, this is shown when 80% of your sales come from only 20% of your product line (the vital few). The quality application of this concept is that IEs should focus on the vital few problems, issues, factors, etc. to improve performance. It can also be used to draw a Pareto chart that shows frequency of occurrence of defects versus causes. Using this chart, one can identify the vital 20% factors causing the majority of the defects.

Cause-and-Effect Diagrams

Fishbone or Ishikawa Diagrams are cause-and-effect diagrams that can be used to identify quality issues. In the graphical technique, a type of defect is placed on a horizontal line; factors that are possible contributors to the problem are put on lines coming off the horizontal line forming a rough approximation of the skeleton of a fish. The chart is used in brainstorming potential factors causing quality concerns. A diagram for a company that has repeatedly had holes drilled to the wrong size would start with the problem and then add general factors (material, equipment, people, etc.). Under each of those, specifics are added.

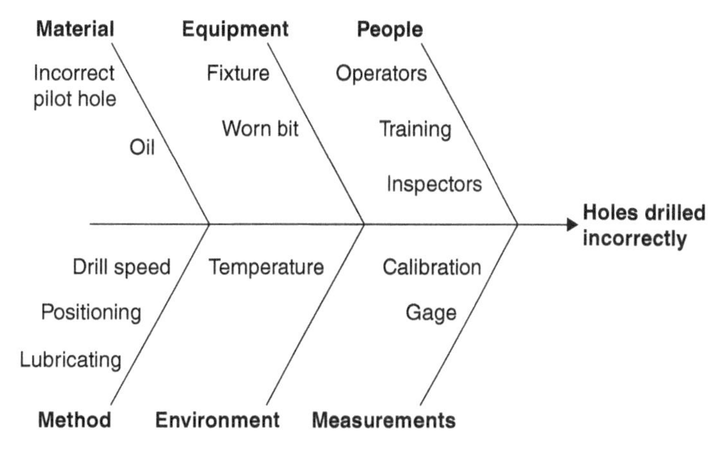

Figure 12.1 Fishbone diagram

STATISTICAL QUALITY CONTROL

The IE chapter of the handbook contains a "Statistical Quality Control" section that gives properties of various control charts—average, range, standard deviation, and moving range—and the tests for detecting an out-of-control condition.

Consider a steady-state process manufacturing pink plastic parts that is being monitored so that corrective action can be taken when a process variable's level or variability goes out of control. The specifications require the color to be such that a colorimeter reading stays within the range of 220 to 280. We might try to center the process so that the average reading is close to 250, but it is important to remember that the statistical condition of being in control or out of control does not depend on the specifications. Instead, it depends on the recent operating history; for instance, if the process has been operating at a mean of 265 and a standard deviation of 5, then an observation less than 250 would be considered as out-of-control because it is more than three standard deviations ("sigmas") below the mean.

Let a sample of n successive parts be tested from time to time, giving a sample average $\bar{X}$ and a sample standard deviation S (or sample range R). For example, if $n = 3$ parts were tested and the colorimeter readings were {258, 262, 245}, then $\bar{X} = 255$ and $S = 8.8882$ (or $R = 17$).

Control Charts

To detect when the level of a process goes out of control, we use an "X-bar" chart, that is, a control chart that tracks the successive values of the grand mean ($\bar{\bar{X}}$), which is defined as the mean of recent $\bar{X}$ values. Similarly we define the grand standard deviation $\bar{S}$ (or the grand range $\bar{R}$) as the average of successive values. We establish an upper-control-limit value, a centerline value, and a lower-control-limit value as follows:

$$UCL_{\bar{X}} = \bar{\bar{X}} + A_3\bar{S} \quad \text{or} \quad \bar{\bar{X}} + A_2\bar{R}$$

$$CL_{\bar{X}} = \bar{\bar{X}}$$

$$LCL_{\bar{X}} = \bar{\bar{X}} - A_3\bar{S} \quad \text{or} \quad \bar{\bar{X}} - A_2\bar{R}$$

Here A_3 is a number tabulated in the "Statistical Quality Control" section of the IE chapter of the handbook. For large n, recall that $\bar{S}$ approaches σ, the true standard deviation. A_3 approaches $3/\sqrt{n}$ for large n, so we see that $A_3\bar{S}$ approaches $3\sigma/\sqrt{n}$, which is 3 times the true standard deviation for a sample of size n. The upper and lower control limits are "3-sigma" limits.

A_2 plays a similar role to that of A_3 when the variability of a sample is expressed as a sample range rather than a sample standard deviation. It is more accurate to use sample standard deviations, but in the past, when computations were performed by hand, ranges were easier to compute, and they remain in common use.

Process level and process variability are two different things. For instance, in the production of pink plastic parts, if the dye were replenished with slightly different dye, the colorimeter readings might move to a new level while maintaining the same variability; on the other hand, if an agitator in the dye vat failed, the readings might become more highly variable (some parts getting more dye than others) while maintaining the same average level.

To detect when the variability of a process goes out of control, we use an S chart (or R chart). This allows us to track the successive values of sample standard deviation S (or sample range R). The IE chapter of the handbook gives formulas for the upper-control-limit value, a centerline value, and a lower-control-limit value for these charts as follows:

$$UCL_S = B_4\bar{S} \quad \text{or} \quad UCL_{\bar{R}} = D_4\bar{R}$$

$$CL_S = \bar{S} \quad \text{or} \quad CL_{\bar{R}} = \bar{R}$$

$$LCL_S = B_3\bar{S} \quad \text{or} \quad LCL_{\bar{R}} = D_3\bar{R}$$

Values of B_3, B_4, D_3, and D_4 are also given. The R chart is more commonly used, although the S chart is more accurate.

For any control chart, the tests for an out-of-control condition are (1) a point falls outside the 3-sigma control limits, (2) two out of three successive points fall outside a 2-sigma limit on the same side, (3) four out of five successive points fall outside a 1-sigma limit on the same side, or (4) eight successive points fall on the same side of the centerline. These tests for an out-of-control condition are given in the handbook.

PROCESS CAPABILITY AND SPECIALITIES

Process capability is a measure of a process's ability to meet its designed purpose. It consists of measuring the variability of the process output and comparing that variability with specified tolerances. It is represented by the variable C_{pk} or C_{pm}. Control charts are used to determine if a process is "in statistical control." If the process is not in statistical control, then capability has no meaning. Process capability involves only common cause variation (also called natural variation) and not special cause variation. Common cause variations are caused by chance, noise, or non-assignable causes. Special cause variations have an assignable cause such as equipment failure or operator error. Capability analysis can be used to predict how many parts will be produced out of specification. This is a new topic for the Industrial Engineering FE exam. The handbook does not currently have formulas for calculating a process capability index. In general it is the ratio of range between the upper specification limit (USL) and lower specification limit (LSL) divided by an estimate of process variability. For a normally distributed process it can be expressed as

$$C_{pk} = (USL - LSL) / 6\sigma$$

Sampling Plans

Acceptance sampling uses statistical sampling to determine whether to accept or reject a lot of material. It is commonly used in the receiving process when shipments arrive from vendors. A sample, a random subset of the total populations (such as a lot or truckload), is tested. If the number of defects in the sample falls below an acceptable quality limit (AQL), the lot is accepted; otherwise additional testing may be required or the lot may be rejected. The rationale behind sampling is that 100% inspection is time consuming, costly, and does not improve quality. Various procedures and tables exist for acceptance sampling.

DESIGN OF EXPERIMENTS FOR QUALITY IMPROVEMENT

The IE chapter of the handbook contains a section that gives formulas for least squares, standard errors, confidence intervals, and sample correction coefficient. Information related to factorial experiments, ANOVA, and randomized block design is also provided. Design of experiments (DOE) uses statistics to explore the effect of some process or intervention (the "treatment") on some objects (the "experimental units"). Design of experiments deals with planning, conducting, analyzing and interpreting controlled tests to evaluate the factors that impact a parameter or group of parameters we are interested in. One example is if we wanted to test whether heat treating a certain metal at a different temperature affects hardness. The factor, the aspect of the process we are testing, is heat treating. If we use different amounts or settings of a factor, they are called "levels." The DOE can be used to test potential causes of quality problems.

RELIABILITY ENGINEERING

Reliability is a device's ability to perform its required function under stated conditions for a specified period of time. It considers the probability of failure and the frequency of failure. This is a new topic for the Industrial Engineering FE exam and such formulas have yet to be added to the handbook. This will likely limit the type of questions that can be asked in this area. Some basic concepts would still be suitable to exam questions.

Reliability in Series

If a set of components are configured in series (one after another), when one component fails, the entire system fails. The reliability of the system is the product of the component reliabilities.

→ Component 1 — Component 2 - - - Component n

$$R_{system} = R_1 \times R_2 \times \ldots R_n$$

So, if you had three identical components in series, each with a reliability of 90%, the system reliability would be $R_{system} = 0.9 \times 0.9 \times 0.9 = 72.9\%$.

Adding one more component in the series with a reliability of 98% would decrease the system reliability to approximately 71.44% (calculated by 72.9 × 0.98). Series is not the best design when seeking high reliability.

Reliability in Parallel

A parallel system is configured such that if not all of the components fail, the entire system still works. This provides redundancy. Redundancy means when one part of a system fails, there is an alternate backup system. An example would be the brake and the emergency brake in your car.

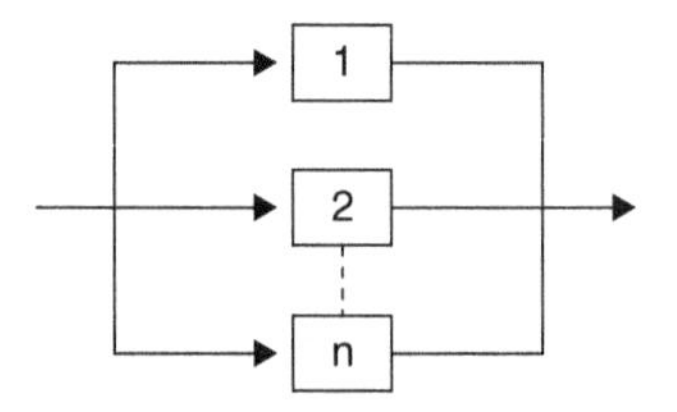

For a parallel system, the reliability formula changes to:

$$R_{system} = 1 - (1 - R_1) \times (1 - R_2) \times \ldots(1 - R_n)$$

If we had three redundant components, each with a reliability of 90%, and we only need one working for the system to work, our reliability would be:

$$R_{system} = 1 - (1 - 0.9) \times (1 - 0.9) \times (1 - 0.9) = 99.9\%$$

Mean time between failures (MTBF) is a common measure of reliability. It is the sum of the operational periods divided by the number of observed failures. If you install six new environmentally-friendly light bulbs in your kitchen and track how long it takes them to burn out (fail), you could calculate MTBF. If the bulbs lasted 6, 9, 18, 21, 23, and 37 months, the MTBF is the average, 19 months.

PROBLEMS

12.1 A control chart does which of the following?

a. Improves process quality
b. Differentiates between common cause and special cause variation
c. Determines if a production lot should be accepted
d. None of the above

12.2 The finished coat thickness given by a painting operation is tested by taking six thickness readings from time to time. The average thickness has been running 3.2 mm, and the standard deviation of thickness has been running 1.1 mm. A range chart for this operation should have *UCL*, *CL*, and *LCL* values most nearly:

a. 6, 3, and 0
b. 5.9, 2.9, and –0.1
c. 7, 3, and 1
d. 7.1, 3.1, and 1.1

12.3 Many tools and techniques are used in total quality management (TQM). Which of the following is *NOT* commonly used in TQM?

a. Employee involvement
b. Statistical process control
c. Inspection quotas
d. Benchmarking

12.4 Beth is the manager of a production line making donuts. She is concerned with the diameter of the donuts. The machine is designed to create donuts that are 4.0 inches wide. Beth reviewed production records of the average donut size for the last ten days and found the following data: 4.0, 4.1, 3.6, 3.7, 4.2, 4.1, 4.0, 3.7, 3.9, and 3.7. She wants to develop a control chart to plot the average of a three donut sample. Which of the following is closest to the values she should use?

a. $CL_x = 3.9$, $UCL_x = 4.0$, $LCL_x = 3.8$
b. CLx = 4.0, $UCL_x = 4.1$, $LCL_x = 3.9$
c. $CL_x = 3.9$, $UCL_x = 4.0$, LCLx = 3.8, and $CL_R = 0.8$, $UCL_R = 2.5$, $LCL_R = 0$
d. More information is needed to calculate the control limits.

12.5 Jose has developed an unmanned submarine that is controlled by a navigation computer (N) and a propulsion computer (P). The reliabilities are 95% and 87% respectively. He has another $5,000 to spend to improve the quality of his design. What should he do to make the greatest improvement to his sub's reliability?

a. Replace the P computer with a 89% reliable system.
b. Add a redundant P computer to the existing P computer.
c. Replace the N computer with a 97% reliable system.
d. Both "a" and "c" are equally good options.

12.6 All of the following costs are likely to decrease as a result of better quality *EXCEPT*:

a. scrap costs
b. inspection costs
c. maintenance costs
d. warranty and service costs

12.7 Pareto charts are used to:

a. identify inspection points in a process
b. organize errors, problems, or defects
c. show an assembly sequence
d. provide guidelines for quality training

12.8 When a sample measurement falls inside the control limits, it means that:

a. each unit manufactured is good enough to sell
b. there is no variability in the process
c. the product will meet or exceeds the customers' requirements
d. if there is no other pattern in the samples, the process is in control

12.9 A store owner notices a strong, positive correlation between in-store promotions and the use of e-coupons. How would an increase in e-coupons affect in-store promotions?

a. There is no association between e-coupons and in-store promotions.
b. When there are more in-store promotions, fewer e-coupons are used.
c. When there are more in-store promotions, e-coupons are not affective.
d. More e-coupons are used when there are more in-store promotions.

12.10 The process capability index uses both ________ and ________.

a. precision; specifications
b. precision; accuracy
c. variability; specification
d. variability; mean

SOLUTIONS

12.1 **b.** Control charts can be used to assess changes to improve quality, but no type of inspection improves quality. Acceptance sampling is used to determine if a lot should be accepted. Control charts track variability and are used to differentiate between common cause and special cause variation (being out of control).

12.2 **a.** We are asked to find values on a range chart (R chart). We are given $\bar{X} = 3.2$, $\bar{S} = 1.1$, and $n = 6$. From tables in the IE chapter of the handbook, the corresponding values are $A_2 = 0.483$, $A_3 = 1.287$, $D_3 = 0$, and $D_4 = 2.004$. We need to find $\bar{R}$. Using the formulas for the $UCL_{\bar{X}}$ we can set $A_3\bar{S} = A_2\bar{R}$. Solving, we have $\bar{R} = 2.931$; this is the centerline value. The upper and lower control limits should be at $D_4\bar{R}$ and $D_3\bar{R}$ respectively; resulting in $UCL = 5.87$ and $LCL = 0$.

12.3 **c.** Inspection and quotas are not common in TQM (review Deming's 14 points). The other answers are.

12.4 **d.** The grand average $\bar{\bar{X}}$ is 3.9. This is found by averaging the 10 averages Beth has. Therefore, the CL for the $\bar{X}$ chart would be 3.9, not the design value of 4.0. To calculate the upper and lower control limits, Beth needs $\bar{R}$ (the average of the range for each of the 10 samples). The data is not given in the problem.

12.5 **b.** We know Jose's computers (N and P) are currently in series. Their combined reliabilities result in a current system reliability of 82.65% ($0.95 \times 0.87 = 0.8265$). Option A changes that to 84.55% ($0.95 \times 0.89 = 0.8455$) and Option C changes that to 84.39% ($0.97 \times 0.87 = 0.8439$). Since option B is adding redundancy, we first calculate that new value $1 - (1 - 0.87)^2 = 98.31\%$ and then find the overall system reliability of approximately 93.4% ($0.95 \times 0.9831 = 0.93394$).

12.6 **c.** With better quality we should have less scrap and warranty and service costs because we are making fewer bad parts. If we improve the quality of our operations, we may be able to inspect less. Maintenance costs will likely go up since control charts and other quality tools will highlight the need to adjust and correct machinery more often.

12.7 **b.** A Pareto chart is concerned with identifying the vital few causes of errors or defects. That most closely matches "organize errors, problems or defects."

12.8 **d.** If a control chart does not have any of the issues listed for tests for out of control given in the handbook, then it is considered to be in control. Being in control does not guarantee that "each unit manufactured is good enough to sell" or "there is no variability in the process." It means that there is no special cause variability in the production process. It doesn't analyze customers' requirements, just the performance on one or more dimensions or aspects of the product.

12.9 d. A strong, positive correlation is an example of a result that can be found from a design of experiment. When there is a correlation between two variables, they are related. A positive correlation means they change in the same way (both improve or both get worse). A negative correction is the opposite (one gets better and the other gets worse).

12.10 c. The capability index evaluates how much the process is varying with respect to specifications.

CHAPTER 13

Systems Engineering

OUTLINE

Expect 8-12 questions on requirements analysis, system design, human systems integration, functional analysis and allocation, configuration management, risk management, verification and assurance, and system life-cycle engineering

Systems engineering is a new topic for the Industrial Engineering FE exam. Let's start by defining some terms. Systems Engineering is an interdisciplinary approach to enable the successful realization of complex engineering projects over the entire life cycle. It considers both the business and the technical needs of all stakeholders with the goal of providing a quality product that meets the users' needs. A system engineer (SE) works to identify and integrate all the varied aspects of a modern complex technical system. Its origins are in the aerospace and defense industries. It has since moved into other industries including transportation systems, electrical power grids, computer networks, wireless communication networks, and healthcare. System Architecture is the arrangement of elements and subsystems and the allocation of functions to them to meet system requirements. Systems engineering should be involved throughout the life of the system, from development through production, deployment, training, support, operation, and disposal. There are several components of systems engineering, including requirements analysis, functional analysis, and synthesis.

In a system, a component is an operating part of the system consisting of input, process, and output. An attribute is a property of the components, which characterize the system. Relationships are links between components and attributes. Actors are others which interact with the system, such as users, other systems, or the environment. For example, imagine a car. A subsystem is the navigation system. The actors are the driver, the GPS network of satellites, the dashboard display system, among others. An attribute would be the frequency the navigation system operates on to communicate with the GPS system outside of the car. The context the car is in would include the weather (rain, temperature, snow, etc.) and location (U.S. vs. other countries, high altitudes, communication "dead zones," etc.). The car must interact in and among all of these, and the SE must insure that everything will work together.

REQUIREMENTS ANALYSIS

A requirement is a physical and functional need that the system must be able to perform. A spacecraft may have requirements to carry a payload weighing *X*, with maximum dimensions *Y*, and maintain a pressure surrounding the payload of *Z*. There are four general categories of requirements:

1. Mission or business requirements—Specifies the context in which the system will operate
2. System/subsystem requirements—What the system's capabilities will be
3. Resource requirements—What the system will use or interface with
4. Performance requirements—What the system will be able to accomplish

System engineers must establish requirements, tie them to testing, and facilitate the process of changing them. A requirement that the braking system must stop a car safely is too vague. A requirement to stop a car going 60 mph on a flat, dry surface being able to come to a complete stop in 100 yards is a testable requirement and a much better written one. If the federal government or a new technology changes how braking systems work in cars, the requirements must be easy to update and a system engineer must track the associate changes to the requirements; this is called configuration management.

SYSTEMS DESIGN

Systems engineers have developed the V-model of the system design process, shown in Figure 13.1.

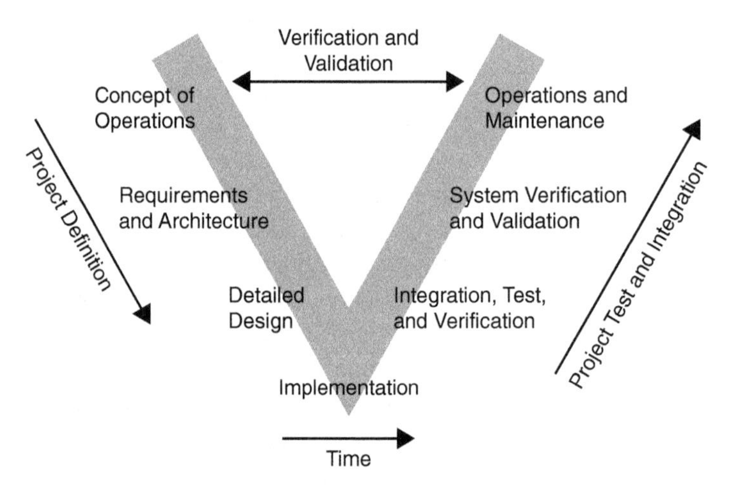

Figure 13.1 V-Model of the systems engineering process
Source: http://ops.fhwa.dot.gov/publications/seitsguide/seguide.pdf

The steps in the V-Model are:

- Concept Operations—Identify and document stakeholders, overall system capabilities, performance measures for the validation at the end of the project.
- System Specification—Develop verifiable system requirements that meet needs defined during concept operations.
- Design (high level and detailed)—Design system architecture that satisfies system requirements.
- Development (hardware and software)—Select and obtain appropriate technology.
- Testing (device, subsystems, system)—Test against requirements and constraints.
- Verifications and Validation—Evaluate that the system does what is expected.
- Operate and Maintain—Use the system.

As the system engineer works through the process (starting at the upper left-hand of the V), the system documentation is created and refined. Requirements are created on the left-hand side that drive the next step in the process (moving down the V) and will be verified (i.e., "Is it doing what we want?"), validated (i.e., "Is it doing it right?"), and tested (i.e., "Is it doing it well enough?") later in the process (on the right-hand side of the V).

HUMAN SYSTEMS INTEGRATION (HSI)

Human Systems Integration has a foundation in human factors but is more than just human factors. It considers how the human interacts with the system. It analyzes the interaction of all humans involved, including operators, maintainers, and operators of other related systems. The HSI considers:

- users' characteristics (size, knowledge, physical abilities, number of operators),
- users' well being (safety, health, ergonomics, habitability), and
- users' abilities (capabilities and limitations, training requirements, human factors).

The HSI specialists consider different domains in their analysis. They include the following:

- Manpower—The number of personnel required to operate, maintain, sustain, and provide training for systems
- Personnel—The cognitive and physical capabilities required to be able to train, operate, maintain, and sustain systems
- Training—The instruction, education, and/or on-the-job training required to provide personnel with the skills and knowledge needed to use and maintain the system
- Human Factors Engineering—Integrating human characteristics, abilities, and limitations into system definition, design, development, and evaluation

- Health Hazard Assessment—Short or long term hazards to health occurring as a result of operating or maintaining the system (i.e., allergic reactions or ergonomic injuries)
- System Safety—Safety risks occurring when the system is functioning in both a normal or abnormal manner
- Survivability—The characteristics of a system that can reduce fratricide, detectability and probability of being attacked and minimize system damage, soldier injury and cognitive and physical fatigue. (This is common in military applications, but is not appropriate for all systems engineering projects.)
- Habitability—Considers the users' personal needs, such as sleeping, eating, and hygiene. (This would be important when designing a tank, airplane, or spacecraft but not appropriate for other applications, such as a wireless communication system or a hand-held entertainment system.)

HSI involves making tradeoffs between these domains. For example, better human factors designs of the user interface may reduce the amount of training required. Adding protective gear for the crew can improve survivability but will potentially increase ergonomic or heat stress risks. These tradeoffs even occur outside of the realm of HSI. A spacecraft designed for human passengers will have numerous technical requirements (temperature, pressure, oxygen, water, etc.) that would not be present in an unmanned vehicle and can easily raise the price by more than an order of magnitude.

FUNCTIONAL ANALYSIS AND ALLOCATION

A function is something that the system must do or accomplish to achieve its purpose. For example a weapon system has loading functions (what kind of ammo, how fast, etc.) and firing functions (distance, accuracy, speed). Specifying "what" must be accomplished to achieve desired objectives (but not "how") within the system/subsystems is the process of functional analysis and allocation. It is accomplished by an iterative process of translating system requirements into detailed design criteria. Specific resource requirements at the subsystem level are also defined at the same time. The following are types of questions which will be answered.

- What functions will the subsystem perform?
- When and for how long will the subsystem be required to perform these functions?
- Where will the system be used? By whom?
- How will the system accomplish its objective?
- What maintenance and support functions are required?

CONFIGURATION MANAGEMENT

Configuration management is a process for establishing and maintaining consistency of a product's performance, functional, and physical attributes with its requirements, design, and operational information throughout its life cycle. It helps to verify that proposed changes are systematically considered to minimize adverse effects. Change management and requirements management are at times used interchangeably with configuration management. Due to the complex nature of systems engineering projects, change will happen. Technological innovation can occur, engineers will refine aspects of the system as the design process progress, customer needs can change, government regulations can be added, and countless other causes of change occur throughout the life cycle.

RISK MANAGEMENT

Risk management is the identification, assessment, and prioritization of risks. Risk is the effect of uncertainty on objectives. It is a function of the magnitude of the potential loss and the probability that the loss will occur. Systems engineers should coordinate an economical application of resources to minimize, monitor, and control the probability and/or impact of unfortunate events. Risks can come from uncertainty in financial markets, project failures (at any phase in design, development, production, or sustainment life-cycle), legal liabilities, accidents, natural causes and disasters as well as deliberate attack from an adversary.

VERIFICATION AND ASSURANCE

Systems engineers use a variety of tests throughout the project's life cycle to verify the system will meet the requirements defined at the beginning of the process. As each component of the system is built, it is tested both by itself and by integrating it with other components to make a subsystem. The components and subsystems are not only tested to see if they work but if they satisfy the requirements allocated to them. The system and subsystems testing process needs to be well tracked. It is typically an iterative process. Engineers and technicians evaluate the functionality, integration, and usability of the designs, prototypes, and products they test.

SYSTEM LIFE-CYCLE ENGINEERING

The system life-cycle engineering is the analysis of the various system phases including: system conception, design and development, production and/or construction, distribution, operation, maintenance and support, retirement, phase-out and disposal.

As the system engineering life cycle begins, key steps in the conceptual design stage include the following:

- Need identification
- Feasibility analysis
- System requirements analysis
- System specification
- Conceptual design review

Once the system concept has been approved, key steps within the preliminary design stage include the following:

- Functional analysis
- Requirements allocation
- Detailed trade-off studies
- Definition of system options
- Preliminary designs
- Development specification

Once preliminary designs are approved, key steps within the detail design and development stage include the following:

- Detailed design
- Detailed synthesis
- Development of prototypes and models
- Revision of specification
- Critical design review

Once the design is finalized and approved, the system is produced based on specification and design; key steps within the product construction stage include the following:

- Production of system components
- Acceptance testing
- System distribution and operation
- Operational testing and evaluation
- System assessment

Once fully deployed and in operation, the key steps within the utilization and support stage include the following:

- System operation in the user environment
- Change management
- Maintenance and logistics support
- System modifications for improvement
- System assessment

Once deployed, the effectiveness and efficiency of the system must be evaluated during the phase out and disposal stage; key steps include the following:

- Determine when the product has met its effective life
- Analysis of operational requirements versus system performance
- Feasibility of system phase-out
- Plan recycling, remanufacturing and/or disposal of system components

PROBLEMS

13.1 The requirements process in systems engineering does all of the following *EXCEPT*:

a. develops a common understanding among users, developers, and acquirers about what is needed
b. defines who will use and maintain the system/subsystem
c. fills in the blanks among the needs and expectations
d. provides a foundation for testing and evaluations

13.2 Which of the following applications is *NOT* well suited for systems engineering techniques?

a. A database of customer information and prior orders
b. A smart TV that allows Internet access, shopping, and downloading games
c. A collision avoidance system that can be added to cars that uses sensors to warn drivers to avoid collisions
d. NASA's potential mission to send an unmanned satellite to Mercury

13.3 Interactions and tradeoff happen frequently in systems engineering. They could be expected among the following pairs *EXCEPT*:

a. human systems integration and project cost
b. the personnel and training domains within human systems integration
c. risk management and test and evaluation
d. survivability domain within human systems integration and requirement documents

13.4 The following statements are true about systems engineering *EXCEPT*:

a. testing should occur throughout the system's life-cycle
b. systems engineering involves various stakeholders (customers, users, management, etc.)
c. once requirements are established in the design process, they should not change to minimize costs
d. systems engineering considers the context the product will operate in

13.5 Systems requirements should be:

a. testable
b. traceable
c. unambiguous
d. all of the above

13.6 Which of the following is a well written requirement?

a. Trip-free circuit breakers should be used to prevent wire damage.
b. The system operating cost should be $1,000 per operating hour or better, where applicable.
c. The system operating cost should be at least as good as the best of the current systems.
d. The circuit breaker subsystem should be user-friendly.

13.7 The process of translating system requirements into detailed design criteria is called:

a. criteria management
b. translation management
c. functional analysis
d. design analysis

13.8 Prototypes would likely be used in which stage of the system engineering design process?

a. Stakeholder requirements
b. System requirements
c. Component requirements & testing
d. Acceptance testing

13.9 Which of the following is a part of the life-cycle of systems engineering?

a. Functional analysis
b. Detailed trade-off studies
c. Phase out and disposal
d. All of the above

13.10 Which of the following is *NOT* a domain in human systems integration?

a. System safety
b. System requirements
c. Human factors engineering
d. Training

SOLUTIONS

13.1 b. Requirements clarifies needs and expectations among stakeholders. It provides a basis for testing systems and subsystems. The requirements state the system must be able to do *X* and the testing and evaluation process is to see if the system does *X*. Considering the humans who will interface with the system is done in the functional analysis and the human systems interaction process, not in the requirements process.

13.2 a. Systems engineering is used for large, complex engineering problems. These types of systems interact with other systems, the environment, and other engineering disciplines. A database should be managed and tested but does not typically have the complexity and interdisciplinary nature that calls for systems engineering.

13.3 c. The HSI will influence the cost (either positively or negatively). For example, adding more displays for the user would increase the cost, while simplifying the maintenance process could reduce costs. The personnel and training domains have significant interaction. Requirements relating to users can increase or decrease training requirements as users have or don't have certain skills or abilities. Survivability needs will appear/influence the requirements and testing of a system. For example, "must stop a 200-pound person moving at 65 mph" is a survivability issue but will become specific requirements and will be tested with crash test dummies. Test and evaluation is a key component of the systems engineering process; risk management is not directly tested or evaluated during this phase of the system development process.

13.4 c. Testing occurs throughout the process from prototypes, to component design, to acceptance testing and beyond. Systems engineering is interdisciplinary and involves all stakeholders. Requirements will change as the understanding of the needs and technology gets better through the development process. It cannot be avoided, just planned for and managed. Understanding the environment the system operates in is key.

13.5 d. Testable and unambiguous requirements are key during verification. Traceable is key in requirement and configuration management.

13.6 a. Response B is ambiguous. Response C is not measureable. Response D is not verifiable.

13.7 c. Function analysis is the process of translating system requirements into detailed design criteria. It specifies "what" must be accomplished to achieve desired objectives, but not the "how."

13.8 c. Stakeholder requirements and system requirements are early in the definition process before a prototype can be built. At the lower portion of the V-model, when components are being designed and tested, a prototype can be used. Acceptance testing is one of the final steps and uses a "more finished" product, not a prototype.

13.9 d. All of these are part of the system engineering life cycle.

13.10 b. System requirements can be modified based on HSI analysis, but it is not a domain.